Quantitative Genetics, Genomics and Plant Breeding

Quantitative Genetics, Genomics and Plant Breeding

Edited by

Manjit S. Kang

*Professor, Quantitative Genetics
Department of Agronomy
Louisiana State University
Baton Rouge
Louisiana
USA*

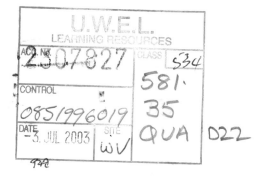
CABI *Publishing*

CABI *Publishing* is a division of CAB *International*

CABI Publishing	CABI Publishing
CAB International	10 E 40th Street
Wallingford	Suite 3203
Oxon OX10 8DE	New York, NY 10016
UK	USA

Tel: +44 (0)1491 832111	Tel: +1 212 481 7018
Fax: +44 (0)1491 833508	Fax: +1 212 686 7993
E-mail: cabi@cabi.org	E-mail: cabi-nao@cabi.org
Web site: www.cabi-publishing.org	

A catalogue record for this book is available from the British Library, London, UK.

Library of Congress Cataloging-in-Publication Data
Quantitative genetics, genomics, and plant breeding / edited by Manjit S. Kang.
    p. cm.
    Papers from the Symposium on Quantitative Genetics and Plant Breeding in the 21st
Century, held at Louisiana State University, Mar. 26–28, 2001.
    Includes bibliographical references.
    ISBN 0-85199-601-9 (alk. paper)
    1. Crops--Genetics--Congresses. 2. Plant breeding--Congresses. 3. Quantitative
genetics--Congresses. I. Kang, Manjit S. II. Symposium on Quantitative Genetics and
Plant Breeding in the 21st Century (2001 : Louisiana State University)

SB123 .Q36 2002
631.5′233--dc21

2001052814

ISBN 0 85199 601 9

Typeset by AMA DataSet Ltd, UK
Printed and bound in the UK by Biddles Ltd, Guildford and King's Lynn

Contents

Contributors

P. Annicchiarico, *Istituto Sperimentale per le Colture Foraggere, viale Piacenza 29, 26900 Lodi, Italy.*

M. Balzarini, *Faculdad Cs. Agropecuarias, Estadistica y Biometria, Universidad Nacional de Córdoba, Av. Valparaiso s/n, 5000 Córdoba, Argentina.*

M. Bänziger, *International Maize and Wheat Improvement Center (CIMMYT), Lisboa 27, Apdo. Postal 6-641, 06600 Mexico DF, Mexico.*

A.S. Basra, *Department of Agronomy and Range Science, University of California, Davis, CA 95616, USA.*

J. Betran, *Department of Soil and Crop Sciences, Texas A&M University, College Station, TX 77843-2474, USA.*

D.S. Brar, *International Rice Research Institute, DAPO 7777, Metro Manila, The Philippines.*

S.C. Chapman, *CSIRO Plant Industry, 120 Meiers Road, Long Pocket Laboratories, Indooroopilly, Qld 4068, Australia.*

M. Cooper, *Pioneer Hi-Bred International Inc., 7300 NW 62nd Avenue, PO Box 1004, Johnston, IA 50131, USA.*

P.L. Cornelius, *Departments of Agronomy and Statistics, University of Kentucky, Lexington, KY 40546-0091, USA.*

J. Crossa, *Biometrics and Statistics Unit, International Maize and Wheat Improvement Center (CIMMYT), Lisboa 27, Apdo. Postal 6-641, 06600 Mexico DF, Mexico.*

B. Cullis, *Wagga Wagga Agricultural Institute, Private Mail Bag, Wagga Wagga, NSW 2650, Australia.*

K. Dreher, *International Maize and Wheat Improvement Center (CIMMYT), Lisboa 27, Apdo. Postal 6-641, 06600 Mexico DF, Mexico.*

J.W. Dudley, *Department of Crop Sciences, University of Illinois, 1102 S. Goodwin Avenue, Urbana, IL 61801, USA.*

G.O. Edmeades, *Pioneer Hi-Bred International Inc., PO Box 609, Waimea, HI 96796, USA.*

S.S. Goyal, *Department of Agronomy and Range Science, University of California, Davis, CA 95616, USA.*

G.L. Hammer, *Agricultural and Production Systems Research Unit (APSRU), Queensland Department of Primary Industries, PO Box 102, Toowoomba, Qld 4350, Australia.*

D. Hoisington, *International Maize and Wheat Improvement Center (CIMMYT), Lisboa 27, Apdo. Postal 6-641, 06600 Mexico DF, Mexico.*

G. Hollamby, *Roseworthy Agricultural College, Roseworthy, South Australia, Australia.*

F. Hospital, *INRA, Station de Génétique Végétale, Ferme du Moulon, 91190 Gif-sur-Yvette, France.*

L.A. Hunt, *Department of Plant Agriculture, University of Guelph, Guelph, Ontario N1G 2W1, Canada.*

J.-L. Jannink, *Department of Agronomy, Iowa State University, Ames, IA 50011, USA.*

N.M. Jensen, *Pioneer Hi-Bred International Inc., 7300 NW 62nd Avenue, PO Box 1004, Johnston, IA 50131, USA.*

C. Jiang, *Monsanto Life Sciences, Research Center, 700 Chesterfield Parkway North, St Louis, MO 63198, USA.*

M.S. Kang, *Department of Agronomy, Louisiana State University, Baton Rouge, LA 70803-2110, USA.*

M.J. Kearsey, *School of Biosciences, University of Birmingham, Birmingham B15 2TT, UK.*

G.S. Khush, *International Rice Research Institute, DAPO 7777, Metro Manila, The Philippines.*

M.D. Krakowsky, *Department of Agronomy, Iowa State University, Ames, IA 50011, USA.*

N.L. Kruger, *School of Land and Food Sciences, University of Queensland, Brisbane, Qld 4072, Australia.*

M. Lee, *Department of Agronomy, Iowa State University, Ames, IA 50011, USA.*

D. Luckett, *Wagga Wagga Agricultural Institute, Private Mail Bag, Wagga Wagga, NSW 2650, Australia.*

D.E. Mather, *Department of Plant Science, McGill University, 21111 Lakeshore Road, Ste-Anne-de-Bellevue, Quebec H9X 3V9, Canada.*

K.P. Micallef, *School of Land and Food Sciences, University of Queensland, Brisbane, Qld 4072, Australia.*

R. Moutiq, *Department of Agronomy, Iowa State University, Ames, IA 50011, USA.*

S.H. Park, *Vegetable and Fruit Improvement Center, 1500 Research Parkway, Suite 120A, Texas A&M University Research Park, College Station, TX 77845, USA.*

D.W. Podlich, *School of Land and Food Sciences, University of Queensland, Brisbane, Qld 4072, Australia.*

G.P. Rédei, *University of Missouri, 3005 Woodbine Ct, Columbia, MO 65203, USA.*

J.-M. Ribaut, *International Maize and Wheat Improvement Center (CIMMYT), Lisboa 27, Apdo. Postal 6-641, 06600 Mexico DF, Mexico.*

A. Smith, *Wagga Wagga Agricultural Institute, Private Mail Bag, Wagga Wagga, NSW 2650, Australia.*

O.S. Smith, *Pioneer Hi-Bred International Inc., 7300 NW 62nd Avenue, PO Box 1004, Johnston, IA 50131, USA.*

R.H. Smith, *Vegetable and Fruit Improvement Center, 1500 Research Parkway, Suite 120A, Texas A&M University Research Park, College Station, TX 77845, USA.*

R. Thompson, *IACR-Rothamsted, Harpenden, Hertfordshire AL5 2JQ, UK.*

N.A. Tinker, *Agriculture and Agri-Food Canada, 960 Carling Avenue, Building 20, Ottawa, Ontario K1A 0C6, Canada.*

F.A. Van Eeuwijk, *Wageningen University, Department of Plant Sciences, Laboratory of Plant Breeding, PO Box 386, 6700 AJ Wageningen, The Netherlands.*

M. Vargas, *International Maize and Wheat Improvement Center (CIMMYT), Lisboa 27, Apdo. Postal 6-641, 06600 Mexico DF, Mexico.*

B. Walsh, *Department of Ecology and Evolutionary Biology, University of Arizona, Tucson, AZ 85721, USA.*

Y. Xu, *RiceTec, Inc., PO Box 1305, Alvin, TX 77512, USA.*

W. Yan, *Department of Plant Agriculture, University of Guelph, Guelph, Ontario N1G 2W1, Canada.*

Foreword

A majority of economically important plant and animal traits, such as grain yield or meat production, can be classified as multigenic or quantitative. The application of statistical methods in the improvement of such traits – applied quantitative genetics – began early in the 20th century, when many researchers questioned whether the inheritance of these continuously distributed traits was Mendelian. During the past century, however, both plant and animal geneticists have obtained convincing evidence that Mendelian principles apply to quantitative as well as qualitative traits. This evidence has also shaped the general model that embraces the multiple-factor hypothesis for quantitative traits (with genes located in chromosomes and hence sometimes linked, and incomplete heritability because of the contribution of environmental factors to total phenotypic variation).

The early efforts by outstanding scientists, such as R.A. Fisher, G.W. Snedecor and J.L. Lush, have been followed over the decades with many avenues of research in both the applied and theoretical aspects of quantitative genetics. More recently, the linking of molecular genetics and genomics with the study, evaluation and improvement of quantitative traits has become the central theme of much exciting research. The recent developments in the areas of quantitative trait locus mapping, bioinformatics, marker-assisted selection and molecular-enhanced breeding strategies have just scratched the surface in the molecular dissection of complexly inherited traits. DNA-based technologies will provide a complete new kit of tools for plant and animal breeders for the improvement of quantitative traits. Genotype–environment interaction will continue to be a major factor as researchers focus on the improvement as well as the stability of such traits.

This book will serve as a valuable research and teaching reference. The collection of papers will provide knowledge from which to launch new research efforts on the nature of the inheritance of quantitative traits as well as the development of new technologies for improving such traits. Hopefully, it will stimulate expanded efforts to exploit the full potential of genomics, bioinformatics and molecular-enhanced breeding technologies in the study and utilization of quantitative variation in plant and animal species.

The efforts of Dr Manjit Kang in organizing the symposium on 'Quantitative Genetics and Plant Breeding in the 21st Century' and the editing of the proceedings are commendable. He has assembled papers for a truly exceptional volume that will serve as an important reference for quantitative geneticists and plant and animal breeders for many years into the future.

Charles W. Stuber
President 2002
American Society of Agronomy

Preface

... simply imagine that new century full of its promise, molded by science, shaped by technology, powered by knowledge. These potent transforming forces can give us lives fuller and richer than we have ever known.

William Jefferson Clinton (May 1997)

This book resulted from an international symposium 'Quantitative Genetics and Plant Breeding in the 21st Century' (www.sigmaxi.org/chapter/LocalEvents/LSU.htm) that was held from 26 to 28 March 2001 in Baton Rouge, Louisiana, under the auspices of the Louisiana State University Chapter of Sigma Xi, Scientific Research Society, of which I was President (July 2000–June 2001). Sigma Xi is an international honour society of scientists and engineers (www.sigmaxi.org).

Quantitative genetics has contributed immensely to the improvement of crops and animals in the 20th century. The field of quantitative genetics has been important for its applications to breeding methodologies. Quantitative genetics (or biometrical genetics, as it was called earlier) came of age with the publication of books by Mather (1949) and Falconer (1960). The field started to receive focused attention through conferences in the 1960s, as evidenced by the publication of *Statistical Genetics and Plant Breeding*, which resulted from a meeting held in Raleigh, North Carolina, during 20–29 March 1961 (Hanson and Robinson, 1963). (Of course, there was a conference specifically devoted to heterosis, which was held in 1950 at Ames, Iowa (Gowen, 1952), followed by another symposium in 1997, which was held in Mexico City (Coors and Pandey, 1999).) In 1976, the first international conference on quantitative genetics was held in Ames, Iowa, proceedings of which have been documented in Pollak *et al.* (1977). Because of the need to document continuing advances in the theory of quantitative genetics and its application to plant and animal breeding, the second international conference on quantitative genetics took place in Raleigh, North Carolina (31 May–5 June 1987). In the 724-page proceedings of the second conference (Weir *et al.*, 1988), restriction fragment length polymorphism (RFLP) was mentioned in two chapters (Schuler, 1988; Soller and Beckmann, 1988), but there was no mention of quantitative trait loci (QTL). Almost every chapter in *Quantitative Genetics, Genomics, and Plant Breeding* has references to QTL. Obviously, it is clear that quantitative genetics has progressed by leaps and bounds, especially in the direction of mapping QTL with molecular markers and marker-assisted selection.

The need for increasing food production is obvious because of projections that the world population will increase from the current 6 thousand million to about 10 thousand million by the mid-21st century, which translates into an increase of 1 thousand million every 10 or

12 years during the next 40–50 years. The challenge before us, therefore, is: 'Can we feed 4 thousand million more people by 2040?' Quantitative genetics, being the bedrock of breeding methodologies, has much to offer in meeting this challenge.

The major issues/areas included in this book are: (i) QTL mapping, genomics, marker-assisted selection, tissue culture and alien introgression for crop improvement; and (ii) advances in genotype–environment interaction (GEI)/stability analyses. The symposium brought together experts in the fields of quantitative genetics, crop improvement, tissue culture and statistics to share their knowledge with colleagues in the profession. The book contains authoritative chapters written by experts in their fields from around the world. Because quantitative genetics itself, without practical applications to breeding, is of limited value, it was deemed necessary to cover other crop-improvement issues, such as tissue culture and its role in improving crops, GEI and stability analyses, and issues such as nitrogen-use efficiency (NUE) to increase crop production. A link between the two major subject areas is suggested by many contributions, highlighting the difficulties that biotechnology-driven breeding programmes may face in breeding across diversified environments.

The book is basically organized into two major sections, excluding the first chapter, which represents the banquet talk by a distinguished geneticist, Dr George P. Rédei, Professor Emeritus, University of Missouri-Columbia. Dr Rédei provides a unique historical perspective on developments in genetics in the 20th century. His narration of the events is tinged with humour but factual. He believes the role of quantitative genetics in basic biology and applied sciences will increase.

Section I relates to issues relative to genomics (including bioinformatics) and QTL. Because tissue culture and molecular genetics protocols are interrelated, two chapters relative to this subject are included in the first section. This section comprises 13 chapters.

Dr Bruce Walsh (Chapter 2) points out that the influence of quantitative genetics spans across plant breeding, animal breeding, evolutionary genetics and human genetics. His view is that genomics will not reduce the importance of quantitative genetics but will increase it. He exhorts plant breeders to consider powerful tools developed in animal breeding, evolutionary genetics and human genetics.

Dr Nicholas A. Tinker (Chapter 3) tells why quantitative geneticists must care about bioinformatics. Bioinformatics tools and concepts can be applied to quantitative genetics, and bioinformatics is an area in which the expertise of the quantitative geneticist is required.

Dr Michael J. Kearsey (Chapter 4) explains the problems encountered in QTL analyses and offers possible solutions. He suggests that initial solutions to some of the problems involve part-chromosome substitution lines and near-isogenic lines and goes on to discuss how a novel set of resources – stepped aligned inbred recombinant strains (STAIRS) – could permit very accurate QTL mapping.

Drs Jean-Luc Jannink and Bruce Walsh (Chapter 5) discuss association-mapping methods reported primarily in the human genetics literature. In mapping QTL, association mapping takes advantage of events that created linkage disequilibrium between DNA markers and QTL in the relatively distant past. The many meioses since those events will have removed association between a QTL and any marker not tightly linked to it. Association mapping thus allows for fine mapping.

Dr John W. Dudley (Chapter 6) establishes an excellent bridge between molecular genetics, quantitative genetics and plant breeding. He points out that combining quantitative genetic methodologies with genomics information is required to maximize the value of investments in genomics research.

Dr Jean-Marcel Ribaut and colleagues (Chapter 7) provide a practical example of the effective use of molecular markers to improve drought tolerance in tropical maize, which represents an important perspective from an international research centre – the International Maize and Wheat Improvement Center (CIMMYT). They strongly suggest that a multidisciplinary effort in the areas of breeding, physiology and biotechnology is needed for understanding plant responses to drought stress.

Dr Diane E. Mather (Chapter 8) provides an overview of genome mapping efforts in barley. She discusses several examples of QTL research in barley to illustrate important insights that genome mapping has provided into the genetic control of a number of quantitative traits in barley.

Dr Yunbi Xu (Chapter 9) describes the use of rice as a model plant in molecular biology research. He covers quantitative traits/QTL relative to bioinformatics and functional genomics. Because rice is a model plant, any success achieved in this crop should benefit other crops as well.

Dr Frederic Hospital (Chapter 10) gives an overview of research achievements in marker-assisted back-cross breeding theory. He also discusses the consequences of marker-assisted selection and genotype building.

Dr Mark Cooper and colleagues (Chapter 11) emphasize that computer simulation will be a central tool for evaluating breeding strategies for their power to improve quantitative traits. They suggest that, by understanding the structure and function of gene networks that determine traits, researchers will be able to describe and predict gene–phenotype relationships and evaluate the power of molecular-enhanced breeding strategies. They describe and apply the $E(N{:}K)$ model, a new class of non-linear quantitative gene network model, as a step towards achieving this objective.

In a companion article (Chapter 12), Dr Scott C. Chapman and colleagues discuss key aspects of a modelling framework for linking gene effects to phenotypes. They remark that the complex coordination of biochemical pathways will mask the ultimate controls of crop yield for some time. This is an attribute of biological systems where the detection of and the response to environmental cues is integrated over time-scales (seconds to weeks) and scales of organism organization (cell compartments to interacting communities of plants).

Dr Roberta H. Smith and Dr Sung Hun Park (Chapter 13) give an overview of the many facets of the use of plant cell, tissue or organ culture techniques for crop improvement. The interface between these techniques and molecular techniques is clearly visible.

Drs Darshan S. Brar and Gurdev S. Khush (Chapter 14) describe tissue culture in the context of transferring alien genes into cultivated rice – one of the world's most important cereals.

Section II includes ten chapters related to GEI. I give an overview of the progress that has been made and future prospects of research related to GEI issues (Chapter 15).

Dr Fred A. Van Eeuwijk and colleagues (Chapter 16) provide an outline of a conceptually simple method of analysing QTL–environment interaction (QEI). Their method is based on regression. They view QEI analysis as a direct elaboration of GEI analysis.

Dr R. Moutiq and colleagues (Chapter 17) present results of an experiment relative to photoperiod response in maize. They describe the elements of GEI for photoperiod sensitivity and suggest that a better understanding of its inheritance should facilitate a rapid exchange of germ-plasm across latitudes.

Drs Amarjit S. Basra and Sham S. Goyal (Chapter 18) discuss mechanisms of NUE. They point out that difficult-to-measure traits or traits of low heritability or high GEI should benefit from molecular-marker technology and that an integration of agronomic and physiological studies with quantitative genetic approaches should allow selection of genotypes with high NUE.

Drs Weikai Yan and L.A. Hunt (Chapter 19) explain the use of the genotype effect and genotype–environment effect (GGE) biplot approach of analysing GEI data. The GGE biplot software is introduced, which also handles genotype–trait data, diallel-cross data, and genotype–pathogen race data.

Drs José Crossa and Paul Cornelius (Chapter 20) discuss linear–bilinear models, such as additive main effects and multiplicative interaction (AMMI) and the shifted multiplicative model (SHMM), for analysing GEI and assessing crossover interaction (COI). They describe a clustering approach using a statistical analysis system (SAS) programme that groups sites (or genotypes) with negligible genotypic COI.

Dr Alison Smith and colleagues present in Chapter 21 methodology for a spatial multiplicative mixed model analysis for GEI data. In the companion chapter (Chapter 22), they illustrate the methodology using two examples.

Dr Monica Balzarini (Chapter 23) discusses applications of mixed model theory to predict cross performance and analyse multienvironment trials. She uses best linear unbiased prediction (BLUP) for prediction of crosses not yet tested and to study GEI.

Dr Paolo Annicchiarico (Chapter 24) considers practical issues of defining adaptation strategies and yield-stability targets in breeding programmes and the contribution of GEI analysis and selection theory applied to multienvironment yield data. Two case-studies from Algeria and Italy are presented.

I trust that breeders/geneticists will find the information in this book stimulating and useful. The authors have done a commendable job of producing most-informative chapters. The book is intended to be of use to practising breeders and geneticists as well as students and teachers of plant breeding, genetics and molecular breeding.

I thank all the authors for making common cause with us and sharing their knowledge and discussing issues with professional colleagues through this vehicle. I also thank Tim Hardwick of CABI *Publishing* for believing in this project and helping to transfer technology to all parts of the world.

Manjit S. Kang
August 2001
Baton Rouge, Louisiana

References

Coors, J.G. and Pandey, S. (eds) (1999) *The Genetics and Exploitation of Heterosis in Crops.* American Society of Agronomy, Crop Science Society of America, and Soil Science Society of America, Madison, Wisconsin.

Falconer, D.S. (1960) *Introduction to Quantitative Genetics.* Oliver and Boyd, London.

Gowen, J.W. (ed.) (1952) *Heterosis.* Iowa State University Press, Ames, Iowa.

Hanson, W.D. and Robinson, H.F. (eds) (1963) *Statistical Genetics and Plant Breeding.* Publication 982, National Academy of Sciences – National Research Council, Washington, DC.

Mather, K. (1949) *Biometrical Genetics,* 1st edn. Methuen, London.

Pollak, E., Kempthorne, O. and Bailey, T.B., Jr (eds) (1977) *Proceedings of International Conference on Quantitative Genetics.* Iowa State University Press, Ames, Iowa.

Schuler, J. (1988) Inserting genes affecting quantitative traits. In: Weir, B.S., Eisen, E.J., Goodman, M.M. and Namkoong, G. (eds) *Proceedings of the Second International Conference on Quantitative Genetics.* Sinauer Associates, Sunderland, Massachusetts, pp. 198–199.

Soller, M. and Beckmann, J.S. (1988) Genomics, genetics and the utilization for breeding purposes of genetic variation between populations. In: Weir, B.S., Eisen, E.J., Goodman, M.M. and Namkoong, G. (eds) *Proceedings of the Second International Conference on Quantitative Genetics.* Sinauer Associates, Sunderland, Massachusetts, pp. 161–188.

Weir, B.S., Eisen, E.J., Goodman, M.M. and Namkoong, G. (eds) (1988) *Proceedings of the Second International Conference on Quantitative Genetics.* Sinauer Associates, Sunderland, Massachusetts.

1 Vignettes of the History of Genetics

George P. Rédei

University of Missouri, 3005 Woodbine Ct, Columbia, MO 65203, USA

Mendelism

Although many people consider the beginning of genetics to be the publication of the 'Versuche über Pflanzenhybriden' by Gregor Mendel in 1866 or the submission of the manuscript during the preceding year, the beginning of genetics goes back to thousands of years before.

All geneticists and practically everybody else agree today that Mendel's discovery was an extraordinary achievement. Fewer people know some interesting details about how Mendel achieved it. Not only had he chosen simple characters of an autogamous plant and counted the segregating offspring, but also it was particularly smart that for some he did not have to grow the second generation because the segregation was already evident by inspecting the pods (Fig. 1.1). The circumstances also taught him common sense since he had about 245 m² nursery space in the monastery garden. It also shows that not only was Mendel a very smart man, he also had great sense for practical matters. During his teaching and priestly duties, he also founded a savings and loan bank and a fire brigade. On a photograph probably taken in 1862, Mendel is shown examining a beautiful *Fuchsia* inflorescence (Fig. 1.2). What intuitive strength that he did

Fig. 1.2. Gregor Mendel examines a *Fuschia* plant. The photo was taken in about 1862. (Courtesy of Dr V. Orel.)

Fig. 1.1. Segregation for wrinkled and smooth seeds within the pea fruits heterozygous for the gene.

©CAB *International* 2002. *Quantitative Genetics, Genomics and Plant Breeding* (ed. M.S. Kang)

not pursue this ornamental plant further! The chromosome numbers of fuchsias vary a great deal – $2n = 22, 55, 66$ and 77 – and this confused other students of inheritance before and after Mendel.

Mendel himself never claimed any laws to his credit. The term (actually rule (*Regel*) rather than laws) was first used by Carl Correns (1900), and he named them: '1. Uniformitäts- und Reziprozitätsgesetz, 2. Spaltungsgesetz, 3. unabhängige Kombination', namely, first law: uniformity of the F_1 (if the parents are homozygous) and the reciprocal hybrids are identical (in the absence of cytoplasmic differences); second law: independent segregation of the genes in F_2 (in the absence of linkage); and third law: independent assortment of alleles in the gametes of diploids. Thomas Hunt Morgan (1919) also recognized three laws of heredity: (i) free assortment of the alleles in the formation of gametes; (ii) independent segregation of the determinants for different characters; and (iii) linkage–recombination. In some modern textbooks only two Mendelian laws are recognized, but this is against the tradition of genetics in which the first used nomenclature is upheld.

Mendel was a former student and teaching assistant of C.J. Doppler, the physicist, and in the laboratory in Vienna they were already teaching some statistics. Mendel was also fortunate in not finding linkage, which might have been confusing. He used seven characters and obtained 128 (2^7) combinations. Peas have seven linkage groups, thus the probability of independence would have been $6!/7^6 = 720/117,649 \approx 0.0061$. Actually some of the genes he studied were syntenic, e.g. *v*, *fa* and *le* in chromosome 4. But the distance between *fa* and *le* is 114 map units and *i* and *a* in linkage group 1 (204 map units) are so far away in the chromosome that they segregate independently. It seems that, among the hybrid combinations he had, *v–le* (12 map units) was not included (Blixt, 1975). This was dubbed appropriately 'Mendel's luck', presumably by J.P. Lotsy, a German geneticist of the early 20th century.

The printer, who introduced numerous small errors, had already abused the classic paper of Mendel. The editor took liberties, too, and changed some of the spellings preferred by Mendel. It is known that Mendel corrected by hand at least some of the 40 reprints he received. Only four of these reprints have survived. One of them, sent by Mendel to the renowned Austrian botanist Anton Kerner, was not opened, as revealed by the uncut edges of the paper (Křiženecký and Němec, 1965).

It was quite unfortunate that his contemporaries failed to recognize the significance of his research. Carl Wilhelm Nägeli, the famous professor of botany at the University of Munich and an internationally renowned authority, felt that it was inconceivable that the plants should obey statistical rules. He advised Mendel: 'You should regard the numerical expressions as being only empirical because they cannot be proved rational.' He went even further and suggested to Mendel the study of *Hieracium* apomicts and raised self-doubts in Mendel as to whether the observations he carefully and conscientiously made would really have general validity (Nägeli, 1867).

One should not be entirely negative about Nägeli. He was probably the first who sighted chromosomes around 1842 and described them in German as *Stäbchen* or little sticks in English (Geitler, 1938).

It was not until 1873 that A. Schneider observed mitosis in Platyhelminthes and, 2 years later, Edouard Strasburger reported chromosome numbers for several plant species. Some counts were correct, some not. The term chromosome was coined in 1888 by the surgeon W. Waldeyer, who was not really an experimental biologist but was very good at pigeon-holing (Rédei, 1974).

Professor Nägeli can really be called an expert by the definition of Henry Ford, who said the expert knows what cannot be done: even when he sees that it has already been accomplished, he can also explain why it should not have been successful. Nägeli almost shot down the Mendelian results. He might also have been influential on Wilhelm Olbers Focke, who in 1881 in his monograph on plant hybrids refers only 15 times to Mendel (nine times in connection with *Hieracium* but only once about the pea experiments) but mentions the name and work of Gärtner 409 times, Kölreuter 214 times and several others dozens of times.

Nägeli evoked the ire of the medical researchers by his ideas on bacterial pleomorphism. Pleomorphy meant that bacteria (he called them *Schizomycetes*) were not supposed to possess hard heredity. He believed that their variability is not hereditary but depends entirely on the culture conditions. Apparently, his laboratory skills were insufficient and he did not understand what pure cultures are. Unfortunately, his influence and 'authority' were a serious impediment to the development of bacteriology.

Dr W. Migula, Professor at the College of Technology in Karlsruhe, Germany, gives a vivid account about the situation in his *System der Bakterien* in 1897:

> When Nägeli says, p. 20, that 'Cohn [the founder of modern bacterial systematics in 1872] had established a system of genera and species, in which each function of the *Schizomycetes* [bacteria] is represented by a particular species; by this he expressed the rather widespread view exclusive to physicians. So far I have not come across any factual ground that could be supported by morphological variations or by pertinent definitive experiments.' When Nägeli still says this in 1877, one must either assume that he was unaware of the work of the preceding 5 years, or that he chose to ignore it on purpose because it did not fit his theory.

Nägeli has also some positive legacies. I have mentioned before that he was probably the first to report seeing chromosomes. In 1884, he published a large volume entitled: *Mechanisch-physiologische Theorie der Abstammungslehre*, which is also the first systematic effort to create a molecular interpretation of the hereditary material.

Mendel's problems did not cease with his death. Anselm Rambousek (Fig. 1.3), who succeeded Mendel as abbot of the monastery, destroyed a large part of the unpublished records and personal notes after the death of his predecessor. There are different ways of leaving a historical legacy.

Fortunately, Mendel did not live to read Sir Ronald Fisher's (1936) devastating criticism. Fisher, one of the greatest statisticians

ever lived, questioned, in good faith, the 'too good to be true' data of Mendel – although Fisher tried to find excuses for Mendel, such as an assistant who was familiar with his expectations and might have deceived him, or that he figured out what he was supposed to find and just wanted to demonstrate the validity of his hypothesis. Nobody will ever find out what happened. Some of the sensation-hungry public media periodically revisit the Fisher paper and question Mendel's integrity. His principles are beyond doubt. I do not wish to go into the details because these are familiar to the majority of the students and workers in genetics. Alfred Sturtevant (1965) points out that Fisher erred in the dates, in the number of years of the experiments and misrepresented some of the statements in Nägeli's letters to Mendel.

F. Weiling (1966), a German statistician, after a thorough analysis arrived at similar conclusions. Weiling also used more technical arguments. He pointed out that the pollen tetrads may clump and then the distribution may be biased and suggests the following calculations for chi-square:

$$\chi^2 = \frac{(x - Np)^2}{Np(1-p)}$$

Fig. 1.3. Anselm Rambousek. (Courtesy of Dr V. Orel.)

where x = the observed, say, recessives, N = the number of individuals in the sample, p = the expected frequency of the phenotype. Weiling provides the following hypothetical example: $x = 152$, $N = 580$, $p = 0.25$:

$$\chi^2 = \frac{\left(152 - [580 \times 0.25]\right)^2}{580\,(0.25) \times (0.75)} = \frac{(152 - 145)^2}{108.75}$$

$$= \frac{49}{108.75} = 0.4505747126$$

This has a probability that is very different from that calculated by Fisher. Weiling also claims that Fisher erred by assuming the identity of the reciprocal crosses and did not take it into account and that might have affected the chi-square, which should have been calculated using a correction factor c:

$$\chi^2 = \frac{(x - Np)^2}{c\left(Np[1-p]\right)}$$

If the distribution is not really binomial but semi-random, this also affects the chi-square value.

Not being a statistician, I do not want to take a position in the dispute. I only wish to provide some food for thought in this case or in general. One point is indisputable: no matter how Mendel reached his conclusions, he was right. Back in the 1950s, I conducted larger experiments with monogenic segregation of auxotrophic mutants of *Arabidopsis* and observed an even better fit to the 3 : 1 under axenic conditions.

Psychologists have a term for problems of judgement: multistability of perception. In layman's words, you see what you want to see. Of course, you do not always get what you see. The British artist Gerald H. Fisher (1968) (I do not know whether he was kin to Sir Ronald) graphically illustrated how these things happen (Fig. 1.4). The upper drawing shows an ugly man, the lower figure displays an undressed woman but if you look long enough both pictures show the same.

Sometimes, failing memory or perhaps a drive for humour distorts the historical facts. In 1949, R.C. Punnett reminisced on the origin of the Hardy–Weinberg law and said:

I was asked why it was that, if brown eyes were dominant to blue, the population was not becoming increasingly brown eyed: yet there was no reason for supposing such to be the case. I could only answer that the heterozygous browns also contributed their quota of blues and that somehow this leads to equilibrium. On my return to Cambridge I at once sought out G.H. Hardy with whom I was then very friendly. For we had acted as joint secretaries to the Committee for the retention of Greek in the Previous Examination and we used to play cricket together. Knowing that Hardy had not the slightest interest in genetics I put my problem to him as a mathematical one. He replied that it was quite simple and soon handed to me the now well-known formula pr = q2 (where p, 2q and r the proportions of *AA*, *Aa* and *aa* individuals in the population varying for the A–a difference). Naturally pleased at getting so neat and prompt an

Fig. 1.4. Gerald Fisher's (1968) graphic illustration of the multistability of perception. Basically the same object (or principle in science) may mean different things depending on when and how one looks at it. Both figures above may appear as a sad male face or a nude. (By permission of the Psychonomic Society.)

answer I promised him that it should be known as 'Hardy's Law' a promise fulfilled in the next edition of my *Mendelism*. Certain it is that 'Hardy's Law' owed its genesis to a mutual interest in cricket.

Punnett might not have ever read the seminal paper of Hugo de Vries in 1900, where he said much earlier:

> Si l'on appelle D les grains de pollen ou les ovules ayant un caractère dominant et R ceux qui ont le caractère récessif, on peut se représenter le nombre et la nature des hybrides par la formule représentative suivante, dans laquelle les nombres D et R sont égaux:
>
> $$(D + R)(D + R) = D^2 + 2DR + R^2$$

This is, of course, no different from what all textbooks call either the Hardy–Weinberg law or the Castle–Hardy–Weinberg theorem.

Why Genetics was a Late Bloomer

The question often emerges why genetics started so late relative to other sciences. Copernicus (1473–1543) centuries earlier had proposed essentially valid ideas about the celestial bodies. Galileo (1564–1642) developed theories on dynamics and astronomy. Newton (1642–1729), who understood something about genetics by being also a sufferer from the complex hereditary disease gout, pioneered in gravitation and energy. Dalton (1766–1844) developed an atomic theory, although he was afflicted by X-linked red–green colour-blindness and, being a physicist, he quite clearly described his malady. In literature, Shakespeare (1564–1616), Molière (1622–1673) and Goethe (1749–1832) preceded Mendel. The latter – besides being an immortal poet – contributed significantly to the understanding of the biology of development. Mozart (1756–1791) and Beethoven (1770–1827) elevated music to an unsurpassable beauty. Strangely, Beethoven was tormented by a hearing deficit and that might have been the reason why he elected not to marry and have offspring, although he was romantically involved with several women.

There were several causes of the late development of genetics. Basic biological mechanisms of reproduction were not understood. Experimental procedures were not used. I cannot tell whether the ancient Egyptians comprehended the consequences of human inbreeding, but the artists of the 14th century BC depict the pharaoh and his wife's offspring like Wilhelm Johannsen's (1857–1927) famous beans (Fig. 1.5).

Fig. 1.5. King Akhenatan, Queen Nefertiti and three of their daughters. Bas-relief from the tomb of Apy at Amarna, c. 1362 BC. (From Aldred (1961) Thames & Hudson, by permission.)

Aristotle (384–322 BC) writes that, in Abyssinia, mice get pregnant if they lick salt. He probably did not believe it, but the 'information' might have come from a respected source so he felt obligated not to dispute it. He also stated that women had fewer teeth than men. It is hard to understand why he never looked into the mouth of his wife or mother; this would not have required a grant or special equipment.

When Aristotle reviews the ancient theories of sex determination, he finds them all unsatisfactory:

> Some suppose that the difference [between sexes] exists in the germs from the beginning; for example, Anaxagoras and other naturalists say that the sperm comes from the male and that the female provides the place [for the embryo], and that the male comes from the right, the female from the left, since in the uterus the males are at the right and the females at the left. According to others, like Empedocles, the differentiation takes place in the mother, because, according to them, the germs penetrating a warm uterus become male, and a cold uterus female.

Several of his other reported cases of heredity seem, however, quite plausible and sensible, while others are utter nonsense. Aristotle states that mutilations are not transmitted to the offspring but blindness and some scars may be. There are more defective males than females. The normal eye colour is black, and blue is a deficiency of the shade. Some of the travelling 'Freiherr Münhausen's'-like stories find their ways into his erudite books. In Libya – he writes – because of drought and heat diverse species of thirsty animals congregate at an oasis and mate. From such misalliances, for example, camel × sparrow → ostrich arises or the wild boar would have its origin by ants mating with lions. The Roman Pliny (AD 23–79) remarks 'si libeat credere' – if we are permitted to believe in tall stories.

On the other hand, even students of Linnaeus – for example, the savant Außtro-Finlandus Johannes J:nis Haartman (1751) – faithfully retell the incredible fantasies. So does practically everybody else through the centuries. The Soviet charlatans during the Lysenko era in the 20th century (Medvedev, 1969), who destroyed genetics and maimed many outstanding geneticists (e.g. Agol, Vavilov and hundreds of others), postulated similar fantastic nonsense (vegetative hybrids, inheritance of environmentally acquired traits, etc.).

Besides the lack of experimentation and the slavish submission to the ancient books, there was another negative force, expressed by Joshua Sylvester in the 16th century. Sylvester answers the 'New objection of Atheists, concerning the capacitie of the Ark':

> O profane mockers! if I but exclude
> Out of this Vessell a vast multitude
> Of since-born mongrels, that derive their birth
> From monstrous medly of Venerian mirth:
> Fantastick Mules, and spotted Leopards
> Of Incest-heat ingendred afterwards:
> So many sorts of Dogs, of Cocks, and Doves
> Since, dayly sprung from strange & mingled loves,
> Where in from time to time in various sort,
> Daedalian Nature seems her to disport:
> If plainer, yet I prove you space by space
> And foot by foot, that all this ample place,
> By subtill judgement made and Symmetrie,
> Might lodge so many creatures handsomely,
> Sith every brace was *Geometricall*:
> Nought resteth (*Momes*) for your reply at all;
> If, who dispute with God, may be content
> To take for current, Reason's argument.

The Reverend Dr Hodge of Princeton University, expressing the opinions of many of his contemporaries about Darwinism, remarked: 'to ignore design as manifested in God's creation is to dethrone God' (Provine, 1971, p. 10).

Dr A.F. Wiegmann, a physician from Brauschweig, Germany, was a prize-winner of the Physical Section of the Royal Prussian Academy of Sciences in 1826; his thesis in the competition sought to shed light on the problem: 'Gibt es ein Bastarderzeugung im Pflanzenreiche?' (Is there any hybridization in the plant kingdom?). On the second attempt he received only half the prize because he could not prove to the distinguished panel's complete satisfaction that plants do form hybrids. In his detailed report, he complains about his deteriorating vision, trembling hand, difficulties in bending and kneeling in his backyard and, above all, he is worried about the neighbours who might think that he is sodomizing plants (Roberts, 1965).

Some of the attempts with animal hybridization (wolf × mastiff) described by George Louis Le Clerc Compte de Buffon (1707–1788) were even more disastrous. The wolf killed the dog and mauled the curious experimenter (Olby, 1966).

During the preceding era, experimentation had not been very popular. All this was changing now with the Enlightenment philosophy of the 18th century. The language may still be Latin but the ideas are revolutionary. In 1759 the St Petersburg Russian Academy of Sciences offered a prize for proving:

> Sexu plantarum argumentis et experimentis novis, praeter adhuc iam cognita, vel corroborare vel impugnare, premissa expositione historica et physica omnium plantae partium, qui aliquid ad fecundationem et perfectionem seminis et fructus conferre creduntur. [Sexuality of plants should be confirmed or refuted by arguments and new experiments, besides those that are already known, by presenting the history and the physical parts of all plants that are believed to have contributed to the seed and fruits.]

Kölreuter, an early plant hybridizer, apparently, stipulated these requirements (Roberts, 1965).

This is a major milestone on the way to experimental science. The Academy wanted to see not just the records of the observations but also the physical evidence, fruits, seeds and all other plant parts. Linnaeus entered and won the contest and later expressed his wishes to spend the rest of his life studying plant hybrids.

Felix Hoppe-Seyler, a not particularly modest editor of the journal *Hoppe-Seylers Medizinische-Chemischen Untersuchungen*, set similar critical requirements. When he received Friedrich Miescher's manuscript of the initial study on nuclein in 1869, the thorough editor did not publish it until 1871, when he himself had a chance to confirm the information along with two separate papers, authored by two of his students, which showed that Miescher was correct. Actually Hoppe-Seyler and his team had proved that nuclein was not a substance unique to pus cells but was present in red blood cells, in yeast and even in milk, and this is also the beginning of the DNA story (Borek, 1965).

The obvious question arises: is such an editorial policy desirable or not? In this case it actually worked well and eventually the priority was posthumously credited to Miescher alone, despite the 'piracy' of his intellectual property. Editorial heavy-handedness does not always have such a happy ending. Hoppe-Seyler rejected the paper of MacMunn dealing with haematin, a pigment present in tissues besides blood. MacMunn's results were thus not appreciated until 1925, when another biochemist, Keilin, showed that MacMunn was right and this pigment was important for respiration (Borek, 1965).

There are several examples of similar poor judgement by experts. The editor of the *Lancet* rejected – for lack of understanding – the seminal manuscript of L. and H. Hirszfeld on the frequencies of the three alleles of the ABO blood group and the article could find its way only into *Anthropologie*, a less widely read journal (Stoneking, 2001). H.J. Muller was fired from the University of Massachusetts shortly before he was awarded the Nobel prize (1946) because the administration was not satisfied with his teaching skills. A graduate student, according to my non-scientific survey, had a completely different view. *Nature (London)* rejected the manuscript of Hans A. Krebs, who became a Nobel laureate for the same work in 1953. In 1970, a distinguished genetics panel declared *Arabidopsis* to be *planta non grata*, but, by 2000, it became the first completely sequenced higher plant and more papers are being published about it than any other plant species.

Hugo de Vries, Carl Correns and Erich von Tschermak-Seysenegg rediscovered Mendel's work in 1900. The circumstances of the rediscovery were also controversial.

H. de Vries, in this first paper, did not refer to Mendel and his explanations regarding whether he had ignored or forgotten him are contradictory. In a letter written to H.F. Roberts (1965), de Vries claimed that he worked out the Mendelian rules all by himself without the help of Mendel's work. A.H. Sturtevant (1965, p. 27) casts some doubt on the truthfulness of this claim:

> In 1954, nineteen years after the death of de Vries, his student and successor Stomps reported that de Vries had told him that he learned of Mendel's work through receiving a reprint of the 1866 paper from Beijerinck, with a letter saying that he might be interested in it. The reprint is still in the Amsterdam laboratory, as has been stated.

Despite these facts, de Vries generally receives more credit in the literature than Correns, whose contributions to genetics are much more substantial. Tschermak's work is the least valuable and the least original.

Bateson, while travelling on a train and reading, came across the Mendelian experiments and the confirmations. He became the most ardent Mendelian and the most diligent public relations man for the new ideas. He encountered stiff resistance from various corners, mainly from the biometricians, students and followers of Sir Francis Galton. Bateson published an enthusiastic book in 1902: *Mendel's Principles of Heredity: a Defence.*

Karl Pearson, a man of enormous intellect, was one of the most vociferous critics of Bateson. According to him, the purity of the gametes theory was 'not elastic enough to account for the numerical values of the constants of heredity hitherto observed' (Pearson, 1904). He requested that the Mendelians provide 'a few general principles . . . which embrace all the facts deducible from the hybridization experiments' (Pearson, 1904). Bateson was ill equipped to deal with the mathematical tasks that would 'form the basis of a new mathematical investigation' (Pearson, 1904). G. Udney Yule (1907) came to the rescue of Batesonism by accepting the compatibility of Mendelism and biometry. Wilhelm Johannsen (1909) wrote a great book with the purpose of demonstrating the need of biometry in understanding genetics. It is regrettable that this monumental work has not been translated into English and is inaccessible to many geneticists due to a language barrier.

The amalgamation of biometry and genetics did not happen readily. In the journal *Genetics*, the statistical papers are still relegated to the back of issues. Many geneticists find the language and concepts obtrusive because of lack of adequate mathematical preparation. Roger Milkman reported several years ago about an international meeting of statistical genetics, that the papers were apparently beautiful, albeit he did not understand them but hoped that the speakers did.

Pearson's confidence in the application of biometry to genetics was well vindicated by the development of the shotgun sequencing of genomes, which could not have been carried out without very powerful computers and computer programs (Sharing the glory not the credit. *Science* 291, 1189 (2001)).

The general acceptance of Mendelism continued after the rediscovery not only by the biometricians but also by the embryologists, evolutionists and zoologists. Nevertheless, at the 6–8 January 1909 meeting of the American Breeders' Association in Columbia, Missouri, Professor T.H. Morgan of Columbia University did not attend personally – maybe because of contempt for the predominantly agricultural audience – but he submitted a paper entitled 'What are "factors" in Mendelian explanations?' A member of the Zoology Department read it:

> In modern interpretation of Mendelism, facts are being transformed into factors at a rapid rate. If one factor will not explain the facts, then two are invoked; if two prove insufficient, three will sometimes work out. The superior jugglery sometimes necessary to account for the results may blind us, if taken too naïvely, to the common-place that the results are often so excellently 'explained' because the explanation was invented to explain them. We work backwards from the facts to the factors, and then, presto! explain the facts by the very factors that we invented to account for

them. I am not unappreciative of the distinct advantages that this method has in handling the facts. I realize how valuable it has been to us to be able to marshal our results under a few simple assumptions, yet I cannot but fear that we are rapidly developing a sort of Mendelian ritual by which to explain the extraordinary facts of alternative inheritance.

The Rise of *Drosophila* and Cytogenetics

By the time this paper and others similar in tone appeared in print, an unusual, strange event took place. (I am relating the story as I heard it from Dr E.G. Anderson, who was at that time a graduate student of R.A. Emerson at Cornell University.)

C.W. Woodworth, an entomology student, introduced *Drosophila* to the Harvard laboratory of William Castle, and Morgan also used it as a tool for his embryology class. One day, he wanted to demonstrate the phototropism of the flies. As Mrs Lillian Morgan opened a matchbox containing *Drosophila*, Professor Morgan went to the window and told the students to watch how the flies would come towards him. Facing the flies, Dr Morgan discovered a white-eyed one. He became interested in it, but, despite the assistance of the students, the fly escaped. Next day, a mutant male was captured and thus the future of genetics was changed.

In 1910 and 1911, Morgan, an embryologist, published the first genetics paper on 'sex-limited' inheritance. This was new for *Drosophila* and Morgan but not for genetics. Four years earlier, Doncaster and Raynor (1906), working with the *Abraxas* moth, discovered criss-cross inheritance and, despite the assistance of William Bateson, the puzzle could not be rationalized. Their hypotheses broke down.

Miss N.M. Stevens and Professor Edmund Wilson each showed in 1905 that the 'unknown' X chromosome of Henking (1891) was actually a sex-determining chromosome. Wilson and Morgan were colleagues at Columbia University and they knew about each other's work. Thus, sex linkage was a simple inference.

There was, by that time, a lot of interest in chromosomes. Before the turn of the century, several authors had published chromosome numbers, including that of humans. Bardeleben observed about 16, while Flemming was sure that there were more than 16 (Sutton, 1903). De Winiwarter (1912) in sectioned testes observed 46 autosomes + an X chromosome but no Y chromosome. The latter is, of course, the smallest: according to the human genome draft (excluding gaps) it contains only 21.8 megabases versus the X chromosome, which has 127.7 megabases (Lander *et al.*, 2001). In the ovaries, de Winiwarter observed, correctly, a total of 46 chromosomes. During the following decades, various numbers were reported even by the same investigators (von Nachtsheim, 1959). In 1952, T.C. Hsu (von Nachtsheim, 1959), using a hypotonic solution, claimed 48, but subsequently Tijo and Levan (1956) showed, by a similar technique, adding also colchicine, beyond any doubt that humans have only 46 (von Nachtsheim, 1959).

A historical irony is that, in 1953, Cyrill Darlington, one of the most renowned cytologists, published a popular book *Facts of Life* with a photomicrograph of Hsu on the cover and showing only 46 chromosomes, but he cited it as evidence for 48 human chromosomes (von Nachtsheim, 1959).

The problem remained controversial, although the majority of cytologists confirmed that 46 was the correct number. M. Kodani in several papers between 1956 and 1958 reported 46, 47 and 48 chromosomes in both Japanese and US white individuals (von Nachtsheim, 1959).

Various banding techniques were developed during the 1970s by Torbjörn Casperson and associates and were expanded by others, which yielded the human chromosome pictures as they are used for cytogenetic maps (Caspersson *et al.*, 1968). By 1996, Speicher *et al.*, using multiplex fluorescence *in situ* hybridization (FISH) technology, distinguished each human chromosome with a distinct colour (Fig. 1.6).

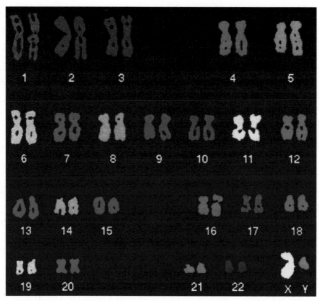

Fig. 1.6. Multiplex FISH staining by 1996 distinguished each human chromosome by their special fluorescent colour. (Courtesy of Michael R. Speicher.)

Let us jump back in time to 1903, when Walter Sutton published an epoch-making paper on chromosomes in heredity. He correctly asserted that the chromosomes are not separated by paternal and maternal groups, although the two groups are equivalent. There are two distinct types of nuclear divisions, equational and reductional (van Beneden, 1883). The chromosomes retain their individuality in the process. He assumed with Bardeleben that there are 16 chromosomes in humans and thus they may produce $16 \times 16 = 256$ gametic combinations. The 256 gametic types can thus produce $256 \times 256 = 65{,}536$ phenotypes. He assumed linkage, but for recombination he suggested 'segmental dominance'. His combinations are not too far from the current estimated human gene numbers.

Carl Correns, who also discovered cytoplasmic (chloroplast) inheritance, observed linkage in 1900 and suggested in 1902 a model for recombination 9 years before Morgan.

The majority of geneticists know that Carl Correns was one of the three rediscoverers of the Mendelian principles in 1900 and reported linkage in *Matthiola* in 1900.

He was also one of the discoverers of cytoplasmic inheritance (Correns, 1909).

In 1902, Correns suggested a mechanism for crossing over 9 years before Morgan's paper appeared in the *Journal of Experimental Zoology*.

> We assume that in the same chromosome the two anlagen of each pair of traits lie next to each other (A next to a and B next to b, etc.) and that the pairs of anlagen themselves are behind each other. A, B, C, D, E, etc. are the anlagen of parent I; a, b, c, d, e, etc. are those of parent II. Through the usual cell and nuclear divisions the same type of products are obtained as the chromosomes split longitudinally . . . When one pair contains antagonistic anlagen, while the rest of the pairs are formed of two identical types of anlagen, or the anlagen are 'conjugated' as they are in *Matthiola* hybrids, which I have described, then further assumptions are necessary . . . Then AbCdE/aBcDe and aBcDe/AbCdE yield both AbCdE and aBcDe; ABcdE/abCDe and abCDe/ABcdE both ABcdE and abCDe, etc.

Another really remarkable paper is slowly sinking into oblivion or is totally misrepresented. On 9 July 1909 (more than two decades earlier than the *Neurospora* work of

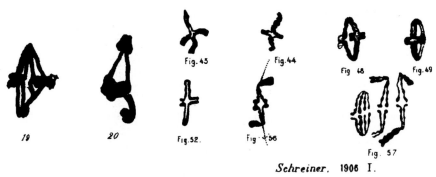

Schreiner. 1906 I.

Fig. 1.7. Meiotic configurations, borrowed from the zoological literature, used by Janssens (1909) to support his idea of chiasmatypy, in current terminology crossing over and recombination. (See text for detailed explanation.)

Carl Lindegren in 1932), F.A. Janssens, Professor at the University of Louvain, Belgium, presented his theory of chiasmatypy (Fig. 1.7) in the journal *La Cellule*:

> In the spermatocytes II, we have in the nuclei chromosomes, which show one segment of two clearly parallel filaments, whereas the two distal parts diverge . . . The first division is therefore reductional for segment A and a and it is equational for segment B and b . . . The 4 spermatids contain chromosomes 1st AB, 2nd Ab, 3rd ab, and 4th aB. The four gametes of a tetrad will thus be different . . . The reason behind the two divisions of maturation is thus explained . . . The field is opened up for a much wider application of cytology to the theory of Mendel.

Elof Carlson (1966) – in his otherwise excellent book – cites this paper and even shows with some drawings that Janssens believed that recombination takes place at the two-strand stage. The drawings of Carlson are, however, nowhere in the publication of Janssens. When Morgan discovered crossing over 2 years later, he acknowledged the priority of Jannsens, who, however, had only cytological evidence.

Morgan's student, Sturtevant, constructed the first genetic map and recognized inversions as crossing-over inhibitors. Morgan, Bridges and Muller revealed the basic mechanics of recombination. Bridges discovered non-disjunction, deletion, duplication and translocation. The list above includes only the most significant

discoveries of the chromosomal theory of inheritance.

Bateson, the great champion of genetics, who coined the term genetics and whom, in 1926, T.H. Morgan eulogized with these words: 'His rectitude was beyond all praise and recognized by friend and foe alike,' concluded a memorial lecture in 1922 at the University of Pennsylvania with the following warning:

> I think we shall do genetical science no disservice if we postpone acceptance of the chromosome theory in its many extensions and implications. Let us distinguish fact from hypothesis. It has been proved that, especially in animals, certain transferable characters have a direct association with particular chromosomes. Though made in a restricted field this is a very extraordinary and most encouraging advance. Nevertheless the hope that it may be safely extended into a comprehensive theory of heredity seems to me ill-founded, and I can scarcely suppose that on wide survey of genetical facts, especially those so commonly witnessed among plants, such an expectation would be entertained. For phenomena to which the simple chromosome theory is inapplicable, save by the invocation of a train of subordinate hypotheses, have been there met with continually, as even our brief experience of some fifteen years has abundantly demonstrated.
>
> (Bateson, 1926)

Morgan very successfully exploited the potentials of his 'fly room' and trained a remarkable series of students (Bridges,

Sturtevant, Muller,* Dobzhansky, Curt Stern, Bonnier, Komai, Gabritchevsky, Olbrycht, Altenburg, Weinstein, Gowen, Lancefield, Mohr, Nachtsheim, E.G. Anderson, Jack Schultz and others), whose work became the foundation of classical genetics and the main menu of textbooks for decades to come. Morgan's association with the California Institute of Technology signalled a more modern trend of genetics and the development of a younger generation of geneticists, such as Beadle,* Tatum,* Ephrussi, Delbrück,* Norman Horowitz, Lindegren, Schrader and E.B. Lewis.* The students of their students, such as Lederberg,* Doerman, Srb and others, made a lasting impact on the future course of genetics.

An interesting episode of the Cal Tech and the preceding period of Morgan has been recorded by Henry Borsook (1956). In the late 1920s, Edwin Cohn, the physical chemist, asked T.H. Morgan, the first Nobel-laureate geneticist, what his research plans were. Morgan's answer was: 'I am not doing any genetics, I am bored with genetics. But I am going out to Cal Tech where I hope it will be possible to bring physics and chemistry to bear on biology.'

Shortly after Morgan arrived at Cal Tech, Albert Einstein visited the laboratory and posed almost the same question. Morgan's answer was about the same as before. Einstein shook his head and said, 'No, this trick won't work. The same trick does not work twice. How on earth are you ever going to explain in terms of chemistry and physics so important a biological phenomenon as first love?' Sure enough, in the 1930s, Morgan could not provide an answer to Einstein's question, but at the current rate of advances of molecular neurogenetics some clues may soon be available.

Mutation

In 1927, H.J. Muller in *Drosophila* and independently L.J. Stadler (1928) in barley and maize proved that X-rays can induce mutations.

The Nobel-laureate immunologist, Peter Medawar, remarked once that wise people may have expectations, but only fools make predictions. Of course, brilliant people may make brilliant errors.

In a somewhat ill-conceived manner, in 1941, at the 9th *Cold Spring Harbor Symposium on Quantitative Biology* (p. 163), H.J. Muller stated:

> We are not presenting . . . negative results as an argument that mutations cannot be induced by chemical treatment . . . It is not expected that chemicals drastically affecting the mutation process while leaving the cell viable will readily be found by our rather hit-and-miss methods. But the search for such agents, as well as the study of the milder, 'physiological' influences that may affect the mutation process, must continue, in the expectation that it still has great possibilities before it for the furtherance both of our understanding and our control over the events within the gene.

Charlotte Auerbach and J.M. Robson might have already solved the problem when belatedly – because of wartime security restrictions – in 1944 they reported successful induction of mutations with radiomimetic chemicals. Muller worked for a period of time along with Auerbach in G. Pontecorvo's laboratory in Edinburgh after his return, via Spain, from his unhappy sojourn in the Soviet Union.

Despite all, H.J. Muller was the well-deserving second geneticist recipient of the Nobel prize for his studies on mutation. *Science* magazine in November 1946 (Vol. 104, p. 483) proudly reported the award, and perhaps appropriately with a printing 'mutation' or typo.

Non-nuclear Inheritance

W. Haacke assumed in 1893 that the waltzing–walking traits of mice are located in cytoplasmic elements (the centrosome),

---

* Nobel laureates.

whereas coat colour (white–grey) segregation is assured by the reductional division of the chromosomes. 'I do not know whether the number of chromosomes present in mice had been recorded, but this number would enable us to establish the possible combinations.' The fact that he was able to obtain experimentally all 16 combinations of these four traits seemed to indicate to him the validity of this interpretation.

C. Correns (1909) and E. Baur (1909), independently, reported genuine cytoplasmic inheritance in various plants, and their findings were abundantly confirmed later.

Professor T.H. Morgan in 1926 expressed the following view: 'except for the rare cases of plastid inheritance, the inheritance of all known characters can be sufficiently accounted for by the presence of genes in the chromosomes. In a word the cytoplasm may be ignored genetically.'

John R. Preer, Jr (1963), an eminent contributor to the field, remarked:

> Cytoplasmic inheritance is a little bit like politics and religion from several aspects. First of all, you have to have faith in it. Second, one is called upon occasionally to give his opinion of cytoplasmic inheritance and to tell how he feels about the subject.

Pleiotropy

The term pleiotropy was coined by Ludwig Plate, a German geneticist, and he wrote in 1913:

> ein Gen in manchen Fällen gleichzeitig mehrere Markmale, die zu ganz verschiedenen Organen gehören können, beinflußt. Eine solche Erbeinheit habe ich . . . pleiotrop genannt [a gene in many instances can influence several traits, which can be involved with different organs].

Interestingly, in Sutton (1959) the following discussion has been recorded:

> *Fremont-Smith*: Can one gene operate only in one highly specified environment and perform only one function? Would any other environment either suppress its activity or be lethal? Or can a gene perform a variety of functions, depending upon the environment to which it is exposed?
> *Lederberg*: There is no qualitative difference in the product, depending on the environment.
> *Wagner*: But that which the gene forms acts differently in different environments.
> *Fremont-Smith*: It has no multiple potentiality at all?
> *Lederberg*: *Pleiotropism non est.*
> *Fremont-Smith*: Did you add, at the 'dogma' level?
> *Lederberg*: In terms of the primary product, that is the doctrine.

By the 1980s and 1990s, mitochondrial functions have been thoroughly studied by many geneticists. The fact that single base-pair mutations in the human mitochondrial tRNA^{Leu} and other tRNAs may cause more than single human disease is clear evidence for pleiotropy (Fig. 1.8).

Definition of the Gene

These and other recent developments may modify the definition of the gene:

> Woltereck (1909): A reaction norm.
> Sturtevant (1965): Mendel usually used the term Merkmal for what we now term gene.
> Suzuki *et al.* (1976): The fundamental physical unit of heredity.
> Klug and Cummings (1983): A DNA sequence coding a single polypeptide.
> Elseth and Baumgardner (1984): a segment of the DNA that codes for one particular product.
> Strickberger (1985): In modern terms, an inherited factor that determines a biological characteristic of an organism is called a gene.
> Russel (1992): The determinant of a characteristic of an organism.
> Gray Lab Internet Glossary (2001): Genes are formed from DNA, carried on the chromosomes and are responsible for the inherited characteristics that distinguish one individual from another. Each human individual has an estimated 100,000 separate genes.

Each of these definitions has some correct elements. Probably the best is still that of Woltereck (1909). The least pleasant one is

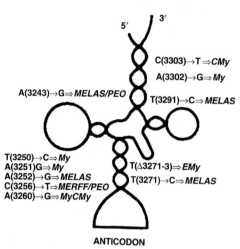

Fig. 1.8. Pleiotropic mutations in a human mitochondrial gene transcribed into UUR-tRNA[Leu]. *CMy*, cardiomyopathy; *My*, myopathy; *MELAS*, mitochondrial myopathy, encephalopthy, lactic acidosis, stroke; *PEO*, progressive external ophthalmoplegia; *EMy*, encephalopathy, myocardia; *MyCMy*, myopathy, cardiomyopathy; *MERFF*, myoclonic epilepsy, ragged-red fibres. (Redrawn and modified after Moraes, 1998.)

The vast majority of the human genes are 'mosaics' containing seven to nine exons of 120 to 150 bp each. In some genes, the exon number may be much larger (e.g. in titin about 200). In between exons, there are 1000–3500 bp introns. The size of the introns may be many times larger. The number of coding nucleotides generally varies between 1100 and 1300 bp, but the larger genes may have much longer coding sequences. The exons + introns + 5′ and 3′ untranslated sequences combined, the genomic genes, in general, extend to 14–27 kb DNA. The human dystrophin gene in the X chromosome extends to about 2300 kb. A large fraction of the human genes are alternatively spliced and thus the same gene may be translated into two, three or more kinds of proteins. Genic sequences (2–3%) are richer in GC nucleotides than the noncoding tracts. In prokaryotes, introns are rare and the genes are much smaller (Rédei, 2002).

Gene Numbers

the last. Genes are not formed from DNA. Genes are either DNA or RNA depending on the organism. Genes are not on the chromosomes but the genes are in the chromosomes; actually the DNA forms the backbone of the chromosomes. The number of human genes is most unlikely to be 100,000; the latest estimates indicate about 35,000.

How would I define briefly the gene today?

Gene: a specific functional unit of DNA (or RNA) potentially transcribed into RNA or coding for protein(s). A group of cotranscribed exons but, due to alternative splicing, exon shuffling, overlapping or using more than one promoter or termination signal, the same DNA sequence may encode more than a single protein. A common structural organization of protein-encoding genes in eukaryotes:

enhancer – promoter – leader – exons – introns – termination signal – polyadenylation signal – downstream regulators

The number of genes per genome of an organism can be estimated by molecular analysis on the basis of mRNA complexity or by total sequencing of the genome. The estimates based on mRNA can be best determined when the entire genome is sequenced. By the latter method, the single-stranded RNA phage, MS2, was found to have four genes. The gene number has also been estimated from mutation frequencies. If the overall induced mutation rate, for example, is 0.5 and the mean mutation rate at selected loci is 1×10^{-5}, then the number of genes is $0.5/(1 \times 10^{-5}) = 50,000$. Although this method is loaded with some errors, the estimates so obtained appear reasonable. On the basis of mutation frequency in *Arabidopsis*, the total number of genes was estimated to be about 28,000 (Rédei *et al.*, 1984). The number of genes of *Arabidopsis* was estimated to be 25,498 after sequencing the genome. In *Drosophila*, ~17,000 genes were claimed on the basis of mRNA complexity. On the basis of the sequenced genome, the estimate is now ~13,600. During the 1930s, C.B. Bridges counted ~5000 bands in the *Drosophila* salivary chromosomes and for many years it was assumed that each band represented a gene. By 1928, John Belling had counted 2193 chromomeres in the pachytene chromosomes of *Lilium pardalinum* and assumed that this number corresponded to the number of genes (Belling, 1928).

Nucleotide sequencing of 69 salivary bands in the long arm of chromosome 2 of *Drosophila* pointed to the presence of 218 protein-coding genes, 11 tRNAs and 17 transposable element sequences within that ~2.9 Mb region. The shotgun sequencing of the *Drosophila* genome identified ~13,600 genes encoding 14,113 transcripts because of alternate splicing. In humans, 75,000–100,000 genes were expected on the basis of physical mapping; of these about 4000 may involve hereditary illness or cancer. The human gene number estimates in 2001 still varied from ~27,000 to ~150,000. In *Saccharomyces*, in the 5885 open reading frames, 140 genes encode rRNA, 40 snRNA and 270 tRNA. About 11% of the total protein produced by the yeast cells (proteome) has a metabolic function; 3% each is involved in DNA replication and energy production; 7% is dedicated to transcription; 6% to translation; and 3% (200) constitutes different transcription factors. About 7% is concerned with transporting molecules and about 4% constitutes structural proteins. Many proteins are involved with membranes. In *Caenorhabditis*, 19,099 protein-coding genes are predicted on the basis of the sequencing of the genome. The minimal essential gene number has also been estimated by comparing presumably identical genes in the smallest free-living cells *Mycoplasma genitalium* and *Haemophilus influenzae*, both completely sequenced. Insertional inactivation mutagenesis indicated the minimal number to be ~265–300. In *Caenorhabditis elegans*, about 20 times more genes are indispensable for survival. In higher organisms, the number of open reading frames may be larger than the number of essential genes (Rédei, 2002).

The gene number may not accurately reflect the functional complexity of a genome or organism because the combinatorial arrangement of proteins may generate great diversity and specificity. A synopsis of how these genes function would be most rewarding if one were able to present it even as a bird's-eye view. The most simplistic views are in the daily newspapers.

This sweeping and selective overview has missed out much important historical development. Fortunately, quantitative and population genetics have been better dealt with by many speakers than I ever could have attempted. I shall deal briefly with an area with which I was especially involved and which may have great significance for the future from the viewpoint of quantitative analysis.

Transformation

Transformation goes back to the late 1920s but it became practical with eukaryotes in the late 1970s and the early 1980s. By the mid-1980s, I was fortunate to be associated with researchers at the Max-Planck-Institut, Cologne, Germany. This effort resulted in the application of *in vivo* transcriptional gene-fusion technology to plants (Fig. 1.9).

In a similar manner, *in vivo* translational gene-fusion vectors can also be constructed in which there are no stop codons in front of the reporter gene and the translation initiation codon is removed, so the plant host protein and the reporter gene fusion would be facilitated. The results of these experiments are illustrated in Figs 1.10–1.13 and the captions provide explanations. Obviously transformation provides unique opportunities to manipulate the genome and facilitates new insights into how in plants indigenous genes and foreign genes are regulated and expressed.

The Future of Genetics

It is customary to finish presentations with some predictions. Why I am shying away

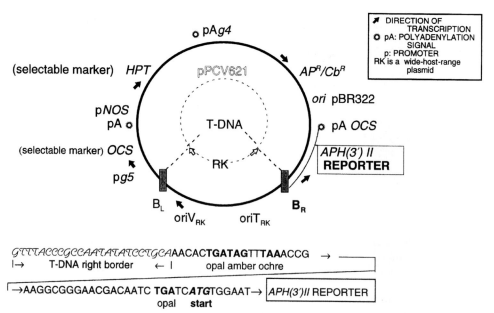

GTTTACCCGCCAATATATCCTGCA**AACAC**CT**GATAGTTTAAA**CCG → ──────────
|→   T-DNA right border   ← | opal amber ochre   ┐
                                                                                  │
┌→AAGGCGGGAACGACAATC **TGA**TC**ATG**TGGAAT→ |APH(3')II REPORTER|
             opal   **start**

Fig. 1.9. The critical feature of this *in vivo* transcriptional gene-fusion vector is that the reporter *(aph(3')II,* luciferase or *gus)* has no promoter and it is fused to the right border of the T-DNA. The structural gene of the reporter can be expressed only if it integrates behind a plant promoter that can provide the promoter function. In front of the structural gene here, there are four nonsense codons to prevent the fusion of the proteins with any plant peptide. (Based on oral communications by Dr Csaba Koncz.)

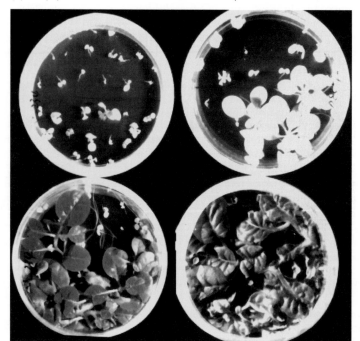

Fig. 1.10. Transformation of tobacco with transcriptional gene-fusion vector. Each Petri plate contains two segregating transgenic progenies (in order to save labour) and the size and colour of the transformed plants reflect the strength and time of function of the promoter. The selective agent, the antibiotic kanamycin, bleaches the non-transformed plants. (G.P. Rédei and Yan Yao, unpublished.)

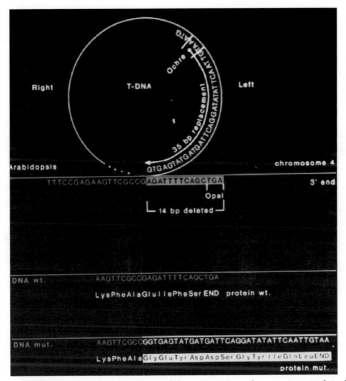

Fig. 1.11. The assumed mechanism of integration of the T-DNA into chromosome 4 of *Arabidopsis* and induction of recessive mutation with normal transmission.

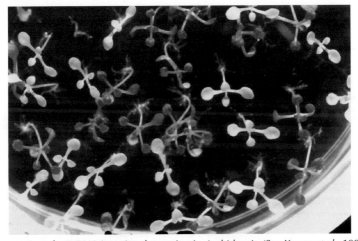

Fig. 1.12. Segregation of a T-DNA insertional mutation in *Arabidopsis*. (See Koncz *et al.*, 1990.)

from general forecasts may be justified by a few more quotes.

Erwin Chargaff, the discoverer of the Chargaff rule, which was one of the cornerstones for the construction of the Watson and Crick model of the double helix, stated in 1955, only 6 years before the nature of the genetic code had been revealed:

> I believe, however, that while the nucleic acids, owing to the enormous number of possible sequential isomers, could contain

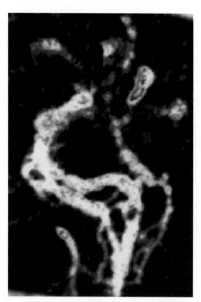

Fig. 1.13. Expression of bacterial luciferase, under the control of a constitutive promoter, in *Arabidopsis* is most intense at the maximal metabolic activity, represented in false colours. (From G.P. Rédei, C. Koncz, and W. Langridge, unpublished.)

enough codescripts to provide a universe with information, attempts to break the communications code of the cell are doomed to failure at the present very incomplete stage of our knowledge. Unless we are able to separate and to discriminate, we may find ourselves in the position of a man who taps all the wires of a telephone system simultaneously. It is, moreover, my impression that the present search for templates, in its extreme mechanomorphism, may well look childish in the future and that it may be wrong to consider the mechanisms through which inheritable characteristics are transmitted or those through which the cell repeats itself as proceeding in one direction only.

J.D. Watson's letter to Max Delbrück on 22 March 1953 sounds quite surprising today:

I have a rather strange feeling about our DNA structure. If it is correct, we should obviously follow it up at a rapid rate. On the other hand, it will, at the same time, be difficult to avoid the desire to forget completely about nucleic acid and to concentrate on other aspects of life.
(Judd, 1979)

The only remark I care to make is that I feel very assured that the role of quantitative genetics in basic biology and applied sciences will increase.

References

Aldred, C. (1961) *The Egyptians. Ancient People and Places.* Thames and Hudson, New York.

Aristotle (384–322 BC) *Generation of Animals*, Book IV, Pt I.

Auerbach, C. and Robson, J.M. (1944) Production of mutations by allylisothiocyanate. *Nature (London)* 154, 81.

Bateson, W. (1926) Segregation: being the Joseph Leidy Memorial Lecture at the University of Pennsylvania, 1922. *Journal of Genetics* 16, 201–235.

Baur, E. (1909) Das Wesen und die Erblichkeitsverhältnisse der 'varietates albomarginatae hort' von *Pelargonium zonale. Z. Ind. Abst. Vererb.-Lehre* 1, 330–351.

Belling, J. (1928) The ultimate chromomeres of *Lilium* and *Aloe* with regard to the numbers of genes. *Univ. Calif. Publ. Bot.* 14, 307–318.

Blixt, S. (1975) Why didn't Gregor Mendel find linkage? *Nature (London)* 256, 206.

Borek, E. (1965) *The Code of Life.* Columbia University Press, New York.

Borsook, H. (1956) Informal remarks 'by way of a summary'. *Journal of Cellular and Comparative Physiology* 47 (Suppl. 1), 283–286.

Carlson, E.A. (1966) *The Gene: a Critical History.* Saunders, Philadelphia.

Caspersson, T., Farber, S., Foley, G.E., Kudynowski, J., Modest, E.J., Simonsson, E., Wagh, U. and Zech, L. (1968) Chemical differentiation along metaphase chromosomes. *Experimental Cell Research* 49, 219–222.

Chargaff, E. (1955) On the chemistry and function of nucleoproteins and nucleic acids. *Istituto Lombardo (Rend. Sci.)* 89, 101–115.

Correns, C. (1900) G. Mendels Regel über das Verhalten der Nachkommenschaft der Rassenbastarde. *Ber. Dtsch. Bot. Ges.* 18, 158–168.

Correns, C. (1902) Über den Modus und den Zeitpunkt der Spaltung der Anlagen bei den Bastarden vom Erbsen-Typus. *Bot. Zeitung* 60(2), 65–82.

Correns, C. (1909) Verebungsversuche mit blaß(gelb)grünen und buntblättrigen Sippen bei Mirabilis, Urtica und Linaria annua. *Z. Ind. Abst. Vererb.-Lehre* 1, 291–329.

de Vries, H. (1900) Sur la loi de disjonction des hybrides. *Comptes Rendus Académie de Science Paris* 130, 845–847.

de Winiwarter, H. (1912) Études sur la spermatogenèse humaine. *Arch. Biol.* 27, 91–189 + VIII plates.

Doncaster, L. and Raynor, G.H. (1906) Breeding experiments with Lepidoptera. *Proceedings of the Zoological Society London* 1, 125–133.

Elseth, G.D. and Baumgardner, K.D. (1984) *Genetics*. Addison-Wesley, Reading.

Fisher, G.H. (1968) Ambiguity of form: old and new. *Perception and Psychophysics* 4, 189–192.

Fisher, R.A. (1936) Has Mendel's work been rediscovered? *Annals of Science* 1, 115–137.

Focke, W.O. (1881) *Die Pflanzenmischlinge. Eine Beitrag zur Biologie der Gewächse*. Borntrager, Berlin.

Geitler, L. (1938) *Chromosomenbau*. Bornträger, Berlin.

Gray Lab Internet Glossary (2001) www.graylab.ac.uk/omd/

Haacke, W. (1893) Die Träger der Vererbung. *Biol. Zbl.* 13, 525–542.

Haartman, J.J. (1751) *Plantae Hybridae*. Uppsala, Sweden.

Henking, H. (1891) Über Spermatogenese bei *Pyrrhocoris apetrus. Zeitschr. Wissensch. Zool.* 51, 685–736.

Janssens, F.A. (1909) La théorie de la chiasmatypie. Nouvelle interprétations des cinèses de maturation. *Cellule* 25, 389–411.

Johannsen, W. (1909) *Elemente der exakten Erblichkeitslehre*. Fischer, Jena.

Judson, H.F. (1979) *The Eighth Day of Creation*. Simon and Schuster, New York.

Klug, W.S. and Cummings, M.R. (1983) *Concepts of Genetics*. Merrill, Columbus.

Koncz, C., Mayerhofer, R., Koncz-Kalman, Z., Nawrath, C., Reiss, B., Rédei, G.P. and Schell, J. (1990) Isolation of a gene encoding a novel chloroplast protein by T-DNA tagging in *Arabidopsis thaliana. EMBO Journal* 9, 1337–1346.

Křiženecký, J. and Němec, B. (1965) *Fundamenta Genetica*. Czechoslovak Academy of Sciences, Prague.

Lander, E.S. with the International Human Genome Sequencing Consortium (2001) Initial Sequencing and Analysis of the Human Genome. Nature (London) 409, 860–921.

Lindegren, C.C. (1932) The genetics of *Neurospora* II. Segregation of the sex factors in asci of *N. crassa, N. sitophila* and *N. tetrasperma. Bulletin of the Torrey Botany Club* 59, 119–138.

Medvedev, Z. (1969) *The Fall and Rise of T.D. Lysenko*. Columbia University Press, New York.

Mendel, G. (1866) Versuche über Pflanzenhybriden. *Verh. Naturforsch. Verein. Brünn* 4, 3–47.

Migula, W. (1897) *System der Bakterien. Handbuch der Morphologie, Entwicklungsgeschichte und Systematik der Bakterien*. Fischer, Jena.

Moraes, C.T. (1998) Characteristics of mitochondrial DNA disease. In: Singh, K.K. (ed.) *Mitochondrial DNA Mutations and Aging, Disease and Cancer*. Springer-Verlag, Berlin, pp. 167–184.

Morgan, T.H. (1909) What are 'factors' in Mendelian explanations? *American Breeders' Association Report* 5, 365–368.

Morgan, T.H. (1910) Sex limited inheritance in Drosophila. *Science* 32, 120–122.

Morgan, T.H. (1911) An attempt to analyze the constitution of the chromosomes on the basis of sex-limited inheritance in Drosophila. *Journal of Experimental Zoology* 11, 365–412.

Morgan, T.H. (1919) *The Physical Basis of Heredity*. Lippincott, Philadelphia, Pennsylvania.

Morgan, T.H. (1926) Genetics and physiology of development. *American Nature* 60, 490–515.

Muller, H.J. (1927) Artificial transmutation of the gene. *Science* 66, 84–87.

Muller, H.J. (1941) Induced mutations in *Drosophila*. Cold Spring Harbor *Symposium on Quantitative Biology* 9, 151–167.

Nägeli, C.W. (1867) Cited in Gregor Mendel's letter to Carl Nägeli on 18 April 1867, p. 5. *Genetics* 35 (Suppl.), 1950.

Nägeli, C.W. (1884) *Mechanisch-physiologische Theorie der Abstammungslehre*. Oldenburg, Munich.

Olby, R.C. (1966) *Origins of Mendelism*. Schocken Books, New York.

Pearson, K. (1904) On a generalized theory of alternative inheritance, with special reference to Mendel's law. *Philosophical Transactions of the Royal Society A* 203, 53–86.

Plate, L. (1913) *Vererbungslehre*. Engelmann, Leipzig.

Pliny, G., II (AD 23–79) *C. Plinii Secundi Naturalis Historia*. Edited by Detlefsen, D. Berolini, Apud Weidmannos, 1866–1882.

Preer, J.R., Jr (1963) Discussion. In: Burdett, W.J. (ed.) *Methodology in Basic Genetics.* Holden-Day, San Francisco, p. 374.

Provine, W.B. (1971) *The Origins of Theoretical Population Genetics.* University of Chicago Press, Chicago.

Punnett, R.C. (1949) Early days of genetics. *Heredity* 4, 1–10.

Rédei, G.P. (1974) Steps in the evolution of genetic concepts. *Biol. Zbl.* 93, 385–424.

Rédei, G.P. (2002) *An Encyclopedia Dictionary of Genetics, Genomics and Proteomics.* John Wiley & Sons, New York.

Rédei, G.P., Acedo, G.N. and Sandhu, S.S. (1984) Mutation induction and detection in *Arabidopsis.* In: Chu, E.H.Y. and Generoso, W.M. (eds) *Mutation, Cancer, and Malformation.* Plenum, New York, pp. 285–313.

Roberts, H.F. (1965) *Plant Hybridization before Mendel.* Hafner, New York.

Russel, P.J. (1992) *Genetics.* Harper Collins, New York.

Speicher, M.R., Ballard, S.G. and Ward, D.C. (1996) Karyotyping human chromosomes by combinatorial multi-fluor FISH. *Nature Genetics* 12, 368–375.

Stadler, L.J. (1928) Mutations in barley induced by X-rays and radium. *Science* 68, 186–187.

Stevens, N.M. (1905) Studies in spermatogenesis with special reference to the 'accessory chromosome'. *Carnegie Institute Publications* 36, 1–32.

Stoneking, M. (2001) Single nucleotide polymorphisms. From the evolutionary past . . . *Nature (London)* 409, 821–822.

Strickberger, M.W. (1985) *Genetics.* Macmillan, New York.

Sturtevant, A.H. (1965) *A History of Genetics.* Harper & Row, New York.

Sutton, E.H. (ed.) (1959) *Genetics: Genetic Control of Protein Structure.* Josiah Macey

Foundation, Madison Press, Madison, Wisconsin, pp. 160–161.

Sutton, W.S. (1903) The chromosomes in heredity. *Biological Bulletin* 4, 231–248.

Suzuki, D.T., Griffiths, A.J.F. and Lewontin, R.C. (1976) *An Introduction to Genetic Analysis.* Freeman, San Francisco.

Sylvester, J. (1880) *The Complete Works,* Vol. I, p. 420. Edited by A.B. Grossart. Printed for private circulation, Edinburgh, UK.

Tijo, J.H. and Levan, A. (1956) The chromosome number of man. *Hereditas* 42, 1–6.

Udney Yule, G. (1907) On the theory of inheritance of quantitative compound characters on the basis of Mendel's laws. In: *Report of the Third International Conference on Genetics.* Spottiswood, London, pp. 140–142.

van Beneden, E. (1883) Recherches sur la maturation de l'oeuf et la fécondation. *Arch. Biol.* 4, 265–640.

von Nachtsheim, H. (1959) Chromosomenaberrationen beim Säuger und ihre Bedeutung für die Entstehung von Mißbildungen. *Naturwiss.* 46, 637–645.

Watson, J.D. (1953) Letter to Max Delbrück on 22 March 1953. Quoted after Judson, H.F. (1979) *The Eighth Day of Creation.* Simon and Schuster, New York, p. 229.

Weiling, F. (1966) Hat J.G. Mendel bei seinen Versuchen 'zu genau' arbeitet? – Der χ^2-Test und seine Bedeutung für die Beurteilung genetischer Spaltungsverhältnisse. *Züchter* 36, 359–365.

Wilson, E.B. (1905) The chromosomes in relation to the determination of sex in insects. *Science* 22, 500–502.

Woltereck, R. (1909) Weitere experimentelle Untersuchungen über Artveränderung, speziell über des Wesen quantitativer Unterschiede der Daphniden. *Verhandl. Dtsch. Zool. Ges.* 110–172.

Section I

Genomics, Quantitative Trait Loci and Tissue Culture

2 Quantitative Genetics, Genomics and the Future of Plant Breeding

Bruce Walsh

Department of Ecology and Evolutionary Biology, University of Arizona, Tucson, AZ 85721, USA

Introduction

Quantitative genetics (in its various guises) has been the intellectual cornerstone of plant breeding for close to 100 years. While the roots of Mendelian genetics, and its rediscovery, are firmly in the hands of plant breeders, it was Fisher's (1918) variance decomposition paper that marks the modern foundation for both quantitative genetics and plant breeding. We are now embarking on the age of genomics, and so it is reasonable to speculate on the implications of both partial and whole genome sequences for quantitative genetics. Likewise, the tools of modern quantitative genetics have been developed in four separate fields: plant breeding, animal breeding, human genetics and evolutionary genetics. Unfortunately, for a variety of reasons, migration of information between these fields has not been what it should be. Thus, it is also an appropriate time to enquire whether useful tools have been developed in these other fields that may be helpful to plant breeders of today and the genomics-based breeders of the near future.

Quantitative Genetics in the Age of Genomics

It took just 100 years to move from the rediscovery of Mendel to the complete sequencing of a higher plant (*Arabidopsis*). Preliminary analysis (*Arabidopsis* Genome Initiative, 2000) of the *Arabidopsis* sequence detected 25,500 genes, almost double the number of detected *Drosophila* genes (13,600) and of the same order as the most recent estimates of the number of human genes (25,000–40,000). Roughly 45% of all detected *Arabidopsis* genes are present in four or more copies, so that the 25,500 genes can be classified into roughly 11,600 protein types. While *Arabidopsis* offers a key portal to the genomes of other higher plants, because of their much larger size, whole-genome sequences of most of the major crops are unlikely to be forthcoming in the near future. Although there is no reason to expect that the basic set of protein types will be any greater than around 12,000, the extensive evolutionary history of polyploidization and segmental duplication in most higher plants (Wendel, 2000) suggests that far more than the 26,000 or so genes of *Arabidopsis* may be present in many of our major crops. While it may seem obvious that genomic information (both current and forthcoming) will have a major impact on quantitative genetics, just how will this information modify quantitative genetics as currently practised and how profound will the change be?

Classical vs. neoclassical quantitative genetics

In the classical (Fisherian) quantitative genetics framework, an observed phenotype (y) is regarded as the sum of genetic (g) and environmental (e) effects plus an interaction between genotype and environmental values ($g \times e$):

$$y = u + g + e + g \times e \qquad (2.1)$$

where u is the population mean. In most plant-breeding situations, this basic model usually involves a further decomposition of the environmental effects (for example, into plot, block, temporal and/or location components), but, for brevity, we ignore these (easily introduced) extensions in our discussion. Resemblance between relatives, line-cross analysis and other approaches are then used to estimate the variance components associated with g, e and $g \times e$ and these are used to predict selection response and the expected values of individuals from defined crosses and other useful quantities. Classical quantitative genetics has been enormously successful in serving the needs of the plant-breeding community, but the use of just a few statistical descriptors (the variance components) to describe the complex underlying genetics tends to leave an uneasy feeling with many of my molecular colleagues. One extreme view put forward by some of these colleagues is that the age of genomics ushers in the death-knell for quantitative genetics, as we can move away from a degree of statistical uncertainly to a framework of known genes. While this view is rather naïve (and often seems to be based more on a general aversion to statistics than solid reasoning), it is quite clear that genomic information does usher in a

transformation of quantitative genetics. The classical framework assumes we only know individual phenotypes and the degree of relationship between these individuals. Genomic information allows for an extended (or neoclassical) framework that also incorporates genetic-marker information. In particular, under the neoclassical framework, genetic markers provide information on the genotypic value of an individual. If m denotes an observed (often multilocus) marker genotype, the basic model now becomes:

$$y = u + G_m + g + e + g \times e \qquad (2.2)$$

where G_m is the genotypic value associated with this genotype. The classical model (Equation 2.1) is a standard random-effects model, where our interest is in estimating variances associated with g, e and $g \times e$. In contrast, the neoclassical approach leads to a mixed model, with G_m regarded as fixed effects. Genotype–environment interactions involving G_m can be incorporated into (Equation 2.2), as can potential epistatic interactions between a particular marker genotype and the background genes ($G_m \times g$). In the neoclassical framework, the marker genotype effect G_m can be directly estimated as a fixed effect. As we detail below, genomic information may be able to suggest particular genes for consideration as candidates, so that G_m is the genotypic value associated with the multilocus genotypes at the candidate loci. The percentage of the total phenotypic variance accounted for by the candidates provides a direct measure of their importance. The hope expressed by my molecular colleagues is that the variance accounted for by the classical part of this model ($g + e + g \times e$) is small relative to the variance accounted for by the marker information (G_m).

Even in situations where the known genotypes do indeed account for a large fraction of trait variance, the importance of particular genotypes may be quite fleeting. Genotypic values for particular loci are potentially functions of the background genotypes and environments, and hence can easily change as crops evolve and as the biotic (pests and pathogens) and abiotic (farming practices) environments change. Further, mutation will generate new quantitative trait loci (QTL), and a candidate locus that works well in one population may be (at best) a very poor predictor in another. Even if only a modest number of QTL influence a trait, then (apart from clones) each individual is essentially unique in terms of its relevant genotypes and the particular environment effects it has experienced. If epistasis and/or genotype–environment interactions are significant, any particular genotype may be a good, but not exceptional, predictor of phenotype. Quantitative genetics provides the machinery necessary for managing all this uncertainty in the face of some knowledge of important genotypes. Indeed, variance components allow one to quantify just how much of the variation is unaccounted for by the known genotypes. A critical feature of quantitative genetics is that it allows for the proper accounting of correlations between relatives in the unmeasured genetic values (g).

We do not mean to paint an overly harsh view of the importance of being able to identify key genotypes. Rather, we simply wish to introduce a little caution to dampen completely unrestrained enthusiasm. It is clear that there are enormous benefits to being able to predict even a fraction of an individual's genotypic value, given a set of genetic markers. For example, even a small increase in the probability of fixation of an advantageous allele during inbreeding to form pure lines can have a dramatic effect. Suppose an F_1 population is segregating favourable alleles at ten loci, and we first inbred to fixation and then select among lines. In the absence of selection, the probability of fixation of any single favourable allele is 0.5. A quick binomial calculation shows that a sample size of 2357 is required to have a 90% probability that at least one line contains all favourable alleles. If we are able to increase the probability of fixation by only 50% (to 0.75), only 40 individuals are required, which is roughly a 60-fold reduction. If 20 favourable alleles are segregating, the reduction is 3330-fold (from 2,414,434 to 725). Likewise, with known genotypes in hand, searches for genotype–environment

interaction ($G \times E$) are much more direct, allowing for the possibility of searching for major genes that are highly adaptive to specific environments.

Genomics and candidate loci

Much of the above discussion of a more generalized view of quantitative genetics has assumed that we know the genotypes (and their effects) at a number of QTL. Given that very few QTL have been fully isolated, we are still far from achieving this goal. At present, the genotypes (m) scored in Equation 2.2 usually consist of anonymous markers shown by statistical association to be linked to QTL. Such marker-assisted selection can result in a significant improvement of the selection response, particularly when the heritability of a character is low (Lande and Thompson, 1990). Given that even a QTL of large effect is typically only initially localized to a region of around 20–50 cM, anonymous markers can easily be 10–25 (or more) cM from the actual QTL. Selecting directly on the QTL genotypes (as opposed to linked markers) increases the efficiency of selection, unless the marker is very tightly linked to the QTL. The relative efficiency of a single generation of selection on linked markers (as opposed to directly selecting on the genotypes) scales as $(1-2c)^2$ or roughly $1-4c$ for a tightly linked marker (where c is the recombination frequency between the marker and QTL).

Hence, we would like to be able to at least localize more tightly linked markers and ideally, to screen potential candidate loci directly to see if they are the QTL. Direct tests of association with a small set of candidates are a more powerful approach than a genome-wide screen using a set of anonymous markers. The difficult issue is selecting the candidates in the first place, and the hope (indeed, often the core assumption) of genomics is that the full genome sequence will, in time, greatly facilitate the selection of candidate genes. A variety of genomics tools (reviewed below) has indeed been suggested to help in the search for candidates. At present, the use of these tools is generally restricted by economic, rather than biological, constraints. One of the major trends expected over the next decade will be to make these tools economically feasible for just about any trait or crop of interest.

Basic genomic tools

Perhaps the single most useful tool is dense marker maps. It is these maps that allow for QTL mapping, association studies and marker-assisted introgression, to name just a few uses. The most obvious tool is the whole-genome sequence. With the complete sequence in hand (or even a partial sequence highly enriched for coding sequences), one can construct any number of DNA chips. These microarrays containing a large number of defined DNA sequences can be used for screening the expression of a large number of genes (via hybridization) in particular tissues (expression-array analysis), for probing a related genome for homologous genes of interest, and for many other interesting possibilities that we are only beginning to consider. Besides faster and cheaper sequencing, a major factor facilitating future genomic projects is the ability to use sequence homology to bootstrap from a model system to a related species. For example, we can use sequences from *Arabidopsis* to probe for homologous genes in related species, and any recovered sequences can, in turn, be used to design probes for even more distantly related species. Given the tendency for many plant genes to exist in large multigene families, the advantage of a full genome sequence is that all members of a particular family can be used as probes, increasing the chance of identifying at least one homologous gene from a related species. Once again, any recovered homologues can themselves be used as probes for other family members within the target species. Thus, centred around a key model system, we can imagine the search for homologous genes spreading out in phylogenetic space like ripples on a pond, reaching ever more phylogenetically distant species.

Prediction of candidate genes

With a genomic sequence in hand, either from the plant of interest or from a sufficiently close relative, how can one use this information to find possible candidate genes? The most straightforward approach is to search the genome for sequences with homologies to known candidate genes from another species. For example, a gene known (say) to create variation in plant architecture in maize (*Zea mays* L.) can be used to probe related grasses. If homologues can be found, association tests between trait values and variation in the potential candidate(s) can be performed. A more brute-force approach is to first limit a QTL to a confidence region (as small as possible) and then use the genomic sequence from that region either to suggest candidates for further testing (see below) or, by simply screening all the genes in this region using expression arrays, to search for those whose expression pattern is consistent with the character of interest. Even in such apparently direct expression studies, some caution is in order. For example, a gene turned on in seed tissue is certainly a candidate for yield, but another gene expressed only in root tips (and hence probably excluded from further consideration) may have a more important effect on yield if it increases the plant's ability to gather and store energy.

More generally, the hope is that the DNA sequence itself may provide clues to its potential as a candidate. The growing field of proteomics has generated an extensive catalogue of known protein motifs, offering the possibility of making some (albeit crude) deductions about the functions of particular open reading frames, such as whether the resulting protein spans a membrane, is involved in transport, is directed to a particular organelle, etc. Even such partial information may be informative in keeping or excluding potential candidates. The hope for the future is that we shall be able to read the regulatory sequences to deduce the expression pattern of a gene directly from its DNA sequence, in essence, during the array expression studies *in silico* (via the computer). Again, the above-mentioned caveat (that expression in very different, and unexpected, tissues may have a dramatic effect on the character of interest) holds.

Transgenics

The tool from biotechnology that perhaps excites breeders the most is the ability to construct transgenics, importing a novel gene into an organism. While transgenics can be constructed for many crop species, their phenotypes are rather unpredictable. Insertion of a new sequence cannot currently be targeted to specific sites, but rather is largely random, with the location of insertion significantly influencing the level of gene expression. Further, plants often suppress multiple-copy genes, which can have a further impact on a transgenic. Even when these issues are resolved, it is clear for the foreseeable future that transgenic technology is restricted to importing genes of major effect. The success in improving a character by importing a modest to large suite of genes of smaller individual effects (but perhaps a great cumulative effect) is less certain, given the above concerns about consistency of expression. However, this could also work in the breeder's favour in that a gene of modest effect may have a more dramatic effect than that expected due to position effect. A further complication is that the introduction of genes of large effect (perhaps generated by using a high-expression promoter on a gene of otherwise modest effect) can often have significant pleiotropic consequences for a number of characters besides the target and hence can reduce crop performance in other aspects. Selection for lines possessing modifiers to reduce any associated deleterious effects is thus a key step in the improvement of an initial transgenic line. Quantitative-genetic machinery can suggest those lines with the greatest potential for modifiers, for example, by searching for lines with large $G_m \times g$ interactions in favour of less deleterious side-effects.

Fishing for useful variation in natural or weakly domesticated populations

The area where genomics may eventually offer the largest pay-off to breeders is in the search for useful genes in natural and/or weakly domesticated populations. The source populations or species, from which modern crops descend, harbour far more genetic diversity than is present in the limited set of highly domesticated lines currently in use for food production. The ability to localize genes of significant effect and, subsequently, to introgress these into cultivars without generating undesirable side-effects on performance is a key aim of breeders. As we detail in the next section, a variety of useful approaches for searching natural populations for genes of interest has been developed in other fields of quantitative genetics.

Useful Tools from Other Fields of Quantitative Genetics

The broad arena of quantitative genetics consists of four rather distinct fields – plant breeding, animal breeding, evolutionary genetics and human genetics (it could be argued that tree breeders form a fifth field). Although all four draw upon the basic foundations of quantitative genetics, each has rather distinct and often non-overlapping literatures, and the information flow across the fields has often been rather restrictive. One consequence of this restricted flow is that approaches for a specific problem are often independently reinvented. A more interesting consequence is that since practitioners in each of the fields are faced with unique issues and constraints, each has developed a number of useful tools that are (unfortunately) often not widely known to outsiders. We (Lynch and Walsh, 1998; Walsh and Lynch, 2003) have recently tried to bring all these tools and approaches together into a unified general framework for quantitative genetics. Since many of these field-limited tools are both largely unknown and yet of potential interest to plant breeders, I conclude by briefly reviewing a few of the more promising approaches (especially those with applications at the quantitative genetics–genomics interface). Extension of some of these approaches, to be of value to plant breeders, may require some non-trivial modifications.

Plant breeding

For a start, it is useful to remind plant breeders of some of the tools they routinely use that are not well known (or at least not widely appreciated) to geneticists outside of the field. As a consequence of having to deal with a diversity of mating systems (most importantly selfing) and sessile individuals, issues that plant breeders tend to focus on more than other quantitative geneticists include the creation and selection among inbred lines and their hybrids, G × E and competition. Some important tools have already migrated from plant breeders to quantitative genetics as a whole. One example is line-cross-based analysis (generation means, diallels), which has seen an increasing use in evolutionary genetics. Somewhat surprisingly, many quantitative geneticists have been a little slow in drawing upon the wealth of field-plot designs, especially analyses for dealing with G × E, that plant breeders have accrued. For example, while additive main effects and multiplicative interaction (AMMI) models (Gollob, 1968; Mandel, 1971; Gauch, 1988, 1992; Gauch and Zobel, 1988; Zobel *et al.*, 1988) and biplots (Gabriel, 1971; Kempton, 1984) have become important tools for plant breeders (as several chapters in this book illustrate), they are generally unknown outside the field. The correct formulation for the covariance between relatives under inbreeding (e.g. Cockerham, 1983) is another important tool developed by plant breeders that has remained largely unappreciated (but see Abney *et al.*, 2000).

Animal breeding

Animal breeders face designs involving complex pedigrees, large half-sib or (more rarely) full-sib families, long lifespans and overlapping generations (many of these same issues are faced, to an even greater extent, by tree breeders). The machinery of predicting breeding values by best linear unbiased prediction (BLUP) (reviewed by Henderson, 1984; Mrode, 1996; Lynch and Walsh, 1998) and the estimation of variance components by restricted maximum likelihood estimation (REML) (reviewed by Searle *et al.*, 1992; Lynch and Walsh, 1998) have been developed by animal breeders to address these concerns. BLUP/REML easily allows for arbitrary pedigrees (through specification of appropriate relationship matrices) and for the estimation of a large number of fixed factors. This BLUP/REML framework is a very appealing one from a genomics standpoint, as scored genotypes of interest can be treated as fixed effects, and complex (fixed and/or random) models with both background genotypes and structured environmental effects can also be introduced.

A second area that may be of interest to plant breeders is the extensive work of animal breeders on maternal effects designs (e.g. Lynch and Walsh, 1998, ch. 23). Although several of these designs are not easily transferred to plant-breeding systems (some are based on cross-fostering offspring), they, none the less, are useful reading when thinking about the importance of maternal effects, a topic that often seems to be overlooked by plant breeders. The widespread availability of cloned individuals can greatly facilitate the estimation of maternal effects, and hence a determination of their importance. Recent theoretical work on the quantitative-genetic implications of endosperm by Shaw and Waser (1994) is a related topic of interest.

Finally, a major push towards the use of Bayesian methods of analysis is coming from the animal breeders (e.g. Gianola and Fernando, 1986). Just as likelihood methods replaced method-of-moments and other estimators when they became computationally feasible in the mid–late 1970s, a variety of Markov chain Monte Carlo simulation approaches (such as the Gibbs sampler) have allowed Bayesian posteriors to be computed for even very complex models (Geyer, 1992; Tierney, 1994; Tanner, 1996). The very appealing feature of a Bayesian analysis is that a marginal posterior distribution incorporates all the uncertainties introduced by having to estimate other parameters of less interest. For example, a model that estimates the additive genetic variance must also estimate a number of other variance components and fixed effects. The marginal posterior for the additive variance naturally incorporates all the uncertainty introduced by having to estimate these additional nuisance parameters. Bayesian analysis provides a powerful framework for analysis for the expected growing complexity of neoclassical models.

Evolutionary genetics

As the search for potentially useful genes moves to natural populations, machinery from evolutionary and population genetics may prove useful. The issues of concern to evolutionary geneticists involve estimating the nature and amount of selection on a defined suite of characters and the population genetics of evolution. Three useful developments from this field may be of interest to plant breeders.

First, methods for estimating the nature of natural selection on any characters of interest have been developed (Lande and Arnold, 1983; Arnold and Wade, 1984a,b; Schluter, 1988; Crespi and Bookstein, 1989; Schluter and Nychka, 1994; Willis, 1996). This machinery allows the breeder to estimate the nature of natural selection on any measurable suite of characters, separating selection into direct and indirect effects (due to selection on correlated characters). A detailed understanding of the nature of natural selection in either wild or domesticated populations can provide the breeder with valuable insight into characters that can further improve performance.

Secondly, there is a rich literature from population genetics dealing with detection of selection from a population sample of DNA sequences (reviewed by Kreitman, 2000). An interesting application of these methods was the finding of reduced levels of polymorphism (consistent with directional selection) in the 5′ control region of the *teosinte-branched 1* gene involved in major morphological differences between teosinte (*Zea mexicana* L.) and domesticated maize (Wang *et al.*, 1999). With a collection of candidate genes in hand, one can search for signatures of selection in homologues from natural populations. Much of the theory underlying tests of selection follows from the explosive development of coalescent theory (reviewed by Hudson, 1991; Tavare and Balding, 1995; Fu and Li, 1999), which describes the genealogy (the distribution of the times to a common ancestor) for a random sample of a particular DNA sequence from the population. There are obvious extensions of this theory to deal with issues of concern to quantitative geneticists, such as estimating the degree of relationship based on molecular data and the fine-mapping of QTL using very tightly linked markers (e.g. Slatkin, 1999; Zollner and von Haeseler, 2000).

Finally, there has been considerable progress in the theoretical analysis of finite locus models (as opposed to the traditional infinitesimal models routinely used by breeders), and these developments are reviewed in Burger (2000). In particular, the response to selection when the underlying distribution of genotypic (or breeding) values is not Gaussian has received significant attention (Barton and Turelli, 1987; Turelli, 1988; Turelli and Barton, 1990, 1994). Such developments in finite-locus models provide a useful framework for predicting selection response when partial genotypic information is available.

Human genetics

The final field of quantitative genetics from which plant breeders may wish to draw upon is developments in human genetics, where small family sizes and a lack of controlled mating designs are common occurrences. Despite these obvious limitations, human geneticists have been rather successful at mapping genes, and some of their tools may prove useful to plant geneticists, especially when trying to isolate genes of interest from natural or weakly domesticated populations for which defined inbred lines may not be available.

One powerful approach has been to use sib pairs to map QTL (reviewed and extended by Abel and Muller-Myhsok, 1998; Monks *et al.*, 1998; McPeek, 1999; Elston and Cordell, 2001), and these approaches can be applied to the offspring from single plants in natural populations (although suitable modifications would have to be introduced to account for selfing). One complication that both human geneticists and plant breeders working with natural populations face when attempting association studies (between candidate genotypes and trait values) is that false positives can be created by population substructure (or stratification). For example, if a marker is very common in a particular subpopulation, and that subpopulation also carries alleles for a trait of interest at high frequencies, then, if the population structure is not accounted for, the marker can show an association with the trait simply by being a predictor of the population from which an individual is drawn. Human geneticists account for any potential population structure by using the transmission–disequilibrium test (TDT), which compares whether an allele is transmitted or not transmitted from a parent to an offspring showing the trait of interest (Spielman *et al.*, 1993; Knapp, 1999a,b). Another powerful tool of human geneticists is fine-mapping of genes by linkage disequilibrium, using the historical recombinations (as reflected in the decay of disequilibrium) that occur between a tightly linked marker and a gene of interest to fine-map that locus

(Hastbacka *et al.*, 1992; Graham and Thompson, 1998; Slatkin, 1999).

A final important tool with its roots in human genetics is random-effects models to map QTL in complex pedigrees (e.g. Amos, 1994; Gessler and Xu, 1996; Xie *et al.*, 1998; Yi and Xu, 1999, 2000). The idea behind a random-effects model is simply to estimate the trait variance associated with any particular genomic region, using anonymous markers that span the genome. As with BLUP/REML, this approach can accommodate both arbitrary pedigrees and numerous fixed effects. It is certainly an approach to consider for QTL mapping in many settings.

Conclusions

The age of genomics is a very exciting time for quantitative geneticists. While the view is often suggested that genomics will reduce the importance of quantitative genetics, in fact the opposite is true. Straightforward modifications of classical quantitative-genetic models provide the natural framework for handling both phenotypic and genotypic information. Equally important for breeders to consider are powerful tools developed in other fields of quantitative genetics, only a few of which have been discussed here.

References

Abel, L. and Muller-Myhsok, B. (1998) Robustness and power of the maximum-likelihood-binomial and maximum-likelihood-score methods, in multiple linkage analysis of affected-sibship data. *American Journal of Human Genetics* 63, 638–647.

Abney, M., McPeek, M.S. and Ober, C. (2000) Estimation of variance components of quantitative traits in inbred populations. *American Journal of Human Genetics* 66, 629–650.

Amos, C.I. (1994) Robust variance-components approach for assessing genetic linkage in pedigrees. *American Journal of Human Genetics* 54, 535–543.

Arabidopsis Genome Initiative (2000) Analysis of the genome sequence of the flowering plant *Arabidopsis thaliana*. *Nature* 408, 796–815.

Arnold, S.J. and Wade, C. (1984a) On the measurement of natural and sexual selection, theory. *Evolution* 38, 709–719.

Arnold, S.J. and Wade, C. (1984b) On the measurement of natural and sexual selection, applications. *Evolution* 38, 720–734.

Barton, N.H. and Turelli, M. (1987) Adaptive landscapes, genetic distances and the evolution of quantitative characters. *Genetical Research* 49, 157–173.

Burger, R. (2000) *The Mathematical Theory of Selection, Recombination, and Mutation.* John Wiley & Sons, New York.

Cockerham, C.C. (1983) Covariances of relatives from self-fertilization. *Crop Science* 23, 1177–1180.

Crespi, B.J. and Bookstein, F.L. (1989) A path-analytic model for measurement of selection on morphology. *Evolution* 43, 18–28.

Elston, R.C. and Cordell, H.J. (2001) Overview of model-free methods for linkage analysis. *Advances in Genetics* 42, 135–150.

Fisher, R.A. (1918) The correlation between relatives on the supposition of Mendelian inheritance. *Transactions of the Royal Society of Edinburgh* 52, 399–433.

Fu, X.-Y. and Li, W.-H. (1999) Coalescing into the 21st century: an overview and prospects of coalescent theory. *Theoretical Population Biology* 56, 1–10.

Gabriel, K.R. (1971) Biplot display of multivariate matrices with applications to principal component analysis. *Biometrika* 58, 453–467.

Gauch, H.G., Jr (1988) Model selection and validation for yield trials with interaction. *Biometrics* 44, 705–715.

Gauch, H.G., Jr (1992) *Statistical Analysis of Regional Yield Trials: AMMI Analysis of Factorial Designs.* Elsevier, Amsterdam, The Netherlands.

Gauch, H.G., Jr and Zobel, R.W. (1988) Predictive and postdictive success of statistical analysis of yield trials. *Theoretical and Applied Genetics* 76, 1–10.

Gessler, D.D.G. and Xu, S. (1996) Using the expectation or the distribution of identical-by-descent for mapping quantitative trait loci under the random model. *American Journal of Human Genetics* 59, 1382–1390.

Geyer, C.J. (1992) Practical Markov chain Monte Carlo (with discussion). *Statistical Science* 7, 473–511.

Gianola, D. and Fernando, R.L. (1986) Bayesian methods in animal breeding theory. *Journal of Animal Science* 63, 217–244.

Gollob, H.F. (1968) A statistical model which combines features of factor analysis and analysis of variance techniques. *Psychometrika* 33, 73–115.

Graham, J. and Thompson, E.A. (1998) Disequilibrium likelihoods for fine-scale mapping of a rare allele. *American Journal of Human Genetics* 63, 1517–1530.

Hastbacka, J., de la Chapelle, A., Kaitila, I., Sistonen, P., Weaver, A. and Lander, E. (1992) Linkage disequilibrium mapping in isolated founder populations, diastrophic dysplasia in Finland. *Nature Genetics* 2, 204–211.

Henderson, C.R. (1984) *Applications of Linear Models in Animal Breeding.* University of Guelph, Guelph, Ontario.

Hudson, R.R. (1991) Gene genealogies and the coalescent process. In: Futuyama, D.J. and Antonovics, J. (eds) *Oxford Surveys in Evolutionary Biology.* Oxford University Press, Oxford.

Kempton, R.A. (1984) The use of biplots in interpreting variety by environment interactions. *Journal of Agricultural Science* 103, 123–135.

Knapp, M. (1999a) The transmission/disequilibrium test and parental-genotype reconstruction: the reconstruction-combined transmission/disequilibrium test. *American Journal of Human Genetics* 64, 861–870.

Knapp, M. (1999b) A note on power approximations for the transmission/disequilibrium test. *American Journal of Human Genetics* 64, 1177–1185.

Kreitman, M. (2000) Methods to detect selection in populations with application to the human. *Annual Review of Genomics and Human Genetics* 1, 539–559.

Lande, R. and Arnold, S.J. (1983) The measurement of selection on correlated characters. *Evolution* 37, 1210–1226.

Lande, R. and Thompson, R. (1990) Efficiency of marker-assisted selection in the improvement of quantitative traits. *Genetics* 124, 743–756.

Lynch, M. and Walsh, B. (1998) *Genetics and Analysis of Quantitative Traits.* Sinauer Associates, Sunderland, Massachusetts.

McPeek, M.S. (1999) Optimal allele-sharing statistics for genetic mapping using affected relatives. *Genetic Epidemiology* 16, 225–249.

Mandel, J. (1971) A new analysis of variance model for non-additive data. *Technometrics* 13, 1–8.

Monks, S.A., Kaplan, N.L. and Weir, B.S. (1998) A comparative study of sibship tests of linkage and/or association. *American Journal of Human Genetics* 63, 1507–1516.

Mrode, R.A. (1996) *Linear Models for the Prediction of Animal Breeding Values*. CAB International, Wallingford, UK.

Schluter, D. (1988) Estimating the form of natural selection on a quantitative trait. *Evolution* 42, 849–861.

Schluter, D. and Nychka, D. (1994) Exploring fitness surfaces. *American Naturalist* 143, 597–616.

Searle, S.R., Casella, G. and McCulloch, C.E. (1992) *Variance Components*. John Wiley & Sons, New York.

Shaw, R.G. and Waser, N.M. (1994) Quantitative genetic interpretations of postpollination reproductive traits in plants. *American Naturalist* 143, 617–635.

Slatkin, M. (1999) Disequilibrium mapping of a quantitative-trait locus in an expanding population. *American Journal of Human Genetics* 64, 1765–1773.

Spielman, R.S., McGinnis, R.E. and Ewens, W.J. (1993) Transmission test for linkage disequilibrium, the insulin gene region and insulin-dependent diabetes mellitus (IDDM). *American Journal of Human Genetics* 52, 506–516.

Tanner, M.A. (1996) *Tools for Statistical Analysis*, 3rd edn. Springer-Verlag, New York.

Tavare, S. and Balding, D.J. (1995) Coalescents and genealogical structure under neutrality. *Annual Review of Genetics* 29, 410–421.

Tierney, L. (1994) Markov chains for exploring posterior distributions (with discussion). *Annals of Statistics* 22, 1701–1762.

Turelli, M. (1988) Population genetic models for polygenic variation and evolution. In: Weir, B.S., Eisen, E.J., Goodman, M.M. and Namkoong, G. (eds) *Proceedings of the Second International Conference on Quantitative Genetics*. Sinauer Associates, Sunderland, Massachusetts, pp. 601–618.

Turelli, M. and Barton, N.H. (1990) Dynamics of polygenic characters under selection. *Theoretical Population Biology* 38, 1–57.

Turelli, M. and Barton, N.H. (1994) Genetic and statistical analyses of strong selection on polygenic traits. What, me normal? *Genetics* 138, 913–941.

Walsh, B. and Lynch, M. (2003) *Evolution and Selection of Quantitative Traits*. Sinauer Associates, Sunderland, Massachusetts.

Wang, R.-L., Stec, A., Hey, J., Lukens, L. and Doebley, J. (1999) The limits of selection during maize domestication. *Nature* 398, 236–239.

Wendel, J.G. (2000) Genome evolution in polyploids. *Plant Molecular Biology* 42, 225–249.

Willis, J.H. (1996) Measures of phenotypic selection are biased by partial inbreeding. *Evolution* 50, 1501–1511.

Xie, C.D., Gessler, D.G. and Xu, S. (1998) Combining data from different line crosses for mapping quantitative trait loci using the identical-by-descent based variance component method. *Genetics* 149, 1139–1146.

Yi, N. and Xu, S. (1999) A random model approach to mapping quantitative trait loci for complex binary traits in outbred populations. *Genetics* 153, 1029–1040.

Yi, N. and Xu, S. (2000) Bayesian mapping of quantitative trait loci under the IBD-based variance component model. *Genetics* 156, 411–422.

Zobel, R.W., Wright, M.J. and Gauch, H.G., Jr (1988) Statistical analysis of a yield trial. *Agronomy Journal* 80, 388–393.

Zollner, S. and von Haeseler, A. (2000) A coalescent approach to study linkage disequilibrium between single-nucleotide polymorphisms. *American Journal of Human Genetics* 66, 615–628.

3 Why Quantitative Geneticists should Care about Bioinformatics

Nicholas A. Tinker

Agriculture and Agri-Food Canada, 960 Carling Avenue, Building 20, Ottawa, Ontario K1A 0C6, Canada

Introduction

Whether you regard it as a scientific discipline or a supporting technology, bioinformatics has gained enough status and recognition to dominate scientific journals, graduate programmes, careers and entire research institutions. Bioinformatics could be defined as 'the storage, retrieval and analysis of information about biological structure, sequence, or function' (Altman, 1998). This is a flexible definition: it relates to information about virtually any biological enquiry – from molecular biology to ecology, and it includes computational biology – an area that sometimes defines itself separately. The term bioinformatics, however, is used most frequently in relation to molecular biology or genomic research. For example, the instructions to authors of the journal *Bioinformatics* state that it is 'a forum for the exchange of information in the fields of computational molecular biology and genome bioinformatics' and historical perspectives of bioinformatics refer exclusively to sequence analysis and other biomolecular research (Roberts, 2000; Trifonov, 2000).

Should quantitative geneticists care about bioinformatics? Regardless of how you define bioinformatics, the answer is 'yes', and the purpose of this discussion is to explain why. In supporting this statement, I shall attempt to provide information about current activities in bioinformatics that might be of interest to researchers in quantitative genetics and plant breeding. This discussion is organized into two sections: first, a general overview of the major components of bioinformatics, followed by a discussion of bioinformatics as it relates to research in quantitative genetics and plant breeding.

Bioinformatics is Characterized by a Set of 'Core Activities'

The broad definition of bioinformatics chosen in the introduction allows for any number of special niches within the field. The narrower genomics-based interpretation of bioinformatics is, however, represented by a well-defined set of activities. A suggested curriculum for bioinformatics students (Altman, 1998) includes a list of 14 'core bioinformatics topics'. I have modified this list by merging these topics into a set of eight general areas, discussed below. The descriptions that follow are superficial, intended only to orientate the reader towards 'classical bioinformatics' and to prepare for further discussion of the relevancy of bioinformatics to quantitative genetics and plant breeding.

Pairwise sequence alignment

I find that the easiest example with which to illustrate the importance of bioinformatics is to say: 'here is a sequence; find some sequences that are similar'. This example can even spark the interest of an English scholar or musician (both of whom may be jealous of the extent to which biological data are organized and publicly available). Depending on the listener, some background explanations may be required:

- Sequences can be either DNA or protein. Since DNA codes for protein, it is possible to match a protein sequence with a DNA sequence by translating the DNA into all possible proteins.
- You may have to search through all known DNA sequences. Genbank currently contains more than 10 million DNA sequences, made up of more than 11 thousand million base pairs.
- It might be best to search for protein matches, since protein sequences are more conserved than DNA sequences. There are currently about 600,000 proteins in Genbank.
- Matches need to be subjected to statistical tests for significance: imperfect matches are useful and interesting, but you will see many partial matches due to random chance. If you are looking hard to find imperfect similarity, you will need to tolerate the risk of finding meaningless matches.
- Matching sequences may require introducing gaps. These gaps may occur anywhere in either of the two sequences being matched.

The above explanations usually suffice to illustrate the size and complexity of the problem. The listener is then assured that the solution to this problem is in better hands than mine and that many good algorithms and computer tools are available. The acronym BLAST (for basic local alignment search tool) is easily remembered and it is easy to demonstrate some Internet BLAST tools that are readily available (e.g. http://www.ncbi.nlm.nih.gov/BLAST/ or http://www2.ebi.ac.uk/blast2/). There are many different BLAST programs and algorithms (not all are called BLAST), and these result from a large body of research into optimization and statistical theory (e.g. Altschul et al., 1997; Pearson et al., 1997; Agarwal and States, 1998; Karplus et al., 1998).

Searching one query sequence at a time is best done through Internet tools and public databases. However, local search capability is needed to search against special or proprietary sequence sets or to submit a large number of query sequences to the same database. Since BLAST searching is a fundamental activity in almost every bioinformatics laboratory, it is useful to put the computational scale and hardware requirements into perspective. Our laboratory routinely searches expressed sequence tags (ESTs), each being a partial DNA sequence derived from a randomly cloned mRNA, against a downloaded version of Genbank. We use the program BLASTALL (Altschul et al., 1997) on a Windows NT machine with two 550 MHz Pentium III processors and 750 Mb RAM. With this configuration, a search of 100 nucleotide sequences (average length 800 bp each) against the non-redundant protein database (600,000 sequences, containing 189,012,571 amino acids) takes approximately 40 min. For major updates of our database, six processors may be busy for most of a week.

Multiple sequence alignment, clustering and phylogenetic inference

When first encountered, the concept of multiple sequence alignment may seem like an extension of pairwise sequence alignment. These two processes, however, exist for entirely different reasons. Pairwise alignment is usually used in conjunction with a search for identity and is fine-tuned to be extremely fast and efficient. Multiple alignment is performed after a set of similar sequences have been identified and the user is concerned with finding the best possible alignment that satisfies the minimum number of alterations from one sequence to the next. This is normally done with protein sequences, but can also be performed with nucleotide sequences. Multiple alignment is used primarily to extract information about gene evolution, to perform hierarchical clustering and to infer phylogeny. The theory and algorithms developed for alignment, clustering and phylogenetic inference are extensive, and will not be discussed here. Recent reviews of this area are provided by Doolittle (1999), Doyle and Gaut (2000) and Phillips *et al.* (2000). Clustering and phylogenetic inference are not restricted to DNA or protein sequence, and much of this area will already be familiar to most classical and quantitative geneticists.

One of the applications of multiple sequence alignment is the discovery of conserved and variable regions within a gene. A highly conserved region of amino acid sequences can indicate an active site within the resulting three-dimensional protein. Thus, sequence alignment is a useful complement to studies of protein structure and function. The alignment of DNA sequences is less useful for this task, because nucleotide changes can have neutral effects. Nucleotide alignment, however, can be very useful for the discovery and design of molecular markers. Our laboratory, in collaboration with Dr Diane Mather of McGill University, is using alignment of DNA sequences from grasses for the discovery and development of molecular markers in oat (*Avena sativa*) and barley (*Hordeum vulgare*). On a much larger scale, studies of DNA sequence alignment have been used to develop a set of 1.4 million single nucleotide polymorphisms (SNPs), which have been placed on the human genome map to facilitate gene discovery through linkage disequilibrium (International SNP Map Working Group, 2001).

Fragment assembly and mapping

Fragment assembly also shows superficial similarity to both pairwise and multiple sequence alignment, but, again, serves a different purpose. In this case, the purpose is to assemble fragments of DNA sequence into contiguous strands. The assumption is made that all fragments come from the same organism, or at least from organisms similar enough for base changes to be rare. Fragment assembly involves the following steps:

- Removal of unwanted sequence (cloning vector or poor-quality sequence).
- Locating pairwise overlaps.
- Resolving overlaps to build larger contiguous strands (contigs).
- Improving alignments in overlapping regions by introducing gaps.
- Manual verification and editing.
- Generation of a consensus sequence.

Depending on the application, the manual verification step can be critical. If raw sequencing files are available (chromatograms showing evidence for each base), these are used to resolve any positions where ambiguity remains.

Fragment assembly is the final step in the generation of complete genomic sequences. In organisms where this has been possible, sequence-based fragment assembly follows the preliminary work of genetic and physical mapping. Genetic mapping of molecular polymorphisms is familiar to many researchers in plant genetics, but the ability to build physical maps of large fragments has been limited to model plant organisms with small genomes. Physical mapping, such as that performed in the public human-genome-sequencing effort, is achieved by dividing the genomic sequence into large (100–200 kb) fragments called bacterial artificial chromosomes (BACs). These BACs are digested and separated by electrophoresis. Overlapping BACs can be identified because they produce significant numbers of identically sized restriction fragments. The assembly of BAC sequences based on overlaps is analogous to sequence-based fragment assembly, but does not require any sequence data. Once a BAC assembly is produced, a set of BACs is chosen to represent a minimum tiling pattern. These BACs are digested into smaller fragments for sequencing and, finally, sequence-based fragment assembly.

Sequence-based fragment assembly is not limited to complete genome-sequencing projects. It also finds use in smaller-scale sequencing projects, such as gene characterization and map-based cloning. Another application has been in the creation of non-redundant sets of ESTs. Most ESTs represent partial genes, but many ESTs are small and poorly characterized. Large-scale fragment assembly provides an automated method to identify EST fragments that belong to the same molecule, and then to generate longer consensus sequences that may represent full genes. We use this method to remove redundancy from our EST collections prior to building expression arrays. By comparing the length of consensus EST sequences with the length of similar proteins (found using BLAST), we are able to identify those that represent complete gene sequences.

Feature prediction and annotation

Feature prediction can refer to the identification of genes within complete genomic sequences or to the identification of motifs, introns, promoters and other features that are diagnostic of gene or genome function. Annotation means recording these identifications in a database. Annotation is also used in general reference to the assignment of putative gene function. Genes are characterized by open reading frames (ORFs) – stretches of DNA without stop codons. However, since eukaryotic genes are interrupted by introns, many true genes are represented by ORFs that are shorter than spurious ORFs. As any statistician will recognize, this is especially true when dealing with a global search through a large set of data, such as the human genome sequence. Thus, gene identification requires additional criteria, such as similarity to known genes or the presence of diagnostic features (e.g. Guigo *et al.*, 2000).

Protein structure, modelling and dynamics

Due to increasing interest in 'proteomics' and 'metabolomics' as the logical succes-

sors to large-scale genomics initiatives, it is inevitable that this area of bioinformatics will become increasingly important. Many additional topics, such as monitoring of protein expression patterns using two-dimension gels, are not captured by this heading, but should be expected to complement efforts to study protein structure and function. Likewise, most of the bioinformatics tools already mentioned as components of genomics will continue to be important in the area of proteomics.

Despite a large amount of interest and effort, bioinformatics has not yet been able to accomplish what some would see as its most important goal: the *ab initio* prediction of tertiary protein structure. Accurate predictions of protein structure can only be achieved by tedious experimental procedures, although bioinformatics plays an important role in the visualization of these experimental predictions and in the discovery of implications. Computer-based modelling of tertiary protein structure is limited to predicting the effects of small changes (site-directed mutagenesis) or to modelling the structure of a protein that has significant sequence homology to a protein with a known structure (a procedure known as threading). Some of the limitations in predicting protein structure may be overcome through brute-force computational power, which may be delivered by massive networks of interconnected computers.

Another application of bioinformatics is in understanding pathways and metabolism. Existing knowledge of metabolic pathways has developed to the extent that very few researchers can be expected to have comprehensive background. Information about biochemical pathways is available at databases such as KEGG (http://star.scl.genome.ad.jp/kegg/) and EXPASY (http://expasy.cbr.nrc.ca/cgi-bin/search-biochem-index). Many seemingly unrelated pathways can interact through common precursors or products, and efforts to modify metabolic processes need to consider both the optimal target (e.g. a rate-limiting step) and the possibility of indirect effects. Information about gene expression that is currently being developed through genomics or proteomics needs to be cross-referenced to appropriate steps in metabolic pathways, so that researchers in diverse fields can access and interpret the implications. Through gene annotations, it is usually possible to link with information about metabolic process. For example, enzyme sequences in the SwissProt database (http://www.expasy.ch/sprot/) are annotated using a standardized enzyme classification (EC) system that can be cross-referenced to known steps in metabolic pathways. Since protein interactions are the basis of metabolic pathways, there is hope that bioinformatics will play an additional role in the detailed analysis of pathway dynamics and, perhaps, the discovery of previously unknown processes. The need for detailed records of protein interactions is recognized in the development of the Biomolecular Interaction Network Database (BIND) (Bader *et al.*, 2001).

Support of laboratory biology

Laboratory support may appear to be the most mundane task in bioinformatics, but it is essential for the operation of high-throughput strategies, such as sequencing and studies of gene expression. For the bioinformaticist engaged in laboratory support, the resulting interactions with biologists can be a rewarding diversion from the task of solving informatics challenges. Laboratory support can range from solving day-to-day computer problems to building new software to collect, store and analyse data.

A major task in laboratory support has arisen from the technology of microarray expression studies (discussed later). These experiments have the capability to rapidly generate hundreds of thousands of data points. Each datum must be identified by meaningful relationships to other databases, and each can be associated with a specific region of a large graphical image. Unlike gene sequencing (where the original image is often discarded), microarray images may be required at a later date to help validate a discovery. Since important information may

be derived through combining information from multiple experiments, there is a need to preserve all data in a common environment. This need serves to introduce the next subject.

Design and implementation of databases

Databases vary in scope and in scale, and the designer must consider whether the database will satisfy a specific local need or whether it is intended to be served publicly on the Internet. Major public databases require consideration of relevancy, scope, database structure and interface, as well as curation and maintenance. The choice of database software or programming language can be influenced by these factors, but may also be influenced by available resources and expertise. The decision may also be based on an existing software licence or the presence of a related database that uses specific software. For example, the plant-genome databases sponsored by the US Department of Agriculture (http://ars-genome.cornell.edu/) use the ACeDB format; therefore, different plant genome databases can share a common set of tools and benefit from cumulative expertise.

Smaller, local databases require most of the considerations described above, but they may not require the programming of a special interface. Our in-house EST database is implemented in Microsoft Access. This choice was made because Access provides a simple interface for database design, and because it requires minimal maintenance. Thus, a single person with limited programming expertise can develop and maintain a database. The Access software also provides a simple interface for users to query the data in sophisticated ways. Rather than providing predefined queries, all of our genomics users take a short course that enables them to understand the database structure and learn to develop their own queries using this visual interface. Disadvantages of this system are that it is limited to use on a local network and it does not scale well.

Mining data from heterogeneous sources

The number of Internet databases related to molecular biology and genomics is impressive. Maintaining a complete listing is a bioinformatics challenge in itself. Each year, the journal *Nucleic Acids Research* devotes its first issue to coverage of new database developments, including an updated directory of databases. The current directory (Baxevanis, 2001) lists almost 300 databases.

Some databases (e.g. the National Centre for Biotechnology Information (NCBI) and European Molecular Biology Laboratory (EMBL)) attempt to provide 'one-stop bioinformatics shopping'. In these two examples, public data from diverse sources are collected and maintained in a standardized format that can be accessed directly through a common set of tools. The NCBI data are accessed though a tool called Entrez, whereas the EMBL databases are accessed using a sequence retrieval system (SRS). Each of these systems has different strengths. For example, complex data-mining questions can be automated in SRS using the PERL scripting language, whereas Entrez is well designed for simple Web-based queries where the user wants to be presented with results as well as links to related information.

Not all questions can be answered from one data source, so the bioinformaticist must be familiar with many diverse and heterogeneous databases. In some cases, it is possible to communicate automatically with multiple databases. For example, relational databases that support a common interface called open database connectivity (ODBC) can be addressed simultaneously using the structured query language (SQL) to filter information based on appropriate relationships between tables. It is even possible to make tables from multiple databases appear as though they belong to a single database. However, these approaches depend on knowledge of the underlying data structure in both databases and on identifying consistent fields in all databases. Sometimes a bioinformaticist must resort to creative strategies, e.g.

- Retrieve data from database no. 1 to text file.
- Parse text file to transform field contents and remove unnecessary fields.
- Retrieve data from database no. 2 to text file.
- Compare text files to identify common elements.

This general strategy can be implemented using a scripting language (e.g. PERL) to produce and parse intermediate text files. An alternate strategy is to build a local relational database using subsets extracted from other databases. This strategy results in an efficient local database that can serve more than one purpose.

Why should Quantitative Geneticists Care about Bioinformatics?

Quantitative geneticists analyse information about biological structure and function. This makes them computational biologists and, in a broad sense, bioinformaticists. Plant breeders, often quantitative geneticists themselves, are also actively engaged in bioinformatics. It may be possible to run a successful breeding programme without a computer, but most breeders need to perform extensive record-keeping, pedigree-tracking and statistical analyses. The purpose of this section is not to extend or argue about the definition of bioinformatics, but rather to discuss the importance of 'classical bioinformatics' (as described in the first part of this chapter) in traditional areas of quantitative genetics or plant improvement and to emphasize the potential role of quantitative geneticists in areas of bioinformatics related to genomics.

Quantitative geneticists are bioinformatics experts

There is some distinction between computational biology and bioinformatics, the former being concerned with computational theory and algorithms and the latter with information systems and data mining. However, the distinction is blurred, since most practitioners require knowledge of both areas. Claverie (2000) believes that both areas have strayed from being bona fide theoretical branches of molecular biology to being merely observational, 'phenomenological' approaches. In supporting this, Claverie (2000) describes the ideal bioinformaticist as a theoretical biologist, analogous to the theoretical physicist, who strives to capture the essence of molecular mechanisms within abstract models. In this sense, the quantitative geneticist already seems to be the ideal bioinformaticist.

A good example of how bioinformatics can be used in the development of biological theory is provided by Mendoza *et al.* (1999, 2000). These authors have developed and tested mathematical models to capture what is known about the genetic control of flower morphogenesis and root-hair development in *Arabidopsis thaliana*. These models lead to predictions of possible effects of external stimuli or altered allelic states. Although this is a mathematical model of developmental circuitry, rather than a genetic analysis of a population of phenotypes, it makes me believe that statistical geneticists possess some of the expertise that is needed to understand the complex interaction of genes at a molecular level. This could include the ability to develop and apply abstract models based on biological assumptions, as well as the ability to understand and account for phenotypic variance, population structure, epistasis, genotype–environment interaction, and other statistical realities.

Our understanding of biological processes comes from observing natural or altered biological systems. The approach of the molecular biologist is often to dissect a single component, whereas the approach of the quantitative geneticist is to discover a model that explains multiple components. One of the roles of bioinformatics in this process will be to give the quantitative geneticist better access to discrete components (genes) from which to build or test complex models. These models might test existing theory about a metabolic process within a background of genetic and phenotypic noise,

or they might lead to the discovery of which genes or pathways are most crucial in the development of an 'economic phenotype'.

Quantitative genetics is becoming quantitative genomics

Since the narrow-sense definition of bioinformatics seems restricted to genomics, it is worth reflecting on the difference between 'genetics' and 'genomics'. Genetics is traditionally defined as the study of inheritance, whereas the term genomics was invented to describe high-throughput or large-scale studies of genome sequence and gene structure. Genomics was then subclassified into 'structural' and 'functional' genomics. At this point, I become confused, especially by the difference between genetics and functional genomics. However, one interpretation is this:

- Genetics means inferring the presence of genes or allelic state based on phenotype.
- Structural genomics means describing the structure of the genome, as well as the location and structure of genes.
- Functional genomics means studying how known genes affect phenotype.

These differences may not be arbitrary, but they invite crossover. The identification of quantitative trait loci (QTL) and measurement of QTL effects are considered to be 'statistical genetics', but these activities have elements of both structural and functional genomics. Efforts to place candidate gene loci on molecular maps are driven partially by the desire to find potential associations with QTL. Efforts to merge genetic maps are being driven partially by the desire to provide additional evidence for the location and effect of QTL. Map-based cloning is a direct route from genetics, through structural genomics, to a functional-genomics result.

In the next two sections, I describe two areas that I believe will be important crossover points between statistical genetics and functional genomics. Both of these areas will require bioinformatics expertise.

Molecular markers, developed through bioinformatics techniques, can identify polymorphism in functional gene candidates

We are currently witnessing the completion of genome sequencing in several model organisms, with promise that complete sets of annotated genes will soon be available. Explaining this to a lay person is much easier than explaining the subtleties of QTL analysis or the complexities of measuring heritability. But, apart from notable results from a handful of transformation events, our food supply is still ensured by the efforts of breeders and quantitative geneticists. Most scientists involved with germ-plasm development probably agree that we shall always rely, to some extent, on the measurement and manipulation of natural genetic variability. This is not a defence of traditional territory, nor is it a sales pitch for molecular marker-assisted selection. It is an attempt to establish a proposition that important genes show natural variability, and that understanding the cause of that variability can be interesting and useful. Whether this information is used for direct genetic manipulation or to enhance traditional selection is not the concern of this discussion.

The discovery and characterization of natural genetic variability can follow two routes: the direct measurement of gene expression (as described in the next section) or inference from phenotypic measurements in a recombinant population. If you can breed for a trait, then the genes must be different. Despite the breeder's dogma that allele combinations are nearly infinite, the success of phenotypic selection implies an important role for a finite set of genes. For argument's sake, assume that there are 30,000 genes in a typical plant. We may soon have names for most of these genes but limited understanding of how they work or interact. Through knowledge of gene function or through other experimental evidence, the number of genes with potential to influence a given trait might be reduced to several hundred. If a tool-box of readily scored markers were available for this set of candidate genes, the breeder/quantitative geneticist could rapidly accumulate data on

associations between genes and phenotype. The power of this strategy would be enhanced by accumulating and coanalysing data from many populations and perhaps many different species. The advantage of basing this analysis on gene candidates rather than on arbitrary markers is that linkage disequilibrium is expected. This is of benefit within species because it may permit the assumption of common parental QTL alleles. For analysis across species, it may permit the assumption of common QTL without the need for detailed mapping of chromosomal rearrangements.

Before straying too far on an argument that can be made better by others, I shall return to the topic statement: that the markers required for QTL-based gene discovery can be developed through bioinformatics techniques. Targeting a marker towards a specific gene requires, at the minimum, a partial gene sequence. Many markers have been generated by designing a semi-arbitrary pair of PCR primers from an EST sequence. Such markers usually rely on serendipitous polymorphism at or between priming sites. Recently, a large set of gene-targeted markers has been developed in maize by designing primers to flank simple sequence repeats that have been discovered within coding sequences (see http://www.agron.missouri.edu/ssr.html). Such markers are more likely to identify polymorphisms than are randomly targeted PCR primers. A recent report (Cato *et al.*, 2001) describes a new approach for the identification of molecular markers based on EST sequences. This technique combines a gene-specific primer with a randomly targeted restriction site that may be outside the coding region. This may prove to be an extremely useful method for generating gene-specific polymorphisms based on a conserved primer that can function across multiple species. Any of the above approaches can be enhanced by comparison of gene sequences across a variety of alleles, both within and across species. Interestingly, all three different types of sequence alignment introduced at the beginning of this discussion are useful in this approach: pairwise sequence alignment (to identify and collect a set of orthologous sequences), sequence assembly (to assemble partial sequences) and multiple sequence alignment (to build consensus and identify conserved or variable regions). Other bioinformatics tools that are important in this process include primer design, identification of genes based on metabolic role, and automation of repetitive steps in the collection and alignment of sequences.

Bioinformatics and quantitative genetics facilitate the understanding of gene expression patterns

Some of the most important tools currently available in functional genomics are microarrays (e.g. Schena *et al.*, 1995), oligonucleotide 'gene chips' (Lipshutz *et al.*, 1995) and serial analysis of gene expression (SAGE) (Velculescu *et al.*, 1995). While different in procedure, these techniques achieve a common objective: the measurement of mRNA samples to identify levels or patterns of gene expression. To the quantitative geneticist, an mRNA sample is simply a phenotype. Despite the fact that each different mRNA traces to a specific gene, the entire collection of mRNA levels, captured at a specific moment in time, is analogous to a set of quantitative trait components. Like quantitative trait components, each mRNA can vary independently or (more probably) it can be correlated with other mRNAs. Also, like trait components, each mRNA level is subject to environmental variance, which can disguise its true genetic value. Finally, like trait components, each mRNA is more simply inherited than is the composite trait.

The analogy described above is not coincidental, but there are some notable differences. Most expression studies have focused on a single individual, with phenotypes measured across different environments, across time or at different stages of development. Quantitative-genetic analysis, while conducted across multiple environments, is seldom conducted across time or across different stages of growth. However, quantitative genetics almost always incorporates multiple individuals and, hence, multiple allelic combinations. Despite these differences, it seems obvious to me that quantitative genetics and expression studies are attempting to accomplish similar goals: both are measuring changes in gene expression, gene interaction and gene–environment interaction. Methods of analysis are also similar. Typical methods of microarray analysis include principal-component analysis and hierarchical clustering. These methods seek to identify genes that are coregulated in a similar pattern across time or environment. Expression studies are often simplified by the absence of allelic variability, but they are complicated by the introduction of a large number of measured components. There is no reason why microarrays cannot be used to investigate patterns of gene expression in segregating populations; however, the analysis of such experiments could be challenging. I do not know how many quantitative geneticists have already been attracted to the challenge of microarray analysis, but I hope and predict that this will be an important area of crossover.

There is an additional use for microarrays, which could provide another crossover point between quantitative genetics and genomics. This is the potential use of oligonucleotide arrays in the high-throughput screening of marker polymorphism (e.g. Radtkey *et al.*, 2000). The ability to score large segregating populations at potentially thousands of polymorphic loci could add new dimensions to the analysis of QTL. With this application comes an urgent need for high-throughput data handling and data analysis.

Further integration of statistical genetics and genomics requires public databases with structured population and phenotypic data

Much of genomics is now concerned with the discovery of associations between genotype and phenotype. With technology to quickly screen alleles at candidate gene loci, the limiting factor becomes phenotype.

Furthermore, with large numbers of data points collected under a variety of experimental conditions, the correct analysis and interpretation of results are also limiting factors. It is predictable that the skills of the quantitative geneticist could be in high demand. But, in my opinion, the field of quantitative genetics has always been lacking in systems to organize and mine primary research results. Anyone who has attempted to reanalyse historical data knows that half of the job can be locating and assembling the original data. There may even be a sentiment of data ownership in this field that has been largely shed in publicly funded genomics initiatives. Sometimes this is justified: the correct interpretation of quantitative data requires knowledge of factors that go beyond the data points, so the data originator has a continued responsibility to ensure correct interpretation.

Because of the success of large, public data banks in genomics, it seems worthwhile to entertain the possibilities of parallel databases for population and quantitative genetics. Such databases are not absent. For example, in databases such as Graingenes (http://wheat.pw.usda.gov/), it is possible to find the phenotypic data sets from which QTL inferences were made. However, a large amount of data is not available or is not structured in a way that permits systematic mining of all information related to a given question. Consider the hypothetical problem of exploring the effect of three potentially epistatic genes. One would want to collect all relevant phenotypic data from every population where those genes were characterized or where extensive mapping had taken place. We may still be at a stage where this can be done through contacts in the research community, but it is difficult for a new researcher or graduate student to assemble this type of data. Many QTL data sets are approaching 10 years of age and, due to statistical limitations, it is unlikely that they will provide new information by themselves. However, when combined with other data or when interrogated with a specific hypothesis, they could continue to provide valuable research results.

The difficulty of standardizing quantitative data is obvious. Buying acceptance of a specific data structure or a specific data warehouse is equally challenging. However, similar challenges are being faced by the DNA microarray community (Brazma *et al.*, 2000). As with genome-sequence databases, there will probably be many different public microarray databases, each sharing and exchanging data. Unlike sequence data, however, microarray data originate from a large variety of very different experiments and are valuable only when the experimental conditions are known. In order for data to be exchanged in a meaningful format, standards must exist. To this end, the international community has formed the Microarray Gene Expression Database collaborative group (http://www.mged.org/), which meets annually to share information and gather consensus. This collaboration contains five working groups, addressing everything from data standards to user interface.

Do quantitative geneticists require this level of organization? Are 'traditional' quantitative data valuable enough to be preserved and organized in a public repository? Some would argue that the intensity of bioinformatics activity in the field of genomics is merely a result of overgenerous funding and that traditional scientists cannot afford the luxury of public institutions with dozens of curators and hundreds of computer programmers. Whatever the reason, the copious bioinformatics activity that has surrounded genomics research has created a wonderful arena of public information, which should be envied and imitated by researchers in other fields.

Summary and Conclusions

The term 'bioinformatics' covers a broad area of research and practice. When restricted to genomics, bioinformatics is characterized by a set of core activities, such as sequence alignment, gene prediction, design of databases and data mining. These bioinformatics activities are essential for the development of an understanding of gene structure and function. Although these activities seem to exclude research in classical quantitative genetics, there are inevitable crossover points. Quantitative genetics will look towards genomics for information to develop more accurate, precise and biologically meaningful models. Genomics will look towards quantitative genetics to develop and validate hypotheses involving complex gene interaction. Bioinformatics will play an important role in facilitating this crossover.

Why should quantitative geneticists care about bioinformatics? In this discussion, the following reasons have emerged: first, because bioinformatics contains useful tools and concepts (core activities) that can be applied in quantitative genetics; secondly, because bioinformatics is an area where the expertise of the quantitative geneticist may be required; thirdly, because genomics and genetics are both becoming high-throughput, information-rich fields with common objectives and a common need for bioinformatics expertise. Finally, because of the public nature of large-scale genomics projects, the bioinformatics community that works in this area has pioneered a highly successful network of information resources that go far beyond simple data warehouses. Other research communities, including the plant-breeding and quantitative-genetics communities, may benefit from this example.

References

Agarwal, P. and States, D.J. (1998) Comparative accuracy of methods for protein sequence similarity search. *Bioinformatics* 14, 40–47.

Altman, R.B. (1998) A curriculum for bioinformatics: the time is ripe. *Bioinformatics* 14, 549–550.

Altschul, S.F., Madden, T.L., Schaffer, A.A., Zhang, J., Zhang, Z., Miller, W. and Lipman, D.J. (1997) Gapped BLAST and PSI-BLAST: a new generation of protein database search programs. *Nucleic Acids Research* 25, 3389–3402.

Bader, G.D., Donaldson, I., Wolting, C., Ouellette, B.F., Pawson, T. and Hogue, C.W. (2001) BIND – the biomolecular interaction network database. *Nucleic Acids Research* 29, 242–245.

Baxevanis, A.D. (2001) The molecular biology database collection: an updated compilation of biological database resources. *Nucleic Acids Research* 29, 1–10.

Brazma, A., Robinson, A., Cameron, G. and Ashburner, M. (2000) One-stop shop for microarray data. *Nature* 403, 699–700.

Cato, S.A., Gardner, R.C., Kent, J. and Richardson, T.E. (2001) A rapid PCR-based method for genetically mapping ESTs. *Theoretical and Applied Genetics* 102, 296–306.

Claverie, J.M. (2000) From bioinformatics to computational biology. *Genome Research* 10, 1277–1279.

Doolittle, W.F. (1999) Phylogenetic classification and the universal tree. *Science* 284, 2124–2129.

Doyle, J.J. and Gaut, B.S. (2000) Evolution of genes and taxa: a primer. *Plant Molecular Biology* 42, 1–23.

Guigo, R., Agarwal, P., Abril, J.F., Burset, M. and Fickett, J.W. (2000) An assessment of gene prediction accuracy in large DNA sequences. *Genome Research* 10, 1631–1642.

International SNP Map Working Group (2001) A map of human genome sequence variation containing 1.42 million single nucleotide polymorphisms. *Nature* 409, 928–933.

Karplus, K., Barrett, C. and Hughey, R. (1998) Hidden Markov models for detecting remote protein homologies. *Bioinformatics* 14, 846–856.

Lipshutz, R.J., Morris, D., Chee, M., Hubbell, E., Kozal, M.J., Shah, N., Shen, N., Yang, R. and Fodor, S.P. (1995) Using oligonucleotide probe arrays to access genetic diversity. *Biotechniques* 19, 442–447.

Mendoza, L. and Alvarez-Buylla, E.R. (2000) Genetic regulation of root hair development in *Arabidopsis thaliana*: a network model. *Journal of Theoretical Biology* 204, 311–326.

Mendoza, L., Thieffry, D. and Alvarez-Buylla, E.R. (1999) Genetic control of flower morphogenesis in *Arabidopsis thaliana*: a logical analysis. *Bioinformatics* 15, 593–606.

Pearson, W.R., Wood, T., Zhang, Z. and Miller, W. (1997) Comparison of DNA sequences with protein sequences. *Genomics* 46, 24–36.

Phillips, A., Janies, D. and Wheeler, W. (2000) Multiple sequence alignment in phylogenetic analysis. *Molecular Phylogenetics and Evolution* 16, 317–330.

Radtkey, R., Feng, L., Muralhidar, M., Duhon, M., Canter, D., DiPierro, D., Fallon, S., Tu, E., McElfresh, K., Nerenberg, M. and

Sosnowski, R. (2000) Rapid, high fidelity analysis of simple sequence repeats on an electronically active DNA microchip. *Nucleic Acids Research* 28, E17.

Roberts, R.J. (2000) The early days of bioinformatics publishing. *Bioinformatics* 16, 2–4.

Schena, M., Shalon, D., Davis, R.W. and Brown, P.O. (1995) Quantitative monitoring of gene expression patterns with a complementary DNA microarray. *Science* 270, 467–470.

Trifonov, E.N. (2000) Earliest pages of bioinformatics. *Bioinformatics* 16, 5–9.

Velculescu, V.E., Zhang, L., Vogelstein, B. and Kinzler, K.W. (1995) Serial analysis of gene expression. *Science* 270, 484–487.

4 QTL Analysis: Problems and (Possible) Solutions

M.J. Kearsey

School of Biosciences, University of Birmingham, Birmingham B15 2TT, UK

Introduction

Many characters of central importance to human health, food production and evolutionary-cum-environmental biology are quantitative in nature, being under the control of several genes plus the environment. Such traits include hypertension, osteoporosis and behaviour in humans, yield and quality in our crop plants and farm animals and competitive ability and fitness in organisms in the wild (Kearsey and Pooni, 1996). In the modern age of the transcriptome, we can add gene expression level, as determined from expression arrays, as an archetypal quantitative trait in that it shows variation at many genes and is affected by variation between samples, between genetically identical individuals and across environments.

Despite their central economic, medical and social importance, these traits are difficult to study because the phenotype does not easily provide an insight into the genotype, unlike most simple single-gene traits with major effects. None the less, considerable theoretical and experimental progress has been made in the past 80 years in measuring the heritability of such traits, predicting their direct and correlated responses to selection and optimizing breeding strategies for their improvement in crop plants and farm animals.

It is now just 100 years since the rediscovery of Mendel's work and over 80 years from Fisher's groundbreaking paper providing a methodology for understanding quantitative traits (Fisher, 1918). During the last century, there was an almost exponential growth in our knowledge and understanding of genetics, which led, appropriately, to the unravelling of the complete human genome sequence in February 2001 (Wolfsberg *et al.*, 2001). However, despite these developments, our understanding of the genes underlying the control of quantitative, polygenic traits is little further advanced than it was when Fisher wrote his seminal paper in 1918. We are able to estimate the statistical effects as means, variances and covariances of groups of genes, but we know very little about the nature of the individual polygenes that underlie the traits (Falconer and Mackay, 1996; Kearsey and Pooni, 1996; Lynch and Walsh, 1998).

Although it is clearly not essential to understand the nature of polygenes or quantitative trait loci (QTL) to estimate heritability or predict selection response, it would be interesting and intellectually satisfying to have some sound understanding of the individual genes involved. Indeed, such knowledge may have a profound influence on the way we tackle theoretical and applied problems related to quantitative traits. So what sorts of questions remain to be answered?

1. How many genes are involved in any given trait and what are the distributions of their effects? For many traits, such as grain yield or individual fitness, it would seem likely that most genes, to some extent, affect the trait. But what about crop quality or human hypertension and intelligence? Are there just a few genes that have a major effect and many more whose effect is individually minor and, if so, how few 'major' genes are there? Answers to such questions would be valuable to breeders optimizing selection strategies and to pharmaceutical companies attempting to find drugs for genetic diseases.
2. What is the nature of the dominance and epistatic properties of these genes and how do they interact with the environment? Answers to such questions will prove very useful to those trying to understand the nature of heterosis (hybrid vigour) and inbreeding depression and how best to exploit it (see Coors and Pandey, 1999). They also impinge on pharmacogenomics.
3. What type of genes are they? Are they largely structural or regulatory and, if the latter, how wide is their sphere of influence? Are they allelic variants of well-known genes or do they belong to that large number of open reading frames for which no function has yet been assigned? To obtain this level of understanding, it is essential that we unambiguously identify individual poly-genes and study them at the sequence and transcriptional level.

The Problems

To obtain answers to all these questions, we require fairly accurate gene location and identification and, for much of the past 10–15 years, this has depended on QTL mapping in segregating populations, such as F_2s, back-crosses, recombinant inbred lines (RILs), doubled haploid (DH) lines or natural populations (Tanksley, 1993). This approach has been made possible by the availability of cheap and simple techniques to identify the extensive natural polymorphism at the DNA level, the so-called molecular markers, such as microsatellites,

restriction fragment length polymorphisms (RFLPs), amplified fragment length polymorphisms (AFLPs), etc. Following the mapping of these molecular 'framework' markers, it is then possible to locate the QTL through their genetic association with particular markers during meiosis, as observed in their progeny. A large number of statistical approaches have been developed to attempt to locate QTL by such methods and we have now probably reached the point where little extra precision can be obtained (Lynch and Walsh, 1998). These approaches have been adopted widely with considerable success, but they have many problems associated with them that cause them to be of little use in answering the detailed questions raised above. So what are these problems?

First, and most importantly, QTL locations obtained from segregating populations have very large confidence intervals (CIs). These CIs are seldom less than 5 cM and often > 30 cM (Van Ooijen, 1992; Darvasi et al., 1993; Hyne et al., 1995). Given that a typical chromosome is about 100 cM long, such intervals amount to between 5% and more than 30% of a chromosome. A typical result obtained by computer simulation is shown in Fig. 4.1, where the range of locations for a QTL are shown in 1000 simulations using an F_2 population of 300 individuals. It is now well established that having more markers beyond about one every 10–20 cM does not reduce the CI and that the only way to reduce it is to increase the population size considerably (Hyne et al., 1995).

The reason for these large CIs is simply the lack of recombination at meiosis (Boehnke, 1994; Kearsey and Pooni, 1996; Guo and Lange, 2000). Table 4.1 illustrates the percentage of chromosomes that pass into gametes with zero, one, two and three crossovers from the parental chromosomes. It is clear that most chromosomes (~80%) survive meiosis with either one crossover or none at all and thus one needs to sample a very large number of meioses to have enough crossovers to map the QTL with any accuracy: the smaller the heritability of the trait, the larger the population required. Further

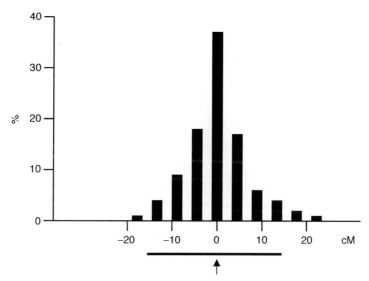

Fig. 4.1. The distribution of estimated QTL locations obtained from 1000 simulated F_2 populations. The actual QTL had a heritability of 20% and was located at the zero point of the distribution. The 95% confidence interval is shown as a horizontal bar.

Table 4.1. The percentage of chromosomes in gametes that have come through meiosis with various numbers of crossovers.

Crossovers per chromosome	Arabidopsis (%)	Rye (%)
0	30	31
1	49	50
2	20	19
3 or more	1	0

recombination can be achieved by randomly mating the population for two or more generations and this is particularly beneficial for closely linked genes. However, both very large populations and several generations of random mating are seldom practical options in a breeding context, particularly when breeders need to have large plots of genetically uniform material (inbred or F_1) to simulate agricultural conditions. Back-crosses and F_2s allow very large populations and hence are efficient, but the spaced plant trials needed to assess them are not very useful for traits of agricultural importance, such as yield, because the conditions are not typical.

One consequence of these large CIs is that they will contain within them a very large number of potential candidate genes. Thus, a 10 cM interval will contain, on average, about 130 rice genes and over 400 *Arabidopsis* genes, while some intervals in which there is little crossing over will contain many more. Hence, while it is not difficult to nominate many potential candidates, it can be very difficult accurately to identify the correct candidate gene.

A second problem concerns multiple QTL on a chromosome. It is difficult to distinguish two QTL that are less than 20 cM apart, even with QTL of moderate heritability, and hence two or more QTL within this interval may be misinterpreted as one (Lebreton *et al.*, 1998). This can result in a large ghost QTL being located between the two true QTL if they are linked in coupling (Martinez and Curnow, 1993) and possibly no QTL being identified if they are linked in repulsion. Either way, one is misled both in the location and in the size of the QTL effect.

A third problem is a statistical one. QTL location by whatever method involves scanning each chromosome for the most likely position of the QTL. This inevitably implies that a large number of possible positions are tested and those whose likelihood of containing a QTL exceeds some critical value are accepted. To avoid too many false positives, the test probability level is adjusted downwards to allow for the multiple tests. Of course, this has the concomitant result of increasing the probability of false negatives. It is a common misconception that, if a QTL is located with some probability of the test statistic under the null hypothesis of α (e.g. 0.001), then the probability of a QTL being there is $1 - \alpha$ (or 0.999). The truth may well be very far from that.

The only real solution to these problems of QTL location in segregating populations is to repeat the experiments using a completely different sample of genotypes derived from the same population. One should then test whether there are QTL in this second population located at the positions identified in the first. This is an approach that the human geneticists realized some time ago and all good human genetics experiments now have an initial identification sample followed by a final testing or confirmation sample. An excellent, and probably unique example of this in plants (maize) is the work of Utz et al. (2000). Of course, such replication comes at a cost. It is very expensive to do, people have limited budgets and limited timescales and, therefore, there is a strong reluctance to do this. Such reluctance tends to encourage people to believe that the QTL that they found on a single trial are genuine and meaningful and so they tend to place considerable, but often undeserved, reliance on this conclusion.

There are several statistical biases involved with these QTL mapping approaches. Only those QTL of sufficiently large effect will be detected in any given trial, either because their true mean effect is large or because they are inflated, by chance, in that environment. For example, consider a QTL whose true, individual size is just at the threshold of detection. Because of environmental variation, it will be below that threshold on 50% of occasions and hence will not be detected. With several such QTL, how many and which ones will be significant on any given repeat of the experiment is entirely a matter of chance, but this will give the false impression of genotype–environment ($G \times E$) interaction

on repetition in space or time. QTL tend to be less likely to be located at markers than between them, even if they are cosegregating with the marker, and truly terminal QTL will tend to be located in subterminal regions of the chromosome (Hyne et al., 1995).

Finally, there is the very real problem that there will be different polymorphisms in each population with different QTL and, of course, different molecular markers segregating.

Possible Solutions

Let us now turn to other approaches to fine-mapping of QTL. Clearly the type of mapping that we have been talking about in segregating populations leads us to rough locations of QTL. They indicate which arm of the chromosome the QTL is on and possibly suggest a more precise location within that arm. To be more precise about the location of these QTL, it is necessary to use some form of chromosome introgression/substitution lines or near-isogenic lines (NILs). The final stage, and the one that I would like to discuss in this chapter, is that of the use of stepped aligned inbred recombinant strains (STAIRS) and I shall be describing these in some detail towards the end. These fine-mapping techniques enable us to narrow down the CIs around a QTL and so focus on a smaller subset of possible candidate genes that might be responsible for the trait in question. Having identified a small number of potential candidate genes, the final stages of this procedure involve approaches such as gene sequencing, expression analysis, transformation and gene silencing, etc., to identify the particular candidate gene that is responsible for the polymorphism identified by the QTL (Albert and Tanksley, 1996).

To illustrate the progress from segregating-population analysis through to substitution lines, I shall use an example of work on Brassica oleracea from my own laboratory. We (Bohuon et al., 1998) have explored the genetics of a cross between a DH line derived from a commercial calabrese hybrid ('Green Duke') and a rapid cycling variety (*B. oleracea* var. *alboglabra*). This has involved mapping markers and QTL for various traits over a number of years and sites in a population of DH lines derived from the F_1 of this cross. QTL analysis in this segregating population identified many potential QTL, among which were six QTL controlling flowering time, one each on chromosomes 2 and 3 and two each on chromosomes 5 and 9 (Fig. 4.2), but their CIs are typically about 20–30 cM. It is well known that *B. oleracea* consists of at least three copies of most genes arising from some ancestral polyploidy and, if we look at chromosomes 2, 3 and 9, which contain QTL from the previous study (Fig. 4.3), we also find that these QTL overlap regions that are syntenous both among themselves and also with regions known to contain the same marker alleles in *Arabidopsis thaliana*. These same chromosomal regions have also been associated with flowering QTL in *Brassica nigra* and also more recently

in *Brassica rapa* (Kowalski *et al.*, 1994; Lagercrantz *et al.*, 1996; Osborn *et al.*, 1997). So there appears to be consistent evidence across chromosomes and species that this particular paralogous region of the chromosome carries a gene or genes that have a major effect on flowering time in *Brassica*. Moreover, the CIs of these QTL locations on all chromosomes and species include a region flanked by the RFLP markers *leu6* and *labi8*. This region is syntenous with a region of the *A. thaliana* genotype that contains the *Constans* gene, which is known to be a gene that controls flowering time in *Arabidopsis*. It is in fact a zinc finger protein, i.e. a transcription factor that controls a complex network of genes relating to flowering time. Therefore, it is perhaps not a coincidence that all of the QTL illustrated in Fig. 4.3 do in fact overlap this region and this may be a very strong indication that *Constans*, or a gene very closely linked to it, is responsible for the flowering-time polymorphism in all of these cases. More recent examples have been explored in other accessions of *B. oleracea* and QTL found in exactly the same locations.

To reduce the large CIs around QTL, a number of workers, and we are included among them, have used the approach of constructing and analysing part-chromosome substitution lines (Howell *et al.*, 1996). The general principle is illustrated in Fig. 4.4. One takes two different parental lines and introgresses parts of each chromosome of one line, the donor line, into the other, the recipient line, by means of marker-assisted back-crossing. Using the *Brassica* lines described in the previous paragraph as an example, we have introgressed sections of donor chromosome from the calabrese parent ('Green Duke') into the recipient *B. oleracea* var. *alboglabra*. In total, we have produced some 70 different substitution lines, each of which has an entirely common recipient background with just a short region of donor chromosome from the calabrese variety (Ramsay *et al.*, 1996). These part-chromosome introgressions vary in length, but it is possible, by comparing the performance for particular traits of each one of these substitution lines with the recipient

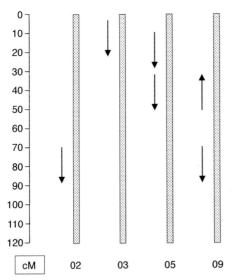

Fig. 4.2. Locations of six QTL for flowering time found in a doubled haploid population of *Brassica oleracea*. The direction of the arrows indicates whether the early-flowering parent carried the early or late allele. The length of the arrows indicate the confidence intervals on the locations. (Redrawn from Bohuon *et al.*, 1998.)

line (var. *alboglabra*), to know whether or not there are QTL in that introgressed region. Furthermore, by comparing different substitution lines, some of which overlap and some of which do not, it is possible to identify fairly specifically the region of the

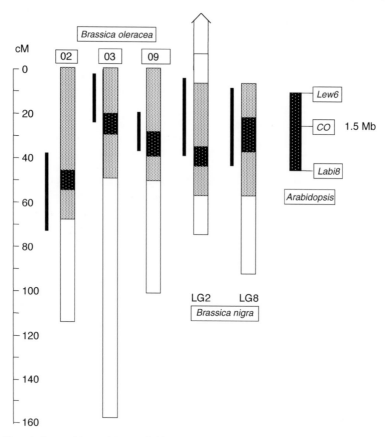

Fig. 4.3. The relative positions of QTL (solid bars) on syntenous regions of *Brassica oleracea* and *Brassica nigra*. Lightly hatched areas of chromosome indicate syntenous regions within and between species. The smaller dark areas indicate the location of DNA syntenous to the region around the *CO* gene in *Arabidopsis*. The 1.5 Mb region of *Arabidopsis* together with the RFLP markers (*Lew6* and *Labi8*) that delineate the region in the brassicas is shown on the right. (Redrawn from Bohuon *et al.*, 1998.)

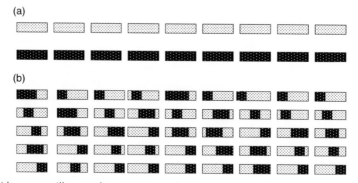

Fig. 4.4. An ideogram to illustrate the constitution of (a) the nine parental and donor chromosomes of *Brassica oleracea* and (b) the associated single part-chromosome substitutions.

chromosome in which the QTL are located (Rae *et al.*, 1999). The problem with this approach, however, is that it is necessary to produce a very large number of substitution lines and, as Fig. 4.4 illustrates, which particular regions you obtain are somewhat random, despite the fact that these lines are produced by marker-assisted selection.

When we compare the locations of QTL found by our previous segregating population with those found in the substitution lines, we observe the relationships shown in Fig. 4.5. We see that some of the QTL from the segregating population are also found in the substitution lines, but we also find a number of other QTL in the substitution lines that were clearly not detected in the segregation populations. So the number of QTL is considerably greater than we had found before. This is due to two factors. First, we have actually separated pairs of QTL that were closely linked in coupling in the original DH population, while we have also been able to detect QTL that were linked in repulsion and hence effectively invisible (Rae *et al.*, 1999).

A different approach to more precise QTL mapping is to use NILs. These are generated by taking an F_1 through several generations of inbreeding (in plants this would normally be by single seed descent) towards creating RILs and eventually identifying individuals in some advanced generation (e.g. the F_6 to F_8) that are entirely homozygous for molecular markers except for one or two loci. By selfing these particular individuals, it is possible to produce two different isogenic lines, which contain either one or the other combination of the alleles that were initially heterozygous in the F_n parent. These are essentially substitution lines but involve very small regions of just a few centimorgans. Such NILs can be used in exactly the same way as substitution lines to delimit very small regions of chromosome within which QTL may be located. Again, the problem with this is that one is looking at a very large number of small regions and to have NILs covering the whole genome would involve a tremendous amount of work. However, the individual regions can be very small and the fact that both NILs

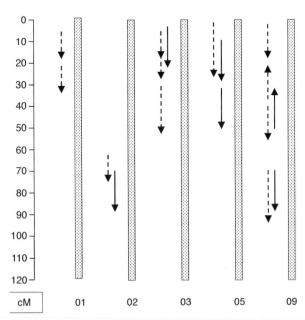

Fig. 4.5. The locations of flowering-time QTLs in *Brassica oleracea*, identified from the substitution lines (dashed arrows) together with those located in the original doubled haploid population as shown in Fig. 4.3.

from a given heterozygote can be obtained from the same F_n parent means that trial designs can ignore maternal effects.

When we get down to the resolution of comparing two or a very few lines for differences due to a short region of chromosome containing a single QTL, another problem of interpretation and experimental design becomes very apparent: that is, we need to ensure that any differences are truly genetic and not due to common environment effects. The conventional approach to genetic experiments with plants is to take seed from a number of relevant families and to raise the progeny across the trial area according to some replicated and randomized design appropriate to the material. This standard approach is designed to remove any bias in the trial site, but ignores biases that may be due to prereplication effects. For example, if there are progeny from two different families (e.g. two NILs), A and B, and all the seed from A comes from one single parent and all the seed from B from another single parent, then any differences among the progeny could be due to both genetic and non-genetic differences between the two parents. Such non-genetic differences might include obvious factors, such as time and location of seed production, but could also include any other, and probably far from obvious, environmental factors that affect the two parents' ability to produce seed. Geneticists working with animals are very well aware of such factors, but plant geneticists tend to assume that they are not important in plants. Our own experience with *Brassica*, where we have replicated and randomized the parents within a uniform glasshouse, indicates that there are frequently significant common environmental effects attributable to non-obvious maternal differences and these effects can persist throughout the life of the progeny (Rae *et al.*, 1999). It is thus essential that the seed parents are also raised in randomized and replicated conditions and that the progeny of different individual seed parents (but putatively the same genetically) are raised and identifiable in the trial. The smaller the QTL effects to be studied, the more important such precautions become.

Although the use of substitution lines and NILs improves the accuracy of QTL location, it is still difficult in most cases to reduce the CI to below 2–5 cM, and the work to achieve even this resolution is time-consuming, while success in finding the appropriate genotypes is somewhat serendipitous. In an attempt to permit a more gene-targeted approach to fine QTL mapping, as well as increasing statistical power, we are moving in a different direction and developing genetic resources in *Arabidopsis* to achieve this. These resources we call STAIRS and they are designed to allow the genetic focus to close in from the whole chromosome to a short 0.1–1 cM interval. The construction and use of STAIRS are described below.

Our approach really falls into two stages. The first stage is to produce whole chromosome substitution strains (CSSs) by taking each chromosome in turn from a donor line and using it to replace the corresponding chromosome in a parental, recipient line. This is a very similar approach to that adopted by Nadeau *et al.* (2000) to produce all the 20 possible chromosome substitution lines for more detailed QTL analysis in mice. We are doing this in *Arabidopsis*, which has only five pairs of chromosomes and can be self-fertilized, so the workload is somewhat simpler than in mice. However, it makes a very good exemplar to illustrate the procedure. So we start by taking two lines, A and B (Fig. 4.6), and introgress each of the chromosomes in turn from B into A to create five whole CSSs. To manipulate chromosomes in this way, it is necessary to inhibit recombination. Previously this was done in model species, such as *Drosophila* and *Aspergillus*, but also in wheat, capitalizing on various features of the meiotic system of these species, which facilitated inhibition of recombination. Thus, in *Drosophila*, inversion stocks were used to prevent the recombination (Kearsey and Kojima, 1967); in wheat monosomic/nullisomic lines were used (Law *et al.*, 1983); and in *Aspergillus* somatic segregation was employed (Varga and Croft, 1994).

In *Arabidopsis*, or in mice, we have none of these techniques available to us. Therefore, we have to identify those particular individuals that have received chromosomes from their parents that have gone through meiosis, naturally, without recombining. Table 4.1 illustrates data on crossover frequencies in chromosomes of *Arabidopsis* and rye, which are good exemplars of dicots and monocots. We see that approximately 30–31% of chromosomes do, indeed, go through meiosis without a single recombinant event. This does not mean that there were no chiasmata on the parental bivalent; indeed, the absence of chiasmata generally results in failure of chromosome disjunction, so every chromosome normally has at least one chiasma at meiosis. However, a single chiasma results in 50% of gametes being non-recombinant, two chiasmata 25% non-recombinant, and so on. Because the majority of chromosomes have one to three chiasmata at meiosis, the frequency of non-recombinant chromosomes remains quite high. Therefore, it is simply a matter of detecting those progeny containing non-recombinant chromosomes, using codominant markers, and selecting them

through marker-assisted back-crossing. The principle is relatively straightforward. We create the F_1 between the donor and the recipient lines. We then back-cross the F_1 to the recipient line, selecting back-cross individuals that have all non-recombinant chromosomes, either recipient chromosomes or donor chromosomes. On finding appropriate individuals, they are self-fertilized and their homozygous progeny selected.

Power calculations indicate that we would need to score about 7000 back-cross individuals to be 90% certain of obtaining all the necessary individuals of the required genotype. However, this does not require a tremendous amount of work in *Arabidopsis* and the procedure we use is to screen batches of 250–400 back-cross individuals at a time, first selecting for terminal markers on chromosome 1 and discarding any recombinant individuals. Of the survivors, which are about 50%, we select for terminal markers on chromosome 2, disregarding any recombination, and so on. By repeating this procedure for all five chromosomes, we finish with approximately 16 individuals. These

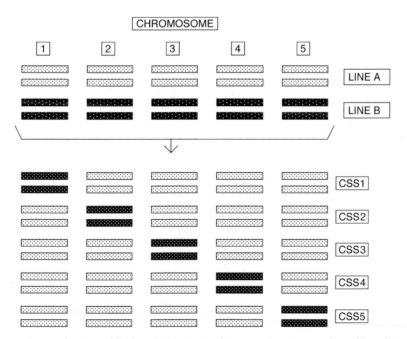

Fig. 4.6. Structure and origins of the five chromosome substitution strains (CSSs) derived by substituting single chromosomes of line B into A.

we screen for intervening markers to make sure that there has not been any recombination within the chromosomes.

Having first derived these CSSs, the next stage is to produce the STAIRS. The STAIRS consist of a set of inbred lines, derived from the CSSs, each of which is homozygous for a chromosome involving a single recombination event between the donor and the recipient chromosome, and these are illustrated in Fig. 4.7. Arranging these single recombinant lines (SRLs) in terms of the location of the recombination event produces the STAIRS, each SRL being a step in the stairs. With 100 random lines and a chromosome of 100 cM in genetic length, each step will, on average, be 1 cM above the previous line, and any difference in phenotype between such SRLs has to be due to genes in this interval.

Such lines are easily constructed and the procedure is illustrated for chromosome

1 in Fig. 4.7. A CSS is crossed to the recipient inbred line and the F_1 back-crossed again to the recipient line. Progeny in this back-cross family are genotyped for 11 well-spaced markers, one at each end and nine interstitial. Individuals that have a single recombination event on that chromosome are selected and self-fertilized to fix the recombinant chromosome. As we saw from Table 4.1, approximately 50% of the chromosomes will arise from a single crossover. Back-cross individuals bearing such chromosomes are easily identified by checking the terminal markers. If the markers are both from the recipient parent or both from the donor parent, these must contain zero, two or four crossovers and are discarded. The remaining ~50% will be mainly (~98%) single recombinants and the approximate location of the recombination event can be identified by the nine internal markers actually scored

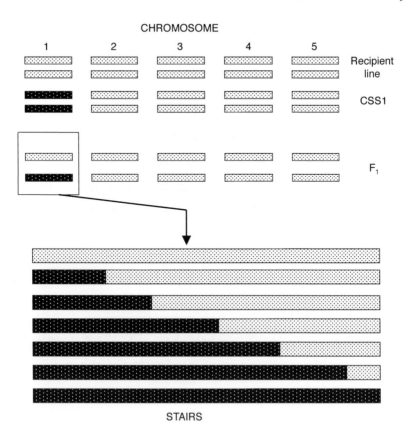

Fig. 4.7. Structure and origins of a set of stepped aligned recombinant inbred strains (STAIRS) in *Arabidopsis*.

at this stage. So we can easily identify a very large number of back-cross individuals that contain single crossovers, and we now know that these crossovers occur in particular regions, as identified by the intervening markers. These individuals are self-fertilized and seed from such families held separately, until needed, in groups ('bins') according to the marker interval within which the crossover had occurred. We also have DNA from all the back-cross individuals selected for selfing. So, if we selected 1000 such back-cross individuals, we would end up with ten bins each of which contains seed (and DNA) from ~100 individuals with recombination in a particular 10 cM interval.

These collections of 'binned' seeds and DNA, plus the residual back-cross seed and CSSs, provide the fine-mapping resource.

To use the resource, one locates a QTL first to a chromosome, using the CSSs, next to a region on that chromosome using a single SRL from each 'bin', and finally to very small regions by further analysis of lines within the 'bin'. This is achieved as follows and is illustrated in detail in Fig. 4.8.

1. All five CSSs plus the recipient line are scored initially for the trait of interest. If we discover, for example, that the CSS with chromosome 1 has a low score while the original recipient is high-scoring, we can be sure that there is a gene or genes on that chromosome controlling the trait in question (Fig. 4.8a).
2. We now score one SRL for the trait of interest from each of the 'bins' – ten in our illustration above – and discover, for example, that the first four are high and the

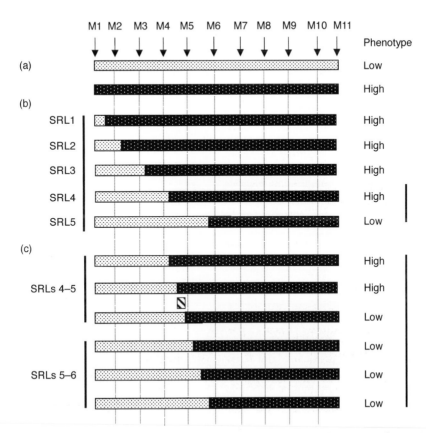

Fig. 4.8. Use of trait information on (a) chromosome substitution lines, (b) wide STAIRS and (c) narrow STAIRS to locate a QTL (▨). (See text for details.)

last six in that series are low (Fig. 4.8b). This proves that the QTL is in the vicinity of marker 5, between markers 4 and 6, but this is still a 20 cM interval. This stage is equivalent to taking large steps up the STAIRS.

3. We genotype the two neighbouring lines, i.e. SRL4 and SRL5, identified in step 2, for interstitial markers to identify the sites of recombination and the size of the differential region. This could well reduce the qualifying QTL search region to 5–10 cM.

4. Residual DNA from back-cross individuals responsible for 'bins' 4 and 5 are screened for single crossovers within the 5–10 cM region identified in step 3, using additional interstitial markers and plants raised from the selfed seed from these individuals to identify homozygotes for the recombinant chromosome.

5. These homozygotes are now scored for the trait and the position of the crossover, where the phenotype changes, is identified, in the same fashion as in step 2 above (Fig. 4.8c). This stage is equivalent to taking small steps up the STAIRS and could locate the QTL to within 0.1–1 cM.

As a result of this series of steps, we have focused in on two homozygous lines that are identical except for a short region of chromosome of less than 1 cM, possibly no more than 0.1 cM, containing a QTL. In *Arabidopsis*, 1 cM is, on average, equivalent to about 45 genes, although some regions may be more or less gene-dense, so it is now possible to consider identifying the candidate gene(s) underlying the QTL. Various approaches are available. Database searches of the published DNA sequence of the region may suggest potential structural genes or transcription factors; differences between the two lines in gene-sequence polymorphism of these candidates can easily be explored. Expression profiling of the two lines is a particularly feasible option. Because the lines are almost identical genetically, the background noise due to variable genotypes is eliminated, while environmental variation can easily be minimized by multiplexing transcript from several individuals. Such analyses may identify differences in the expression profile of genes within the differential region of the two lines. Alternatively, they may indicate up- or down-regulation of genes outside the region, but due to polymorphism in regulatory genes within the region. Such analyses may benefit from identifying environments that alter the relative phenotypic differences between the two lines and from looking for corresponding changes in the expression profile. We can also examine other crosses to see whether the same polymorphism in sequence of the structural gene results in the same change in phenotype. It may also be possible, using single nucleotide polymorphisms (SNPs), to identify two SRLs in the STAIRS that only differ by one gene yet maintain the QTL difference, so confirming the candidate exactly. Transformation and gene-silencing procedures can also be used, once strong potential candidates are known.

The advantages of the STAIRS approach are that it provides a generally applicable and yet focused, approach to precise QTL location, it allows one to 'zoom in' on the genes responsible to almost any degree of accuracy and, because only a few lines are required at any one time, it permits very large-scale replication to enhance the statistical power of both trait- and gene-expression analyses. We have so far generated several CSSs and 'bins' of STAIRS for two different crosses in *Arabidopsis*, 'Landsberg' × 'Columbia' and 'Niederzenz' × 'Columbia', and these will be made available through the Nottingham *Arabidopsis* Stock Centre (NASC) in the UK. We plan to use these resources to develop some understanding of the range and nature of genetic effects for a number of QTL for different traits.

Summary

Quantitative, multifactorial traits are becoming a major area of genetic research in the post-genomics era because of their central importance in medicine, plant and animal breeding and evolutionary biology. However, they present a major challenge because of difficulties in accurately locating and identifying the underlying genes or QTL. There are a variety of questions to be answered about particular QTL. These include their number, distribution of size of effects, gene action and interaction (dominance and epistasis) and, most important, their identity and nature, i.e. whether they are structural or regulatory genes.

The problems of QTL location using molecular markers in segregating populations include large CIs, low statistical power and biased estimates. This chapter summarizes these and explains their causes and possible solutions. Partial solutions to these problems have involved part-chromosome substitution lines and NILs, and their advantages and disadvantages are discussed. A novel set of resources for fine-mapping, STAIRS, is described and its development and application described.

The solutions proposed are aimed more for the use of the geneticist wishing to obtain a fundamental understanding of the nature of the genetic control of quantitative traits, rather than as an everyday tool for the practical plant breeder. However, they could well be judiciously applied in a breeding context and will reveal important information that could be of value to a breeder.

References

Albert, K.D. and Tanksley, S.D. (1996) High-resolution mapping and isolation of a yeast artificial chromosome containing *fw2.2*: a major fruit weight quantitative trait locus in tomato. *Proceedings of the National Academy of Sciences, USA* 93, 15503–15507.

Boehnke, M. (1994) Limits of resolution of linkage studies: implications for the positional cloning of human disease genes. *American Journal of Human Genetics* 55, 379–390.

Bohuon, E.J.R., Ramsay, L.D., Craft, J.A., Arthur, A.E., Lydiate, D.J., Marshall, D.F. and Kearsey, M.J. (1998) The association of flowering time QTL with duplicated regions and candidate loci in *Brassica oleracea*. *Genetics* 150, 393–401.

Coors, J.G. and Pandey, S. (eds) (1999) *The Genetics and Exploitation of Heterosis in Crops*. American Society of Agronomy, Crop Science Society of America and Soil Science Society of America, Madison, Wisconsin.

Darvasi, A., Weintreb, A., Minke, V., Weller, S. and Soller, M. (1993) Detecting marker QTL linkage and estimating QTL gene effect and map location using a saturated genetic map. *Genetics* 134, 943–951.

Falconer, D.S. and Mackay, T. (1996) *Introduction to Quantitative Genetics*, 4th edn. Longman, New York.

Fisher, R.A. (1918) The correlation among relatives on the supposition of Mendelian inheritance. *Transactions of the Royal Society of Edinburgh* 52, 399–433.

Guo, S.W. and Lange, K. (2000) Genetic mapping of complex traits: promises, problems and prospects. *Theoretical Population Biology* 57, 1–11.

Howell, P.M., Marshall, D.F. and Lydiate, D.J. (1996) Towards developing intervarietal substitution lines in *Brassica napus* using marker-assisted selection. *Genome* 39, 348–358.

Hyne, V., Kearsey, M.J., Pike, D.J. and Snape, J.W. (1995) QTL analysis: unreliability and bias in estimation procedures. *Molecular Breeding* 1, 273–282.

Kearsey, M.J. and Kojima, K. (1967) The genetic architecture of body weight and egg hatchability in *Drosophila melanogaster*. *Genetics* 56, 23–37.

Kearsey, M.J. and Pooni, H.S. (1996) *The Genetical Analysis of Quantitative Traits*. Chapman & Hall, London.

Kowalski, S.P., Lan, T.H., Feldman, K.A. and Paterson, A.H. (1994) Comparative mapping of *Arabidopsis thaliana* and *Brassica oleracea* chromosomes reveals islands of conserved variation. *Genetics* 138, 499–510.

Lagercrantz, U., Putterill, G., Coupland, G. and Lydiate, D. (1996) Comparative mapping in *Arabidopsis* and *Brassica*, fine scale genome collinearity and congruence of genes controlling flowering time. *The Plant Journal* 9, 13–20.

Law, C., Snape, J.W. and Worland, A.J. (1983) Aneuploidy in wheat and its use in genetical analysis. In: Lupton, F.G.H. (ed.) *Wheat Breeding and its Scientific Basis*. Chapman & Hall, London, pp. 77–108.

Lebreton, C.M., Visscher, P.M., Haley, C.S., Semikhodskii, A. and Quarrie, S.A. (1998) A nonparametric bootstrap method for testing close linkage vs. pleiotropy of coincident quantitative trait loci. *Genetics* 150, 931–943.

Lynch, M. and Walsh, B. (1998) *Genetics and Analysis of Quantitative Traits.* Sinauer Associates, Sunderland, Massachusetts.

Martinez, O. and Curnow, R.N.C. (1993) Estimating the locations and sizes of the effects of quantitative trait loci using flanking markers. *Theoretical and Applied Genetics* 85, 480–488.

Nadeau, J.H., Singer, J.B., Martin, A. and Lander, W.S. (2000) Analysing complex genetic traits with chromosome substitution strains. *Nature Genetics* 24, 221–225.

Osborn, T.C., Kole, C., Parkin, I.A.P., Sharpe, A.G., Kuiper, M., Lydiate, D.J. and Trick, M. (1997) Comparison of flowering time genes in *Brassica rapa*, *B. napus* and *Arabidopsis thaliana*. *Genetics* 156, 1123–1129.

Rae, A.M., Howell, E.C. and Kearsey, M.J. (1999) More QTL for flowering time revealed by substitution lines in *Brassica oleracea*. *Heredity* 83, 586–596.

Ramsay, L.D., Jennings, D.E., Bohuon, E.J.R., Arthur, A.E., Lydiate, D.J., Kearsey, M.J. and Marshall, D.F. (1996) The construction of a substitution library of recombinant backcross lines in *Brassica oleracea* for the precision mapping of quantitative trait loci. *Genome* 39, 558–567.

Tanksley, S.D. (1993) Mapping polygenes. *Annual Reviews of Genetics* 27, 205–233.

Utz, H.F., Melchinger, A.E. and Schon, C.C. (2000) Bias and sampling error of the estimated proportion of genotypic variance explained by quantitative trait loci determined from experimental data in maize using cross validation and validation with independent samples. *Genetics* 154(4), 1839–1849.

Van Ooijen, J.W. (1992) Accuracy of mapping quantitative trait loci in autogamous species. *Theoretical and Applied Genetics* 84, 803–811.

Varga, J. and Croft, J.H. (1994) Assignment of RFLP, RAPD and isozyme markers to *Aspergillus nidulans* chromosomes using chromosome substituted segregants of a hybrid of *A. nidulans* and *A. quadrilineatus*. *Current Genetics* 25, 311–317.

Wolfsberg, T.G., McEntyre, J. and Schuler, G.D. (2001) Guide to the draft human genome. *Nature* 409, 824–826.

5   Association Mapping in Plant Populations

Jean-Luc Jannink[1] and Bruce Walsh[2]

[1]Department of Agronomy, Iowa State University, Ames, IA 50010, USA;
[2]Department of Ecology and Evolutionary Biology, University of Arizona, Tucson,
AZ 85721, USA

Introduction

The objective of genetic mapping is to
identify simply inherited markers in close
proximity to genetic factors affecting quantitative traits (quantitative trait loci (QTL)).
This localization relies on processes that
create a statistical association between
marker and QTL alleles and processes that
selectively reduce that association as a
function of the marker distance from the
QTL. When using crosses between inbred
parents to map QTL, we create in the F_1
hybrid complete association between all
marker and QTL alleles that derive from the
same parent. Recombination in the meioses
that leads to doubled haploid, F_2 or recombinant inbred lines reduces the association
between a given QTL and markers distant
from it. Unfortunately, arriving at these generations of progeny requires relatively few
meioses, such that even markers that are far
from the QTL (e.g. 10 cM) remain strongly
associated with it. Such long-distance associations hamper precise localization of the
QTL. One approach for fine-mapping is to
expand the genetic map – for example,
through the use of advanced intercross
lines, such as F_6 or higher generational
lines derived by continual generations of
outcrossing the F_2 (Darvasi and Soller,
1995). In such lines, sufficient meioses have
occurred to reduce disequilibrium between

moderately linked markers. When these
advance generation lines are created by
selfing, the reduction in disequilibrium is
not nearly as great as that under random
mating.

The central problem with any of the
above approaches for fine-mapping is the
limited number of meioses that have
occurred and (in the case of advanced
intercross lines) the cost of propagating lines
to allow for a sufficient number of meioses.
An alternative approach is 'association mapping', taking advantage of events that created
association in the relatively distant past.
Assuming many generations, and therefore
meioses, have elapsed since these events,
recombination will have removed association between a QTL and any marker not
tightly linked to it. Association mapping
thus allows for much finer mapping than
standard biparental cross approaches. In
our review of this topic, we first define
association quantitatively and describe
mechanisms that generate it. To motivate
our discussion of rigorous methods to test for
marker association with a quantitative trait
allele, we then discuss in some detail an
example from the plant-breeding literature.
Next, we review an analysis frequently used
in human genetics to find marker associations with disease-susceptibility alleles, the
transmission/disequilibrium test (TDT). We
touch upon work to extend the TDT to

quantitative traits and to identify QTL–
environment (QTL × E) interactions. We
describe recent developments making use of
multiple-marker haplotypes to locate QTL
and conclude with some points concerning
the power of association mapping.

Association Between a Neutral Mendelian Marker and the Phenotype

A statistical association between a neutral
marker allele and the phenotype occurs
when marker alleles are in gametic-phase
disequilibrium (GPD) with alleles at a QTL.
Two alleles at distinct loci are in positive
GPD if they occur together more often than
predicted on the basis of their individual
frequencies. This definition of association
says nothing concerning the physical position of the loci or of the alleles' joint effects
on the phenotype. The term GPD is used
synonymously with the term 'linkage
disequilibrium', but we use the former
term since it avoids reference to linkage
(as unlinked markers can still be in GPD)
and emphasizes that associated alleles must
co-occur in gametes.

In the example of Table 5.1, the combination of alleles (or haplotype) QM is
observed with frequency $p_{QM} = 0.4$, while its
predicted frequency is only $p_Q p_M = 0.3$. The
alleles Q and M are in GPD with the disequilibrium coefficient $D = p_{QM} - p_Q p_M = $ cov
(Q,M) $= 0.1$. Note that, since D can be
expressed as a covariance, we can bound
its possible values by considering the case
when the correlation is +/− 1, giving

$$|D| \le \sigma_Q \sigma_M = \left[p_Q (1 - p_Q) p_M (1 - p_M) \right]^{1/2} \quad (5.1)$$

Table 5.1. Haplotype, and marginal marker and QTL frequencies.

	QTL allele		
Marker allele	Q	q	
M	0.4	0.2	0.6
M	0.1	0.3	0.4
	0.5	0.5	

For a pair of diallelic loci, the expected
value of the estimate of D is equal in
magnitude irrespective of the haplotype
frequencies used and can be calculated as
$D = p_{QM} p_{qm} - p_{Qm} p_{qM}$. For each generation
of random mating, D decays by a factor
of $(1 - r)$, where r is the recombination
rate between the two loci considered. Thus,
after t generations, only $(1 - r)^t$ of the initial
disequilibrium remains.

A variety of mechanisms generate linkage disequilibrium, and several of these can
operate simultaneously. Some of the more
common mechanisms are:

1. Populations expanding from a small
number of founders. The haplotypes present
in the founders will be more frequent than
expected under equilibrium. Three special
cases are noteworthy. First, genetic drift
affects GPD by this mechanism in that a
population experiencing drift derives from
fewer individuals than its present size. Secondly, by considering an individual with a
new mutation as a founder, we see that its
descendants will predominantly receive the
mutation and loci linked to it in the same
phase. Linked marker alleles will therefore
be in GPD with the mutant allele. Finally,
an extreme case arises in the F_2 population
derived from the cross of two inbred lines.
Here, all individuals derive from a single F_1
founder genotype and association between
loci can be predicted based on their mapping
distance (e.g. Lynch and Walsh, 1998).
2. GPD arises in structured populations
when allelic frequencies differ at two loci
across subpopulations, irrespective of the
linkage status of the loci. Admixed populations, formed by the union of previously
separate populations into a single panmictic
one, can be considered a case of a structured
population where substructuring has
recently ceased.
3. Negative GPD will occur between loci
affecting a character in populations under
stabilizing or directional selection as a result
of the Bulmer effect.
4. Positive GPD will occur between loci
affecting a character under disruptive
selection.

5. When loci interact epistatically, haplotypes carrying the allelic combination favoured by selection will also be at higher than expected frequencies.

Effects of Population Admixture and Selection on Association: an Illustration

Studies to determine association between a marker allele and the phenotype can take two forms. In one form, groups are distinguished on the basis of their divergent phenotypes (diseased vs. healthy; low vs. high trait value) and allele frequencies are compared across groups. Such studies are often referred to as case–control studies in the human genetics literature, since they contrast disease-affected individuals (cases) with unaffected (control) individuals. The second type of study uses groups distinguished on the basis of their marker genotypes, and phenotypic means are compared across groups. An example of this is Beer *et al.* (1997), who analysed 13 quantitative traits on 64 North American oat varieties and landraces grouped according to restriction fragment length polymorphism (RFLP) genotype at 48 loci. Significant associations between RFLP fragments and group means occurred for 11.2% of fragments when testing at a 1% type I error rate, indicating many more associations than expected by chance alone. Some caution is in order, because (as the authors point out) the observed marker–trait association does not necessarily imply that markers showing a significant effect on the phenotype are linked to QTL. Rather, the marker–trait disequilibrium may exist in the absence of linkage and, instead, may have arisen simply as a consequence of population structure.

A classic example from humans of this population-stratification effect is Knowler *et al.* (1988), who examined candidate haplotypes for type 2 diabetes in members of the Pima and Tohono O'odham tribes of southern Arizona. Individuals with one particular haplotype had only an 8% rate of diabetes, while those lacking this haplotype had a 30% rate of diabetes. However, this particular haplotype is much more common in Caucasian populations than in full-heritage native American populations. When correcting for this population difference by only considering individuals of full heritage, 59% of individuals with the haplotype had diabetes, while 60% of the individuals lacking the haplotype had diabetes. In a similar fashion, the marker alleles associated with significantly different trait values observed by Beer *et al.* (1997) may have become associated with the phenotype through admixture of genetically divergent populations (for both markers and QTL) or through the effects of selection on both marker frequency and phenotype. In the former case, we can conceptualize the association between marker allele and phenotype as arising from the allele's association with the polygenic effect. If two populations diverge in phenotypic mean and in frequency of a marker allele, then admixture of these populations will create such an association. Under random mating, an unlinked marker allele's association with the phenotypic variance will be divided by four in each generation. Unfortunately, this rule only applies to outbreds that may conceivably random-mate. It will be more difficult to predict the decay of marker association with phenotype in a germ-plasm pool of self-pollinators, such as Beer *et al.*'s (1997) oat data. One obvious population structure in the Beer *et al.* (1997) data is the distinction between spring and winter oat varieties, which differ in both phenotype and marker frequencies (Souza and Sorrells, 1991). Beer *et al.* (1997) did not take these two divergent subpopulations into account in their analysis.

Another potential level of population structuring is a temporal one: Beer *et al.* (1997) analysed germ-plasm spanning about four decades of genetic improvement. Varieties grouped by year of release are expected to differ in mean for traits such as grain yield and harvest index. Under selection, the frequency of favourable QTL alleles at all loci increases and covariances among marker alleles across generations arise. These covariances hamper the estimation of the phenotypic effect associated with any single marker (Kennedy *et al.*, 1992). In effect, we may consider the germ-plasm pool analysed by Beer *et al.* (1997) as an admixture of old and modern subpopulations, the one having undergone less selection than the other. We would then expect to find fewer associations between marker alleles and phenotypes within each subpopulation than in the combined pool. Beer *et al.* (1997) performed this analysis and found only 6.5% and 4.9% of allele–trait associations were significant in the subpopulations of old and modern varieties, respectively. Some of the decline in the frequency of significant results would be due to the difference in power between tests on the combined pool versus within each subpopulation. It seems likely, however, that the difference in the results also indicates that the partition of the combined pool into old and modern varieties successfully separates subpopulations that are divergent both in phenotypic mean and in allele frequencies at certain markers.

The obvious weakness of group-comparison studies is that the grouping method may result in groups that contain predominantly individuals from different subpopulations. To eliminate this weakness, family-based control methods seek case and control individuals or marker alleles within the same family.

The Transmission/Disequilibrium Test

The problem of population admixture is ubiquitous in human-disease mapping, promoting considerable work to develop unbiased association estimators. Perhaps the most successful is the TDT of Spielman *et al.* (1993) to identify loci contributing to disease susceptibility in humans in the presence of population structure. For outbred species, the test employs family trios consisting of both parents and a progeny that is affected by disease (or, in general, that belongs to one category of a dichotomous trait). One of the parents must be heterozygous and carry one copy of the focal marker allele putatively linked to the disease-susceptibility allele. The test consists of determining the frequency of transmission of the focal allele to affected

progeny. A chi-square or binomial test can determine whether that frequency deviates from the expectation of 0.5. Two conditions are necessary for a significant deviation: the marker allele must be both in GPD with and also linked to a disease-susceptibility allele. In the TDT, both case and control marker alleles are in effect within the same heterozygote parent. Random Mendelian segregation therefore ensures that the distribution of the TDT statistic under the null hypothesis is unaffected by population structure or selection within the pedigree (Spielman and Ewens, 1996).

No TDT tests have been developed for predominantly selfing species. The extension, however, should be straightforward. A selfing TDT could employ marker information on F_1 hybrid/selfed progeny pairs, where the F_1 is heterozygous at a putatively linked marker locus and the progeny is affected. In this situation, transmission frequencies have the same expectations as for the TDT test, even if several generations of selfing occur between F_1 and inbred progeny. If the F_1 itself was not genotyped, its genotype may be inferred either from the known genotypes of its inbred parents or by pooling DNA from a number of its progeny derived by selfing. A potential complication (especially in hybrids) is gametic selection, which can bias transmission ratios. Hence, when using a TDT, one should always also perform a test of equal allelic transmission when phenotypic value is ignored.

While the TDT is always a valid test of linkage, researchers have devoted substantial effort to inferring in what cases the TDT is a valid test of population-wide association (Spielman and Ewens, 1996). In particular, when the family trios used are related, the test may detect association that exists solely in the pedigree from which those families derive but not in the general population (Martin *et al.*, 2000). We view the problem as one of determining the correct inference space for the test result. When the test uses multiple related families, the correct inference space for association is the pedigree from which they derive, not the general population. Asserting broader inference would be an example of pseudoreplication

(Hurlbert, 1984). Further, while the TDT remains a valid test for linkage, the critical interest in using association mapping is in finding tightly linked markers. A TDT based on multiple related families may detect association based on fairly distant marker–QTL pairs simply because recombination within the confines of the single pedigree evaluated will fail to reduce their association.

Extensions of the TDT to Quantitative Traits

As developed, the TDT only applies to traits that can be scored as dichotomously in the progeny, though these traits may be influenced by more than one underlying genetic factor. For populations undergoing artificial selection on a quantitative trait, Bink *et al.* (2000) take advantage of the insight that 'selected' versus 'not selected' constitutes a dichotomous trait. All families with selected progeny are therefore genotyped and the standard TDT is applied to those data. In the case of recurrent selection, the observed families will generally not be independent of each other, related as they are through cycles of intermating. As discussed, care must be taken in determining the inference space for positive association results. Data sets containing genotype information on current and previously released varieties of inbred crops could be analysed using the TDT in this way. Indeed, variety pedigrees are generally known (though some pedigrees may contain errors (e.g. Lorenzen *et al.*, 1995)). We can assume that a derived variety was selected from its parental varieties because of its agronomically favourable traits. Thus, a preferentially transmitted marker allele could be inferred to be in GPD with an agronomically favourable QTL allele.

Allison (1997) proposed five extensions of the TDT for quantitative traits. These extensions either compare the means of progeny, conditional on whether they received the putatively associated allele, or examine the frequency of inheritance of the allele among progeny whose trait values are above or below specified thresholds. In this latter case, we see that the use of thresholds reduces quantitative traits to dichotomous traits, bringing us back to the standard TDT. Unfortunately, these tests impose restrictive conditions on usable family trios: one heterozygous and one homozygous parent, and only one offspring. In practice, one family may have multiple progeny and/or the parents may lack genotypic data. To gain power from such data, Monks and Kaplan (2000) present a parametric procedure that relaxes family restrictions, allowing families of different types and several progeny per family to be used. The test defines a statistic, T_{MK}, based on the mean cross product between the deviation of the progeny phenotype from the population mean and the transmission of the focal marker allele from heterozygous parents. For large sample sizes, T_{MK} is approximately distributed as a unit normal [$T_{MK} \sim N(0,1)$]. To apply the test to small sample sizes or when multiple markers or marker alleles are used, Monks and Kaplan (2000) describe permutation procedures to obtain empirical distributions for T_{MK}. Finally, to account for environmental covariates that affect the quantitative trait of interest, the population mean can be adjusted by regression of the trait on the environmental covariates (Rabinowitz, 1997). A cross product is then calculated using the progeny deviation from this adjusted mean.

While plant geneticists have long been interested in genotype–environment interaction, efforts to account for it within human genetics and in association tests in particular are more recent (Schaid, 1999b; Guo, 2000a,b). In the standard TDT, QTL × E would lead to environmental influences on the transmission frequency of the focal marker allele from a heterozygotic parent to affected progeny. Such an effect could be detected by grouping family trios according to their environment or level of exposure to a risk factor. Heterogeneity of transmission frequency across groups would provide evidence in favour of QTL × E (Schaid, 1999b). Similarly, for the Monks and Kaplan (2000) test, environments would affect the magnitude of T_{MK} in the presence of QTL × E. Existence of QTL × E could then be inferred if the variance of T_{MK} across environments is significantly greater than zero. In observational studies where environments cannot be randomized across family trios, interpretation of such a result would need to be treated carefully: an association between environments and different subpopulations could also lead to heterogeneity of transmission or of T_{MK} in the absence of QTL × E.

Association Mapping with Multiple Markers

Given data on multiple linked markers, each particular combination, or haplotype, can be considered an allele at a 'supralocus'. Extensions to the TDT for multiple marker alleles can then be applied to this supralocus (Spielman and Ewens, 1996; McIntyre et al., 2000). A drawback to these methods is that they fail to make full use of all the haplotype information, as some haplotypes are more closely related (i.e. fewer mutational/recombinational steps away) than others. This potentially induces a correlation structure among haplotypes that needs to be considered. Several approaches have been developed to use the full haplotype information to pinpoint more precisely the location of mutations affecting disease status or the value of a quantitative trait. These methods are like typical linkage methods of QTL mapping in that, for specified map locations, they relate identity by descent (IBD) probabilities with phenotypic resemblance among individuals. For this task, however, linkage methods can calculate exact IBD probabilities based on meiotic events recorded in a pedigree. Association methods cannot rely on a recorded pedigree and so use haplotype similarities either to infer IBD probabilities directly or to create cladograms, which can be considered as approximate pedigrees. We describe three approaches.

Templeton and co-authors (Templeton et al., 1987; Templeton and Sing, 1993) use the haplotype marker profiles to construct a cladogram that estimates the evolutionary history and relationships among haplotypes.

Assume that a mutation occurred at some point in this history on one branch of the cladogram. Haplotypes along that branch will be IBD for the mutation and distinct from haplotypes along other branches. The branches of the cladogram therefore define nested sets of haplotypes that should have related associations with the phenotype. Templeton et al. (1987) present a nesting algorithm to group haplotypes hierarchically, enabling a nested analysis of variance. We note that the cladogram could also be used to define a covariance matrix among haplotype effects that would enable a mixed-model analysis of variance to detect QTL on the basis of significant among-haplotype variance. While this approach does not localize the mutation within the set of markers used to define haplotypes, it will increase the power to detect QTL in linkage disequilibrium with those markers.

Meuwissen and Goddard (2000) use an approach to estimate the covariance matrix among haplotype effects that does help predict QTL position within the set of markers. Starting from assumptions concerning the population history since mutation caused polymorphism at the QTL (i.e. effective population size and number of generations since mutation), the algorithm repeatedly simulates haplotype evolution and samples the probability of IBD status across specified categories of identity by state (IBS) among markers within haplotypes. The covariance among haplotype effects is then based on their IBS and its inferred relation to IBD. Since the probability function $P(\text{IBD} \mid \text{IBS})$ depends on QTL location within the set of markers, the assumed QTL location affects the haplotype covariance matrix. A maximum-likelihood QTL position is inferred from the covariance matrix most consistent with the observed phenotypes. Note that Meuwissen and Goddard's (2000) approach assumes a single (monophyletic) polymorphism at the QTL while Templeton et al.'s (1987) cladogram approach does not. Finally, the simulation approach may require information about the population history that is unavailable in practice: the population size and possible substructure over time, the age of the mutation, admixture through migration and selection on the mutation. While Meuwissen and Goddard (2000) show that the approach is, within limits, robust to population-size and mutation-age assumptions, applying the analysis to real (rather than simulated) data would be of interest in testing it.

The two methods discussed above apply to a random sample of individuals chosen independently of their phenotypic value. In mapping human disease, individuals are not sampled randomly, but rather chosen precisely because they are affected. Assume now that all affected individuals carry a disease-susceptibility allele from a mutation that occurred only once in the population. For a hypothesized QTL position, the likelihood problem is now turned on its head: rather than seeking the likelihood of the phenotype measurements given marker haplotypes, we seek the likelihood of observing this sample of marker haplotypes given that all individuals share a phenotype. The likelihood of the haplotypes depends on their genealogy, which cannot be known. In this situation, a marginal likelihood is calculated by averaging likelihoods over different possible genealogies, weighted by the genealogy probabilities (Graham and Thompson, 1998; Rannala and Slatkin, 1998, 2000). This high-dimensional integration over genealogies can be performed using random samples of genealogies. The algorithms to obtain samples use recent developments in the field of coalescent theory (Hudson, 1993; Donnelly and Tavaré, 1995). In essence, coalescent theory considers the haplotypes in a sample to be the tips of the genealogical tree and then defines probability distributions for the time in the past when branches were joined through common ancestor haplotypes. These steps are iterated until the whole genealogy coalesces into a single common ancestor. Having defined the coalescent genealogy, different events can be placed along its branches, such as recombination events (Graham and Thompson, 1998) or marker or QTL mutations (Zöllner and von Haeseler, 2000). The resultant coalescent genealogy represents one possible path generating the currently observed sample. Again, the hypothesized QTL position will affect the distribution of samples obtained through this process and will produce different Monte Carlo estimates of the observed sample likelihood. Differences in likelihood resulting from QTL position then allow for fine-scale disequilibrium mapping using multiple markers (Graham and Thompson, 1998).

Power of Association Mapping

Risch and Merikangas (1996) discussed power in the context of a complete human-genome scan to detect disease-susceptibility alleles of relatively small effect. Given their assumptions about the disease genetics, they showed substantial benefits of association mapping over linkage methods available in humans. Here we briefly discuss factors that affect the power of association mapping to detect QTL and reproduce selected guidelines for sample sizes (Schaid, 1999a; Monks and Kaplan, 2000).

The first determinant of power is the magnitude of GPD itself, which depends on mechanisms previously discussed. The different mechanisms also lead to different relationships between GPD and genetic distance (Jorde *et al.*, 1994; Laan and Pääbo, 1997). Ideally, a marker based on polymorphism in the causal locus itself is used, ensuring maximum marker–QTL GPD (Risch and Merikangas, 1996). Either candidate gene approaches or, with improving genotyping capability, exhaustive genome scan approaches, makes this ideal feasible. Even with complete disequilibrium, maximum GPD depends on the frequencies of the marker and QTL alleles, as shown above (Equation 5.1). Consequently, alleles with a frequency of 0.5 are most easily detected and detection power decreases for more extreme allele frequencies of either marker or QTL. These considerations further indicate that, when multiple alleles exist that affect the trait (a condition called genetic heterogeneity), association mapping loses power (Ott, 1999, pp. 290–291). A more subtle consequence of Equation (5.1) is that a marker more distant from a QTL may actually display a higher level of disequilibrium than a more closely linked marker.

A second determinant of power is the effect size of the QTL allele. In human-disease mapping, a joint measure of recombination fraction between marker and QTL and QTL effect size is given by the genotype relative risk (GRR) of disease for different marker genotype classes (Schaid and Sommer, 1993). Considering marker classes mm, mM and MM, $GRR_1 = P(\text{disease} | \text{mM}) / P(\text{disease} | \text{mm})$ and $GRR_2 = P(\text{disease} | \text{MM}) / P(\text{disease} | \text{mm})$. Higher GRRs indicate larger QTL effect size. Under a multiplicative model of gene mode of action where $GRR_2 = (GRR_1)^2$, disease susceptibility loci where $GRR_1 = 2$ are considered to have relatively small effects (Risch and Merikangas, 1996). For comparison with quantitative traits, these GRRs are similar to the relative 'risk' for individuals carrying a favourable allele of being selected under an intensity of 20% when the marker is in complete disequilibrium with an additive QTL that explains 10% of the phenotypic variance and $P_q = P_Q = 0.5$.

Finally, the mode of gene action (dominant, recessive, additive or multiplicative) greatly influences the power of QTL detection (Schaid, 1999a). Mode of gene action determines the relationship between GRR_1 and GRR_2 as follows: dominant $GRR_1 = GRR_2$; recessive $GRR_1 = 1$; additive $GRR_2 = 2GRR_1 - 1$; multiplicative $GRR_2 = (GRR_1)^2$. The TDT makes no assumptions concerning this mode of action when it looks at the transmission of marker alleles to affected offspring. But, because only affected offspring are sampled, the distribution of marker genotypes that they carry depends on the GRR. Consequently, the power of the TDT will depend on those GRRs (Schaid, 1999a). Likelihood methods of analysing TDT data that are general across modes of gene action or specific to an assumed mode and that are more powerful than the TDT were presented by Schaid and Sommer (1993, 1994). Tables 5.2 and 5.3 reproduce results from Schaid (1999a) and Monks and Kaplan (2000), respectively, to provide a general idea of the sample sizes required to detect with 80% power QTL that affect disease

Table 5.2. Number of family trios (last two columns) required to obtain 80% power of detecting association with a type I error rate of 5×10^{-8}. This error rate allows for a genome-wide scan over the whole human genome. (Results from Schaid, 1999a.)

Mutant allele frequency	Disequilibrium coefficient D	GRR_1	GRR_2	Statistical test	
				TDT	General likelihood
0.01	0.01	2	4	5,730	6,480
		1	2	3.86×10^7	7.72×10^5
0.10	0.09	2	4	687	776
		1	2	44,800	8,516
0.50	0.25	2	4	337	381
		1	2	946	710

Table 5.3. Number of families required to obtain 80% power of detecting association with a type I error rate of 0.01. Multiple offspring may be used per family. The segregating QTL causes 10% of the phenotypic variance and has an additive mode of action. (Results from Monks and Kaplan, 2000.)

Mutant allele frequency	Disequilibrium coefficient D	Number of progeny per family	
		1	5
0.10	0.02	7130	1600
	0.05	1810	404
0.50	0.10	808	182
	0.25	202	45

susceptibility or a quantitative trait measured on a continuous scale.

Final Remarks

The reader will no doubt notice the heavy influence of human genetics in much of the above discussion of association mapping. Plant breeders will do well in the future to continue to follow the human-genetics literature for continued developments and refinements. As stressed by Walsh (Chapter 2, this volume), it behoves all practitioners of quantitative genetics to follow developments in other subfields outside their own.

References

Allison, D.B. (1997) Transmission–disequilibrium tests for quantitative traits. *American Journal of Human Genetics* 60, 676–690.

Beer, S.C., Siripoonwiwat, W., O'Donoughue, L.S., Sousza, E., Matthews, D. and Sorrells, M.E. (1997) Associations between molecular markers and quantitative traits in an oat germplasm pool: can we infer linkages? *Journal of Agricultural Genomics* 3. [online] URL: http://www.ncgr.org/research/jag/papers97/paper197/indexp197.html

Bink, M.C.A.M., Te Pas, M.F.W., Harders, F.L. and Janss, L.L.G. (2000) A transmission/disequilibrium test approach to screen for quantitative trait loci in two selected lines of Large White pigs. *Genetical Research* 75, 115–121.

Darvasi, A. and Soller, M. (1995) Advanced intercross lines, an experimental population for fine genetic mapping. *Genetics* 141, 1199–1207.

Donnelly, P. and Tavaré, S. (1995) Coalescents and genealogical structure under neutrality. *Annual Review of Genetics* 29, 401–421.

Graham, J. and Thompson, E.A. (1998) Disequilibrium likelihoods for fine-scale mapping of a rare allele. *American Journal of Human Genetics* 63, 1517–1530.

Guo, S.W. (2000a) Gene–environment interactions and the affected-sib-pair designs. *Human Heredity* 50, 271–285.

Guo, S.W. (2000b) Gene–environment interaction and the mapping of complex traits: some statistical models and their implications. *Human Heredity* 50, 286–303.

Hudson, R.R. (1993) The how and why of generating gene genealogies. In: Takahata, N. and Clarck, A.G. (eds) *Mechanics of Molecular Evolution*. Sinauer, Sunderland, Massachusetts.

Hurlbert, S.H. (1984) Pseudoreplication and the design of ecological field experiments. *Ecological Monographs* 54, 187–211.

Jorde, L.B., Watkins, W.S., Carlson, M., Groden, J., Albersen, H., Thliveris, A. and Leppert, M. (1994) Linkage disequilibrium predicts physical distance in the adenomatous polyposis coli region. *American Journal of Human Genetics* 54, 884–898.

Kennedy, B.W., Quinton, M. and van Arendonk, J.A.M. (1992) Estimation of effects of single genes on quantitative traits. *Journal of Animal Science* 70, 2000–2012.

Knowler, W.C., Williams, R.C., Pettitt, D.J. and Steinberg, A.G. (1988) Gm 3;5,13,14 and type 2 diabetes mellitus: an association in American Indians with genetic admixture. *American Journal of Human Genetics* 43, 520–526.

Laan, M. and Pääbo, S. (1997) Demographic history and linkage disequilibrium in human populations. *Nature Genetics* 17, 435–438.

Lorenzen, L.L., Boutin, S., Young, N., Specht, J.E. and Shoemaker, R.C. (1995) Soybean pedigree analysis using map-based molecular markers: I. Tracking RFLP markers in cultivars. *Crop Science* 35, 1326–1336.

Lynch, M. and Walsh, B. (1998) *Genetics and Analysis of Quantitative Traits*. Sinauer, Sunderland, Massachusetts.

McIntyre, L.M., Martin, E.R., Simonsen, K.L. and Kaplan, N.L. (2000) Circumventing multiple testing: a multilocus Monte Carlo approach to testing for association. *Genetic Epidemiology* 19, 18–29.

Martin, E.R., Monks, S.A., Warren, L.L. and Kaplan, N.L. (2000) A test for linkage and association in general pedigrees: the pedigree disequilibrium test. *American Journal of Human Genetics* 67, 146–154.

Meuwissen, T.H.E. and Goddard, M.E. (2000) Fine mapping of quantitative trait loci using linkage disequilibria with closely linked marker loci. *Genetics* 155, 421–430.

Monks, S.A. and Kaplan, N.L. (2000) Removing the sampling restrictions from family-based tests of association for a quantitative-trait locus. *American Journal of Human Genetics* 66, 576–592.

Ott, J. (1999) *Analysis of Human Genetic Linkage.* Johns Hopkins University Press, Baltimore, Maryland, 382 pp.

Rabinowitz, D. (1997) A transmission disequilibrium test for quantitative trait loci. *Human Heredity* 47, 342–350.

Rannala, B. and Slatkin, M. (1998) Likelihood analysis of disequilbrium mapping and related problems. *American Journal of Human Genetics* 64, 1728–1738.

Rannala, B. and Slatkin, M. (2000) Methods for multipoint disease mapping using linkage disequilibrium. *Genetic Epidemiology* 19, S71–S77.

Risch, N. and Merikangas, K. (1996) The future of genetic studies of complex human diseases. *Science* 273, 1516–1517.

Schaid, D.J. (1999a) Likelihoods and TDT for the case–parents design. *Genetic Epidemiology* 16, 250–260.

Schaid, D.J. (1999b) Case–parents design for gene–environment interaction. *Genetic Epidemiology* 16, 261–273.

Schaid, D.J. and Sommer, S.S. (1993) Genotype relative risks: methods for design and analysis of candidate gene association studies. *American Journal of Human Genetics* 53, 1114–1126.

Schaid, D.J. and Sommer, S.S. (1994) Comparison of statistics for candidate-gene association studies using cases and parents. *American Journal of Human Genetics* 55, 402–409.

Souza, E. and Sorrells, M.E. (1991) Relationships among 70 North American oat germplasms: I. Cluster analysis using quantitative characters. *Crop Science* 31, 599–605.

Spielman, R.S. and Ewens, W.J. (1996) The TDT and other family-based tests for linkage disequilibrium and association. *American Journal of Human Genetics* 59, 983–989.

Spielman, R.S., McGinnis, R.E. and Ewens, W.J. (1993) Transmission test for linkage disequilibrium: the insulin gene region and insulin-dependent diabetes mellitus (IDDM). *American Journal of Human Genetics* 52, 506–516.

Templeton, A.R. and Sing, C.F. (1993) A cladistic analysis of phenotypic associations with haplotypes inferred from restriction endonuclease mapping. IV. Nested analyses with cladogram uncertainty and recombination. *Genetics* 134, 659–669.

Templeton, A.R., Boerwinkle, E. and Sing, C.F. (1987) A cladistic analysis of phenotypic associations with haplotypes inferred from restriction endonuclease mapping. I. Basic theory and an analysis of alcohol dehydrogenase activity in *Drosophila. Genetics* 117, 343–351.

Zöllner, S. and von Haeseler, A. (2000) A coalescent approach to study linkage disequilibrium between single-nucleotide polymorphisms. *American Journal of Human Genetics* 66, 615–628.

6 Integrating Molecular Techniques into Quantitative Genetics and Plant Breeding

John W. Dudley

Department of Crop Sciences, University of Illinois, 1102 S. Goodwin Avenue, Urbana, IL 61801, USA

Introduction

This chapter deals with the relationships between molecular genetic techniques, quantitative genetics and plant breeding. Molecular genetic techniques include the array of methods now available at the molecular level, including molecular-marker technology, the technologies available for studying gene structure and function and the bioinformatics tools necessary for extracting information from the data being generated from sequencing and functional genomics studies. Plant breeding is the science and art of genetic improvement of crop plants. Quantitative genetics is the study of genetic control of traits that show a continuous distribution in segregating generations and is concerned with inheritance of differences between individuals that are of degree rather than kind (Falconer, 1989). An understanding of quantitative genetic principles is generally considered critical to the design of efficient plant-breeding programmes (Baker, 1984), and plant breeding has been considered as applied quantitative genetics (Kempthorne, 1977).

Because the only way either molecular genetic techniques or quantitative genetics contribute to plant improvement is through plant breeding, this chapter discusses the steps involved in a plant-breeding programme. For each step, the contributions of quantitative genetics to the plant-breeding process will be discussed, along with a discussion of where molecular-level techniques intercept with plant breeding and quantitative genetics.

The Plant Breeding Process

Plant breeding consists of the creation of genetic variability, selection of élite types from that variability and synthesis of a stable cultivar from the élite selections. The history of plant breeding precedes that of the understanding of genetics, dating back to the times when primitive people saved seed to plant in succeeding years. Many major plant-breeding discoveries precede the rediscovery of Mendel's laws – including the development of major crops, such as maize (*Zea mays* L.), wheat (*Triticum aestivum* L.) and barley (*Hordeum vulgare* L.) through selection from primitive races. Mass selection for sucrose concentration in the sugar beet (*Beta vulgaris* L.) root began in 1786 and resulted in the first beet sugar factory being built in 1802 (Smith, 1987), 100 years before the rediscovery of Mendel's laws. The basic principles underlying the breeding of hybrid maize were known prior to 1900 (Zirkle, 1952).

With the rediscovery of Mendel's laws, genetic principles began to be applied to plant breeding. Smith (1966) traces the developments from 1901 to 1965. Because most of the traits of economic importance were under quantitative genetic control, quantitative genetics became an important contributor to plant-breeding theory. The contributions of quantitative genetics to plant breeding have been discussed in detail (Dudley, 1996). Lee (1999) summarizes the current and potential contributions of molecular genetics to plant breeding, particularly with regard to understanding heterosis and breeding hybrid plants. Moose (2000) summarizes the potential impact of maize genomics on breeding improved maize hybrids.

Quantitative-genetics Tools

Quantitative genetics approaches the understanding of genetic control of phenotype in a deductive manner: that is, it attempts to provide a genetic model for observed phenotypic variation. A major contribution of quantitative genetics has been to provide tools for separating genetic variation from environmental variation. As part of that separation, tools for measuring genotype–environment interaction (GE) and its importance have made a major contribution. Because the ramifications of GE are discussed in Chapters 15–25 in this volume and a symposium volume discussing this topic is available (Kang and Gauch, 1996), a detailed discussion is not given here. Suffice it to say that, when environment affects a trait, genes may act differently in different environments.

Because quantitative traits are those for which the effects of genotype and environment cannot be readily distinguished, a major contribution of quantitative-genetic theory was to provide methods for separating genetic effects from environmental effects. The work of Fisher (1918) and the elaborations by Cockerham (1954) and Kempthorne (1954) provided procedures for describing genetic variation in a population. These developments allowed the division of genetic variability into additive, dominance and epistatic variation. A major assumption in these derivations was that gene effects at individual loci were small and equal.

The general procedure for estimating genetic components of variance is to devise a mating design that will estimate covariances between relatives (such as the covariance of full-sibs or half-sibs). The mating design is then grown in an environmental design. The environmental design includes the choice of environments (usually locations and years) and environmental stresses (such as plant population, irrigation or lack thereof, fertility levels, etc.), as well as the experimental design (such as a randomized complete block, incomplete block or other type of design). From the appropriate analysis of variance, design components of variance are estimated and equated to covariances between relatives. Estimates of covariances between relatives are then equated to expected genetic variance components and genetic variances are estimated (Cockerham, 1963). Such estimates have limitations. Assumptions usually include linkage equilibrium in the population from which the parents of the mating design were obtained and negligible higher-order epistatic effects.

Using estimates of genetic variances and of genotype–environment variances, gain from selection can be predicted. In fact, one of the major contributions of quantitative genetics to plant breeding was the development of theory for prediction of gain from selection. A generalized prediction equation for gain per year can be written as follows (Empig et al., 1972):

$$R = cis_g^2/y\sigma_P \qquad (6.1)$$

where c is a pollen control factor (1/2 if selection is after pollination, 1 if selection is prior to pollination and 2 if selfed progenies are recombined), y is the number of years per cycle, i is the selection differential expressed as number of σ_P, s_g^2 is the appropriate genetic variance for the type of selection being practised and σ_P is the appropriate phenotypic standard deviation for the progenies being evaluated in the selection programme. This equation is critical for comparing selection procedures. Examples of its use are given by Hallauer and Miranda (1988) and Fehr (1987).

Selection by plant breeders results in changes, not only in the trait for which selection is being practised, but in other traits as well (correlated response). The extent of correlated response is a function of the heritabilities of the primary and correlated traits, as well as the genetic correlation between the traits. Falconer (1989) presents the correlated response equation, a corollary to the predicted gain equation, as:

$$CR_Y = ih_x h_y r_A \sigma_{PY} \qquad (6.2)$$

where CR_Y is the correlated response in trait Y when selection is based on trait X, i is the standardized selection differential for X, h_x and h_y are the square roots of heritability of traits X and Y, respectively, r_A is the additive genetic correlation between X and Y and σ_{PY} is the appropriate phenotypic standard deviation for Y. Equation 6.2 becomes important not only in determining the type of correlated response that may occur under selection, but also in determining effectiveness of indirect selection. If $r_A h_x > h_y$ indirect selection for X will be more effective for improving Y than direct selection for Y, all other factors being equal. If, in addition, selection for X allows progress in an environment where Y cannot be measured, as may be true for marker-assisted selection, additional benefits accrue from indirect selection. Molecular genetics and genomics, which aim to more precisely determine the degree of relationship between genes and phenotypes, may help improve plant breeding by providing more direct estimates of r_A. These relationships may work in both ways, where knowledge of positive correlations between two desirable traits can be exploited and, conversely, where knowledge of negative correlations may suggest inefficient selection methods. The genomics-based knowledge may be at an individual gene level or at the level of protein expression. In either case, methods of combining information over several genes or proteins will be required.

The interest of plant breeders in selection for net worth, a function of several traits, has led to the development of methods for the simultaneous improvement of a population for multiple traits (Falconer, 1989). Three general procedures – tandem selection, independent culling levels and index selection – have been used to approach the question. Smith (1936) was the first to present the concept of index selection. Smith presented an index of the form:

$$I = b_1 X_1 + b_2 X_2 + \ldots b_m X_m$$

where I is an index of merit of an individual and $b_1 \ldots b_m$ are weights assigned to phenotypic trait measurements represented as $X_1 \ldots X_m$. The b values are the product of the inverse of the phenotypic variance–covariance matrix, the genotypic variance–covariance matrix and a vector of economic weights. A number of variations of this index have been developed. These include the base index of Williams (1962), the desired-gain index of Pesek and Baker (1969) and, more recently, retrospective indexes proposed by Johnson et al. (1988) and Bernardo (1991). The emphasis in the retrospective-index developments is on quantifying the knowledge of experienced breeders. Although breeders may not use a formal selection index in making selections, every breeder, either consciously or unconsciously, assigns weights to different traits when making selections.

The tools quantitative genetics provides for plant breeding are those that allow for a statistical description of genetic variability, for separation of genetic variability from environmental variability and prediction of gain from various types of selection. Although the tools are crude and the assumptions made are often considered unrealistic, they provide predictive power and have served the plant breeder well.

Molecular Genetic Tools

Molecular markers

The widespread use of molecular markers in plants is largely because of the implications they have for helping solve problems common to quantitative genetics and plant breeding. The use of markers as a potential aid in selection dates back to Sax (1923), who found seed colour related to seed size in beans. Stuber and Edwards (1986) were pioneers in the use of molecular markers (isozymes) in plant breeding. Stuber (1992) reviewed this work. The use of markers for selection in plant-breeding programmes is the application of a form of indirect selection and was reviewed in detail by Dudley (1993). The availability of molecular markers provides an additional dimension to the use of quantitative genetics in plant breeding. Potential applications of molecular markers include marker-assisted selection, identification of the number of genes controlling quantitative traits, grouping germ-plasm into related groups, selection of parents and marker-assisted back-crossing (Lee, 1995). A detailed discussion of the ways a major seed company has attempted to use molecular markers is given by Johnson and Mumm (1996). New technologies, such as single nucleotide polymorphisms (SNPs), provide the capability of developing densely saturated genetic maps. Further, this technology may be automated and conducted on an industrial scale, thus dramatically shortening the time and cost required to genotype a breeding population. Eventually, maize marker maps may be dense enough and the cost of marker analysis low enough for it to become cheaper to genotype populations than to collect the phenotypic data necessary for identifying quantitative trait loci (QTL) (Moose, 2000).

Transformation

The ability to insert a single gene into a plant cell or tissue and to regenerate an entire plant containing a functional version of that gene has opened the door to transferring genes across species barriers. Widespread use of this technology produced such new crops as Roundup Ready® soybeans and *Bacillus thuringiensis* (Bt) maize. Limitations of the technology in the past depended upon the species. In some species, transformation was limited to certain genotypes having the capability of regenerating whole plants from callus tissue or from cell suspensions. This limitation is now being removed. Regardless of the ease of transformation, however, the general practice is to back-cross a transgene into élite lines or cultivars rather than transforming each individual line, because of the cost of getting regulatory approval to use a particular transgenic event. Thus, molecular-marker technology is needed to hasten recovery of the recurrent parent in the back-crossing procedure.

Genomics

Although Liu (1998) includes molecular-marker technology in his definition of genomics, the term is used here to include the extensive research effort aimed at sequencing the genomes of plants and animals and understanding the functions of the sequenced genes. The emerging area of bioinformatics, which is providing the statistical and computer-programming tools necessary to mine the rapidly expanding databases being produced by major sequencing efforts, is included in this definition. Genomics under this definition is a reductionist approach to understanding phenotype: that is, the genotype is reduced to gene sequences, the function of the sequences is determined and, finally, attempts are made to determine the effects of the genes on the phenotype. The hope is that, by obtaining such an understanding, a plant can be assembled with an improved phenotype.

Progress in sequencing the genomes of crop plants is being made at an ever-increasing rate. Currently, the complete sequence of the wild mustard *Arabidopsis* has been obtained. The rice (*Oryza sativa* L.) genome has been sequenced and large numbers of expressed sequence tags (ESTs) are being generated from crops such as maize and soybean (*Glycine max* L.). Progress is being made in developing methods for studying the functions of the sequenced genes. The potential use of this information in plant-breeding programmes and its relationship to quantitative genetics will be discussed in a later section.

Creation or Assembly of Genetic Variability

The choice of parental germ-plasm with which to begin a breeding programme is the most important decision a breeder makes and is the first step, after establishment of objectives, in any plant-breeding programme. It is only recently, however, that quantitative-genetic theory has been applied to this question (Dudley, 1996) and even more recently that molecular techniques have had an impact. Essentially, the contributions of quantitative genetics relate to choice of parents to produce segregating generations, whereas those of molecular techniques, while useful in the choice of parents, relate more to classifying germ-plasm into related groups and broadening the gene pool that can be considered in creating genetically variable populations. To date, this broader gene pool has been the source of single genes controlling characteristics expressed in a manner not readily available in the usual gene pool.

Contributions of quantitative genetics

The contributions of quantitative genetics to selection of parents are greatest for breeding programmes where the objective is to develop an improved line or population for a quantitative trait or a series of such traits. A common procedure for such circumstances in both self-pollinated species, such as wheat or soybeans, and in cross-pollinated species, where a hybrid between inbreds is the desired end-product, is to cross two homozygous lines, self from the F_1 and select out an élite type from segregating generations. In self-pollinated species, these lines are usually evaluated for *per se* performance. In cross-pollinated species, where hybrids are the end-product, similar breeding procedures are used, with the exception that the end-product will be a hybrid. Thus, the criterion for selection is combining ability of some form, rather than line *per se* performance.

In either case, the general axiom, supported by quantitative-genetic theory, is to cross good with good. When choosing parents, the objective is to maximize the probability of generating new lines that will perform better than the best pure line currently in use. The parents chosen should generate a population for selection that will meet the criterion of usefulness described by Schnell (1983), as discussed in Lamkey *et al.* (1995). The usefulness of a segregating population was described by Schnell as the mean of the upper $\alpha\%$ of the distribution expected from the population. Mathematically, $U_\alpha = Y \pm G_\alpha$, where U_α is usefulness, Y is the mean of the unselected population and G_α is gain from selection of the upper $\alpha\%$ of the population. This statistic takes into account both the mean and the genetic variability, thus emphasizing a basic axiom in plant breeding: a population that will produce an improved cultivar will have both a high mean and an adequate genetic variability.

Panter and Allen (1995) suggested using best linear unbiased prediction (BLUP) methods to predict the mid-parent value – a good predictor of the mean of lines from a cross – of soybean crosses. BLUP methods take into consideration the performance of lines related to the line for which performance is being predicted. They found the coefficient of parentage between a pair of lines was related to genetic variance in the progeny. Based on these results, they suggested that an effective method of choosing parents would be to identify pairs of lines with high mid-parent values estimated from BLUP and to select among such pairs those that were the most genetically diverse based on the genetic relationship matrix. Their suggestion is supported by the results of Toledo (1992). With the availability of genetic markers, degree of relationship between lines can be established from molecular-marker data (Lee, 1995). This provides an alternative method of determining relatedness when pedigree information is unavailable or of uncertain accuracy.

The same general principles apply to choice of parents in a hybrid breeding programme. The basic question in choosing parents is identification of those lines or populations that contain favourable alleles not present in a hybrid being improved. Dudley (1984a) framed the following questions relative to choice of parents for a hybrid maize-breeding programme. Which hybrid should be improved? Which lines should be chosen as donors to improve the target hybrid? Which parent of the target hybrid should be improved? Should selfing begin in the F_2 or should back-crossing be used prior to selfing?

Procedures for answering these questions were developed based on the concept of classes of loci (Dudley, 1982). Using this concept, methods of identifying donors with the greatest numbers of loci carrying favourable alleles not present in a hybrid to be improved were devised (Dudley, 1984b,c, 1987a,b). Modifications of these methods were proposed by Gerloff and Smith (1988), Bernardo (1990a,b) and Metz (1994). Evidence for their effectiveness in selecting superior parents and identifying heterotic relationships was presented by Dudley (1988), Misevic (1989), Zanoni and Dudley (1989), Hogan and Dudley (1991) and Pfarr and Lamkey (1992).

Contributions of molecular genetics technology

There are two major contributions of molecular technology to the creation of genetic variability. First, molecular-marker technology has provided a method of determining relationships among germ-plasm sources. Secondly, the potential gene pool for improving cultivated crops has been greatly broadened by the use of transgenic technology. A third contribution relates to choice of parents.

Molecular markers have been used in a variety of crops to determine relationships among germ-plasm sources. In maize, clustering based on molecular-marker-estimated relationships has been shown to agree with

known heterotic relationships among inbreds (Mumm and Dudley, 1994). In soybeans, clustering based on molecular markers has been useful in identifying germ-plasm accessions from China with minimal relationship to US germ-plasm (Brown-Guedira et al., 2000). Thus, combining molecular techniques with quantitative-genetic and statistical procedures has provided a useful tool for cataloguing genetic variability.

Although interspecific crosses have been sources of genetic variability available to plant breeders for many years, the development of gene sequencing coupled with transformation has allowed exploitation of genetic variation in species, such as bacteria, for which crosses with cultivated species are impossible. A classic example is the isolation of the Bt gene from Bacillus thuringensis, modifying it to make it work in a maize plant, insertion of the gene along with a promoter and a selectable marker into maize tissue culture and development of maize with resistance to European maize borer (Ostrinia nubilalis) at a level not previously known (Koziel et al., 1993; Armstrong et al., 1995). Another potential of molecular technology is the development of genes for the production of drugs or polymers and their insertion into plants. This development greatly broadens the potential usefulness of a species. In addition, molecular-marker technology is being used to monitor recovery of the recurrent parent in back-crossing programmes aimed at incorporating transgenes into adapted cultivars.

Although transgenic technology allows the tapping of genetic resources hitherto unavailable to the plant breeder, it is still primarily a one-gene-at-a-time approach, even though a number of groups have successfully introduced multiple genes into one line by transformation. A good example is golden rice (Ye et al., 2000). Not every insertion of a transgene provides a transformed plant and not every transgenic event will be expressed in every genetic background. In addition, the genotype limitations on transformation in some species require that transgenes be bred into élite lines. Thus, attention to quantitative genetic principles and hard-headed pragmatic plant breeding in the form of extensive testing must be coupled with the transgenic approaches.

Molecular-marker technology has been proposed for use in selecting parents of crosses. Hanafey *et al.* (1998) obtained a patent for a procedure in which markers were used to genotype élite lines and their early progenitors. Because the élite lines had been selected for yield for a number of generations, changes in marker allele frequency were presumed to be the result of selection for favourable linked genes. Élite lines would be chosen for crossing in which progeny would have the maximum number of marker loci containing favourable marker alleles.

Molecular-marker technology has a role in identifying the best parents to use to produce a single-cross hybrid. Choice of parents to produce a cultivar directly is usually the result of extensive testing of a number of combinations of potential parents. One of the major problems facing breeders is reducing the number of possible hybrids to be tested to a reasonable number. In general, maize breeders work with heterotic groups, and crosses likely to be successful as cultivars are usually between inbreds from different heterotic groups (Hallauer *et al.*, 1988). Even if breeding is restricted to two heterotic groups, however, thousands of potential hybrids are possible.

Bernardo (1994) proposed applying BLUP to this problem. In this procedure, information on hybrid performance of a subset of lines is combined with information on the genetic relationship between the lines tested and an untested set of lines to predict the performance of untested hybrids. Bernardo (1994), using a limited number of hybrids, found correlations between observed and predicted performance to range from 0.65 to 0.80. He compared restriction fragment length polymorphism (RFLP)-based estimates of relationship with pedigree-based estimates and found higher correlations for the RFLP-based estimates. In a more recent study (Bernardo, 1996) involving 600 inbreds and 4099 tested single crosses, correlations between predicted and observed yields ranged from 0.426 to 0.762.

Bernardo concluded that BLUP was useful for routine identification of single crosses prior to testing.

Selection

Selection from the genetic variability produced in the first step in the plant-breeding process is the stage most often thought of when the words 'plant breeding' are mentioned. A large body of literature on selection theory and results is available. Responses to selection have been reported for almost every species (Hallauer, 1985). Knowledge of the relative importance of different types of gene action is important in designing effective selection programmes.

Quantitative genetics and selection

Comstock (1978) suggested that the development of a theoretical basis for comparing response to selection from different breeding methods was one of the most significant contributions of quantitative genetics to maize breeding. Baker (1984) concluded that this statement could be extended to all economically important crops. Of the contributions of quantitative genetics to plant breeding, the predicted-gain equation, the correlated-response equation and index-selection theory are all contributions to increasing selection efficiency. Although plant breeders may not use these formal equations in their selection experiments, they are guided by the principles embodied in the equations. Selection procedures can be roughly divided into those to be used during selfing after the cross between inbreds and those that fall under the heading of recurrent selection.

As Hallauer *et al.* (1988) point out, the methods used to select during inbreeding and recurrent selection procedures are complementary parts of a breeding programme. In fact, because one result of selection during inbreeding is the development of improved lines that are crossed and another round of selection carried out, selection during inbreeding is one form of recurrent selection.

The appropriate generation in which to select during inbreeding is a major question when developing improved inbreds, whether for use directly as cultivars (in breeding self-pollinators) or as parents of hybrids (as in breeding maize or sorghum). Because of the rapid approach to homozygosity within lines and the increased variability between lines as inbreeding progresses, Brim (1966) suggested using a modified pedigree (single-seed descent) procedure, in which yield testing was delayed until homozygosity was nearly complete. Other workers have proposed the use of doubled haploids to provide 'instant inbreds'.

In maize breeding, there is an extensive literature on the question of whether to test for combining ability in early generations or to wait until homozygosity is more nearly complete. Bernardo (1992) developed theory for the genetic and phenotypic correlations between test-cross values of lines tested in a given selfed generation and their selfed progeny. As selfing advances, the correlation increases. Heritability of test-cross means also affects the correlation between early-generation phenotypic values and the expected genetic values of the progeny. Based on theory and simulation results, Bernardo (1992) suggested saving approximately 25% of lines based on S_1 or S_2 testing if heritability is 0.25 or 0.5 in the S_1 generation. He also presented tables showing the probability of retaining lines in the upper $\alpha\%$ of a distribution of homozygous lines, given that a line selected in a preceding generation (S_n) was in the upper $\alpha\%$ of lines in the S_n generation. Empirical results (Jensen et al., 1983; Hallauer and Miranda Fo, 1988) agree with this theory.

Recurrent selection is one of the most extensively studied forms of selection in plant breeding (Coors, 1999). Much of the work on recurrent selection was initiated in maize because of the controversy over the causes of heterosis (Gowen, 1952). Two types of recurrent selection had been devised based on differing views as to the cause of heterosis. Recurrent selection for general combining ability (g.c.a.) (Jenkins, 1940) was presumed to take advantage of average effects. Recurrent selection using an inbred tester (i.e. recurrent selection for specific combining ability) was proposed by Hull (1945) as a means of taking advantage of overdominance, a phenomenon that he suggested was of primary importance relative to heterosis. Comstock et al. (1949) proposed reciprocal recurrent selection with the idea that it would take advantage of both dominant and overdominant types of gene action. Although the studies initiated as a result of the debate over the causes of heterosis did not resolve the causes of heterosis, they did provide a great deal of information on the effectiveness of different selection procedures (Coors, 1999).

Quantitative-genetic studies to determine the relative importance of dominance and overdominance can usually be summarized by the statement 'although the results suggest additive and partial to complete dominance effects are most important, over-dominance cannot be eliminated as having an effect on heterosis for maize grain yield'. Quantitative-genetic studies have been even less informative as to the importance of epistasis as a cause of heterosis. Because of the regression basis of methods for estimating types of genetic variance, estimates of epistatic variances have large standard errors and typically are found to be non-significant (Hallauer and Miranda Fo, 1988). Procedures for detecting the presence of epistasis (as opposed to measuring epistatic variance), however, generally support the presence of epistasis for grain yield in maize. Based on the concept that a plant is the end-product of a collection of biosynthetic and chemical reactions, epistasis must play a role in the performance of plants and in causing heterosis.

Quantitative genetics has played a major role in refining selection procedures. It has provided general clues as to the types of gene action responsible for improved performance, but it has not proved to be a good vehicle for identifying and measuring types of gene action and their relative importance.

Molecular genetic techniques and selection

As with quantitative-genetic techniques, molecular genetic techniques have a role in selection and in identifying important types of gene action. Very few selection experiments based on molecular markers are reported in the literature. Such experiments are expensive and often not considered as 'basic enough' to be funded by granting agencies. Thus, their primary development is in the private sector. Two examples of the successful use of selection based on molecular markers have been reported. Edwards and Johnson (1994) showed selection based on markers to be effective for improving sweetcorn. Johnson and Mumm (1996) reported the effective use of markers as an aid in early testing for combining ability in maize. They also reported the effective use of markers for recurrent selection for specific combining ability. The selection was effective, however, because of the elimination of low-yielding types associated with markers, rather than because it identified types that were higher yielding than the best lines that might have been identified by phenotype selection. Work by Eathington *et al.* (1997) agreed with that of Johnson and Mumm (1996) in demonstrating the effectiveness of markers as an aid in early testing programmes.

A major contribution of molecular techniques has been in gaining an understanding of types of gene action. Although marker-associated effects are not necessarily single-gene effects, a few key points have come out of the large number of studies aimed at identifying QTL. First, unlike the assumption in classical quantitative genetics of small and equal gene effects for genes controlling a trait, gene effects have a distribution from small to large. For quantitative traits, individual chromosome segments with effects large enough to be found significant in experiments of reasonable size have been identified. Careful analysis of experimental data using sophisticated quantitative-genetic techniques has shown that chromosome segments originally thought to be showing overdominance were, in fact,

carrying favourable alleles linked in repulsion and the apparent overdominance was pseudo-overdominance (Cockerham and Zeng, 1996). Development of near-isogenic lines (NILs) from these regions has allowed the isolation of chromosome segments responsible for the apparent overdominance (Graham *et al.*, 1997).

The importance of epistasis for control of heterosis for grain yield (Yu *et al.*, 1997) and grain-yield components (Li *et al.*, 1996) in rice (*O. sativa* L.) was demonstrated using QTL analysis. Recent work has demonstrated the importance of epistasis for QTL controlling heading date in rice using NIL analysis (Lin *et al.*, 2000). Work in tomato (*Lycopersicon esculentum* L.) (Eshed and Zamir, 1996) documented the importance of epistasis among introgressed chromosome segments using NILs. Of particular importance is the finding that epistatic effects of a few regulatory genes or factors may significantly influence phenotypic variation and the identification of QTL (Lin *et al.*, 2000). Thus, molecular techniques, combined with quantitative genetics, are enhancing our understanding of gene action controlling quantitative traits.

The Future

Where do we go from here? The hopes for quantitative genetics included an understanding of heterosis, a separation of genetic effects from environmental effects and an understanding of gene action. Certainly the concepts of quantitative genetics are an important part of any plant breeder's toolkit. However, quantitative genetics has not led to a clear understanding of the genetic basis for heterosis (Coors and Pandey, 1999). Molecular-marker technology, when combined with quantitative-genetic methodology, has provided additional insights into gene action. The work of Cockerham and Zeng (1996) showed that Stuber's marker information could be analysed to demonstrate repulsion–phase linkage in a chromosome region. Graham *et al.* (1997) isolated chromosome segments showing the importance of dominance of linked genes as an explanation for heterosis within a chromosome segment. The work in rice (Lin et al., 2000) demonstrating the importance of epistasis between chromosome segments is another example of increased under-standing of gene action relating to a quantitative trait. These types of studies, combined with quantitative-genetic princi-ples, help elucidate some of the questions for which quantitative-genetic methods alone have lacked precision.

Molecular-marker technology has con-tributed to plant breeding in a number of areas. A major contribution, however, has been in combination with transformation technology to allow rapid incorporation of transgenes having major effects for desirable traits, such as insect and herbicide resis-tance. Work with NILs in tomato has isolated chromosome segments with large, economi-cally significant effects from crosses of culti-vated tomato with wild species (Tanksley et al., 1996; Fulton et al., 1997). Similarly, chromosome segments allowing significant improvement in yield have been found in rice (Xiao et al., 1996, 1998). Stuber (1998) and Stuber et al. (1999) demonstrated the usefulness of marker techniques in isolating chromosome segments with large favourable effects from apparently unadapted germ-plasm, either inbreds or exotics, in maize. These studies, in three different species, document the potential for identifying useful alleles in exotic, unadapted, germ-plasm sources.

To date, genome sequencing and gene expression analysis have not contributed directly to quantitative genetics or plant breeding. At the cell-biology level, Schulze and Downward (2000) suggest that micro-arrays can be used to gain detailed informa-tion about specific metabolic pathways and their response to specific stimuli. They point out, however, that the use of microarrays will generate huge amounts of expression data. Whether such data lead to advances in the understanding of biological problems will depend on the development of method-ologies, in experimental biology and bioin-formatics, that allow meaningful knowledge to be extracted. Miflin (2000) considers that

the promise of recombinant DNA technology is in the potential for molecular analyses of genetic systems designed to improve crop productivity. Miflin (2000) suggests that this goal may be approached in three ways, start-ing at the beginning by generating complete sequences of the plant genome; starting at the end by genetic analysis of phenotypes using genetic-marker technology; or starting in the middle by metabolic analysis. He points out that the onset of genomics will provide massive amounts of information, but the success of genetic analysis will depend on using that information to improve crop phenotypes. Thus, information from gene sequencing and from functional genomic analysis will be useful only when the effects of reassortment of genomic sequences are measured at the phenotypic level. Bernardo (2001) considered the question of the bene-fits of knowing all the genes for a quantita-tive trait in a hybrid crop. Using computer simulation, he concluded that such informa-tion is most useful in selection when only a few loci (e.g. ten) control the trait.

To take into account the complexities of the combinations of gene sequences and the interaction between those sequences will require the coordinated efforts of a number of scientists with different backgrounds. Plant breeders are the individuals most able to grow plants in the field and measure pheno-type. They have the knowledge of phenotype necessary to determine when changes in phenotype have been made and whether those changes are positive or negative. Their skills and abilities will be essential if the developments of the basic molecular biolo-gists are to be useful. To organize the vast amounts of unorganized sequence and func-tion data being produced by the molecular biologists will require not only the bioinfor-matics specialist to organize the data but the quantitative geneticist to interpret the infor-mation and organize the results into systems that can be used by the plant breeder.

Integration of information into a system

A plant-breeding programme is an integrated, highly time-dependent system. Thus, as molecular markers were integrated into transgenic-conversion programmes, rapid-advancement systems making use of off-season nurseries and minimizing the time from planting to harvest became essential. Coupled with that was the necessity for high, accurate throughput in molecular-marker laboratories. The results of marker analysis must often be obtained for hundreds or thousands of plants within the time from planting to flowering (at least in maize). These procedures must be done in a cost-efficient manner. Thus, a great deal of effort has been put into mechanizing the whole process, in the field and in the laboratory, to save time. The necessity for rapid turnover and timely results means that planning of a particular transgene conversion, for example, requires a detailed outline of the time from the first cross to the production of seed for wide-scale testing.

An integrated system for incorporating transgenes into a breeding programme has been accomplished in some crops. The promise of more sophisticated marker systems with increased reliability and throughput provides hope for even more effective use of marker technology. Of major importance is the question of allocation of resources to transgenic development as opposed to continued development of increased base performance for yield and other agronomic traits. Because the incorporation of transgenes is a back-crossing process and the basic genotype resulting from the incorporation of a transgene is not improved except for the single trait affected by the transgene, continued improvement of other traits must be accomplished in programmes separate from the transgenic programme (Messmer, 1997).

How will gene sequence information and functional genomic information be incorporated into breeding programmes? The results from functional genomic studies in humans were targeted to understanding the genetic control of large differences (for example, tumorous vs. healthy cells; cells treated with a given drug vs. untreated) (Schulze and Downward, 2000). Thus comparisons of the genes expressed in tumorous tissue vs. healthy tissue of a particular organ afflicted with a particular disease were the bases for inference. In plants, there are relatively few cases where these types of comparisons are of primary importance. Gene libraries are being developed for different tissues from plants at different maturities and under differing types of stress. In the short run, it seems unrealistic to expect that such studies will have application to traits such as grain yield in maize except where the trait is a defensive one with a major effect. On the other hand, a clearer understanding of biochemical pathways affecting kernel quality in maize or grain quality in soybean may allow genetic intervention into a particular pathway and the development of designer genes for grain with enhanced nutritional quality for food or feed. Such interventions may take the shape of designing genes or combinations of genes to be created artificially or cloned and then transformed into other maize or soybean plants. In this case, the incorporation of the technology into plant-breeding systems will probably take the back-crossing route that has been followed with current transgenic approaches. The development of golden rice is an example of such an approach. Potentially, with the development of transformation technology for élite germ-plasm, transgenes might be inserted and evaluated in breeding procedures that simultaneously improve line performance and optimize expression of the transgene phenotype. This presumes that regulatory procedures will change to allow the approval of new transgenic events without the current costly, time-consuming procedures.

If microarray procedures can be used to develop assays that identify desirable patterns of gene expression in a particular tissue that can be reliably measured, the problems of timely data acquisition and analysis pervasive in transgenic procedures will again be encountered. The whole question of interaction between genes (epistasis) and the interactions of genes with environments has yet to be addressed in a meaningful fashion at the gene sequence and function level. Thus, for the immediate future, the application of these technologies to many important plant-breeding problems is limited.

Plant breeders and quantitative geneticists have developed high-throughput procedures to measure phenotypes at the whole-plant level and computational methods to quantify the relationships between phenotype, genotype and environment. Combining these existing approaches with genomics information will be required to maximize the value obtained from the significant public and private investments in genomics research with crop species. Without public research investment in plant breeding and quantitative genetics training, the potential benefits of genomics research will not be realized. Without quantitative geneticists and plant breeders, the public sector cannot educate the personnel the commercial companies, now heavily involved in genomics research, will need to take advantage of the results of their genomic investment. Miflin (2000) points out the large gap between gene-sequence information and effects on phenotype. Unless the scientific community and the sources of funding for public research recognize this gap and involve public plant breeders and quantitative geneticists with the process, the gap will only grow larger.

References

Armstrong, C.L., Parker, G.B., Pershing, J.C., Brown, S.M., Sanders, P.R., Duncan, D.R., Stone, T., Dean, D.A., DeBoer, D.L., Hart, J., Howe, A.R., Morrish, F.M., Pajeau, M.E., Petersen, W.L., Reich, B.J., Rodriguez, R., Tarochione, L.J. and Fromm, M.E. (1995) Field evaluation of European maize borer control in progeny of 173 transgenic maize events expressing an insecticidal protein from *Bacillus thuringensis*. *Crop Science* 35, 550–557.

Baker, R.J. (1984) Quantitative genetic principles in plant breeding. In: Gustafson, J.P. (ed.) *Gene Manipulation in Plant Improvement*. Plenum Press, New York and London, pp. 147–175.

Bernardo, R. (1990a) An alternative statistic for identifying lines useful for improving parents of elite single crosses. *Theoretical and Applied Genetics* 80, 105–109.

Bernardo, R. (1990b) Identifying populations useful for improving parents of a single cross based on net transfer of alleles. *Theoretical and Applied Genetics* 80, 349–352.

Bernardo, R. (1991) Retrospective index weights used in multiple trait selection in a maize breeding program. *Crop Science* 31, 1174–1179.

Bernardo, R. (1992) Retention of genetically superior lines during early-generation testcrossing of maize. *Crop Science* 32, 933–937.

Bernardo, R. (1994) Prediction of maize single-cross performance using RFLPs and information from related hybrids. *Crop Science* 34, 20–25.

Bernardo, R. (1996) Best linear unbiased prediction of the performance of crosses between untested maize inbreds. *Crop Science* 36, 872–876.

Bernardo, R. (2001) What if we knew all the genes for a quantitative trait in hybrid crops? *Crop Science* 41, 1–4.

Brim, C.A. (1966) A modified pedigree method of selection in soybeans. *Crop Science* 6, 220.

Brown-Guedira, G.L., Thompson, J.A., Nelson, R.L. and Warburton, M.L. (2000) Evaluation of genetic diversity of soybean introductions and North American ancestors using RAPD and SSR markers. *Crop Science* 40, 815–823.

Cockerham, C.C. (1954) An extension of the concept of partitioning hereditary variance for analysis of covariances among relatives when epistasis is present. *Genetics* 39, 859–882.

Cockerham, C.C. (1963) Estimation of genetic variances. In: Hanson, W.D. and Robinson, H.F. (eds) *Statistical Genetics and Plant Breeding*. Publication 982, NAS-NRC, Washington, DC, pp. 53–94.

Cockerham, C.C. and Zeng, Z.B. (1996) Design III with marker loci. *Genetics* 143, 1437–1456.

Comstock, R.E. (1978) Quantitative genetics in maize breeding. In: Walden, D.B. (ed.) *Maize Breeding and Genetics*. John Wiley & Sons, New York, pp. 191–206.

Comstock, R.E., Robinson, H.F. and Harvey, P.H. (1949) A breeding procedure designed to make maximum use of both general and specific combining ability. *Agronomy Journal* 41, 360–367.

Coors, J.G. (1999) Selection methodology and heterosis. In: Coors, J.G. and Pandey, S. (eds) *The Genetics and Exploitation of Heterosis in Crops*. American Society of Agronomy, Crop Science Society of America and Soil Science

Society of America, Madison, Wisconsin, pp. 225–245.

Coors, J.G. and Pandey, S. (eds) (1999) *The Genetics and Exploitation of Heterosis in Crops.* American Society of Agronomy, Crop Science Society of America and Soil Science Society of America, Madison, Wisconsin.

Dudley, J.W. (1982) Theory for transfer of alleles. *Crop Science* 22, 631–637.

Dudley, J.W. (1984a) Identifying parents for use in a pedigree breeding program. In: *Proceedings of the 39th Annual Maize and Sorghum Research Conference.* American Seed Trade Association, Washington, DC, pp. 176–188.

Dudley, J.W. (1984b) A method of identifying lines for use in improving parents of a single cross. *Crop Science* 24, 355–357.

Dudley, J.W. (1984c) A method for identifying populations containing favorable alleles not present in elite germplasm. *Crop Science* 24, 1053–1054.

Dudley, J.W. (1987a) Modification of methods for identifying populations to be used for improving parents of elite single crosses. *Crop Science* 27, 940–944.

Dudley, J.W. (1987b) Modification of methods for identifying inbred lines useful for improving parents of elite single crosses. *Crop Science* 27, 945–947.

Dudley, J.W. (1988) Evaluation of maize populations as sources of favorable alleles. *Crop Science* 28, 486–491.

Dudley, J.W. (1993) Molecular markers in plant improvement: manipulation of genes affecting quantitative traits. *Crop Science* 33, 660–668.

Dudley, J.W. (1996) Quantitative genetics and plant breeding. *Advances in Agronomy* 59, 1–23.

Eathington, S.R., Dudley, J.W. and Rufener, G.K., II (1997) Usefulness of marker–QTL associations in early generation selection. *Crop Science* 37, 1686–1693.

Edwards, M. and Johnson, L. (1994) *RFLPs for Rapid Recurrent Selection.* Proceedings of Joint Plant Breeding Symposia Series, American Society of Horticultural Science and Crop Science Society of America, Corvallis, Oregon.

Empig, L.T., Gardner, C.O. and Compton, W.A. (1972) *Theoretical Gains for Different Population Improvement Procedures.* Miscellaneous Publication 26 (revised), Nebraska Agricultural Experiment Station, Lincoln, Nebraska.

Eshed, Y. and Zamir, D. (1996) Less-than-additive epistatic interactions of quantitative trait loci in tomato. *Genetics* 143, 1807–1817.

Falconer, D.S. (1989) *Introduction to Quantitative Genetics.* John Wiley & Sons, New York.

Fehr, W.R. (1987) *Principles of Cultivar Development, Theory and Technique.* Macmillan, New York.

Fisher, R.A. (1918) The correlation between relatives on the supposition of Mendelian inheritance. *Transactions Royal Society of Edinburgh* 52, 399–433.

Fulton, T.M., Beck, B.T., Emmatty, D., Eshed, Y., Lopez, J., Petiard, V., Uhlig, J., Zamir, D. and Tanksley, S.D. (1997) QTL analysis of an advanced backcross of *Lycopersicon peruvianum* to the cultivated tomato and comparisons with QTLs found in other wild species. *Theoretical and Applied Genetics* 95, 881–894.

Gerloff, J.E. and Smith, O.S. (1988) Choice of method for identifying germplasm with superior alleles. I. Theoretical results. *Theoretical and Applied Genetics* 76, 209–216.

Gowen, J.W. (ed.) (1952) *Heterosis.* Iowa State College Press, Ames, Iowa.

Graham, G.E., Wolff, D.W. and Stuber, W. (1997) Characterization of a yield quantitative trait locus on chromosome five of maize by fine mapping. *Crop Science* 37, 1601–1610.

Hallauer, A.R. (1985) Compendium of recurrent selection methods and their application. *Critical Reviews in Plant Sciences* 3, 1–30.

Hallauer, A.R. and Miranda Fo, J.B. (1988) *Quantitative Genetics in Maize Breeding.* Iowa State University Press, Ames, Iowa.

Hallauer, A.R., Russell, W.A. and Lamkey, K.R. (1988) Corn breeding. In: Sprague, G.F. and Dudley, J.W. (eds) *Corn and Corn Improvement.* American Society of Agronomy, Madison, Wisconsin, pp. 463–564.

Hanafey, M.K., Sebastian, S.A. and Tingey, S.V. (1998) Method to identify genetic markers that are linked to agronomically important genes. US Patent 5 746 023. Date issued 5 May 1998.

Hogan, R.M. and Dudley, J.W. (1991) Evaluation of a method for identifying sources of favorable alleles to improve an elite single cross. *Crop Science* 31, 700–704.

Hull, F.H. (1945) Recurrent selection and specific combining ability in maize. *Journal of the American Society of Agronomy* 37, 134–135.

Jenkins, M.T. (1940) The segregation of genes affecting yield of grain in maize. *Journal of the American Society of Agronomy* 32, 55–63.

Jensen, S.D., Kuhn, W.E. and McConnell, R.L. (1983) Combining ability studies in elite US maize germplasm. In: *Proceedings of the 38th Annual Maize and Sorghum Research Conference*. American Seed Trade Association, Washington, DC, pp. 87–96.

Johnson, B., Gardner, C.O. and Wrede, K.C. (1988) Application of an optimization model to multi-trait selection programs. *Crop Science* 28, 723–728.

Johnson, G.R. and Mumm, R.H. (1996) Marker assisted maize breeding. In: *Proceedings of the Fifty-first Annual Maize and Sorghum Research Conference*. American Seed Trade Association, Washington, DC, pp. 75–84.

Kang, M.S. and Gauch, H.G., Jr (eds) (1996) *Genotype-by-environment Interaction*. CRC Press, Boca Raton, Florida.

Kempthorne, O. (1954) The correlations between relatives in a random mating population. *Proceedings of the Royal Society, London, B* 143, 103–113.

Kempthorne, O. (1977) The international conference on quantitative genetics, introduction. In: Pollak, E., Kempthorne, O. and Bailey, T.B., Jr (eds) *Proceedings of the International Conference on Quantitative Genetics*. Iowa State University Press, Ames, Iowa, pp. 3–18.

Koziel, M.G., Beland, G.I., Bowman, C., Carozzi, N.B., Crenshaw, R., Crosland, L., Dawson, J., Desai, N. and Kadwell, S. (1993) Field performance of transgenic maize plants expressing an insecticidal protein derived from *Bacillus thuringensis*. *Bio-Technology* 11, 194–200.

Lamkey, K.R., Schnicker, B.J. and Melchinger, A.E. (1995) Epistasis in an elite maize hybrid and choice of generation for inbred line development. *Crop Science* 35, 1272–1281.

Lee, M. (1995) DNA markers and plant breeding programs. *Advances in Agronomy* 55, 265–344.

Lee, M. (1999) Towards understanding and manipulating heterosis in crop plants – can molecular genetics and genome projects help? In: Coors, J.G. and Pandey, S. (eds) *The Genetics and Exploitation of Heterosis in Crops*. American Society of Agronomy and Crop Science Society of America, Madison, Wisconsin, pp. 185–194.

Li, Z.K., Pinson, S.R.M., Park, W.D., Paterson, A.H. and Stansel, J.W. (1996) Epistasis for three grain yield components in rice (*Oryza sativa* L.). *Genetics* 145, 453–465.

Lin, H.X., Yamamoto, T., Sasaki, T. and Yano, M. (2000) Characterization and detection of epistatic interactions of 3 QTLs, *Hd1*, *Hd2*, and *Hd3*, controlling heading date in rice using nearly isogenic lines. *Theoretical and Applied Genetics* 101, 1021–1028.

Liu, B.H. (1998) *Statistical Genomics, Linkage, Mapping, and QTL Analysis*. CRC Press, Boca Raton, Florida.

Messmer, M. (1997) Twenty-first century plant breeders, seamlessly integrating the lab and the field. In: *Proceedings of the 52nd Annual Corn and Sorghum Research Conference*, Chicago, Illinois, pp. 64–72.

Metz, G. (1994) Probability of net gain of favorable alleles for improving an elite single cross. *Crop Science* 34, 668–672.

Miflin, B. (2000) Crop improvement in the 21st century. *Journal of Experimental Botany* 51, 1–8.

Misevic, D. (1989) Identification of inbred lines as a source of new alleles for improvement of elite maize single crosses. *Crop Science* 29, 1120–1125.

Moose, S. (2000) Maize genomics, present and future impacts on breeding improved hybrids. In: *Proceedings of the 37th Illinois Corn Breeders School*. University of Illinois Urbana, Illinois, pp. 60–66.

Mumm, R.H. and Dudley, J.W. (1994) A classification of 148 US maize inbreds, I. Cluster analysis based on RFLPs. *Crop Science* 37, 1475–1481.

Panter, D.M. and Allen, F.L. (1995) Using best linear unbiased predictions to enhance breeding for yield in soybean. I. Choosing parents. *Crop Science* 35, 397–404.

Pesek, J. and Baker, R.J. (1969) Desired improvement in relation to selection indices. *Canadian Journal of Plant Science* 49, 803–804.

Pfarr, D.G. and Lamkey, K.R. (1992) Comparison of methods for identifying populations for genetic improvement of maize hybrids. *Crop Science* 32, 670–676.

Sax, K. (1923) The association of size differences with seed coat pattern and pigmentation in *Phaseolus vulgaris*. *Genetics* 8, 552–560.

Schnell, F.W. (1983) Probleme der Elternwahl-Ein Uberblick. In: *Arbeitstagung der Arbeitsgemeinschaft der Saatzuchtleiter in Gumpenstein, Austria, 22–24 Nov. 1983*. Verlag und Druck der Bundesanstalt fur alpenlandische Landwirtschaft, Gumpenstein, Austria, pp. 1–11.

Schulze, A. and Downward, J. (2000) Analysis of gene expression by microarrays, cell biologist's gold mine or minefield. *Journal of Cell Science* 113, 4151–4156.

Smith, D.C. (1966) Plant breeding – development and success. In: Frey, K.J. (ed.) *Plant

Breeding. Iowa State University Press, Ames, Iowa, pp. 3–54.

Smith, G.A. (1987) Sugar beet. In: Fehr, W.R. (ed.) *Genetic Contributions to Yield Gains of Five Major Crop Plants.* CSSA Special Publication No. 7, ASA, CSSA and SSSA, Madison, Wisconsin, pp. 577–625.

Smith, H.F. (1936) A discriminant function for plant selection. *Annals of Eugenics* 7, 240–250.

Stuber, C.W. (1992) Biochemical and molecular markers in plant breeding. In: Janick, J. (ed.) *Plant Breeding Reviews.* Wiley, New York, pp. 37–61.

Stuber, C.W. (1998) Case history in crop improvement: yield heterosis in maize. In: Paterson, A.H. (ed.) *Molecular Dissection of Complex Traits.* CRC Press, Boca Raton, Florida, pp. 197–206.

Stuber, C.W. and Edwards, M.D. (1986) Genotypic selection for improvement of quantitative traits in maize using molecular marker loci. In: *Proceedings of the 41st Annual Maize and Sorghum Research Conference.* American Seed Trade Association, Chicago, Illinois, pp. 40–83.

Stuber, C.W., Polacco, M. and Lynn, M. (1999) Synergy of empirical breeding, marker-assisted selection, and genomics to increase crop yield potential. *Crop Science* 39, 1571–1583.

Tanksley, S.D., Grandillo, S., Fulton, T.M., Azmir, D., Eshed, Y., Petiard, V., Lopez, J. and Beck, B.T. (1996) Advanced backcross QTL analysis in a cross between an elite processing line of tomato and its wild relative *L.*

pimpinellifolium. Theoretical and Applied Genetics 92, 213–224.

Toledo, J.F.F. (1992) Mid parent and coefficient of parentage as predictors for screening among single crosses for their inbreeding potential. *Brazilian Journal of Genetics* 15, 429–437.

Williams, J.S. (1962) The evaluation of a selection index. *Biometrics* 18, 375–393.

Xiao, J., Ahn, S.N., McDouch, S.R., Tanksley, S.D., Lin, J. and Yuan, L.P. (1996) Genes from wild rice improve yield. *Nature* 384, 223–224.

Xiao, J., Li, J., Grandillo, S., Ahn, S.N., Yuan, L., Tanksley, S.D. and McCouch, S.R. (1998) Identification of trait-improving quantitative trait loci alleles from a wild rice relative, *Oryza rufipogon. Genetics* 150, 899–909.

Ye, X.D., Al-Babili, S., Kloti, A., Zhang, J., Lucca, P., Beyer, P. and Potrykus, I. (2000) Engineering the provitamin A (beta-carotene) biosynthetic pathway into (carotenoid-free) rice endosperm. *Science* 287, 303–305.

Yu, S.B., Li, J.X., Xu, C.G., Tan, Y.F., Gao, Y.J., Li, X.H., Zhang, Q. and Saghai-Maroof, M.A. (1997) Importance of epistasis as the genetic basis of heterosis in an elite rice hybrid. *Proceedings of the National Academy of Sciences, USA* 94, 9226–9231.

Zanoni, U. and Dudley, J.W. (1989) Comparison of different methods of identifying inbreds useful for improving elite maize hybrids. *Crop Science* 29, 577–587.

Zirkle, C. (1952) Early ideas on inbreeding and crossbreeding. In: Gowen, J.W. (ed.) *Heterosis.* Iowa State College Press, Ames, Iowa, pp. 1–13.

7 Use of Molecular Markers in Plant Breeding: Drought Tolerance Improvement in Tropical Maize

J.-M. Ribaut,[1] M. Bänziger,[1] J. Betran,[2] C. Jiang,[3] G.O. Edmeades,[4] K. Dreher[1] and D. Hoisington[1]

[1]International Maize and Wheat Improvement Center (CIMMYT), Lisboa 27, Apdo. Postal 6-641, 06600 Mexico DF, Mexico; [2]Department of Soil and Crop Sciences, Texas A&M University, College Station, TX 77843-2474, USA; [3]Monsanto Life Sciences, Research Center, 700 Chesterfield Parkway North, St Louis, MO 63198, USA; [4]Pioneer Hi-Bred Int. Inc., PO Box 609, Waimea, HI 96796, USA

Introduction

The application of molecular markers to plant breeding can be divided into three main categories: (i) the characterization of germ-plasm, known as fingerprinting; (ii) the genetic dissection of the target trait – actually the identification and characterization of genomic regions involved in the expression of the target trait; and (iii) following the identification of the genomic regions of interest, crop improvement through marker-assisted selection (MAS). The first two applications have proved themselves by generating knowledge about the genetic diversity of germ-plasm, thereby allowing placement into heterotic groups and a better understanding of the genetic basis of agronomic traits of interest. For simply inherited traits – those that have high heritability and are regulated by only a few genes – the use of molecular markers to accelerate germ-plasm improvement has been well documented (e.g. Johnson and Mumm, 1996; Mohan et al., 1997; Young, 1999). Such work has proved successful in: (i) tracing favourable alleles in the genomic

background of genotypes of interest; and (ii) identifying individual plants in large segregating populations that carry the favourable alleles. Moreover, with the recent development of PCR-based markers, for example, simple sequence repeats (SSRs) (Chin et al., 1996; Powell et al., 1996) and single nucleotide polymorphisms (SNPs) (Gilles et al., 1999), a substantial improvement in the capacity to efficiently screen large populations has been achieved, thereby increasing the efficiency of MAS experiments.

Traits controlled by single genes or major quantitative trait loci (QTL) are easy to transfer from a donor line to a recipient line via line conversion. Such line conversions, achieved through a back-cross (BC) approach, are conducted at the International Maize and Wheat Improvement Center (CIMMYT), for example, to introgress favourable alleles of the opaque-2 gene or a major QTL identified on the short arm of maize (Zea mays L.) chromosome 1 that is associated with maize streak virus (MSV) resistance. Although it is easy to lay out a BC-MAS experiment, but given the number of parameters involved, designing the most appropriate and efficient strategy is generally not a straightforward task. This chapter presents some theoretical and practical guidelines for identifying the most appropriate BC-MAS strategy based on the objectives of different types of applied breeding experiments.

For improvement of polygenic traits, the efficiency of using molecular markers in plant-breeding programmes remains questionable, with only a few success stories published to date (Mohan *et al.*, 1997; Ribaut and Hoisington, 1998; Young, 1999). The difficulty of manipulating quantitative traits is related to their genetic complexity – principally the number of genes involved in their expression and interactions between genes (epistasis). Since several genes are involved in the expression of polygenic traits, they generally have smaller individual effects on the plant phenotype and are cross-dependent. This implies that several regions, or QTL, must be manipulated simultaneously to have a significant impact, and that the effect of individual regions is not easily identified. In addition, the evaluation of the QTL–environment interaction $(Q \times E)$ remains a handicap for the efficiency of MAS, since QTL identification can be strongly affected by environmental factors (Beavis and Keim, 1996). To illustrate the potential and the limitations of using DNA markers to improve crops for complex traits or specific environments, we present work conducted at CIMMYT on improving drought tolerance in tropical maize, including the genetic dissection of target traits of interest, a report on MAS experiments and some innovative ideas on new MAS strategies that might be successful. Based on experience, the limitations of polygenic trait improvement through QTL manipulation have been clearly identified. One of the weaknesses of the quantitative-genetic approach that limits its use in plant breeding is that it provides very little information about the mechanisms and pathways involved in drought tolerance or about the multitude of genes involved in the plant's response. Recent developments in functional genomics should help in overcoming this problem, because these new approaches

allow the simultaneous study of the expression of several thousand genes. At the end of this chapter, we also present our views on the potential of functional genomics and the role of physiological/biochemical pathways as the link between functional genomics and plant phenotype.

Cost and Efficiency of SSRs for High-throughput MAS Experiments

The first application using a DNA marker system, the restriction fragment length polymorphism (RFLP) analysis, was reported in 1980, in relation to the construction of a genetic linkage map in humans (Botstein *et al.*, 1980). Since then, RFLP markers have been widely used to construct linkage maps for several crop species, including maize (Helentjaris *et al.*, 1986), tomato (Paterson *et al.*, 1988) and rice (McCouch *et al.*, 1988). To date, many RFLP markers have been identified with tight linkage to the genes controlling economically important traits in various crop species. RFLPs are reliable markers, and the same probe can usually be hybridized on different crop genomes, making RFLP markers useful for comparative mapping studies as well. However, RFLP analysis requires large quantities of quality DNA, and detection of RFLPs by Southern blot hybridization may be laborious and time-consuming, which make this assay undesirable for plant-breeding projects with high-throughput requirements. Beginning in 1990 (Williams *et al.*, 1990), the development of diverse PCR-based markers has been vigorous and has provided the basis for a large number of innovative methods for recognizing DNA polymorphisms among individuals. PCR-based techniques are robust and amenable to automation and, therefore, widely applied to large-scale marker development or implementation procedures. Among the many different types of PCR-based DNA markers available for use in plant breeding, SSR markers are often preferred for reasons of cost and simplicity. Moreover, SSRs are reliable, codominant, abundant and uniformly dispersed in plant genomes. From a practical point of view, SSR assays can be reliably applied on a large scale at an early stage of plant development, due to the small amount of tissue required to extract an adequate amount of DNA for PCR amplification. These qualities make SSRs attractive for high-throughput MAS experiments.

Recently, the cost-effectiveness of using SSRs in MAS experiments was estimated at CIMMYT, using a spreadsheet-based budgeting approach (Dreher *et al.*, 2000). Although the basic laboratory protocol used at CIMMYT for SSR analysis remains the same, the number of samples processed and the number of markers analysed varies according to the application. Table 7.1 shows the cost per data point of SSR analysis under several scenarios. Because of the indivisibility of several key inputs, especially labour, the cost of SSR analysis is high when the number of samples and/or the number of markers is small, but, because of economies of scale, the cost decreases rapidly as the number of samples and/or the number of markers increases. For combinations involving ten or more markers and sample sizes above 100, the cost per data point falls below US$1.35. These cost estimates are sensitive to changes in laboratory protocols, and they can rise or fall significantly if the technical parameters of the protocols are altered.

These results demonstrate that the data-point cost for SSRs is quite independent of the number of genotypes screened when a large number of markers is considered, for instance, during the construction of a linkage map or for a BC-MAS experiment when SSRs are used to identify genotypes with lower contributions of a donor allele at non-selected loci. However, when only a few SSRs are used, for instance, in the selection of a favourable allele at target loci, this type of selection is cost-effective when screening several hundred genotypes. A typical example of this kind of MAS experiment is the screening of a single SSR on several genotypes to select for a mutant allele at the *opaque-2* locus in segregating populations. Maize seed-protein quality can be improved by selecting for mutations in the *opaque-2* gene (Mertz *et al.*, 1964), which is located on the short arm of chromosome 7. The presence of the homozygous mutant *o2* allele at the *opaque-2* locus is correlated with changes in the amino acid balance within the endosperm, specifically, a favourable increase in the proportion of lysine and tryptophan. The *opaque-2* locus has been cloned (Schmidt *et al.*, 1990), and its sequence has been published. Three SSRs have been detected within the sequence of the gene itself (phi057, phi112 and umc1066). For several years, CIMMYT has routinely screened thousands of genotypes, using one of three available polymorphic SSRs, in segregating populations to identify,

Table 7.1.   Data-point costs for SSR molecular marker analysis (US$ per data point).

Sample size	Number of markers analysed					
	1 marker	10 markers	50 markers	100 markers	200 markers	500 markers
2	33.55	4.37	1.83	1.53	1.38	1.30
10	7.79	1.85	1.35	1.31	1.27	1.25
100	2.26	1.35	1.26	1.25	1.25	1.24
250	2.00	1.32	1.26	1.25	1.24	1.24
500	1.96	1.31	1.26	1.25	1.24	1.24
1000	1.94	1.31	1.26	1.25	1.24	1.24
5000	1.91	1.31	1.26	1.25	1.24	1.24

Note: These unit costs assume that leaf samples are harvested from the field and DNA is extracted using a sap extractor. The DNA present in each sample is quantified using a spectrophotometer and the samples are PCR-amplified with 22-base-pair custom-made markers. Amplified fragments are separated using an agarose gel with 2% Metaphor and 1% Seakem. When there are fewer than 500 samples, a 110 ml gel is used; when 500 or more samples are analysed, a 280 ml gel is used. Explicit travel costs associated with leaf harvest are not included. Any change in protocol will change these representative costs.

at an early stage of recombination, genotypes that have one copy of the mutant allele at the *opaque-2* locus (BC strategy) and those that have two copies (self-pollination strategy). Selection is conducted before flowering to allow the pollination of only the selected plants. Using DNA markers to select for the *opaque-2* gene typifies the use of MAS as an efficient substitute for phenotypic selection, considering the recessive nature of the gene, the absence of obvious visual selection due to the interaction of this gene with modifiers involved in kernel hardness and the cost of $3.54 per sample (when processed through chemical analysis) to quantify total nitrogen and tryptophan levels at CIMMYT's Soils and Plant Analysis Laboratory.

BC-MAS for Simply Inherited Traits

The efficiency of MAS experiments for the transfer of a single target region has been reported for several plant genomes, the integration of the *Bacillus thuringiensis* (*Bt*) transgene into different genetic backgrounds being a good example (Ragot *et al.*, 1995). When the expression of a target trait is regulated by a single gene or by a gene responsible for a high percentage of the phenotypic variance of the trait, the transfer of a single genomic region from a donor to a recipient line, or line conversion, can produce significant trait improvement. By making an allelic map of the genome with DNA markers, plants possessing a 'better' genome composition can be efficiently identified, i.e. the donor allele at the target segment plus the largest proportion of the recurrent genome in the rest of the genome process (Tanksley *et al.*, 1989). The use of DNA markers, which permits the genetic dissection of the progeny at each generation, increases the speed of selection when compared with phenotypic selection.

Before any BC-MAS experiment gets under way, the number of target genes involved in selection and the expected level of line conversion must be defined. Then, one must identify, in each generation, the size of the population to be screened, the number, position and nature of molecular markers used and the number of genotypes selected. The expected level of conversion is closely related to the number and distribution of the DNA markers at non-target loci and the recombination frequencies between the target gene and flanking markers. All of these parameters influence the number of generations required to achieve a specific level of BC-MAS, while offering different alternatives for defining a strategy. Simulation results indicate that the selection response in the BC_1 could be increased significantly when the selectable population size is less than 50; a diminished return is observed when this number exceeds 100 (Ribaut *et al.*, 2002). Selectable population size is defined as the number of individuals with favourable alleles at the target genes from which selection with markers can be carried out on the rest of the genome at non-target loci. Simulations demonstrate that this recommendation is independent of the number of selection regions considered. For an introgression at one or only a few target genes in a partial line conversion and using only one generation of MAS at non-target loci, such a selection conducted at BC_3 would be more efficient than if it were conducted at BC_1 or BC_2. With selection only for the presence of the donor allele at one locus in BC_1 and BC_2 and MAS at BC_3, lines with less than 5% of the donor genome can be obtained with a selectable population size of ten in BC_1 and BC_2 and 100 in BC_3. To illustrate this approach, two schemes are presented in detail in Fig. 7.1. In Fig. 7.1A, the complete MAS step is conducted in all three BC generations, while, in Fig. 7.1B, the complete MAS step is conducted only at the third BC. The scheme that resulted in the least amount of the donor genome (1.5% after three BCs) utilized a complete MAS step at each generation (Fig. 7.1A). Obtaining this 1.5% required screening a population of 200 individuals at a single locus, followed by the screening of 109 markers (11 per chromosome) at non-target loci at each BC on the selectable population size ($N_{sl} = 100$). When the single complete MAS step is conducted only at the third BC (Fig. 7.1B), the remnant donor genome

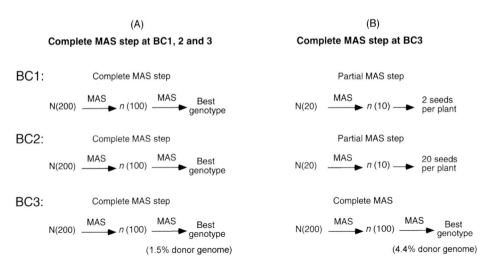

Fig. 7.1. Donor genome contribution for the allelic introgression at one target gene when (A) a complete MAS step (*n* = 100) is conducted during the three BC cycles, and (B) partial MAS steps are conducted during the first two BC cycles (*n* = 10) and a complete MAS step is conducted once at the third BC (*n* = 100).

contribution at non-target loci is 4.4%. In this scheme, screening of only 20 plants at the target locus in BC_1 and BC_2 is required. The strategy employing a single complete MAS cycle at an advanced BC is an attractive option, especially if allelic introgression at a few target genes is considered concomitantly in several recipient lines. Indeed, the small population required for the first generations, in which selection is only conducted at target loci, represents a major logistical advantage. Moreover, if one target gene is linked to a phenotypic marker or is a transgene with a selectable marker gene, such as herbicide resistance included in the gene construct, selection for this gene can be conducted phenotypically, thereby reducing the cost of selection. If this is the case, DNA extraction is not required to conduct the selection during the first generations. The 'penalty' for this strategy is the retention of some donor genome contribution at non-target loci, most of it flanking the target genes on the carrier chromosomes. Possible negative impacts from this remnant donor genome on plant performance can be minimized if the donor line is élite germ-plasm, because the probability of having bad agronomic characteristics 'dragged' into the selection at non-target loci is reduced. The

approach that uses a single complete MAS step only at the third BC has been adopted at CIMMYT for the introgression of MSV-resistance alleles into several élite lines. A major QTL involved in resistance to this African virus has been identified in four different segregating populations on the short arm of chromosome 1, between the fourth and the fifth bin (artificial chromosome regions bounded by core markers). Since the phenotypic variance expressed by this region in the four QTL studies was greater than 60%, we are concentrating our ongoing selection exclusively on this single genomic region. This MAS will be complemented by a phenotypic screening after the first self-pollination.

MAS for Polygenic Traits: Drought Tolerance in Maize as an Example

Most traits of agronomic importance are complex and regulated by several genes, with yield being among the most polygenic and complex. Regulation of plant responses to stresses (abiotic and biotic) is even more complex, due to the interaction between the environment and the plant genotype, especially when plants are grown in marginal environments. Among abiotic stresses, breeding for drought tolerance is one of the most challenging tasks, because selected material should be outstanding not only under water-limited conditions but also in cases when rainfall is adequate. Drought is a major cause of yield loss in cereals, with sub-Saharan Africa being the region receiving the most negative impact in the world. The Food and Agriculture Organization (FAO) estimates that 44% of the land surface in sub-Saharan Africa is subject to a high risk of meteorological drought, which principally affects maize, sorghum, rice, wheat and pearl millet.

Establishing optimal environments to select for improved performance under drought is complicated by environmental variation; there is generally only one drought generation per year in the tropics and conventional selection for drought tolerance requires careful irrigation management, which is a constraint for many breeding programmes. In addition, there is clearly a yield barrier for any plant and, under drought, this yield barrier is highly related to water supply. Grain production requires a plant to support and produce the floral structure(s) and, without a minimum of water, there is little or no plant development. This limitation makes it difficult to predict the level of the yield barrier under drought conditions. Germ-plasm improvement for drought tolerance is a high priority for the CIMMYT maize programme because most tropical maize is produced under rain-fed conditions in areas where drought is widely considered to be the most important abiotic constraint to production (Heisey and Edmeades, 1999). For example, 21% of the total maize area in sub-Saharan Africa often experiences drought, resulting in a 33% decrease in yield. Drought at any stage of plant development affects production, but maximum damage is inflicted on maize when it occurs around flowering. Farmers may respond to drought at the seedling stage by replanting their crop and, at later stages, some yield may yet be salvaged because maize is relatively drought-tolerant at the grain-filling stage. Drought at flowering, however, can be mitigated only

by irrigation (Pingali, 2001). CIMMYT scientists, therefore, have devoted considerable effort during the past three decades to improving drought tolerance in maize for the period before and during flowering. Extensive research has been conducted in the areas of breeding, physiology, agronomy and, most recently, biotechnology. Today, significant progress has been achieved through conventional breeding (Bänziger *et al.*, 2000), but this approach is slow, time-consuming and with uncertain potential for further progress. Although challenging, biotechnology approaches that combine QTL studies and functional genomics should provide useful information and tools to effectively complement conventional selection for drought-tolerance improvement.

Genetic dissection of drought components

CIMMYT research on drought tolerance using biotechnology was initiated about 10 years ago and has focused on the genetic dissection of drought tolerance, identifying QTL for yield components, secondary morphological traits of interest, e.g. the anthesis–silking interval (ASI) (Bolaños and Edmeades, 1996) and, more recently, physiological parameters. To date, the genetic dissection of yield components and secondary traits of interest has been conducted in four different segregating crosses, under diverse water regimes in several environments and at different inbreeding levels (Table 7.2). In addition, to identify the QTL for key physiological components (e.g. hormones, proteins, etc.) involved in drought-tolerance mechanisms, a recombinant inbred line (RIL) population was developed by single-seed descent from F_3 families obtained by crossing Ac7643 with Ac7729/TZSRW, one of the four crosses mentioned above. The same morphological traits as measured in the F_3 families have been evaluated in this RIL population. In addition to physiological parameters measured in-house, such as relative water content, osmotic adjustment and chlorophyll content, we collaborated with other research groups to determine abscisic acid

(ABA) concentration in target tissues, root parameters and parameters involved in dehydration tolerance to low-temperature stress. Recently a mapping project following the Tanksley and Nelson (1996) approach has been initiated with the

Table 7.2. Segregating populations analysed for yield components, morphological traits and physiological parameters, under different stress regimes, in different locations and at different inbreeding levels.

Populations	Trials	Traits
Ac7643 × Ac7729/TZSRW (236 $F_{2/3}$ families)	92A WW (TL) 94A IS, SS (TL) 96A/B LowN (PR) 96B HiN (PR)	MFLW, FFLW, ASI, LNO, EHT, PHT, ENO, GY, HK, KNO, CHL
Ac7643 × CML247 (236 $F_{2/3}$ families)	96A SS (TL)	MFLW, FFLW, ASI, LNO, EHT, PHT, ENO, GY, HK, KNO
Ac7643 × Ac7729/TZSRW (236 RIL families)	96A IS, SS (TL) 96B WW (TL) 99A WW, IS, SS (TL)	MFLW, FFLW, ASI, LNO, EHT, PHT, EPO, ENO, GY, HK, KNO
K64R × H16 (280 F_3 topcross families)	99B IS, SS (ZW) 00A IS, SS (KY)	MFLW, FFLW, ASI, PHT, ENO, GY, SEN
K64R × H16 (170 F_4 families)	00B SS (ZW)	MFLW, FFLW, ASI, PHT, EHT, EPO, SEN, ENO, GY
CML444 × SC-Malawi (234 F_3 families)	00B WW, IS, SS (ZW) 01A IS, SS (TL)	MFLW, FFLW, ASI, PHT, SEN, TBNO, ENO, GY
Jalisco (teosinte) × LPC21 (200 BC_3F_2 families)	01A IS, SS (TL)	MFLW, FFLW, ASI, ENO, GY
Ac7643 × Ac7729/TZSRW (236 RIL families)	96A IS, SS (TL) 96B WW (TL) 99A WW, IS, SS (TL) 01A IS, SS (TL)	RWC, OP, OA, RCT, CHL(J), CHL(E), ABA(E), ABA(S), ABA (EL), EW, EGR
Ac7643 × Ac7729/TZSRW (140 RIL families)	98 (laboratory test)	Root parameters under hydroponics
Ac7643 × Ac7729/TZSRW (220 RIL families)	00 (Phytotron) Low temperature	Pigments, photosystem parameters, RWC, SW, RW
Ac7643 × Ac7729/TZSRW (220 RIL families)	01 (Phytotron) Drought, drought and low temperature	Pigments, photosystem parameters, RWC, SW, RW

Stress regime: WW, well-watered; IS, intermediate water stress; SS, severe water stress; LowN, low-nitrogen trial; HiN, high-nitrogen trial.
Location: KY, Kenya; TL, Tlaltizapan, Mexico; PR, Poza Rica, Mexico; ZW, Zimbabwe.
Morphological traits: MFLW, male flowering; FFLW, female flowering; ASI, anthesis–silking interval; LNO, number of leaves; EHT, ear height; PHT, plant height; EPO, ear position; SEN, senescence; TBNO, number of tassel branches; EW, ear weight; EGR, ear growth rate for 1 week; SW, shoot weight; RW, root weight.
Physiological parameters: RWC, relative water content; OP, osmotic potential; OA, osmotic adjustment; CHL, chlorophyll content in a young leaf (J) and in the ear leaf (E); RCT, root conductivity; ABA, abscisic acid content, in the ear (E), in the silk (S) and in the ear leaf (EL).
Yield components: GY, grain yield; ENO, number of ears; HK, hundred-kernel weight; KNO, number of kernels.

Mexican National Agricultural Research Program (INIFAP) to identify exotic favourable alleles for drought tolerance in teosinte populations.

Our QTL identification effort first focused on yield components and morphological traits such as the ASI (Table 7.2), because it was important to have some information on the genetic complexity of the components/traits involved directly in the phenotypic selection conducted by breeders. During the last few years, we intensified our efforts aimed at identifying the QTL involved in the expression of physiological parameters, to further our understanding of the genetics of plant responses under drought. Indeed, a plant phenotype is the result of a differential expression of several physiological/biochemical pathways; a short ASI phenotype, for instance, might involve carbohydrate and hormone metabolism and translocation, as well as water parameter regulation and/or membrane stability. An understanding of the genetic basis of these key physiological parameters is very important, because it will be a key link between the results emerging from functional genomics and morphological plant responses and, therefore, should allow the identification of major pathways.

MAS experiments

After identifying a set of QTL in the first cross (Ac7643 × Ac7729/TZSRW) (Ribaut *et al.*, 1996, 1997), an initial BC-MAS project was launched in 1994. The line Ac7643 was used as the drought-tolerant donor and CML247 was used as the recurrent parent. CML247 is an élite tropical inbred line developed by CIMMYT, with outstanding combining ability and good yield *per se* under well-watered conditions. It is susceptible to drought, in part because its ASI under drought is large. Five genomic regions involved in the expression of a short ASI were selected for transfer from Ac7643 into CML247. After two BCs and two self-pollinations (Fig. 7.2), the best

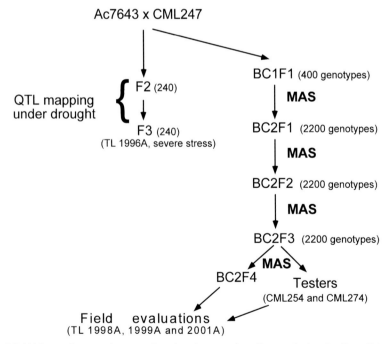

Fig. 7.2. BC-MAS experiment to improve drought tolerance of an élite tropical maize line, CML247, by introgressing five selected regions derived from Ac7643, a drought-tolerant line characterized by a very short ASI.

genotype was fixed from the donor line for the five target regions (12% of the genome), as well as for 7% of the genome lying outside the QTL regions. The 70 best BC_2F_3 (i.e. S_2) lines were identified and crossed with two CIMMYT testers, CML254 and CML274. These hybrids and the BC_2F_4 families (S_3 lines) derived from the selected BC_2F_3 plants were evaluated during the dry winter season in 1998, 1999 and 2001 in Tlaltizapan, Mexico, under several water regimes. Results show that, under stress conditions that induce a yield reduction of at least 80%, the mean of the 70 selected genotypes performed better than the control crossed with CML254 and CML274. In addition, the best genotype among the 70 selected ($BC_2F_3 \times$ testers) performed two to four times better than the control. This difference became less marked when the intensity of stress decreased and, for a stress inducing less than 40% of yield reduction, hybrids resulting from the MAS or developed with the 'original' version of CML247 performed the same. Although the genotypes that performed the best depended on the stress intensity, few genotypes performed always significantly better compared with the controls across the six water-limited trials. No yield reduction was observed under well-watered conditions (Table 7.3).

Although the BC-MAS experiment described above was successful, germ-plasm improvement for polygenic traits restricted to QTL manipulation has several well-identified limitations, the most critical being the inability to predict the phenotype of any given genotype based on the allelic constitution. This limitation implies that, to increase the probability of success of a MAS experiment, QTL identification must be achieved on a cross basis; only a few such experiments, including the one presented here, have proved successful and have demonstrated that this approach works (Ragot *et al.*, 2000). From an economic perspective, this recommendation has some significant consequences, because it implies construction of a linkage map and the phenotypic evaluation of segregating families for each cross, to identify the target genomic regions to be manipulated. Using the spreadsheet

developed at CIMMYT, construction of a linkage map alone costs an estimated US$25,000 (estimated for CIMMYT-Applied Biotechnology Center (ABC)). When considering the entire experiment (Fig. 7.2), including all the different selection steps, the cost increases to US$98,000. Such costs obviously limit the application of MAS for polygenic trait analysis in breeding programmes at institutions such as CIMMYT, where there is no direct financial return from the germ-plasm. In addition, considering the number and the diversity of the environments targeted by CIMMYT in developing countries, a major investment to improve a single line will rarely provide an adequate return. To try to overcome these limitations, new strategies have been developed at CIMMYT that are aimed at reducing the cost of MAS experiments and delivering new germ-plasm instead of improved versions of existing lines (Ribaut and Betrán, 1999). Some of these new strategies are now being tested at CIMMYT, and resources have been allocated to: (i) the construction of a consensus map that combines information related to QTL characterization and gene expression; and (ii) the identification through functional genomics of a set of key genes/pathways involved in maize drought response that will be used as selection tools in breeding programmes.

Consensus map

The objective of the consensus map considered here is to compile information that is or will soon be available at CIMMYT (Table 7.2) and in the public domain on the mechanisms of drought tolerance in maize at the genetic and genomic levels. To construct the consensus map, anchor molecular markers that are common to the four segregating crosses developed at CIMMYT are used to position all the markers on a single map through linear regression. In a second phase, the QTL information generated for each cross (QTL for yield components, morphological traits and physiological parameters) will be compiled on the consensus map, assigning a 'weight' for each QTL set. The

Table 7.3. Grain yields per plot (kg ha^{-1} ± SD) under different water regimes (WW, well-watered; IS, intermediate stress; SS, severe stress conditions) of the original CML247 version, the donor line (Ac7643) and the MAS-improved versions of CML247 when crossed with two CIMMYT tester inbred lines (CML254 and CML274). The mean of the control was calculated based on ten entries for CML247 and five entries for Ac7643, while the mean of the improved genotypes represents 70 different entries selected after MAS. 'Best genotype' is the yield of the highest-yielding genotype cross out of the entries.

Trial	Genotype	Hybrids CML254			Hybrids CML274		
		Mean	Yield loss (%)	Best genotype	Mean	Yield loss (%)	Best genotype
1998A WW	CML247 Orig	6153 ± 780	0	7401	7638 ± 689	0	8964
	CML247MAS	7173 ± 846	0	8712 (gen. 2)	7674 ± 928	0	9483 (gen. 21)
1998A IS	CML247 Orig	748 ± 475	87.8	1303	1854 ± 665	75.7	2640
	CML247MAS	2028 ± 982	71.7	4529 (gen. 57)	2166 ± 862	71.8	4057 (gen. 10)
1998A SS	CML247 Orig	578 ± 550	90.6	1443	464 ± 597	93.9	1696
	CML247MAS	904 ± 1399	95.5	4169 (gen. 24)	829 ± 898	89.2	3006 (gen. 49)
1999A WW	CML247 Orig	10705 ± 186	0	10935	10637 ± 194	0	10842
	Ac7643	10505 ± 302	0	10814	10722 ± 196	0	11048
	CML247MAS	10527 ± 248	0	11101 (gen. 57)	10851 ± 291	0	11607 (gen. 47)
1999A IS	CML247 Orig	6421 ± 231	40.0	6812	6248 ± 309	41.3	6711
	Ac7643	5936 ± 447	43.5	6523	4863 ± 340	54.6	5495
	CML247MAS	5980 ± 407	43.2	6709 (gen. 15)	5443 ± 435	49.8	6616 (gen. 43)
1999A SS	CML247 Orig	6156 ± 398	42.5	6862	5472 ± 247	48.6	6084
	Ac7643	5387 ± 381	48.7	5930	4900 ± 529	54.3	5378
	CML247MAS	5730 ± 477	45.6	6560 (gen. 14)	5229 ± 405	51.8	6057 (gen. 12)
2001A IS	CML247 Orig	4709 ± 486		5233	3264 ± 497		3977
	Ac7643	4072 ± 234		4417	2109 ± 268		2415
	CML247MAS	4900 ± 614		6166 (gen. 26)	3500 ± 711		5188 (gen. 70)
2001A SS	CML247 Orig	3889 ± 356		4410	3006 ± 519		3604
	Ac7643	4221 ± 309		4629	2606 ± 246		2919
	CML247MAS	4311 ± 551		5761 (gen. 64)	3265 ± 656		5144 (gen. 70)

weight will have two major components: (i) the nature of the trait; and (ii) the threshold value of the QTL identified under a specific environment – likelihood ratio (LR). Once all the QTL and gene expression information is integrated into a consensus map, we hope to identify noteworthy regions involved in maize drought tolerance. Those regions might be related to the expression of the same trait (different crosses or environments) and/or a combination of different target traits (same cross and/or different crosses or environments). Therefore, a consensus map should highlight common QTL and gene expression among populations as well as among environments. The stability of a QTL, i.e. the frequency with which it is identified across experiments, might also be taken into account when assigning a weight to each QTL to further reinforce this important characteristic.

A drought-tolerant phenotype is the result of the accumulation of favourable alleles at a large number of genes. The fact that some of those genes form clusters within the maize genome, as suggested by the genetic information available in the maize database (Khavkin and Coe, 1997), might be essential for the identification of those 'common drought regions' across crosses. The different components and therefore the nature of this consensus map will evolve over time, as more information from QTL and gene discovery studies is incorporated, and also as the information generated at the gene expression level is analysed and integrated with available QTL/gene information. If genomic regions involved in the expression of drought tolerance can be identified using the approach described here, new MAS strategies that do not necessarily require the construction of linkage maps for every new cross under consideration might be developed. A MAS experiment based on the regions of interest reported on the consensus map will not be the most efficient approach, because only some, rather than all, of those regions will have a significant impact on the plant phenotype, depending on different allelic composition of the crosses. Nevertheless, this type of MAS can be conducted at a very low cost (several

thousand dollars per cross), with a potentially large throughput.

Gene discovery and characterization

The identification of genes involved in plant responses under drought has been widely explored (Skriver and Mundy, 1990; Bray, 1993; Ingram and Bartels, 1996; Cushman and Bohnert, 2000) and, to date, if one tries to establish a list of candidate genes for drought tolerance based on the gene function, hundreds of genes can easily be listed. The question is how to prioritize research aimed at validating those genes putatively involved in the drought-tolerance process and how to characterize their impact on plant response under a given set of experimental conditions.

One option is to prioritize the gene-expression study for the genes/expressed sequence tags (ESTs) that mapped – when this mapping position is known – into a genomic region of interest previously identified through a QTL study. This correlation at the mapping level should increase the probability that a differential expression of these genes will have an impact on the plant phenotype. Another option is to consider the chronological sequence of gene expression within a target pathway. In this case, the characterization of the genes involved in the initiation phase of the stress response (e.g. genes encoding for stress-induced transcription factors) represent a logical priority, since they represent the 'upstream keys' to global genomic responses that might involve hundreds of genes. Moreover, once they have been identified, expression of these key genes should serve as a 'timing reference' to identify expression products from downstream genes involved in stress responses. Examples of these 'upstream genes' include the recent discovery of the promoter regulatory elements DRE (dehydration-responsive element) and ABRE (ABA-responsive element) associated with dehydration- and low-temperature-induced gene expression in *Arabidopsis* (Shinozaki and Yamaguchi-Shinozaki, 1997), as well as the identification of transcription factors interacting with these promoters (Liu *et al.*, 1998). Due to their potential impact on a plant phenotype, a large effort is dedicated to the identification and characterization of transcriptional factors involved in the plant response under abiotic stress conditions. This effort goes towards the identification of the mechanisms involved in drought-perception and signal-transduction pathways that are still poorly understood.

Functional genomics: a new source of knowledge

Through the development of emerging technologies that provide functional information at the gene level, gene characterization under stress conditions (i.e. expression framework) has received a significant boost in recent years (e.g. Bohnert *et al.*, 2000). Maize cDNA libraries have been generated from stressed roots and leaves from immature 2-week-old plantlets. Many of these transcripts have been sequenced and the sequences deposited in the maize database (http://www.zmdb.iastate.edu/). These sequences (or part of them), based on random or functional selection, have been arrayed to provide gene-expression screening tools, such as microarray (EST sequences) (Schena *et al.*, 1995) or DNA chip (oligonucleotides) (Fodor, 1997). To date, microarrays generated with cDNA sequences from maize are available as a product of a National Science Foundation plant genome research project – maize gene discovery, DNA sequencing and phenotypic analysis – led by Virginia Walbot (Stanford University). So the tools are available, but the challenges that still remain are: (i) how to design the most suitable profiling experiments; and (ii) how to efficiently normalize and organize the generated information to make it useful. The design of a suitable profiling experiment is not a straightforward task, because the germ-plasm, the target tissue, the time line, the experimental conditions and the repeatability of the experiment need to be carefully considered. The RNA to hybridize on a microarray can

come from contrasting germ-plasm (line, hybrids, families) for a target trait or from the same germ-plasm but under different experimental conditions, or both. If working with fixed lines or hybrids to identify key pathways involved in the plant response proves satisfactory, going to a 'finer' level of understanding of plant response will require access to segregating populations, such as RILs or near-isogenic lines (NILs). The phenotypic and genetic (QTL) characterization of such segregating populations under several environments and conditions would be a significant advance, because this information would allow the study of differential expression of genotypes from the two tails of the phenotypic distribution of a target trait. In addition, and as already mentioned above, the QTL location can be used as a validation step for the candidate genes that display differential expression and map at the same genomic region as a QTL. The use of segregating families with the same genetic background will allow the elimination of many false-positive genes, i.e. genes that present differential expression but are not related to the target trait, as is the case when working with contrasting lines and hybrids. Due to the large potential of functional genomics, several suitable approaches can be considered depending on the question one tries to answer. Undoubtedly, this new area of investigation will generate a large amount of information that should help bridge the gap between gene function and plant phenotype.

Conclusions

Projections indicate that the extent and severity of the crop-production problem stemming from water-limited conditions will, unfortunately, increase in both developed and developing countries during the coming years. That this is recognized by policy-makers and scientists alike is reflected in the increasing number of teams working on bettering our understanding of drought tolerance and improving plant

response under drought conditions, in model plants such as *Arabidopsis* and in food crops (for a review, see Edmeades *et al.*, 2001). During the last 10 years, the development of molecular genetics and QTL analysis has allowed us to identify genomic regions involved in drought tolerance in several cereal species (e.g. maize, sorghum, wheat and rice). In maize, this genetic dissection has been conducted for yield components and morphological and physiological traits. The weakness of this quantitative-genetic approach is that it provides very little information about the mechanisms and pathways involved in drought tolerance or about the multitude of genes involved in the plant's responses. The recent development of functional genomics should help overcome this limitation, because it will allow a simultaneous study of the expression of several thousand genes.

Based on progress to date, it is clear that a multidisciplinary approach – combining breeding, physiology and biotechnology – is required for an effective understanding of a plant's response to drought stress. Given the complexity of the problem, a better understanding of the genes and the pathways involved in plant responses will be crucial to accelerating, sustaining and complementing conventional breeding programmes. The QTL-characterization efforts initiated several years ago provide a powerful base of information and germ-plasm for the genetic dissection of physiological drought-adaptive traits. This physiological genetic approach, combined with the potential of functional genomics, including gene-expression profiling and proteomics, will allow us to identify the key pathways involved in drought stress and to know how they interact. This, in turn, will lead to efficient and effective strategies for developing cereals with higher productivity under water-limited conditions.

Acknowledgements

The authors thank Dr Jose Crossa for his help in analysing the data presented herein and David Poland for his helpful editorial review of the manuscript.

References

Bänziger, M., Pixley, K.V., Vivek, B. and Zambezi, B.T. (2000) *Characterization of Elite Maize Germplasm Grown in Eastern and Southern Africa: Results of the 1999 Regional Trials Conducted by CIMMYT and the Maize and Wheat Improvement Research Network for SADC (MWIRNET)*. CIMMYT, Harare, Zimbabwe, pp. 1–44.

Beavis, W.D. and Keim, P. (1996) Identification of quantitative trait loci that are affected by environment. In: Kang, M.S. and Gauch, H.G. (eds) *Genotype-by-environment Interaction.* CRC Press, Boca Raton, Florida, pp. 123–149.

Bohnert, H., Fischer, R., Kawasaki, S., Michalowski, C., Wang, H., Yale, J. and Zepeda, G. (2000) Cataloging stress-inducible genes and pathways leading to stress tolerance. In: Ribaut, J.-M. and Poland, D. (eds) *Molecular Approaches for the Genetic Improvement of Cereals for Stable Production in Water-limited Environments. A Strategic Planning Workshop held at CIMMYT, El Batan, Mexico, 21–25 June 1999.* CIMMYT, Mexico, DF, Mexico, pp. 156–161.

Bolaños, J. and Edmeades, G.O. (1996) The importance of the anthesis–silking interval in breeding for drought tolerance in tropical maize. *Field Crops Research* 48, 65–80.

Botstein, D., White, R.L., Skolnick, M. and Davis, R.W. (1980) Construction of a genetic linkage map in man using restriction fragment length polymorphisms. *American Journal of Human Genetics* 32, 314–331.

Bray, E.A. (1993) Molecular response to water deficit. *Plant Physiologist* 103, 1035–1040.

Chin, E.C.L., Senior, M.L., Shu, H. and Smith, J.S.C. (1996) Maize simple repetitive DNA sequences: abundance and allele variation. *Genome* 39, 866–873.

Cushman, J.C. and Bohnert, H. (2000) Genomic approaches to plant stress tolerance. *Current Opinion in Plant Biology* 3, 117–124.

Dreher, K., Morris, M.L., Ribaut, J.-M., Khairallah, M., Pandey, S. and Srinivasan, G. (2000) *Is marker-assisted selection cost-effective compared to conventional plant breeding methods? The case of quality protein maize.* Conference paper, Third Annual Conference of the International Consortium for

Agricultural Biotechnology Research, 25–28 August, Ravello, Italy.

Edmeades, G.O., Cooper, M., Lafitte, R., Zinselmeier, C., Ribaut, J.-M., Habben, J.E., Löffler, C. and Bänziger, M. (2001) Abiotic stresses and staple crops. In: Nosberger, J., Geiger, H.H. and Struik, P.C. (eds) *Crop Science: Progress and Prospects. Proceedings of the Third International Crops Science Congress, 17–21 August 2000.* CAB International, Wallingford, UK, pp. 137–154.

Fodor, S.P.A. (1997) Massively parallel genomics. *Science* 277, 393–395.

Gilles, P.N., Wu, D.J., Foster, C.B., Dillon, P.J. and Chanock, S.J. (1999) Single nucleotide polymorphic discrimination by an electronic dot blot assay on semiconductor microchips. *Nature Biotechnology* 17, 365–370.

Heisey, P.W. and Edmeades, G.O. (1999) Part 1. Maize production in drought stressed environments: technical options and research resource allocation. In: CIMMYT (ed.) *World Maize Facts and Trends 1997/98.* CIMMYT, Mexico, DF, Mexico, pp. 1–36.

Helentjaris, T., Slocum, M., Wright, S., Schaefer, A. and Nienhuis, J. (1986) Construction of genetic linkage maps in maize and tomato using restriction fragment length polymorphisms. *Theoretical and Applied Genetics* 72, 761–769.

Ingram, J. and Bartels, D. (1996) The molecular basis of dehydration tolerance in plants. *Annual Review Plant Physiology Plant Molecular Biology* 47, 377–403.

Johnson, G.R. and Mumm, R.H. (1996) Marker assisted maize breeding. In: *Proceedings of the 51st Annual Corn and Sorghum Research Conference, Chicago, Illinois, 11–12 December 1996.* American Seed Trade Association, Washington, DC, pp. 75–84.

Khavkin, E. and Coe, E. (1997) Mapped genomic locations for developmental functions and QTLs reflect concerted groups in maize (*Zea mays* L.). *Theoretical and Applied Genetics* 95, 343–352.

Liu, Q., Kasuga, M., Sakuma, Y., Abe, H., Miura, S., Shinozaki-Yamaguchi, K. and Shinozaki, K. (1998) Two transcription factors, DREB1 and DREB2, with an EREBP/AP2 DNA binding domain separate two cellular signal transduction pathways in drought- and low-temperature-responsive gene expression, respectively, in *Arabidopsis. Plant Cell* 10, 1391–1406.

McCouch, S.R., Kochert, G., Yu, Z.H., Wang, Z.Y. and Khush, G.S. (1988) Molecular mapping of rice chromosomes. *Theoretical and Applied Genetics* 76, 815–829.

Mertz, E.T., Bates, L.S. and Nelson, O.E. (1964) Mutant that changes protein composition and increases lysine content of maize endosperm. *Science* 145, 279–280.

Mohan, M., Nair, S., Bhagwat, A., Krishna, T.G., Yano, M., Bhatia, C.R. and Sasaki, T. (1997) Genome mapping, molecular markers and marker-assisted selection in crop plants. *Molecular Breeding* 3, 87–103.

Paterson, A.H., Lander, E.S., Hewitt, J.D., Peterson, S., Lincoln, S.E. and Tanksley, S.D. (1988) Resolution of quantitative traits into Mendelian factors by using a complete linkage map of restriction fragment length polymorphisms. *Nature* 335, 721–726.

Pingali, P.L. (ed.) (2001) *CIMMYT 1999–2000 World Maize Facts and Trends. Meeting World Maize Needs: Technological Opportunities and Priorities for the Public Sector.* CIMMYT, Mexico, DF.

Powell, W., Machray, G.C. and Provan, J. (1996) Polymorphism revealed by simple sequence repeats. *Trends in Plant Science* 1, 215–222.

Ragot, M., Biasiolli, M., Delbut, M.F., Dell'orco, A., Malgarini, L., Thevenin, P., Vernoy, J., Vivant, J., Zimmermann, R. and Gay, G. (1995) Marker-assisted backcrossing: a practical example. In: *Techniques et utilisations des marqueurs moléculaires.* Les Colloques, no. 72, INRA, Paris, pp. 45–56.

Ragot, M., Gay, G., Muller, J.-P. and Duroway, J. (2000) Efficient selection for adaptation to the environment throught QTL mapping in manipulation in maize. In: Ribaut, J.-M. and Poland, D. (eds) *Molecular Approaches for the Genetic Improvement of Cereals for Stable Production in Water-Limited Environments. A Strategic Planning Workshop held at CIMMYT, El Batan, Mexico, 21–25 June 1999.* CIMMYT, Mexico, DF, Mexico, pp. 128–130.

Ribaut J.-M. and Betrán, F.J. (1999) Single large-scale marker-assisted selection (SLS-MAS). *Molecular Breeding* 5, 531–541.

Ribaut, J.-M. and Hoisington, D. (1998) Marker-assisted selection: new tools and strategies. *Trends Plant Science* 3, 236–239.

Ribaut, J.-M., Hoisington, D.A., Deutsch, J.A., Jiang, C. and González-de-León, D. (1996) Identification of quantitative trait loci under drought conditions in tropical maize: 1. Flowering parameters and the anthesis-silking interval. *Theoretical and Applied Genetics* 92, 905–914.

Ribaut, J.-M., Jiang, C., González-de-León, D., Edmeades, G.O. and Hoisington, D.A. (1997) Identification of quantitative trait loci under drought conditions in tropical maize: 2. Yield components and marker-assisted selection strategies. *Theoretical and Applied Genetics* 94, 887–896.

Ribaut, J.-M., Jiang, C. and Hoisington, D. (2002) Efficiency of a gene introgression experiment by backcrossing. *Crop Science* (in press).

Schena, M., Shalon, D., Davis, R.W. and Brown, P.O. (1995) Quantitative monitoring of gene expression patterns with a complementary DNA microarray. *Science* 20, 467–470.

Schmidt, R.J., Burr, F.A., Aukerman, M.J. and Burr, B. (1990) Maize regulatory gene *opaque-2* encodes a protein with a 'leucine-zipper' motif that binds to zein DNA. *Proceedings National Academy of Sciences* 87, 46–50.

Shinozaki, K. and Yamaguchi-Shinozaki, K. (1997) Gene expression and signal transduction in water-stress response. *Plant Physiology* 115, 327–334.

Skriver, K. and Mundy, J. (1990) Gene expression in response to abscisic acid and osmotic stress. *Plant Cell* 2, 503–512.

Tanksley, S.D. and Nelson, J.C. (1996) Advanced backcross QTL analysis: a method for the simultaneous discovery and transfer of valuable QTLs from unadapted germplasm into elite breeding lines. *Theoretical and Applied Genetics* 92, 191–203.

Tanksley, S.D., Young, N.D., Paterson, A.H. and Bonierbale, M.W. (1989) RFLP mapping in plant breeding: new tools for an old science. *Biotechnology* 7, 257–264.

Williams, J.G.K., Kubelik, A.R., Livak, K.J., Rafalski, J.A. and Tingey, S. (1990) DNA polymorphisms amplified by arbitrary primers are useful as genetic markers. *Nucleic Acids Research* 18, 6531–6535.

Young, N.D. (1999) A cautiously optimistic vision for marker-assisted breeding. *Molecular Breeding* 5, 505–510.

8 Explorations with Barley Genome Maps

Diane E. Mather

*Department of Plant Science, McGill University, 21111 Lakeshore Road,
Ste-Anne-de-Bellevue, Quebec H9X 3V9, Canada*

Introduction

Barley (*Hordeum vulgare* L.) is an economically important grain crop and has long been an interesting and useful model for genetic analysis. Classical genetic analyses in barley were facilitated by the availability of many morphological and biochemical variants (Lundqvist *et al.*, 1997), by diploid inheritance and by a high degree of self-pollination. Cytogenetic studies of the barley genome were facilitated by its relatively small number ($2n = 2x = 14$) of large and visually distinct chromosomes. With translocation stocks, trisomics and wheat–barley addition lines (Ramage, 1985; Shepherd and Islam, 1992), genes were assigned to chromosomes, linkage maps were orientated and centromere positions were estimated.

With the development of molecular-marker technology, it became possible to construct detailed whole-genome linkage maps within mapping populations derived from individual crosses. Marker polymorphism within *H. vulgare* was sufficient to permit the construction of intraspecific genome maps (e.g. Heun, 1992; Kleinhofs *et al.*, 1993; Qi *et al.*, 1996). Inbred lines or doubled haploid lines could be used as mapping parents. Crosses among parents could be easily made and random samples of recombinant-inbred or doubled-haploid progeny could be easily derived, providing 'immortal' mapping populations. Seed of these populations could be readily reproduced and shared among researchers, permitting genotyping at many marker loci and phenotyping of many traits under a range of environmental conditions. Thus, barley was among the first crop species for which quantitative trait loci (QTL) were mapped (Heun, 1992; Hayes *et al.*, 1993). QTL mapping provided a major new tool for the study of quantitative traits in barley, traits for which genetic analysis had previously been limited to the estimation of population parameters, such as heritability, combining ability and the effective number of genes (Hockett and Nilan, 1985).

Positions, effects and interactions of QTL have now been estimated for many barley traits and some of these QTL have been experimentally validated and/or been the object of marker-assisted selection efforts. Information on the more than 750 barley QTL reported in the literature has been summarized by Hayes *et al.* (2000). Here, I shall present and discuss several examples of barley QTL research to illustrate some of the insights that genome mapping has been able to provide regarding the genetic control of a diversity of quantitative traits in barley.

QTL Affecting Barley Grain Yield

Crop–plant yield can be viewed as an integrative trait that reflects the overall vigour
and physiological efficiency of a plant genotype, its responses to environmental factors,
including abiotic and biotic stress, and its
patterns of biomass partitioning. Therefore,
it seems reasonable to expect barley grain
yield to be affected directly and indirectly
by many genes and to be subject to environmental effects and genotype–environment
interactions. Given the inherently low power
of QTL experiments for low-heritability
traits, most QTL analyses of barley grain
yield have detected only a few loci.

Some individual QTL apparently have
quite large effects on grain yield and, in some
cases, it has been possible to attribute these
to specific major genes or gene clusters
and/or to loci that affect growth habit and/or
specific yield components. In six-rowed
barley, Hayes et al. (1993) found that yield
QTL often coincided with QTL for plant
height and lodging and that one important
QTL coincided with a previously mapped
Ea maturity locus. In two-rowed barley,
Thomas et al. (1995) and Bezant et al. (1997)
both detected significant QTL effects for
grain yield in the region of the major plant
stature gene sdw1 (denso). In a winter–
spring barley mapping population, Oziel
et al. (1996) found loci affecting vernalization requirement, photoperiod reaction and
low-temperature tolerance to be important
determinants of grain yield. In populations
derived from crosses between two-rowed
and six-rowed parents, the region of chromosome 2 that contains the vrs1 locus (the
determinant of two-rowed vs. six-rowed
spike morphology) has been found to be an
important determinant of grain yield (Kjaer
and Jensen, 1996; Marquez-Cedillo et al.,
2001).

In multiple-environment experiments,
QTL for grain yield have exhibited
interactions with environments. Based
on six-rowed barley yield data from five
environments, Hayes et al. (1993) observed
QTL–environment interactions that were
due to differences in magnitudes of QTL

effects. Based on two-rowed barley yield
data from 28 environments, Tinker et al.
(1996) observed QTL–environment interactions involving changes of both sign and
magnitude across environments. Further
analysis of QTL–environment interaction
patterns for grain yield and other agronomic
traits may provide insights into the patterns
and mechanisms underlying genotype–
environment interaction in barley.

QTL Affecting Developmental Traits in Barley

Traits related to the rate of plant development (often assessed as heading date and/or
maturity date) are of major importance
in determining the suitability of barley
cultivars for particular production environments. Loci involved in the determination
of days to heading in barley include the
sdw1 gene, major and minor genes for vernalization response and photoperiod sensitivity and 'earliness per se' genes (Laurie
et al., 1995; Hay and Ellis, 1998). Yin et al.
(1999) demonstrated that crop-development
models can be used in QTL analyses of
traits that vary with developmental stage.
Using a simple ecophysiological model,
they integrated daily-temperature data into
the assessment of specific leaf area and
were able to gain insight into how different
QTL affect specific leaf area at different
phenological stages. For example, they
found that an apparent effect of the sdw1
locus on specific leaf area is its indirect
effect on preflowering duration.

QTL Affecting Barley Grain- and Malt-quality Traits

Malting of barley for use in brewing
involves a steeping of the grain in water,
followed by germination in a controlled
environment. The resulting 'green malt' is
then dried by kilning. Assessment of malting quality involves a series of grain-quality
assays conducted on barley grain samples
and a series of malt-quality assays conducted

on malt samples obtained by 'micromalting' barley samples. The biochemistry and physiology of barley malting are relatively well understood and the genetic variation for malting quality is quite well documented, but the genetic control of malting quality is apparently complex and is not well understood. The high cost of assaying malting quality for large numbers of samples has limited the number and scope of QTL mapping experiments for malting quality. Nevertheless, QTL for malting-quality traits have been mapped in several major germplasm groups of malting barley, European two-rowed spring barley (Thomas *et al.*, 1996; Bezant *et al.*, 1997; Powell *et al.*, 1997), North American two-rowed barley (Hayes *et al.*, 1996; Mather *et al.*, 1997), Australian two-rowed barley (Langridge *et al.*, 1996) and North American six-rowed barley (Hayes *et al.*, 1993; Han *et al.*, 1995; Oberthur *et al.*, 1995; Ullrich *et al.*, 1997) and in crosses between two-rowed and six-rowed barley (Hayes *et al.*, 1996; Marquez-Cedillo *et al.*, 2000) and between winter and spring barley (Oziel *et al.*, 1996).

Most of these mapping experiments involved crosses between a malting-quality cultivar and a cultivar or line not suitable for malting. At most malting-quality QTL, the favourable allele came from the malting-quality parent, however, there appears to be some scope for the identification of favourable alleles in non-malting germ-plasm. Certain malting-quality traits (e.g. kernel weight and malt quality, grain protein and diastatic power, beta-glucan and soluble total protein) tend to exhibit unfavourable associations. These may be due to genetic and/or environmental causes. In some cases, QTL experiments have provided evidence of genetic causes for such associations (i.e. coincident QTL, indicating pleiotropic genes or repulsion-phase linkage blocks) but have also provided some evidence of independent genetic control (i.e. QTL that affect one trait but not the other). In at least some germ-plasm groups, the QTL results indicate that it should be possible to achieve some improvements in these traits, despite the unfavourable associations (Igartua *et al.*, 2001).

Some of the QTL detected for malting-quality traits coincide with amylase loci (*Amy1*, *Amy2*, *Bmy1*, *Bmy2*), with hordein loci (*Hor1*, *Hor2*) or with loci affecting grain, spike and plant morphology (*vrs1*, *int-c*, *sdw1*). In two-rowed × six-rowed crosses, the *vrs1* region has important QTL effects on multiple grain- and malt-quality traits (indicating pleiotropy of *vrs1* and/or a linkage block of different genes) and exhibits epistatic interactions with other regions of the genome (Hayes *et al.*, 1996; Marquez-Cedillo *et al.*, 2000). Effects of this region have been found even in two-rowed material that is not segregating at the *vrs1* locus (Langridge *et al.*, 1996), suggesting that there may be other malting-quality genes in this region. In two-rowed barley, important effects on malting quality have also been found in the *sdw1* region. In most cases, the favourable effects in this region are associated with the dwarfing allele, which is present in many European malting-quality cultivars. There are also reports, however, of favourable effects being associated with the alternate allele of a malt-quality QTL in a cross not segregating at the *sdw1* locus. Thus, there may be other malting-quality QTL in this region of chromosome 3.

The results of several mapping experiments indicate that chromosome 7 (5H) is of particular importance for malting quality (Hayes *et al.*, 1993; Langridge *et al.*, 1996; Mather *et al.*, 1997; Marquez-Cedillo *et al.*, 2000). Different regions of that chromosome are important in different crosses.

QTL Affecting Traits Assessed by Digital-image Analysis

In malting, the sizes, shapes and uniformity of barley kernels can be of critical importance. Conventional methods of assessing these characters relied mostly upon the determinations of the mean weight of kernel and the proportions of the kernels that passed over or through sieves of specific sizes. Digital-image analysis has provided new ways of assessing size and shape characteristics, providing data on the length, width, area, perimeter and roundness of individual kernels and on the variability among kernels within samples. Similarly, digital-image analysis can be used to study the sizes, shapes and uniformity of starch granules within barley endosperm, and these characteristics may also influence malting, given that starch degradation is a key process in malting. Image analysis of samples from a mapping population permits QTL analysis of kernel or granule size and shape characteristics and of the uniformity of these characteristics.

For starch-granule traits in a six-rowed barley population, Borém *et al.* (1999) reported that the major QTL effects were in a chromosome region that also affected days to heading and plant height but contained no QTL for grain or malt quality. That region affected the overall mean starch-granule volume via effects on the average diameter (but not thickness or roundness) of type A (large) granules and the proportion of type A granules. The same QTL region affected the roundness but not the size of type B (small) granules.

In a two-rowed × six-rowed mapping population, Ayoub *et al.* (2002) applied QTL analysis to image-analysis data on kernel size and shape characteristics, and detected QTL affecting only the means of these characteristics, QTL affecting only the variability of these characteristics and QTL affecting both means and variability. The *vrs1* region had the most important effects, and there were also QTL effects detected at or near *int-c*, a locus that affects the development of lateral spikelets. QTL analysis of two-rowed and six-rowed subpopulations gave differential results, providing evidence of epistasis between the *vrs1* region and other QTL. Numerous significant QTL were detected in the two-rowed subpopulation, but, surprisingly, none was detected in the six-rowed subpopulation.

QTL Affecting Quantitative Disease Resistance in Barley

When it is possible to classify disease responses as either resistant or susceptible,

the resistance genes responsible for these responses can be mapped in the same manner as marker loci. In barley, resistance genes have been mapped in this way for many diseases. For some crop–plant diseases and under many environmental conditions, plants exhibit partial resistance to disease. In these cases, disease responses are assessed as quantitative traits (usually as measurements or ratings of symptom severity) and disease-resistance genes are mapped using QTL analysis.

Using simple rating-scale data for the symptom severity of several naturally occurring diseases in the field, Spaner *et al.* (1998) were able to map QTL affecting up to 45% of observed variation. For powdery mildew (caused by *Blumeria graminis* f. sp. *hordei*) and stem rust (caused by *Puccinia graminis* f. sp. *tritici*), QTL were mapped to within a few centiMorgans of positions that had been estimated based on the results of classification data from carefully inoculated trials. Similarly, for powdery mildew, Falak *et al.* (1999) showed that, in a two-rowed mapping population, a previously known *Mlg* locus and a new locus tentatively designated *Ml(TR)* could be mapped based on classification data obtained after inoculation with specific isolates of the pathogen and that the same loci could be mapped as QTL using symptom-severity results from naturally infected field trials. In that study, QTL analysis also detected a third locus, and an analysis based on both the classification data and the quantitative data suggested a fourth locus with minor effects.

For stripe rust (caused by *Puccinia striiformis* f. sp. *hordei*), several resistance genes have been mapped based on observations of qualitative responses in barley seedlings and several QTL have been mapped based on observations of symptom severity in seedlings and adult plants (Chen *et al.*, 1994; Thomas *et al.*, 1995; Hayes *et al.*, 1996; Toojinda *et al.*, 2000). QTL contributing to stripe-rust resistance have been used successfully in marker-assisted selection (Toojinda *et al.*, 1998).

Currently, the most important disease challenge facing barley production is *Fusarium* head blight (FHB) (caused by *Fusarium graminearum*). Through effects on grain yield and through contamination of the grain with mycotoxins, this disease has devastating effects on barley production. As no sources of complete (qualitative) resistance have been identified, this disease problem has been the subject of intensive QTL mapping, and marker-assisted selection efforts are under way. De la Pena *et al.* (1999) and Zhu *et al.* (1999b) have published QTL results for six-rowed and two-rowed barley, respectively. De la Pena *et al.* (1999) identified several QTL that they considered useful targets for marker-assisted selection, and one for which the resistance allele had an association with late heading. Zhu *et al.* (1999b) detected QTL for FHB resistance on six barley chromosomes, and noted that most of these coincided with QTL determining plant height and/or spike morphology. They suggested that the development of FHB-resistant barley cultivars will require an understanding of the biology underlying coincident QTL between plant-architecture traits and FHB resistance.

QTL Validation in Barley

Detection of QTL and estimation of QTL positions and effects are subject to various sources of experimental error and bias. Mapping experiments may detect spurious QTL or may fail to detect real QTL. They may over- or underestimate the true effects of QTL and they are unable to provide precise estimates of QTL position. In barley, several attempts have been made to experimentally validate the existence, location and positions of putative QTL.

For the six-rowed mapping population 'Steptoe'/'Morex', Larson *et al.* (1996) used marker-assisted back-crossing to transfer alleles at two yield QTL from 'Steptoe' into 'Morex'. They detected significant effects on grain yield for one QTL but not for the other. For the same population, Romagosa *et al.* (1999) verified the effects of four QTL on grain yield but found that effects were more consistent for some QTL than for others. Similarly, Han *et al.* (1997) applied marker-assisted selection for two 'Steptoe'/'Morex' malting-quality QTL and found that it was more effective than phenotypic selection, but only for one of the two QTL. Zhu *et al.* (1999a) pyramided alleles for which favourable grain-yield effects had been detected in 'Steptoe'/'Morex'. Although most of the loci showed significant effects, the significance, magnitude and direction of these effects varied across environments, and epistatic interactions were detected among QTL. Kandemir *et al.* (2000) used marker-assisted back-crossing to transfer large chromosome fragments from grain-yield QTL regions from 'Steptoe' into 'Morex'. They found effects on yield-related traits but no significant effects on grain yield itself.

Using marker-based selection within a new sample of 'Harrington'/'TR306' progeny (i.e. lines not used in the original mapping experiment), Igartua *et al.* (2000) confirmed the existence of favourable grain- and malt-quality QTL alleles on chromosome 7(5H) of 'Harrington'. In the same material, Spaner *et al.* (1999) confirmed the effects of these QTL on agronomic traits. Igartua *et al.* (2000) were unable to definitively confirm the existence of two smaller-effect malting-quality QTL that had been mapped on chromosomes 3 and 6. Nevertheless, selection based on marker genotypes on all three chromosomes was effective in selecting progeny lines with superior malting quality.

The mixed results obtained in these validation and marker-assisted selection experiments demonstrate that, while some of the QTL detected in mapping experiments have repeatable and verifiable effects, others may be spurious, poorly mapped or subject to strong interactions with environmental conditions and/or other loci. The results of QTL analysis do not provide a sufficient basis to definitively predict the optimum genotypic constitution for a quantitative trait, but they do provide a basis for formulating hypotheses to be tested in marker-assisted selection experiments and for the design of breeding programmes that involve both genotypic and phenotypic selection.

Acknowledgements

I am grateful to the Natural Sciences and Engineering Research Council of Canada, the Brewing and Malting Barley Research Institute, Canada Malting Company and Agriculture and Agri-Food Canada for their support of my research on QTL mapping in barley, and to all of the researchers, support staff and funding sources that have participated in and supported barley genome mapping efforts internationally.

References

Ayoub, M., Symons, S., Edney, M.J. and Mather, D.E. (2002) QTL affecting kernel size and shape in a two-rowed by six-rowed barley cross. *Theoretical and Applied Genetics* (in press).

Bezant, J., Laurie, D., Pratchett, N., Chojecki, J. and Kearsey, M. (1997) Mapping QTL controlling yield and yield components in a spring barley (*Hordeum vulgare* L.) cross using marker regression. *Molecular Breeding* 3, 29–38.

Borém, A., Mather, D.E., Rasmusson, D.C., Fulcher, R.G. and Hayes, P.M. (1999) Mapping quantitative trait loci for starch granule traits in barley. *Journal of Cereal Science* 29, 153–160.

Chen, F., Prehn, D., Hayes, P.M., Mulrooney, D., Corey, A. and Vivar, H. (1994) Mapping genes for resistance to barley stripe rust (*Puccinia striiformis* f. sp. *hordei*). *Theoretical and Applied Genetics* 88, 215–219.

De la Pena, R.C., Smith, K.P., Capettini, F., Muehlbauer, G.J., Gallo-Meagher, M., Dill-Macky, R., Somers, D.A. and Rasmusson, D.C. (1999) Quantitative trait loci associated with resistance to *Fusarium* head blight and kernel discoloration in barley. *Theoretical and Applied Genetics* 99, 561–569.

Falak, I., Falk, D.E., Tinker, N.A. and Mather, D.E. (1999) Resistance to powdery mildew in a doubled haploid barley population and its association with marker loci. *Euphytica* 107, 185–192.

Han, F., Ullrich, S.E., Chirar, S., Menteur, S., Jestin, L., Sarrafi, A., Hayes, P.M., Jones, B.L., Blake, T.K., Wesemberg, D.M., Kleinhofs, A. and Kilian, A. (1995) Mapping of β-glucan content and β-glucanase activity loci in barley grain and malt. *Theoretical and Applied Genetics* 91, 921–927.

Han, F., Romagosa, I., Ullrich, S.E., Jones, B.L., Hayes, P.M. and Wesenberg, D.M. (1997) Molecular marker-assisted selection for malting quality traits in barley. *Molecular Breeding* 3, 427–437.

Hay, R.K.M. and Ellis, R.P. (1998) The control of flowering time in wheat and barley, what recent advances in molecular genetics can reveal. *Annals of Botany* 82, 541–554.

Hayes, P.M., Liu, B.H., Knapp, S.J., Chen, F., Jones, B., Blake, T., Franckowiak, J., Rasmusson, D., Sorrells, M., Ullrich, S.E., Wesenberg, D. and Kleinhofs, A. (1993) Quantitative trait locus effects and environmental interaction in a sample of North American barley germplasm. *Theoretical and Applied Genetics* 87, 392–401.

Hayes, P.M., Prehn, D., Vivar, H., Blake, T., Comeau, A., Henry, I., Johnston, M., Jones, B., Steffenson, B., St Pierre, C.A. and Chen, F. (1996) Multiple disease resistance loci and their relationship to agronomic and quality loci in a spring barley population. *Journal of Quantitative Trait Loci* 2, 2.

Hayes, P.M., Castro, A., Corey, A., Marquez-Cedillo, L., Jones, B., Mather, D., Matus, I., Rossi, C. and Sato, K. (2000) A summary of published barley QTL reports. http //www.css.orst.edu/barley/nabgmp/qtlsum.htm

Heun, M. (1992) Mapping quantitative powdery mildew resistance of barley using a restriction fragment length polymorphism map. *Genome* 35, 1019–1025.

Hockett, E.A. and Nilan, R.A. (1985) Genetics. In: Rasmusson, D.C. (ed.) *Barley*. American Society of Agronomy–Crop Science Society of America–Soil Science Society of America, Madison, Wisconsin, pp. 187–230.

Igartua, E., Edney, M., Rossnagel, B.G., Spaner, D., Legge W.G., Scoles, G.J., Eckstein, P.E., Penner, G.A., Tinker, N.A., Briggs, K.G., Falk, D.E. and Mather, D.E. (2000) Marker-based selection of QTL affecting grain and malt quality in two-row barley. *Crop Science* 40, 1426–1433.

Igartua, E., Hayes, P.M., Thomas, W.T.B., Meyer, R. and Mather, D.E. (2001) Genetic control of quantitative grain and malt quality traits in barley. *Journal of Crop Production* (in press).

Kandemir, N., Jones, B.L., Wesenberg, D.M., Ullrich, S.E. and Kleinhofs, A. (2000) Marker-assisted analysis of three grain yield QTL in barley (*Hordeum vulgare* L.) using near isogenic lines. *Molecular Breeding* 6, 157–167.

Kjaer, B. and Jensen, J. (1996) Quantitative trait loci for grain yield and yield components in a cross between a six-rowed and a two-rowed barley. *Euphytica* 90, 39–48.

Kleinhofs, A., Kilian, A., Saghai Maroof, M.A., Biyashev, R.M., Hayes, P.M., Chen, F.Q., Lapitan, N., Fenwick, A., Blake, T.K., Kanazin, V., Ananiev, E., Dahleen, L., Kudrna, D., Bollinger, J., Knapp, S.J., Liu, B., Sorrells, M., Heun, M., Franckowiak, J.D., Hoffman, D., Skadsen, R. and Steffenson, B.J. (1993) A molecular, isozyme and morphological map of the barley (*Hordeum vulgare*) genome. *Theoretical and Applied Genetics* 86, 705–712.

Langridge, P., Karakousis, A., Kretschmer, J., Manning, S., Chalmers, K., Boyd, R., Li, C.D., Islam, R., Logue, S., Lance, R. and SARDI (1996) RFLP and QTL analysis of barley mapping populations. http: //greengenes.cit. cornell.edu/WaiteQTL/

Larson, S.R., Kadyrzhanova, D., McDonald, M., Sorrells, M. and Blake, T.K. (1996) Evaluation of barley chromosome-3 yield QTLs in a backcross F$_2$ population using STS-PCR. *Theoretical and Applied Genetics* 93, 618–625.

Laurie, D.A., Pratchett, N., Bezant, J.H. and Snape, J.W. (1995) RFLP mapping of five major genes and eight quantitative trait loci controlling flowering time in a winter × spring barley (*Hordeum vulgare* L.) cross. *Genome* 38, 575–585.

Lundqvist, U., Franckowiak, J.D. and Konishi, T. (1997) New and revised descriptions of barley genes. *Barley Genetics Newsletter* 26, 22–43.

Marquez-Cedillo, L.A., Hayes, P.M., Jones, B.L., Kleinhofs, A., Legge, W.G., Rossnagel, B.G., Sato, K., Ullrich, S.E., Wesenberg, D.M. and the North American Barley Genome Mapping Project (2000) QTL analysis of malting quality in barley based on the doubled haploid progeny of two elite North American varieties representing different germplasm groups. *Theoretical and Applied Genetics* 101, 173–184.

Marquez-Cedillo, L.A., Hayes, P.M., Kleinhofs, A., Legge, W.G., Rossnagel, B.G., Sato, K., Ullrich, S.E., Wesenberg, D.M. and the North American Barley Genome Mapping Project (2001) QTL analysis of agronomic traits in barley based on the doubled haploid progeny of two elite North American varieties representing different germplasm groups. *Theoretical and Applied Genetics* 103, 625–637.

Mather, D.E., Tinker, N.A., LaBerge, D.E., Edney, M., Jones, B.L., Rossnagel, B.G., Legge, W.G., Briggs, K.G., Irvine, R.B., Falk, D.E. and

Kasha, K.J. (1997) Regions of the genome that affect grain and malt quality in a North American two-row barley cross. *Crop Science* 37, 544–554.

Oberthur, L., Blake, T.K., Dyer, W.E. and Ullrich, S.E. (1995) Genetic analysis of seed dormancy in barley (*Hordeum vulgare* L.). *Journal of Quantitative Trait Loci* 1, 5.

Oziel, A., Hayes, P.M., Chen, F.Q. and Jones, B. (1996) Application of quantitative trait locus mapping to the development of winter habit malting barley. *Plant Breeding* 115, 43–51.

Qi, X., Stam, P. and Lindhout, P. (1996) Comparison and integration of four barley genetic maps. *Genome* 39, 379–394.

Powell, W., Thomas, W.T.B., Baird, E., Lawrence, P., Booth, A., Harrower, B., McNicol, J.W. and Waugh, R. (1997) Analysis of quantitative traits in barley by the use of the amplified fragment length polymorphisms. *Heredity* 79, 48–59.

Ramage, R.T. (1985) Cytogenetics. In: Rasmusson, D.C. (ed.) *Barley*. American Society of Agronomy–Crop Science Society of America–Soil Science Society of America, Madison, Wisconsin, pp. 127–154.

Romagosa, I., Han, F., Ullrich, S.E., Hayes, P.M. and Wesenberg, D.M. (1999) Verification of yield QTL through realized molecular marker-assisted selection responses in a barley cross. *Molecular Breeding* 5, 143–152.

Shepherd, K.W. and Islam, A.K.M.R. (1992) Progress in the production of wheat–barley addition and recombination lines and their use in mapping the barley genome. In: Shewry, P.R. (ed.) *Barley: Genetics, Biochemistry, Molecular Biology and Biotechnology*. CAB International, Wallingford, UK, pp. 99–114.

Spaner, D., Shugar, L.P., Choo, T.M., Falak, I., Briggs, K.G., Legge, W.G., Falk, D.E., Ullrich, S.E., Tinker, N.A., Steffenson, B.J. and Mather, D.E. (1998) Mapping of disease resistance loci in barley on the basis of visual assessment of naturally occurring symptoms. *Crop Science* 38, 843–850.

Spaner, D., Rossnagel, B.G., Legge, W.G., Scoles, G.J., Eckstein, P.E., Penner, G.A., Tinker, N.A., Briggs, K.G., Falk, D.E., Afele, J.C., Hayes, P.M. and Mather, D.E. (1999) Verification of a quantitative trait locus affecting agronomic traits in two-row barley. *Crop Science* 39, 248–252.

Thomas, W.T.B., Powell, W., Waugh, R., Chalmers, K.J., Barua, U.M., Jack, P., Lea, V., Forster, B.P., Swanston, J.S., Ellis, R.P., Hanson, P.R. and Lance, R.C.M. (1995) Detection of quantitative trait loci for agronomic, yield, grain and disease characters in spring barley (*Hordeum vulgare* L.). *Theoretical and Applied Genetics* 91, 1037–1047.

Thomas, W.T.B., Powell, W., Swanston, J.S., Ellis, R.P., Chalmers, K.J., Barua, U.M., Jack, P., Lea, V., Forster, B.P., Waugh, R. and Smith, D.B. (1996) Quantitative trait loci for germination and malting quality characters in a spring barley cross. *Crop Science* 36, 265–273.

Tinker, N.A., Mather, D.E., Rossnagel, B.G., Kasha, K.J., Kleinhofs, A., Hayes, P.M., Falk, D.E., Ferguson, T., Shugar, L.P., Legge, W.G., Irvine, R.B., Choo, T.M., Briggs, K.G., Ulrich, S.E., Franckowiak, J.D., Blake, T.K., Graf, R.J., Dofing, S.M., Saghai Maroof, M.A., Scoles, G.J., Hoffman, D., Dahleen, L.S., Kilian, A., Chen, F., Biyashev, M., Kudrna, D.A. and Steffenson, B.J. (1996) Loci that affect agronomic performance in two-row barley. *Crop Science* 36, 1053–1062.

Toojinda, T., Baird, E., Booth, A., Broers, L., Hayes, P., Powell, W., Thomas, W., Vivar, H. and Young, G. (1998) Introgression of quantitative trait loci (QTLs) determining stripe rust resistance in barley, an example of marker-assisted line development. *Theoretical and Applied Genetics* 96, 123–131.

Toojinda, T., Broers, L.H., Chen, X.M., Hayes, P.M., Kleinhofs, A., Korte, J., Kudrna, D., Leung, H., Line, R.F., Powell, W., Ramsay, L., Vivar, H. and Waugh, R. (2000) Mapping quantitative and qualitative disease resistance genes in a doubled haploid population of barley (*Hordeum vulgare*). *Theoretical and Applied Genetics* 101, 580–589.

Ullrich, S.E., Han, F. and Jones, B.L. (1997) Genetic complexity of the malt extract trait in barley suggested by QTL analysis. *Journal of the American Society of Brewing Chemists* 55, 1–4.

Yin, X., Kropff, M.J. and Stam, P. (1999) The role of ecophysiological models in QTL analysis, the example of specific leaf area in barley. *Heredity* 82, 415–421.

Zhu, H., Briceño, G., Dovel, R., Hayes, P.M., Liu, B.H., Liu, C.T., Toojinda, T. and Ullrich, S.E. (1999a) Molecular breeding for grain yield in barley: an evaluation of QTL effects in a spring barley cross. *Theoretical and Applied Genetics* 98, 772–779.

Zhu, H., Gilchrist, L., Hayes, P., Kleinhofs, A., Kudrna, D., Liu, Z., Prom, L., Steffenson, B., Toojinda, T. and Vivar, H. (1999b) Does function follow form? QTLs for *Fusarium* head blight (FHB) resistance are coincident with QTLs for inflorescence traits and plant height in a doubled haploid population of barley. *Theoretical and Applied Genetics* 99, 1221–1232.

9   Global View of QTL: Rice as a Model

Yunbi Xu

RiceTec, Inc., PO Box 1305, Alvin, TX 77512, USA

Introduction

Most traits of agronomic importance are quantitatively inherited. Quantitative traits (QTs) are genetically controlled by effects of polygenes that are greatly modified by environments. Each of these genes potentially has a relatively small effect. With the advent of molecular markers, such as restriction fragment length polymorphisms (RFLPs), simple sequence repeats (SSRs) and single nucleotide polymorphisms (SNPs), high-density genetic maps can be constructed and QTs can be associated with molecular markers. A chromosomal region that is associated with molecular markers and with a QT is defined as a quantitative trait locus (QTL). Plant populations with different genetic structures have been created for genetic mapping, including F_2/F_3, back-cross (BC), doubled haploids (DHs), recombinant inbred lines (RILs), near-isogenic lines (NILs), back-cross inbred lines (BILs) and various mutants. Development of advanced statistical methods, such as composite interval mapping (Zeng, 1993, 1994; Jansen and Stam, 1994) and Bayesian and Markov chain Monte Carlo analyses (Satagopan *et al.*, 1996; Uimari *et al.*, 1996), along with powerful software, has contributed to genetic mapping using data collected for multiple traits in different environments.

In QTL analysis, rice (*Oryza sativa* L.) has been receiving exceptional attention, because it has been used as a model plant in molecular biology. As a primary food source for more than a third of the world's population, rice has one of the most compact genomes among cereals, with the smallest genome of any monocots known. It contains about 3.5 times as much DNA as *Arabidopsis*, but only about 20% as much as maize (*Zea mays* L.) and about 3% as much DNA as wheat (*Triticum aestivum* L.). The first RFLP map was constructed in the 1980s (McCouch *et al.*, 1988), and two high-density maps were subsequently developed (Causse *et al.*, 1994; Kurata *et al.*, 1994), which have been widely applied to the mapping of genes controlling traits of agronomic importance. Since the first QTL studies (Ahn *et al.*, 1993; Xu *et al.*, 1993), more than 80 articles with over 1000 QTL had been documented by 2000. Rice is the first example in monocot species of the successful cloning of a major gene (the bacterial blight resistance gene, *Xa21*) (Song *et al.*, 1995) and a QTL (heading date, *Hd1*) (Yano *et al.*, 2000) through map-based cloning. In this chapter, QTL will be viewed globally from different perspectives, using rice as a model. Several review articles on rice QTL are also available elsewhere (McCouch and Doerge, 1995; Yano and Sasaki, 1997). For general reviews, the reader may refer to Xu (1997), Liu (1998),

Lynch and Walsh (1998), Paterson (1998), Flint and Mott (2001), Kearsey (Chapter 4, this volume) and Dudley (Chapter 6, this volume).

QTL Across Environments

QTL analysis involves extracting a genetic signal from many sources of 'noise', such as those from external environments and internal genetic backgrounds. For accurate QTL analysis, the 'noise' must be minimized or eliminated. 'Controlled' environments or genetic backgrounds are usually created for filtering the 'noise'. Plant populations used for QTL analysis can be evaluated in either natural or controlled environments or both. Controlled environments can be compared with each other or with natural environments. If two environments mainly differ in one macro-environmental factor, they are considered to be contrasting or near-iso-environments (NIEs) and the standard plot-to-plot variation and other residual microenvironmental effects can be neglected. If the two environments are from experiments of different years or locations, we assume that location and year effects do not confound the effect of the macroenvironmental factor.

Near-iso-environments

Some traits need to be measured under NIEs, where plants respond differently. In this case, the first environment imposes much less stress on plants than the second, e.g. two environments with normal and high temperature, respectively. The effect of the stress environment can be measured using the much-less-stress or normal environment as a control. A relative trait value is then derived from two direct trait values measured in each environment to ascertain the sensitivity of plants to the stress (see, for example, Ni *et al.*, 1998). If different plants have an identical phenotype under the much-less-stress environment, the direct trait value in the stress environment can be used to measure sensitivity. When both environments impose little stress on plants, however, one should use relative trait values instead. A typical example is the photoperiod sensitivity that can only be measured in NIEs, one with a short day length and the other with a long day length. A relative measure for this type of trait (sensitivity) should be: sensitivity = the difference of measures in the NIEs, divided by the measure in one of the NIEs or in the normal environment when the other is stressful.

Xu *et al.* (1997) provided an example of how rice plants respond to photoperiod and temperature. Using Zhaiyeqing 8/Jingxi 17 DHs, days-to-heading (DTH) and photo-thermosensitivity (PTS) were measured in two environments (Beijing and Hangzhou) that mainly differ in day length and temperature. At the photo-thermosensitive stage, Beijing has long day length (14.5–15 h) and low temperature (20–27°C), whereas Hangzhou has short day length (13–13.5 h) and high temperature (25.5–30°C). Rice is considered a short-day plant, and development from vegetative to reproductive stages is promoted under short-day-length and high-temperature conditions. Differences in photoperiod and temperature in the two locations resulted in differences in DTH of 0 to 39 days for individual DH lines (Fig. 9.1). Using the relative difference, [(DTH in Beijing – DTH in Hangzhou)/DTH in Beijing × 100], genes associated with photo-thermosensitivity were mapped with 155 RFLP and 92 SSR markers. Four chromosomal regions were identified as significantly associated with DTH (Fig. 9.1) in either or both locations, whereas LOD scores for the PTS in these regions were much lower than 2.4. A region on chromosome 7 (G397A–RM248) was significantly associated with PTS (LOD = 4.47), which LOD scores for DTH in both locations were much lower than 2.4 (Fig. 9.1), indicating that this PTS QTL is independent of the QTL for heading date.

A second example is from CO39/Moroberekan RILs grown under greenhouse conditions and exposed to two different photoperiod regimes (Maheswaran *et al.*, 2000). Days-to-flowering (DTF) of individual lines were evaluated under 10 h and 14 h

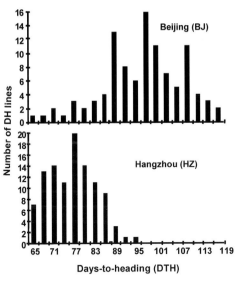

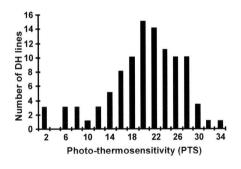

LODs for DTH and PTS

Chr	Marker interval	DTH(BJ)	DTH(HZ)	PTS
1	RG400-RM84	2.41*	1.68	0.87
7	G379A-RM248	1.21	0.56	4.47*
8	RG885-RM44	7.35*	6.56*	1.07
10	C16-RM228	2.67*	3.04*	0.41
12	RG463-RG323	2.30	2.55*	0.82

Fig. 9.1. QTL mapping for photo-thermosensitivity (PTS) in rice under two environments (Beijing and Hangzhou). Left: Days-to-heading (DTH) distribution in Zhaiyeqing 8/Jingxi 17 DH population planted in Beijing and Hangzhou. Top right: PTS distribution in the population when PTS was measured by the difference of DTH in the two environments divided by the DTH in Beijing. Bottom right: QTL identified for DTH in Beijing and Hangzhou and for PTS (*LOD > 2.4).

day lengths, and loci associated with photo-period sensitivity were identified based on the delay in flowering under the 14 h photoperiod (DTF at 14 h – DTF at 10 h). In total, 15 QTL were associated with DTF. Only four of them were also identified as influencing response to photoperiod. None of these QTL is allelic to the PTS QTL on chromosome 7.

Different QTL are identified using direct and relative trait values and, in rice, DTH and photoperiod are controlled by different QTL. On the other hand, direct and relative traits share some QTL. This means that DTH and photoperiod sensitivity are genetically related to some extent. This is because both traits are related to the basic vegetative growth that rice plants must achieve in order to flower. There are QTL mapping studies undertaken in NIEs, but QTL were mapped using trait values scored in each environment rather than using relative measures. The traits themselves were mapped rather than the relative response measured under the NIEs.

Heterogeneous environments

QTL have been studied in multiple environments with many factors being different (heterogeneous environments). When the same mapping population is phenotyped in different environments, some QTL could be detected in one environment but not in others. For the convenience of comparison, rice QTL mapped in two environments were selected for sharing analysis (Table 9.1). A total of 159 QTL were identified in ten QTL mapping reports for 11 categories of quantitative traits. For different traits, QTL-sharing frequencies between two environments range from 9.5% for drought avoidance to 52.9% for 1000-grain weight, and, for all traits, on the average, 46 (30%) of them are shared or common between two environments. Mean variances explained by QTL identified for each category of traits range from 7.1% for panicle number per plant to 24.4% for flood tolerance. For all shared QTL, mean variance explained is 16.7%, whereas for the unshared QTL,

Table 9.1. Comparison of QTL mapped in two environments using the same populations.

| Trait | Number of QTL | | Mean VE | | | |
	Total	Shared (%)	Total	Shared (%)	Unshared	Reference
Yield	15	2 (13.3)	8.7	12.8	8.1	Yu et al., 1997
Panicle per plant	7	3 (42.9)	7.1	6.7	7.4	Yu et al., 1997; Li et al., 2000
Grain per panicle	16	4 (25.0)	11.7	12.9	11.3	Yu et al., 1997; Li et al., 2000
1000-Grain weight	17	9 (52.9)	12.7	14.0	11.1	Yu et al., 1997; Li et al., 2000
Root	30	9 (30.0)	11.8	15.0	10.4	Ali et al., 2000
Drought avoidance	21	2 (9.5)	9.8	10.2	9.8	Courtois et al., 2000
Flood tolerance	12	3 (25.0)	24.4	48.8	16.3	Sripongpangkul et al., 2000
Aluminium tolerance	4	2 (50.0)	12.5	16.0	9.0	Wu et al., 2000
Disease resistance	17	7 (41.2)	10.4	11.0	10.1	Tang et al., 2000; Zou et al., 2000
Seedling vigour	13	3 (23.1)	16.0	19.5	14.9	Redoña and Mackill, 1996
Paste viscosity	7	2 (28.6)	19.3	37.7	11.9	Bao et al., 2000
Total	159	46 (30.0)	12.6	16.7	10.9	

VE, Variance explained.
Traits in each category: Yield – grains (t ha^{-1}); Root – root number, root length and thickness; Drought avoidance – leaf rolling and relative water content; Flood tolerance – initial plant height, plant-height increment, internode increment and leaf-length increment; Seedling vigour – shoot length, root length, coleoptile length and mesocotoyl length; Paste viscosity – peak viscosity, hot-paste viscosity and cool-paste viscosity.

it is 10.9%. Major-gene-related QTL (for flooding tolerance and paste viscosity) had the highest QTL-sharing frequency. QTL with large effect (higher proportion of the variance explained) are shared more frequently.

When compared across three or more environments, QTL-sharing frequencies become lower. For example, a total of 22 QTL for six agronomic traits were identified in Zhaiyeqing 8/Jingxi 17 DHs, only seven of which were shared in all three tested environments (Lu et al., 1996). In three trials using Tesanai 2/CB F$_2$ and its two equivalent F$_3$s, eight QTL were identified, two of which were detected in all three trials (Zhuang et al., 1997). In another report, three of 11 QTL identified for leaf rolling were shared in the three trials with different drought-stress intensities (Courtois et al., 2000).

In QTL mapping, permanent populations, such as DHs and RILs, are now used more often because of their inherent advantages of providing permanent DNA supply and phenotyping opportunities for many different studies. In rice, there are 14 permanent populations reported so far (Table 9.2), including two DH, nine RIL and one BIL populations, with population sizes of 65 to 315. Genetic maps used for QTL studies consist of 113 to 399 molecular markers. These populations have been used in up to 15 research projects for 161 trial-trait combinations, with 682 QTL reported. One of the DH populations, IR64/Azucena, has been shared internationally for the mapping of many agronomic traits, disease resistance, cold tolerance and water-stress tolerance. Two other widely used populations are Zhaiyeqing 8/Jingxi 17 DHs and Lemont/Teqing RILs. Allelic differences and genetic polymorphism, however, are limited among these populations, since each cross has only two alleles segregating at each polymorphic locus.

Table 9.2. Rice permanent populations used in genetic mapping.

Population	Population size	No. of markers	No. of traits	No. of QTL
1. IR64/Azucena DH	105–135	146–175	56	215
2. Zhaiyeqing 8/Jingxi 17 DH	132	137–243	35	115
3. 9024/LH422 RIL	194	141	25	74
4. CO39/Moroberekan RIL	143–281	127	14	121
5. Lemont/Teqing RIL	255–315	113–217	8	46
6. IR58821/IR52561 RIL	166	399	5	28
7. IR74/Jalmagna RIL	165	144	5	18
8. Nipponbare/Kasalath BIL	98	245	4	19
9. Zhenshan 97/Minghui 63 RIL	238	171	3	6
10. Asominori/IR24 RIL	65	289	2	17
11. Acc8558/H359 RIL	131	225	1	11
12. IR1552/Azucena RIL	150	207	1	4
13. IR74/FR13A RIL	74	202	1	4
14. IR20/IR55178-3B-9-3 RIL	84	217	1	4
Overall	65–315	113–399	161	682

References
1: Lorieux *et al.*, 1996; Ghesquière *et al.*, 1997; Huang *et al.*, 1997; Taguchi-Shiobara *et al.*, 1997; Yadav *et al.*, 1997; Alam and Cohen, 1998; Albar *et al.*, 1998; Pressoir *et al.*, 1998; Wu *et al.*, 1998; Yan *et al.*, 1998a,b, 1999; Courtois *et al.*, 2000; Hemamalini *et al.*, 2000; Zheng *et al.*, 2000. 2: Lu *et al.*, 1996; He *et al.*, 1998, 1999; Zhang *et al.*, 1999; Bao *et al.*, 2000; Qian *et al.*, 2000a,b; Gong *et al.*, 2001; Liu *et al.*, 2001. 3: Xiao *et al.*, 1995, 1996. 4: Wang *et al.*, 1994; Champoux *et al.*, 1995; Lilley *et al.*, 1996; Ray *et al.*, 1996; Maheswaran *et al.*, 2000. 5: Li *et al.*, 1995a,b, 1997, 1999; Tabien *et al.*, 2000. 6: Ali *et al.*, 2000. 7: Sripingpangkul *et al.*, 2000. 8: Taguchi-Shiobara *et al.*, 1997; Lin *et al.*, 1998. 9: Tan *et al.*, 2000. 10: Sasahara *et al.*, 1999. 11: Tang *et al.*, 2000. 12: Wu *et al.*, 2000. 13: Nandi *et al.*, 1997. 14: Ni *et al.*, 1998.

QTL × E interaction

QTL can be studied under adverse environments (abiotic stress), NIEs or a uniform environment by replicating DH or RIL populations and splitting tillers or ratooning a segregating population. When two or more environments are involved, QTL– environment (QTL × E) interaction can be estimated from a complete analysis of variance (ANOVA), $QTL_i + E_j + QTL \times E_{ij}$, where significant QTL × E interactions are assessed from the significance or lack of significance of $QTL \times E_{ij}$ interaction terms. However, QTL × E interaction has been predicted by comparing the QTL detected separately in different environments in many crops, including rice (Lu *et al.*, 1996; Zhuang *et al.*, 1997). That a QTL can be detected in one environment but not in others, as discussed earlier, could result from experimental noise, sampling error or experimental error, and thus does not

necessarily indicate QTL × E interaction. As indicated by Jansen *et al.* (1995), the chance for simultaneous detection of QTL in multiple environments is small. On the other hand, sharing QTL among environments does not necessarily mean no QTL × E interaction. This is supported by the fact that QTL × E interaction was identified for some sharing QTL by incorporating $QTL \times E_{ij}$ into QTL analysis (Yan *et al.*, 1999) and by the fact that QTL effects estimated across environments could be very different.

QTL Across Genetic Backgrounds

Homogeneous genetic backgrounds

Populations developed for QTL analysis can be very heterogeneous in genetic backgrounds, with hundreds or thousands of genes segregating simultaneously, or very homogeneous, with only a target gene segregating. Homogeneous or isogenic backgrounds can be created through one of the following five approaches.

1. Back-cross-derived NILs. NILs are generated by introgression, and the resulting inbred lines differ at the targeted locus or region. Introgression is accomplished by repeatedly back-crossing one line carrying a gene of interest (donor parent) to another line that has other desirable properties (recurrent parent).

2. Selfing-derived NILs. NILs are derived through continuous selfing while keeping the target trait locus heterozygous. Once other genetic backgrounds are almost all fixed, an additional generation of selfing will result in a pair of NILs that differ only at the target locus (Xu and Zhu, 1994). Selfing-derived NILs, however, can be any combination of parental genotypes, whereas the back-cross-derived NILs have the same genetic constitution as the recurrent parent.

3. Whole-genome selection of permanent populations. With the availability of permanent mapping populations, such as RILs and DHs, it is possible to find two of them that, except for one or a few marker loci, are almost genotypically identical for the whole genome.

4. Mutation. The creation of a collection of single-locus mutants is a quick approach to producing a large number of NILs. For most mutants, mutation only happens at one or few genetic loci. These mutants can be considered near-isogenic to their 'wild type' and are thus called isomutagenic lines (IMLs).

5. Chromosome substitution. Through chromosome engineering and/or marker-assisted selection (MAS), whole or partial chromosome-substitution lines can be created, so that each line has one chromosome or partial chromosome replaced.

Genetic materials, such as NILs with homogeneous backgrounds, have been used in many different investigations. If NILs are used, interaction between the target QTL and other major genes/QTL can be eliminated and only epistasis between multiple target QTL needs to be considered. With removal of noise from heterogeneous backgrounds, the proportion of variance explained by the target QTL will increase and minor QTL can be identified. Without disturbance from the background effect, multiple QTL can be easily partitioned. Since all genotypic variation comes from the target loci, environmental effects can be estimated. In QTL cloning, NILs have been used to map the target QTL precisely by using all of these advantages.

Heterogeneous genetic backgrounds

Although the genetic distances and order of DNA markers are comparable among very different rice crosses (Antonio *et al.*, 1996), QTL mapping using different populations derived from the same cross has identified very different QTL. Only some QTL are common across populations of different structures, such as DHs and RILs derived from a single cross (He *et al.*, 2001), where there is an identical set of genes segregating. Heterogeneous genetic backgrounds can also come from various crosses derived from different varieties, subspecies, species and families. Genetic materials with heterogeneous genetic backgrounds can be used to estimate epistasis, detect non-allelic QTL, discover multiple alleles and identify paralogous and orthologous QTL. Molecular markers developed in rice have been selected as anchor markers for cross-mapping in cereals. The use of anchor markers has enabled the detection of possible orthologous QTL by comparing QTL across cereals or the construction of phylogenetic relationships. Although it is unclear how many claimed orthologous QTL are real, detection of QTL that are common across cereals at least indicates that the same QTL could be identified from very different genetic backgrounds.

As a contribution to complicated genetic backgrounds, many quantitative traits *per se* are a complex consisting of several components or subtraits. In rice, for example, polygenic sterility can be partitioned into several components, including male and female sterility, or ovary and pollen abortion, so that polygenes can be divided into several components with different functions and thus can be handled with ease. Genetic backgrounds in a population can also be complicated by the contribution of other related traits. Most QTL reported in the literature are based on separate single-trait analyses. Joint analysis of multiple traits will improve QTL mapping power and precision and provide the capability of testing linkage versus pleiotropy, where QTL in apparently the same region affect two distinct traits (Jiang and Zeng, 1995). That these methods have not been widely adopted is probably a reflection of their relative statistical complexity. Use of multitrait analysis should expand as suitable software becomes more widely available.

Epistasis

The importance of epistasis to the genetic control of QTs has been debated. There are two major reasons for contradictory reports or infrequent discovery of QTL interactions, especially at the early stage of QTL mapping. First, contradictory reports may result from the fact that QTL mapping studies and analytical methods have not been able to detect epistasis and thus the conclusions could be biased, preferentially identifying genes that have large effects and/or act independently (Xu, 1997). This argument is supported by the results that QTL with large effects are detected in very different crosses and environments. The second reason is that ordinal QTL analysis was made with populations segregating for the whole genome simultaneously, so that it may be difficult to detect an interaction in a specific combination of QTL genotypes. For example, Yano *et al.* (1997) predicted an interaction between the two largest QTL, *Hd1* and *Hd2*, for heading date. But the existence of another QTL, *Hd6*, and its interaction could not be detected in their primary population (F$_2$), where many epistatic interactions could exist in so-called minor QTL. Successful examples of the detection of epsitatic interactions by using

primary populations seem to be related to population sizes and structures, QTs, the number of existing QTL and QTL effects. The more QTL involved, the more difficult is the detection of significant differences for individual QTL. Although using a large population size may help to detect epistatic interactions, it increases experimental errors, due to increasing difficulty in managing such a population effectively. To improve confidence, different types of plant materials have been constructed. A series of chromosomal substitution lines or NILs with QTL (QTL-NILs) have been developed, and the gene actions of QTL have been analysed in detail.

Alleles at multiple loci

When multiple QTL control a trait, their alleles of positive or negative effect (increasing or decreasing trait value) tend to be dispersed between parents, each with positive alleles at one or some loci but negative alleles at other loci. These dispersed alleles can be cryptic transgressive, which can be found even in parents with similar phenotypes. For example, genotypes *AAbb* and *aaBB* have the same phenotype when a trait is determined by two unlinked loci with additive effect. If these two genotypes were mated to each other and F$_1$ progeny selfed, the resulting F$_2$ progeny array would span the entire range of phenotypes, from some genotypes (*AABB*) with higher phenotypic values than either parent to others (*aabb*) with lower phenotypic values than either parent. It is such extreme transgressive individuals that are often of greatest value in breeding. In rice, dispersed alleles for tiller angle were identified in four varieties with similar phenotype. Association of these alleles was realized by selection of phenotypic extremes in their F$_2$s, resulting in fixed transgression (Xu *et al.*, 1998). In QTL mapping, phenotypic difference between parents is not necessary for the detection of QTL. In most cases where no parental difference is found, QTL are still detected, which could be due to the

Table 9.3.   Plant-height QTL identified across rice populations and corresponding major genes.

QTL	Major gene	Linked marker	Population	Reference
qPH1-1	sd-1	RZ730-RZ801	Zhaiyeqing 8/Jingxi17 DH	Lu et al., 1996
			IR64/Azucena DH	Yan et al., 1999
		RZ730-RG381	Palawan/IR24 F2	Wu et al., 1996
		RZ649-RG374	Tesanai 2/CB F2	Zhuang et al., 1997
		OSR23	IR64/TOG5681//IR64	Lorieux et al., 2000
		RG331	CO39/Moroberekan RIL	Huang et al., 1996
qPH1-2		RZ776-RG375	9042/LH422 RIL	Xiao et al., 1996
		RZ744	CO39/Moroberekan RIL	Huang et al., 1996
qPH1-3	d-18	RG612	CO39/Moroberekan RIL	Huang et al., 1996
qPH2-1		RZ213	IR64/Azucena DH	Yan et al., 1999
qPH2-2	d-5	RG654-RZ260	Lemont/Teqing RIL	Li et al., 1995a
		RG256-RG324	Tesanai 2/CB F2	Zhuang et al., 1997
			Tesanai 2/CB F3	Zhuang et al., 1997
		Amy1C-RG95	IR64/Azucena DH	Yan et al., 1999
		RZ213-RG95	Waiyin/CB F2	Huang et al., 1996
qPH2-3	d-32	RZ166	Palawan/IR24 F2	Wu et al., 1996
qPH2-4	d-30	RG157	Waiyin/CB F2	Huang et al., 1996
qPH2-5		RZ599-RG544	9042/LH422 RIL	Xiao et al., 1996
			9024/LH422 RIL//9042	Xiao et al., 1995
qPH3-1		RG418-RM148	IR64/TOG5681//IR64	Lorieux et al., 2000
		RG418-RG910	IR64/Azucena DH	Yan et al., 1999
qPH3-2		G62-G144	Zhaiyeqing 8/Jingxi17 DH	Lu et al., 1996
		XNbp249-RZ16	9042/LH422 RIL//9042	Xiao et al., 1995
			9024/LH422//LH422	Xiao et al., 1995
qPH3-3	d-56	RG348-RZ329	IR64/Azucena DH	Yan et al., 1999
		RG348-RG944	Lemont/Teqing RIL	Li et al., 1995a
		RG348-RG409	Tesanai 2/CB F2	Zhuang et al., 1997
			Tesanai 2/CB F3	Zhuang et al., 1997
		RG104	CO39/Moroberekan RIL	Huang et al., 1996
qPH4-1	d-31	RG143-RG620	Tesanai 2/CB F2	Zhuang et al., 1997
qPH4-2	d-11	RG163-RZ590	IR64/Azucena DH	Yan et al., 1999
		G513-G271	Zhaiyeqing 8/Jingxi17 DH	Lu et al., 1996
		RG864	CO39/Moroberekan RIL	Huang et al., 1996
qPH5-1	sdg	RG182-RG9	Tesanai 2/CB F2	Zhuang et al., 1997
qPH5-2		RG480-RG697	9024/LH422 RIL//9042	Xiao et al., 1995
			9024/LH422 RIL//LH422	Xiao et al., 1995
			9042/LH422 RIL	Xiao et al., 1996
qPH6-1	d-58	RZ682-RG653	9024/LH422 RIL//9042	Xiao et al., 1995
			9024/LH422 RIL//LH422	Xiao et al., 1995
		RZ682-CDO544	9042/LH422 RIL	Xiao et al., 1996
qPH6-2	d-9	RG64	CO39/Moroberekan RIL	Huang et al., 1996
		RZ667-RG648	Waiyin/CB F2	Huang et al., 1996
qPH7-1		RG351	CO39/Moroberekan RIL	Huang et al., 1996
qPH7-2		RG146-CDO497	9042/LH422 RIL	Xiao et al., 1996
qPH7-3		C285-G20	Zhaiyeqing 8/Jingxi17 DH	Lu et al., 1996
qPH7-4		RG528	CO39/Moroberekan RIL	Huang et al., 1996
qPH8-1		RG333-RZ562	9024/LH422 RIL//9042	Xiao et al., 1995
			9024/LH422 RIL//LH422	Xiao et al., 1995
			9042/LH422 RIL	Xiao et al., 1996
		RG20-RG1034	Lemont/Teqing RIL	Li et al., 1995a
		RG885-RZ617	Zhaiyeqing 8/Jingxi17 DH	Lu et al., 1996
		RZ562-RG978	Tesanai 2/CB F2	Zhuang et al., 1997
qPH8-2		G1073-RG1	Zhaiyeqing 8/Jingxi17 DH	Lu et al., 1996
qPH8-3		Amy3D-RZ66	IR64/Azucena DH	Yan et al., 1999

Table 9.3. *Continued.*

QTL	Major gene	Linked marker	Population	Reference
qPH9-1		RZ9106-RZ777	Lemont/Teqing RIL	Li *et al.*, 1995a
		RZ422-*Amy3ABC*	IR64/Azucena DH	Yan *et al.*, 1999
qPH10-1		G1082-G291	Zhaiyeqing 8/Jingxi17 DH	Lu *et al.*, 1996
		RZ625-CDO93	IR64/Azucena DH	Yan *et al.*, 1999
		RG257	CO39/Moroberekan RIL	Huang *et al.*, 1996
qPH11-1	d-27	RG118	Tesanai 2/CB F3	Zhuang *et al.*, 1997
qPH12-1	d-33	RG869	CO39/Moroberekan RIL	Huang *et al.*, 1996
qPH12-2		RG235-RG341	Tesanai 2/CB F2	Zhuang *et al.*, 1997
		RG574	CO39/Moroberekan RIL	Huang *et al.*, 1996

The same QTL locus name is designated to the linked markers located within a 15 cM region. The plant-height QTL is named '*qPH*' plus chromosome numbers (the first number after '*qPH*') and locus number (the second number after '*qPH*'). The corresponding major genes are determined based on genetic linkage between these genes and molecular markers (Huang *et al.*, 1996; Kamijima *et al.*, 1996).

complementary patterns of positive and negative allelic effects.

As observed in QTL mapping, on the average, about four QTL are identified for each trait (Tables 9.1 and 9.2), the same as the average obtained for 176 trial-trait combinations as reviewed by Kearsey and Farquhar (1998). When QTL identified for the same trait are summarized over different projects/populations, this number becomes much larger. For example, plant height has been mapped, using 13 populations, with 63 QTL reported. Some of the QTL are allelic to each other, i.e. they were mapped to the same chromosomal region or intervals of less than 15 cM. After elimination of possible allelic QTL, the total number of QTL for plant height is reduced to 29, with up to five QTL existing on a chromosome (Table 9.3). The QTL *qPH1-1*, which corresponded to a major semidwarf gene *sd-1*, and *qPH8-1* were each detected in six populations. QTL *qPH2-2* and *qPH3-3* were each detected in five populations. Over 50 major genes for dwarf and semidwarf mutants have been found (Kinoshita, 1995), and 14 of them have been linked to molecular markers (Huang *et al.*, 1996; Kamijima *et al.*, 1996), with 13 of them (93%) colocalized with plant-height QTL. More plant-height QTL will probably be colocalized with major loci, as more major loci are linked with molecular markers. These colocalizations support Robertson's (1985) hypothesis that alleles for qualitative mutants are simply 'lost-function' alleles at the same loci underlying quantitative

variation. Until QTL are mapped to higher degrees of precision and/or cloned, however, it would be difficult to prove that the particular QTL actually correspond to known loci defined by macromutant alleles and which QTL are allelic to each other. The QTL allelism test and the determination of the major-gene and QTL correspondence depend on the availability of high-density molecular maps with a common set of markers shared among researchers.

Multiple alleles at a locus

Two-parent derived populations in crops like rice usually have only two alleles segregating at each locus. Identification of multiple alleles requires comparison of populations derived from different crosses. To distinguish QTL alleles identified in one cross from those in another, all mapped alleles must be accurately sized and documented.

As an example of multiple alleles at a locus, rice amylose content, mainly controlled by the *wx* gene, will be discussed. Wide variation in amylose content occurs and varieties with different amylose content, varying from waxy (0–2%), very low (3–9%), low (10–19%) and intermediate (20–25%) to high (> 25%), have been selected in breeding programmes. Conventional genetic studies using varieties with different amylose content revealed transgressive segregation in F_2s in almost all possible parental combinations

(Pooni *et al.*, 1993). Recently, a polymorphic microsatellite was identified in the *wx* gene (Bligh *et al.*, 1995), located 55 bp upstream of the putative 5' leader-intron splice site. Ayres *et al.* (1997) determined the relationship between polymorphism at that locus and variation in amylose content. Eight *wx* microsatellite alleles were identified from 92 long-, medium- and short-grain US rice cultivars, which explained 85.9% of the variation. The amplified products ranged from 103 to 127 bp in length and contained $(CT)_n$ repeats, where *n* ranged from 8 to 20. Average amylose content in varieties with different alleles varied from 14.9% to 25.2%. Using more diverse rice germ-plasm accessions (*n* = 243), Zeng *et al.* (2000) identified 15 alleles at the *wx* locus, using microsatellite class and G–T polymorphism, resulting in a total of 16 alleles identified so far. Now the question is whether the multiple alleles identified at the waxy locus can be associated with QTL alleles and whether the case can be extended to other traits or genetic loci.

Using molecular markers with multiple alleles in QTL mapping will help identify multiple QTL alleles. QTL studies using different populations have identified some common QTL. It is necessary, however, to further clarify whether they identified common or different alleles at these QTL. Reporting the sizes of associated alleles and using allele-rich markers in QTL studies will provide the information required for clarification, with the assumption that each marker allele has a corresponding QTL allele.

QTL Across Growth and Developmental Stages

Measures of agricultural productivity usually reflect the effects of many genes acting at different times during the period of growth and development of an organism. Genetic expression of QTs varies greatly with development stage, and some QTL may be turned on or off at specific stages or may respond to environmental changes over different stages (Xu, 1997). The developmental genetics of

QTs has been studied using conventional quantitative genetics. In rice, for example, genetic analysis of tiller number was made at different growth stages (Xu and Shen, 1991). Using six indica rice varieties and their diallel F_1s, tiller number was counted at a 10-day interval after transplanting. An identical polygenic system was found to be responsible for tillering ability at different growth stages. For the terminal character, productive tiller number, gene effects changed with growth stages. Non-additive gene action and environmental effect decreased, but additive gene action increased with the progressive development of plants.

Most QTL studies have so far been focused on trait values measured at a specific stage or the final growth stage. This static mapping strategy could not fully reveal the action of genes during the development of traits. To understand genetic expression at different developmental stages, dynamic mapping has been proposed (Xu, 1994, 1997; Xu and Zhu, 1994). There are three approaches to dynamic analysis or time-related mapping. One is based on the analysis of trait values measured at each observation time (Bradshaw and Settler, 1995; Plomion *et al.*, 1996; Price and Tomos, 1997; Verhaegen *et al.*, 1997), from which the accumulated effect of a QTL, from the beginning of ontogenesis to each observation time, can be estimated. This is called effect–accumulation analysis or unconditional QTL mapping (Yan *et al.*, 1998a). The second approach is to analyse trait-value increments observed at sequential time intervals (Bradshaw and Settler, 1995; Plomion *et al.*, 1996; Verhaegen *et al.*, 1997), from which the incremental or net effect of a QTL at each time interval can be estimated. This is called effect-increment analysis or conditional QTL mapping (Yan *et al.*, 1998a). Phenotypic data collected at different growth stages or time intervals can be analysed either separately or jointly. Compared with separate analysis, joint analysis can synthesize all the information from different times or time intervals to give a comprehensive estimate of each QTL position, according to which a corresponding complete expression (or expression rate) curve of each QTL can be estimated (Wu *et al.*, 1999). In practice, both separate and joint analyses should be conducted. A third approach to looking at a QTL over time is to do a multivariate analysis based on fitting the parameters of the growth curve (animal breeders call this general approach 'random regression').

Using IR64/Azucena DHs, Yan *et al.* (1998a,b) studied the developmental characteristics of QTL for tiller numbers and plant height by conditional and unconditional interval mapping, in combination with phenotyping these traits every 10 days after transplanting. They concluded that many QTL identified at early stages were undetectable at the final stage. Conditional mapping identified more QTL than unconditional mapping. Temporal patterns of gene expression changed with developmental stages. Genes at a specific genomic region might have opposite genetic effects at various growth stages. For chromosomal regions significantly associated with plant height, conditional QTL were found only at one to several specific periods and no QTL for plant height was continually active during the entire period of growth.

QTL in Association Genetics

The breadth of genetic information from thousands of DNA polymorphisms and the depth of phenotypic measure hold promise for identifying marker–trait correlation. Allele association between marker loci and association between marker alleles and phenotypes can be designated as marker–marker association and marker–trait association, respectively. In humans, genetic-association studies have been used to assess correlations between variant genotypes and trait phenotypes on a 'population' scale (or a group of human beings). The power of association analysis to detect genetic contribution to complex diseases is considered to be much greater than that of linkage studies (Risch, 2000). At a fundamental level, both genetic association and linkage rely on the coinheritance of adjacent DNA variants, with linkage capitalizing on this by identifying haplotypes that are inherited intact

over several generations and association relying on the retention of adjacent DNA variants over many generations. Thus, association studies can be regarded as very large linkage studies of unobserved, hypothetical pedigrees (Cardon and Bell, 2001). Recombination is the primary force that eliminates linkage and association over generations.

In theory, marker–trait association can be established based on either the phenotypic differences associated with alternative marker genotypes (Lander and Botstein, 1989) or on the difference of allele frequencies between phenotypic extremes in a derived population (Lebowitz *et al.*, 1987). Differences in both phenotype and allele frequency can be identified in a group of cultivars that are derived from a common ancestral gene pool (Xu and Zhu, 1994). The procedure is regarded as an initial screening for identification of QTL (Bar-Hen *et al.*, 1995; Virk *et al.*, 1996). The development of saturated linkage maps and highly informative microsatellite markers in rice makes it possible to systematically survey marker–trait association on a whole-genome scale. Compared with transmission-based genetic mapping, association-based mapping provides more opportunities for breeding applications, since hundreds of germ-plasm accessions that are useful as parents in breeding are involved. It is worth determining whether this mapping strategy will be useful in agricultural species, such as inbred plants, where pedigree information is widely available but inbreeding is not strictly controlled, resulting in genetic impurities in putative inbred lines. Attempts have been made in rice to detect marker–trait association based on the use of unmapped randomly amplified polymorphic DNA (RAPD) markers (Virk *et al.*, 1996). As more and more germ-plasm accessions are evaluated with molecular markers and phenotyped for agronomic traits, it is essential to consider using the association-based approach to map genes or at least to provide a prescreen for linkage-based genetic mapping.

As an example, association genetics was used in rice to reveal marker–marker association and marker–trait association (Y. Xu and S.R. McCouch, Cornell University, unpublished data). Based on genetic diversity and ancestral relationship, 237 rice accessions collected from around the world were classified as all rice accessions, US rice varieties, Cypress-pedigree-related varieties and worldwide complex. With genotypic data for 100 RFLP and 60 SSR marker loci and phenotypic data for 12 traits, a stronger marker–marker association was found in the varietal groups that had greater genetic variation or closer pedigree relationship, as revealed by correlation coefficients between allele sizes of markers. Markers within linkage groups showed stronger allelic association than markers between linkage groups, indicating that marker–marker association in rice germ-plasm was influenced by genetic linkage to some extent. The statistical associations, however, could not be interpreted solely from genetic linkage. Comparison of marker–trait association in different varietal groups demonstrated that both phenotypic variation and pedigree relationship among rice accessions strongly influenced the association detection. A highly consistent allele–trait association was revealed among multiple alleles at a given locus. Marker–trait associations identified were compared with the markers genetically linked with major genes and QTL. Although more work is needed for accurate correspondence among classical genetic loci, reported QTL and marker–trait associations, several chromosomal regions as hot spots for marker–trait associations have been assigned to QTL clusters.

Evidence of allelic association or marker–trait association does not always imply that two loci are linked. A spurious association can be generated by random genetic drift, founder effect, mutation, selection and/or population admixture and stratification (Sham, 1997). Human geneticists account for any potential population substructure/stratification by using a transmission–disequilibrium test (TDT) (Lynch and Walsh, 1998). Recombination tends to erode linkage disequilibrium, and the erosion is slow between closely linked loci. For example, for loci that are 1 cM apart, more than 50% of the initial disequilibrium remains after 50 generations (Falconer and Mackay,

1996). Therefore, putative associations require confirmation based on analysis of multiple samples from genetically isolated groups or relatively homogeneous groups of germ-plasm or by using 'linkage genetics' with a standard population derived from two known varieties.

For several reasons, there is great enthusiasm at present about the promise of association studies for uncovering the genetic components of complex traits in humans: dense SNP maps across the genome, elegant, high-throughput genotyping techniques, simultaneous comparison of groups of loci, statistical measures for assessing genome-wide significance and phenotypic insights that might accompany comparative genomic studies among different human groups. All these conditions have already been or will be satisfied and association studies will inevitably proliferate in plants like rice. Now it is the time to consider critically the design of such studies. In molecular breeding, there is increasing demand for the establishment of molecular profiles for each germ-plasm accession, so that specific germ-plasm accessions can be selected based on breeding purposes. In this process, a large number of germ-plasm accessions will be genotyped with hundreds of molecular markers. This becomes feasible with the development of highly informative DNA markers and high-throughput genotyping technology. Several institutions have started to profile rice germ-plasm on a large scale and a huge amount of data is being generated. It is apparent that molecular marker-based germ-plasm evaluation will produce a large data set that can be explored for association-genetics or linkage-disequilibrium study.

QTL in Breeding Programmes

There is every reason to believe that plant breeding in the 21st century will still depend, to a great extent, on conventional methods for phenotypic selection. Molecular biology could help improve recombinant frequency for favourable alleles and select the traits that are not measurable under normal environments with conventional methods. Using molecular markers in plant-breeding programmes has been discussed elsewhere (Beckmann and Soller, 1986; Paterson *et al.*, 1991; Dudley, 1993; Stuber, 1994; Xu and Zhu, 1994; Lee, 1995; Hospital and Charcosset, 1997; Xu, 1997; Mackill and Ni, 2001). In this section, discussion will focus on how genetic mapping will improve our breeding process for some special traits.

Selection without test crossing and/or a progeny test

In plant breeding, many traits need test crossing and a progeny test for unambiguous identification. Typical examples in rice include male-sterility restorability, wide compatibility, heterosis, combining ability, outcrossing ability and a recently named trait in plants, loss of heterozygosity (Wang *et al.*, 1999). In test crossing, each candidate plant will be crossed to testers and then its genotype will be inferred from a progeny test in the next season. Each candidate plant must be harvested and maintained separately and only the plants with the target trait will advance to the next level. Test crossing may continue for several generations until the selected plants reach a certain level of homozygosity. In back-crossing programmes handling recessive traits, an additional selfing generation is required for the progeny test in order to make sure that the plants used for back-crossing contain the target gene. Using MAS, test crossing and/or a progeny test can be eliminated, since the target trait can be identified from the candidate plant itself, based on DNA-marker analysis, saving laborious test crossing and time-consuming progeny tests.

Selection independent of environments

Many traits must be screened in specific or controlled environments where they can be fully expressed. For example, photoperiod or temperature sensitivity can only be identified by comparison of their phenotypes

in two distinct photoperiod or temperature conditions, as discussed earlier. For identification of insect/disease resistance, plants must be inoculated artificially or naturally. For abiotic resistance, such as salinity and submergence tolerance and lodging resistance, selection in traditional breeding programmes can only be done when the specific stress is present. To measure responses to agrochemicals, such as herbicides and plant-growth regulators, these chemicals must be applied to plants at the right stage under suitable environments. MAS has made it possible to perform indirect selection for these traits, using tightly linked molecular markers.

Selection without laborious fieldwork or intensive laboratory work

Many important traits are phenotypically invisible or unscorable by visual observation and must be measured in the laboratory using sophisticated equipment or facilities, or a large number of samples is required, which means that it cannot be measured until late generations when a relatively large amount of seed becomes available for each selection entry. Grain chemical and physical properties are examples that fall under this category. Traits such as tissue culturability need laborious laboratory work for testing each sample. Using MAS, a piece of leaf harvested at any growth stage of plants will be enough for accurate measurement of all the traits mentioned above, once closely associated markers have been identified.

Selection at an early breeding stage

Traits that are only measurable at or after the reproductive stage would be good candidates for MAS. For example, grain quality can only be tested using mature seeds. Yield heterosis and yield potential must be measured after harvest and/or in advanced generations. MAS can be made at any stage and in any generation, so that breeders do not need to maintain a large number of candidate plants generation after generation.

Selection for multiple genes and/or multiple traits

In some cases, multiple pathogen races or insect biotypes must be used to identify plants for multiple resistances, but, in practice, this may be difficult or impossible, because different genes may produce similar phenotypes that cannot be distinguished from each other. Marker–trait association can be used to select multiple resistances simultaneously and transfer them into a single line.

Consider selection for multiple traits – for example, temperature-sensitive genic male sterility (TGMS), amylose and wide compatibility. Candidate plants must be tested under two different environments where TGMS can be identified. Each plant must be test-crossed with wide-compatibility testers, following up with a progeny test in the next season. At the same time, a large amount of seed must be harvested for amylose measurement. Thus, using conventional selection methods, we must wait until a large number of seeds are available and a reasonable level of homozygosity is reached. For all these traits, in MAS, one just needs a piece of leaf harvested at any growth stage in any segregating generation.

Whole-genome selection

MAS can also be practised at the whole-genome level. Whole-genome selection can be used to eliminate the donor genome in back-cross breeding or to get rid of linkage drag when a wide cross is involved. Compared with a back-cross programme, which usually takes five to seven generations to recover most recurrent parental backgrounds, MAS may save two to four back-cross generations in the transfer of a single target allele (Tanksley et al., 1989; Hospital et al., 1992; Fisch et al., 1999). Combined with MAS for multiple traits, on the other hand, whole-genome selection allows the breeder to transfer multiple traits through back-crossing simultaneously. QTL mapping precision is now good enough for elimination of most linkage drag that conventional methods cannot achieve efficiently. As a practical consideration in MAS, molecular markers used in MAS should have less requirement for DNA quality and quantity and be highly informative, replicable and easy to use (Mackill and Ni, 2001). To reduce the false positives in MAS, markers must be tightly linked to the target trait, and flanking markers or multiple markers around the region could be used simultaneously. Currently, the cost for MAS is still too high and only very few genes are finely mapped. The development of highly informative molecular markers, such as SNP, and high-throughput technology will finally overcome these limitations and accelerate fine-mapping and MAS. QTs are greatly affected by genetic backgrounds (gene interaction) or environmental factors or both, i.e. gene–environment (G × E) interaction (Fig. 9.2). Because of environmental influence, different genotypes may have a very similar phenotype, whereas different phenotypes may come from an identical genotype. Continuous variation could result either from the segregation caused by the interaction of a major gene/QTL with compound environmental factors or from a combining effect of many minor QTL. On the contrary, discrete variation could result either from the segregation of a major gene or from a minor QTL with well-controlled environments. Partition of a complex environment into single independent and measurable components can be very helpful. The levels of QT expression among individuals can be maximized by exposing the population to appropriate environments.

The effective utilization of molecular-marker technology and QTL management in breeding programmes depends on tight linkage between markers and QTL (Dudley, 1993). Multiple QTL must be manipulated simultaneously to have a significant impact. The QTL × E interaction and epistasis need to be localized and quantified. Otherwise, the genomic region involved in the

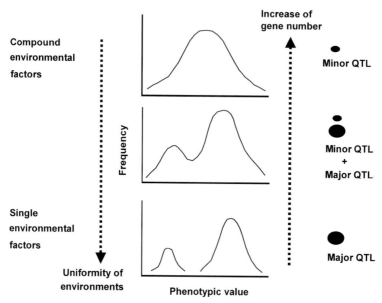

Fig. 9.2. Relationship between phenotypes, genes and environments. Discrete phenotypic distribution for qualitative traits arises from major genes, bimodal distribution for qualitative–quantitative traits from the joint effect of a major QTL (with dominant effect) and some minor QTL, and normal distribution for typical quantitative traits from many minor QTL. With partition and uniformity of environments, some continuously distributed traits can be converted into a bimodal distribution of discretely distributed traits (after Xu, 1997).

interactions cannot be incorporated in the selection scheme and hence the selection process will be biased. Furthermore, QTL information obtained from one population may not mirror that from another population, since they may have different QTL and/or QTL alleles.

QTL and Bioinformatics

With hundreds of QTL mapping reports published every year, bioinformatics tools are needed to manipulate huge amounts of QTL-related data. It is important for all researchers to follow general rules for QTL reporting, such as the standardization of QTL nomenclature, as proposed by McCouch *et al.* (1997). A standard reporting system is critical for comparative genomics, QTL allelism tests, data sharing and mining and the association between major genes and QTL. This system should include: (i) associated alleles and allele characterization, such as allele sizes; (ii) QTL effect;

(iii) variation explained by single and total QTL or all QTL in the model; (iv) QTL interaction if more than one QTL is identified; and (v) the QTL × E interaction if more than one environment is involved. QTL information should be shareable and combined with data generated in genetics and plant breeding, e.g. germ-plasm diversity, mapping populations, pedigrees, gene positions and effects, graphical genotypes, mutants and other genetic stocks.

The explosion of interest in QTL mapping has led to numerous studies in plants, each based on its own experimental population(s). Each experiment is limited in size and usually restricted to a single population, or a cross, planted in specific environment(s). Therefore, QTL effects that can be detected are also limited. One direction for QTL analysis is to combine information from several studies – for example, by meta-analysis of the results of QTL studies (Goffinet and Gerber, 2000) or joint analysis of the raw data (Haley, 1999). Joint analysis using the raw data potentially allows more information to

be extracted than meta-analysis. Walling *et al.* (2000) did a joint analysis of data from seven pig QTL mapping populations from the UK, France, Germany, Sweden, the USA and The Netherlands, totalling over 3000 animals. In rice, the RiceGene database is accessible to the public. Extension of Rice-Gene to include raw data from QTL mapping projects will stimulate this effort. Many permanent populations have been shared internationally for mapping. The raw data should be shared, too. A rice RFLP map constructed by using IR64/Azucena DHs has been saturated with more than 500 SSR markers (Chen *et al.*, 1997; Temnykh *et al.*, 2000, 2001). Researchers involved in QTL mapping, however, have been using the first version of the molecular map, consisting of only 175 RFLP markers. Sharing marker and phenotype information will make more comprehensive and conclusive analyses possible.

For joint or pooled-data analysis, more powerful QTL mapping procedures need to be developed, which use all information available, including marker linkage, phenotypic correlation, $G \times E$ interaction, multilocation phenotype, phenotype from multiple populations (the more populations used, the more alleles will be identified), DNA profiling of germ-plasm and marker–trait association established by germ-plasm fingerprinting. The corresponding software should be available publicly, with flexibility for different population structures, generations and breeding systems.

QTL and Functional Genomics

Recent advances in genome mapping have made it possible to map and determine the magnitude of the effect of individual loci controlling QTs. In a conventional approach to positional cloning, once very tightly linked markers are identified, the genetic map is related to a physical map and sequencing can begin. When the target region is sequenced, the genes within the region become candidate genes and are then used in complementation tests. This approach has been successfully used to clone QTL

in tomato (*Lycopersicon esculentum* L.) for fruit weight (Frary *et al.*, 2000) and sugar content (Fridman *et al.*, 2000) and in rice for heading-date/photoperiod sensitivity (Yano *et al.*, 2000).

The story is the same for all successful examples currently available for QTL cloning. Three cloned QTL have relatively large effects, with over 40% variation explained in near-isogenic backgrounds, which is comparable to major genes. Both crops have a large number of molecular markers available, so QTL mapping using primary populations could narrow the target region down to less than 1 cM. Usually, several thousand individuals from the secondary populations, such as those from advanced back-cross or NILs, are used for fine-mapping. Each crop has a well-established genetic-transformation system. Several years of hard work, along with luck, are needed for the cloning of a gene.

In complementation tests for the function of the candidate genomic region of a QTL, transformation for creating quantitative variation will depend heavily on the expression and inheritance of the introduced QTL, because QTs are usually controlled by a number of genes, each with relatively small effect, and mimicked by internal genetic backgrounds and external environmental factors. Therefore, it would be a great challenge for molecular geneticists to verify the effect of minor QTL (Xu, 1997). Although most studies have identified QTL that explain more than 10% of phenotypic variance (Table 9.1), minor QTL can be identified if larger population sizes are used. On the other hand, minor QTL can be detected with greater power by first removing the effects from detected QTL of larger effects – for example, by including their genotypes as cofactors in the regression. In maize, a QTL contributing as little as 0.3% of phenotypic variance was reported, from an F_2 population of 1700 individuals with a probability threshold of 0.05 (Edwards *et al.*, 1987). For such a minor QTL, however, it is very difficult to use the approach discussed above to do a complementation test.

A typical higher-plant genome encodes 10,000–100,000 genes, scattered through a

total of 10^8–10^{10} bp of DNA. Consequently, 0.1% of the genome would include an average of ten to 100 genes. Due to environmental and epistatic influences, current procedures allow the placement of QTL with only low resolution (10–30 cM intervals of the genome) (Kearsey and Farquhar, 1998). Such large segments probably contain millions of base pairs of DNA and a multitude of genes, so it is unclear whether individual effects are caused by a single gene or a set of linked genes. Low-resolution mapping also makes it a daunting task to sort through large segments of DNA to clone the gene(s) responsible for the QTL effects. Map-based cloning must narrow down the candidate genome region to several to tens of kilobases. The ultimate resolution of QTL mapping, however, is limited by the number of meioses. This number can be increased by using larger sample sizes or by accumulating meioses over a number of generations. Fine-mapping of a QTL to a small chromosomal region involves the identification of unique recombinants that differ in genome composition near the QTL and the phenotypic evaluation of numerous progeny from these recombinants to obtain a reliable measure of the true QTL genotype. High-throughput analysis, combined with highly informative molecular markers, such as SSRs and SNPs, enables us to manage populations with thousands of plants using thousands of markers in fine-mapping. With the development of chip technology, it is possible to make massive parallel data acquisition and analysis, providing the potential for miniaturization and multiplexing required for high-throughput analysis. Array-based genotyping of SNPs will accelerate positional cloning and the high-throughput identification of both monogenic and polygenic traits.

Using a QTL mapping strategy, the quantitative variation of proteins or protein quantity separated by two-dimensional polyacrylamide gel electrophoresis can be localized. This genetic location can be called a protein-quantity locus (PQL). The PQL strategy can also result in the identification of candidate proteins. Proteins whose genetic factors control quantity and/or

activity can be colocalized with QTL defined for the agronomic trait. In this connection, de Vienne *et al.* (1999) showed that three PQLs controlling the quantity of a single leaf protein and three QTL for height growth in maize were colocalized. Because protein loci sample the genome differently from most PCR-based techniques, they provide a different level of information with respect to the diversity questions being asked. Common proteins were found when studying related species belonging to the same family. If a protein appears to be of interest in a cultivated species and if the gene coding an electrophoretically identical protein is already sequenced in another species, this may dramatically shorten the time taken to establish it as a gene of agronomic interest (Thiellement *et al.*, 2001). One can envisage the development in the coming years of such comparative proteomics. The use of protein analysis in the candidate-gene approach is still in its infancy and it will be developed predominantly for the benefit of the plant breeders.

As nucleotide sequencing of the *Arabidopsis* and rice genomes have been completed, and large amounts of expressed sequence tag (EST) information have been obtained for many other plants, there are many opportunities to use this wealth of sequence information to accelerate progress towards a comprehensive understanding of the genetic mechanisms that control plant growth and development and responses to the environment (Somerville and Somerville, 1999). With a complete sequence for the target organism (or even a closely related organism that has been aligned with the target by comparative mapping), genetic mapping may remain the most convenient to process in scanning the genome for target loci. The information provided by genetic mapping about the approximate location of a target gene would equate to having a list of open reading frames within the target region and putative functions for many of them. Further inferences about likely candidate genes might be based on the presence of tissue-specific promoters or other clues regarding when and where specific transcripts might be expressed. A large number

of molecular polymorphisms, such as SNPs and small insertions/deletions or large insertions, discovered with genome sequencing, also provides an opportunity for identifying the nucleotide change associated with QT variation. The nucleotide change that contributes to quantitative variation has been referred to as a QT nucleotide (QTN) (Lyman *et al.*, 1999). Fine-mapping combined with sequence analysis could narrow the chromosomal region associated with quantitative variation (QTL) down to a specific nucleotide change.

The wealth of information from various genome-sequencing projects and sequence-analysis tools provides the biologist with information for gene prediction. Chief in the arsenal of gene prediction are coding exon prediction, BLASTN searches of dbEST and BLASTX searches of protein databases. Bioinformatics-based functional genomics utilizes these sophisticated search tools to compare sequences of the candidate genes with the existing gene sequence and look for homology (similarity) between the candidate gene and known genes (for reviews, see Higgins and Taylor, 2000; Mount, 2001). The basis of homology searching is that related genes have similar sequences and so a new gene can be discovered by virtue of its similarity to an equivalent, already sequenced, gene from a different organism. Prediction tools are used to predict the sequence of proteins that would be produced by the candidate genes and to compare the predicted protein sequence with known protein sequences. When the function of genes in the databases is known, information can be used to further qualify a candidate gene. Those biologists working with less genetically endowed organisms might be able to lever the genetic information from model organisms, such as rice, by taking advantage of homology. In this way, a reverse quantitative-genetics approach could be fruitful in that one could ask how much phenotypic variation in a non-model organism is explained by the homologue of a gene with a similar phenotype in a model organism (Mauricio, 2001).

Experimental approaches to functional genomics include analyses at the levels of genetics, expression, protein and metabolism. The main roadblock on the journey from QTL to gene is that tests for quantitative complementarity can only be done if a stock containing a mutant allele at the candidate locus exists. Much of the *Drosophila* genome consists of loci with known function but no mutant allele (Mackay, 2001). Therefore, the most direct approach could be one that utilizes knockouts to 'turn off' a gene and analyse the phenotype. The general principle of this conventional analysis is that the genes responsible for a phenotype can be identified by determining which genes are inactivated in organisms that display a mutant version of the phenotype. If the starting-point is the gene, rather than the phenotype, then the equivalent strategy would be to mutate the gene and identify the phenotypic change that results. This approach requires large-scale mutagenesis. As an initiative in functional genomics, the International Rice Research Institute (IRRI) has begun the systematic production of rice mutants with the goal of creating a collection of 40,000 independent deletion lines, each carrying an average of ten mutations per genome, resulting in a 90% probability of detecting all possible genes in rice (http://www.cgiar.org/irri/genomics/index.htm). As of January 2001, about 47,000 M_3 or M_4 lines were obtained from ~17,000 independent M_1 plants from treatment with diepoxybutane (DEB) (0.006%), fast neutrons (FN) (33 Gy) and gamma rays (GR) (250 Gy). Large collections of insertion mutants are available for *Arabidopsis*, maize, petunia and snapdragon, and collections of insertion mutants are being created in several other species. Three types of insertional-mutant libraries in rice are being constructed based on random insertions of a DNA (a T-DNA, an endogenous transposon or a plasmid containing a transposable element that can be introduced and result in transposition) into the genome (for a review, see Wu, 1999). Both deletion and insertional mutants will provide IMLs for major genes as well as QTL. A further advantage of a collection of single-deletion and insertional mutation in a co-isogenic background is that they can be used to determine how interlocus epistatic

interactions shape the phenotype, by crossing mutated lines in all possible pairwise combinations and assessing the behaviour in the doubly heterozygous progeny (Mackay, 2001). Numerous IMLs derived from a single genetic background will also open the door to genetic and functional analyses of QTL with minor effects. In order to assign a function to an unknown gene, however, examining the phenotype of the mutant can be much more difficult than it sounds, especially for a quantitative phenotype. For some genes, inactivation may not have any apparent effect.

With the development of microarrays, gene-expression profiling is emerging as an effective way of finding candidate genes. It is not clear how applicable expression profiling will prove to be to the analyses of QTL, as there are three potential pitfalls to this approach (Flint and Mott, 2001). First, QTL might produce only modest changes (that is, less than a twofold difference) in gene expression, which cannot be distinguished from changes caused by non-genetic factors using the microarrays currently available. However, improved quantification of mRNA levels might overcome this problem. Secondly, there might be difficulties in identifying the correct tissue and developmental stage on which to carry out gene-expression analysis. As discussed previously, however, QTL mapping across different developmental stages could provide some clues to this issue. Thirdly, we are ignorant of the extent to which quantitative variation is regulated at the transcriptional level; mRNA differences might be a secondary consequence of genetic action.

The methodologies and strategies currently used are not entirely adequate for assigning functions to the vast numbers of unknown genes that could be discovered by sequencing projects. In the future, gene functional analysis will be made at different levels with different methods, which may include complete sets of mRNA molecules (transcriptome level), protein molecules (proteome level) and metabolites (low-molecular-weight intermediates) (metabolome level) present in a cell, tissue or organ (Hunt and Livesey, 2000; Pennington and Dunn, 2001). It can be expected that technology development in all these areas will bring up new tools that are more suitable for the functional analysis of QTL.

Conclusions

Molecular-marker technology has revolutionized our understanding of quantitative traits under different backgrounds at different levels of genomics. In the 21st century we are starting with numerous developments in science and technology to manipulate genes for both qualitative and quantitative traits. Now we can use every possible genetic tool to improve QTs. Using marker–trait association, for example, MAS can now help breeders manipulate traits that are difficult or impossible to handle in conventional breeding. As human population increases yearly and because food supply in the world must keep pace, improvement of plants continues to be a challenge to geneticists and plant breeders. As a major food source for human beings and a model plant for molecular biology, rice has been moving well ahead of other plants in QTL analysis. Any success in this crop will benefit others in certain ways. In the future, more attention should be paid to pooled-data analysis, transgenic (gene plus) and knockout (gene minus) mutant analysis, G × E interaction, epistasis and developmental genetics for complicated traits, trait complexes and trait components. With the techniques currently available, it takes years to clone a gene using map-based cloning. The lack of significant phenotypic effect for numerous QTL will greatly challenge our current systems. Additional research is urgently needed to map, clone and better characterize QTL. In particular, we need to be able to map mutant genes to a small interval more efficiently and we need a high-throughput detection system to identify the molecular lesion in the critical region more quickly. Therefore, the development of an ideal system for QTL analysis in the era of functional genomics is critical for all plants. Considering interaction of all types that could happen among numerous genes with minor effects and interaction with external environments that could change from time to time and from location to location, we need to develop much more efficient systems for separating, pyramiding and packaging all these genes and to make them functional at a full scale in specific environments. It seems that these systems would be too complicated to be practical. It is reasonable to believe that this could be a major objective for many generations of geneticists and plant breeders. Highly informative markers, isogenic-mutation libraries, high-throughput technology and powerful statistical methods and computational software are key tools for the genetic manipulation of minor QTL. With all the technical developments, manipulating quantitative traits in the near future could be as easy as manipulating qualitative traits controlled by major genes.

Acknowledgements

I thank Drs Manjit S. Kang, Bruce Walsh, John W. Dudley and Mark Walton for their valuable comments and suggestions on the manuscript, and Dr Junjian Ni for his critical reading and help in collecting QTL references. I also thank my collaborators/supervisors for their encouragement and support, including Professor Zongtan Shen at Zhejiang University and Professors Lihuang Zhu and Ying Chen at the Institute of Genetics, Chinese Academy of Sciences, China, and Dr Susan McCouch at Cornell University, USA.

References

Ahn, S.N., Bollich, C.N., McClung, A.M. and Tanksley, S.D. (1993) RFLP analysis of genomic regions associated with cooked-kernel elongation in rice. *Theoretical and Applied Genetics* 87, 27–32.

Alam, S.N. and Cohen, M.B. (1998) Detection and analysis of QTLs for resistance to the brown planthopper, *Nilaparvata lugens*, in a doubled-haploid rice population. *Theoretical and Applied Genetics* 97, 1370–1379.

Albar, L., Lorieux, M., Ahmadi, N., Rimbault, I., Pinel, A., Sy, A.A., Fargette, D. and Ghesquière, A. (1998) Genetic basis and mapping of the resistance to rice yellow mottle virus. I. QTLs identification and relationship between resistance and plant morphology. *Theoretical and Applied Genetics* 97, 1145–1154.

Ali, M.L., Pathan, M.S., Zhang, J., Bai, G., Sarkarung, S. and Nguyen, H.T. (2000) Mapping QTLs for root traits in a recombinant inbred population from two indica ecotypes in rice. *Theoretical and Applied Genetics* 101, 756–766.

Antonio, B.A., Inoue, T., Kajiya, H., Nagamura, Y., Kurata, N., Minobe, Y., Yano, M., Nakagahra, M. and Sasaki, T. (1996) Comparison of genetic distance and order of DNA markers in five populations of rice. *Genome* 39, 946–956.

Ayres, N.M., McClung, A.M., Larkin, P.D., Bligh, H.F.J., Jones, C.A. and Park, W.D. (1997) Microsatellites and a single-nucleotide polymorphism differentiate apparent amylose classes in an extended pedigree of US rice germ plasm. *Theoretical and Applied Genetics* 94, 773–781.

Bao, J.S., Zheng, X.W., Xia, Y.W., He, P., Shu, Q.Y., Lu, X., Chen, Y. and Zhu, L.H. (2000) QTL mapping for the paste viscosity characteristics in rice (*Oryza sativa* L.). *Theoretical and Applied Genetics* 100, 280–284.

Bar-Hen, A., Charcosset, A., Bourgoin, M. and Guiard, J. (1995) Relationship between genetic markers and morphological traits in a maize inbred lines collection. *Euphytica* 84, 145–154.

Beckmann, J.S. and Soller, M. (1986) Restriction fragment length polymorphisms in plant genetic improvement. *Oxford Surveys of Plant Molecular and Cell Biology* 3, 196–250.

Bligh, H.F.J., Till, R.I. and Jones, C.A. (1995) A microsatellite sequence closely linked to the waxy gene of *Oryza sativa*. *Euphytica* 86, 83–85.

Bradshaw, H.D., Jr and Settler, R.F. (1995) Molecular genetics of growth and development in populus. IV. Mapping QTLs with large effects on growth, form, and phenology traits in a forest tree. *Genetics* 139, 963–973.

Cardon, L.R. and Bell, J.I. (2001) Association study designs for complex diseases. *Nature Reviews Genetics* 2, 91–99.

Causse, M.A., Fulton, T.M., Cho, Y.G., Ahn, S.N., Chunwongse, J., Wu, K.S., Xiao, J.H., Yu, Z.H., Ronald, P.C., Harrington, S.E., Second, G., McCouch, S.R. and Tanksley, S.D. (1994)

Saturated molecular map of the rice genome based on an interspecific backcross population. *Genetics* 138, 1251–1274.

Champoux, M.C., Wang, G., Sarkarung, S., Mackill, D.J., O'Toole, J.C., Huang, N. and McCouch, S.R. (1995) Locating genes associated with root morphology and drought avoidance in rice via linkage to molecular markers. *Theoretical and Applied Genetics* 90, 969–981.

Chen, X., Temnykh, S., Xu, Y., Cho, Y.G. and McCouch, S.R. (1997) Development of microsatellite framework map providing genome-wide coverage in rice (*Oryza sativa* L.). *Theoretical and Applied Genetics* 95, 553–567.

Courtois, B., McLaren, G., Sinha, P.K., Prasad, K., Yadav, R. and Shen, L. (2000) Mapping QTLs associated with drought avoidance in upland rice. *Molecular Breeding* 6, 55–66.

de Vienne, D., Leonardi, A., Damerval, C. and Zivy, M. (1999) Genetics of proteome variation as a tool for QTL characterization: application to drought stress responses in maize. *Journal of Experimental Botany* 50, 303–309.

Dudley, J.W. (1993) Molecular markers in plant improvement: manipulation of genes affecting quantitative traits. *Crop Science* 33, 660–668.

Edwards, M.D., Stuber, C.W. and Wendel, J.F. (1987) Molecular-marker-facilitated investigations of quantitative trait loci in maize. I. Numbers, genomic distribution and types of gene action. *Genetics* 116, 113–125.

Falconer, D.S. and Mackay, T.F.C. (1996) *Introduction to Quantitative Genetics*, 4th edn. Longman, Harlow, UK.

Fisch, M., Bohn, M. and Melchinger, A.E. (1999) Comparison of selection strategies for marker-assisted backcrossing of a gene. *Crop Science* 39, 1295–1301.

Flint, J. and Mott, R. (2001) Finding the molecular basis of quantitative traits: successes and pitfalls. *Nature Reviews Genetics* 2, 437–445.

Frary, A., Nesbitt, T.C., Frary, A., Grandillo, S., van der Knaap, E., Cong, B., Liu, J., Meller, J., Elber, R., Alpert, K.B. and Tanksley, S.D. (2000) *fw2.2*: a quantitative trait locus to the evolution of tomato fruit size. *Science* 289, 85–88.

Fridman, E., Pleban, T. and Zamir, D. (2000) A recombination hotspot delimits a wild-species quantitative trait locus for tomato sugar content to 484 bp within an invertase gene. *Proceedings of the National Academy of Sciences USA* 97, 4718–4723.

Ghesquière, A., Albar, L., Lorieux, M., Ahmadi, N., Fargette, D., Huang, N., McCouch, S.R. and Notteghem, J.L. (1997) A major quantitative trait locus for rice yellow mottle virus resistance maps to a cluster of blast resistance genes on chromosome 12. *Phytopathology* 87, 1243–1249.

Goffinet, B. and Gerber, S. (2000) Quantitative trait loci: a meta-analysis. *Genetics* 155, 463–473.

Gong, J., Zheng, X., Du, B., Qian, Q., Chen, S., Zhu, L. and He, P. (2001) Comparative study of QTLs for agronomic traits of rice (*Oryza sativa* L.) between salt stress and nonstress environment. *Science in China* 44, 73–82.

Haley, C. (1999) Advances in quantitative trait locus mapping. In: Dekkers, J.C.M., Lamont, S.J. and Rothschild, M.F. (eds) *From Jay Lush to Genomics: Visions for Animal Breeding and Genetics.* Animal Breeding and Genetics Group, Department of Animal Science, Iowa State University, Ames, Iowa, pp. 47–59.

He, P., Shen, L., Lu, C., Chen, Y. and Zhu, L. (1998) Analysis of quantitative trait loci which contribute to anther culturability in rice (*Oryza sativa* L.). *Molecular Breeding* 4, 165–171.

He, P., Li, S.G., Qian, Q., Ma, Y.Q., Li, J.Z., Wang, W.M., Chen, Y. and Zhu, L.H. (1999) Genetic analysis of rice grain quality. *Theoretical and Applied Genetics* 98, 502–508.

He, P., Li, J.Z., Zheng, X.W., Shen, L.S., Lu, C.F., Chen, Y. and Zhu, L.H. (2001) Comparison of molecular linkage maps and agronomic trait loci between DH and RIL populations derived from the same rice cross. *Crop Science* 41, 1240–1246.

Hemamalini, G.S., Shashidhar, H.E. and Hittalmani, S. (2000) Molecular marker assisted tagging of morphological and physiological traits under two contrasting moisture regimes at peak vegetative stage in rice (*Oryza sativa* L.). *Euphytica* 112, 69–78.

Higgins, D. and Taylor, W. (2000) *Bioinformatics: Sequence, Structure and Databanks.* Oxford University Press, New York, 249 pp.

Hospital, F. and Charcosset, A. (1997) Marker-assisted introgression of quantitative trait loci. *Genetics* 147, 1469–1485.

Hospital, F., Chevalet, C. and Mulsant, P. (1992) Using markers in gene introgression breeding programs. *Genetics* 132, 1199–1210.

Huang, N., Courtois, B., Khush, G.S., Lin, H., Wang, G., Wu, P. and Zheng, K. (1996) Association of quantitative trait loci for plant height with major dwarfing genes in rice. *Heredity* 77, 130–137.

Huang, N., Parco, A., Mew, T., Magpantay, G., McCouch, S., Guiderdoni, E., Xu, J., Subudhi, P., Angeles, E.R. and Khush, G.S. (1997) RFLP mapping of isozymes, RAPD and QTLs for grain shape, brown planthopper resistance in a doubled haploid rice population. *Molecular Breeding* 3, 105–113.

Hunt, S.P. and Livesey, F.J. (2000) *Functional Genomics*. Oxford University Press, New York, 253 pp.

Jansen, R.C. and Stam, P. (1994) High resolution of quantitative traits into multiple loci via interval mapping. *Genetics* 136, 1447–1455.

Jansen, R.C., van Ooijen, J.W., Stam, P., Lister, C. and Dean, C. (1995) Genotype-by-environment interaction in genetic mapping of multiple quantitative trait loci. *Theoretical and Applied Genetics* 91, 33–37.

Jiang, C. and Zeng, Z.-B. (1995) Multiple trait analysis of genetic mapping for quantitative trait loci. *Genetics* 140, 1111–1127.

Kamijima, O., Tanisaka, T. and Kinoshita, T. (1996) Gene symbols for dwarfness. *Rice Genetics Newsletter* 13, 19–25.

Kearsey, M.J. and Farquhar, A.G.L. (1998) QTL analysis in plants: where are we now? *Heredity* 80, 137–142.

Kinoshita, T. (1995) Report of committee on gene symbolization, nomenclature and linkage groups. *Rice Genetics Newsletter* 12, 9–153.

Kurata, N., Nagamura, Y., Yamamoto, K., Harushima, Y., Sue, N., Wu, J., Antonio, B.A., Shomura, A., Shimizu, T., Lin, S.Y., Inoue, T., Fukuda, A., Shimano, T., Kuboki, Y., Toyama, T., Miyamoto, Y., Kirihara, T., Hayasaka, K., Miyao, A., Monna, L., Zhong, H.S., Tamura, Y., Wang, Z.X., Momma, T., Umehara, Y., Yano, M., Sasaki, T. and Minobe, Y. (1994) A 300 kilobase interval genetic map of rice including 883 expressed sequences. *Nature Genetics* 8, 365–372.

Lander, E.S. and Botstein, D. (1989) Mapping Mendelian factors underlying quantitative traits using RFLP linkage maps. *Genetics* 121, 185–199.

Lebowitz, R.L., Soller, M. and Beckmann, J.S. (1987) Trait-based analysis for the detection of linkage between marker loci and quantitative trait loci in cross between inbred lines. *Theoretical and Applied Genetics* 73, 556–562.

Lee, M. (1995) DNA markers and plant breeding programs. *Advances in Agronomy* 55, 265–344.

Li, J.X., Yu, S.B., Xu, C.G., Tan, Y.F., Gao, Y.J., Li, X.H. and Zhang, Q. (2000) Analysing quantitative trait loci for yield using a vegetatively replicated F2 population from a cross between the parents of an elite rice hybrid. *Theoretical and Applied Genetics* 101, 248–254.

Li, Z.K., Pinson, S.R., Stansel, J.W. and Park, W.D. (1995a) Identification of quantitative trait loci (QTL) for heading date and plant height in cultivated rice (*Oryza sativa* L.). *Theoretical and Applied Genetics* 91, 374–381.

Li, Z.K., Pinson, S.R., Marchetti, M.A., Stansel, J.W. and Park, W.D. (1995b) Characterization of quantitative trait loci contributing to field resistance to sheath blight (*Rhizoctonia solani*) in rice. *Theoretical and Applied Genetics* 91, 382–388.

Li, Z., Pinson, S.R.M., Park, W.D., Paterson, A.H. and Stansel, J.W. (1997) Epistasis for three grain yield components in rice (*Oryza sativa* L.). *Genetics* 145, 453–465.

Li, Z.-K., Luo, L.J., Mei, H.W., Paterson, A.H., Zhao, X.H., Zhong, D.B., Wang, Y.P., Yu, X.Q., Zhu, L., Tabien, R., Stansel, J.W. and Ying, C.S. (1999) A 'defeated' rice resistance gene acts as a QTL against a virulent strain of *Xanthomonas oryzae* pv. *oryzae*. *Molecular and General Genetics* 26, 58–63.

Lilley, J.M., Ludlow, M.M., McCouch, S.R. and O'Toole, J.C. (1996) Locating QTL for osmotic adjustment and dehydration tolerance in rice. *Journal of Experimental Botany* 47, 1427–1436.

Lin, S.Y., Sasaki, T. and Yano, M. (1998) Mapping quantitative trait loci controlling seed dormancy and heading date in rice, *Oryza sativa* L., using backcross inbred lines. *Theoretical and Applied Genetics* 96, 997–1003.

Liu, B.H. (1998) *Statistical Genomics: Linkage, Mapping, and QTL Analysis*. CRC Press, Boca Raton and New York, 611 pp.

Liu, Y.S., Zhu, L.H., Sun, J.S. and Chen, Y. (2001) Mapping QTLs for detective female gametophyte development in an inter-subspecific cross in *Oryza sativa* L. *Theoretical and Applied Genetics* 102, 1243–1251.

Lorieux, M., Petrov, M., Huang, N., Guiderdoni, E. and Ghesquière, A. (1996) Aroma in rice, genetic analysis of a quantitative trait. *Theoretical and Applied Genetics* 93, 1145–1151.

Lorieux, M., Ndjiondjop, M.-N. and Ghesquière, A. (2000) A first interspecific *Oryza sativa* × *Oryza glaberrima* microsatellite-based genetic linkage map. *Theoretical and Applied Genetics* 100, 593–601.

Lu, C., Shen, L., Tan, Z., Xu, Y., He, P., Chen, Y. and Zhu, L. (1996) Comparative mapping of QTLs for agronomic traits of rice across environments by using a doubled haploid population. *Theoretical and Applied Genetics* 93, 1211–1217.

Lyman, R.F., Lai, C. and Mackay, T.F. (1999) Linkage disequilibrium mapping of molecular polymorphisms at the scabrous locus associated with naturally occurring variation in bristle number in *Drosophila melanogaster*. *Genetical Research* 74, 303–311.

Lynch, M. and Walsh, B. (1998) *Genetics and Analysis of Quantitative Traits*. Sinauer Associates, Sunderland, Massachusetts, 980 pp.

McCouch, S.R. and Doerge, R.W. (1995) QTL mapping in rice. *Trends in Genetics* 11, 482–487.

McCouch, S.R., Kochert, G., Yu, Z.H., Wang, Z.Y., Khush, G.S., Coffman, W.R. and Tanksley, S.D. (1988) Molecular mapping of rice chromosomes. *Theoretical and Applied Genetics* 76, 815–829.

McCouch, S.R., Cho, Y.G., Yano, M., Paul, E., Blinstrub, M., Morishima, H. and Kinoshita, T. (1997) Report on QTL nomenclature. *Rice Genetics Newsletter* 14, 11–13.

Mackay, T.F.C. (2001) Quantitative trait loci in *Drosophila*. *Nature Reviews Genetics* 2, 11–20.

Mackill, D.J. and Ni, J. (2001) Molecular mapping and marker-assisted selection for major-gene traits in rice. In: *Rice Genetics IV*. IRRI and Science Publisher, Los Baños, The Philippines, pp. 131–151.

Maheswaran, M., Huang, N., Sreerangasamy, S.R. and McCouch, S.R. (2000) Mapping quantitative trait loci associated with days to flowering and photoperiod sensitivity in rice (*Oryza sativa* L.). *Molecular Breeding* 6, 145–155.

Mauricio, R. (2001) Mapping quantitative trait loci in plants: uses and caveats for evolutionary biology. *Nature Reviews Genetics* 2, 370–381.

Mount, D.W. (2001) *Bioinformatics: Sequence and Genome Analysis*. Cold Spring Harbor Laboratory Press, New York, 564 pp.

Nandi, S., Subudhi, P.K., Senadhira, D., Manigbas, N.L., Sen-Mandi, S. and Huang, N. (1997) Mapping QTLs for submergence tolerance in rice by AFLP analysis and selective genotyping. *Theoretical and Applied Genetics* 255, 1–8.

Ni, J.J., Wu, P., Senadhira, D. and Huang, N. (1998) Mapping QTLs for phosphorus deficiency tolerance in rice (*Oryza sativa* L.). *Theoretical and Applied Genetics* 97, 1361–1369.

Paterson, A.H. (1998) *Molecular Dissection of Complex Traits*. CRC Press, Boca Raton, Florida, 305 pp.

Paterson, A.H., Tanksley, S.D. and Sorrells, M.E. (1991) DNA markers in plant improvement. *Advances in Agronomy* 46, 39–90.

Pennington, S.R. and Dunn, M.J. (2001) *Proteomics: From Protein Sequence to Function*. Springer-Verlag, New York, 313 pp.

Plomion, C., Durel, C.-E. and O'Malley, D.M. (1996) Genetic dissection of height in maritime pine seedlings raised under accelerated growth conditions. *Theoretical and Applied Genetics* 93, 849–858.

Pooni, H.S., Kumar, I. and Khush, G.S. (1993) Genetical control of amylose content in selected crosses of *indica* rice. *Heredity* 70, 269–280.

Pressoir, G., Albar, L., Ahmadi, N., Rimbault, I., Lorieux, M., Fargette, D. and Ghesquière, A. (1998) Genetic basis and mapping of the resistance to rice yellow mottle virus. II. Evidence of a complementary epistasis between two QTLs. *Theoretical and Applied Genetics* 97, 1155–1161.

Price, A.H. and Tomos, A.D. (1997) Genetic dissection of root growth in rice (*Oryza sativa* L.): II. Mapping quantitative trait loci using molecular markers. *Theoretical and Applied Genetics* 95, 143–152.

Qian, Q., He, P., Zheng, X., Chen, Y. and Zhu, L. (2000a) Genetic analysis of morphological index and its related taxonomic traits for classification of indica/japonica rice. *Science in China* 43, 113–119.

Qian, Q., Zeng, D., He, P., Zheng, X., Chen, Y. and Zhu, L. (2000b) QTL analysis of the rice seedling cold tolerance in a double haploid population derived from anther culture of a hybrid between *indica* and *japonica* rice. *Chinese Science Bulletin* 45, 448–453.

Ray, J.D., Yu, L., McCouch, S.R., Champoux, M.C., Wang, G. and Nguyen, H.T. (1996) Mapping quantitative trait loci associated with root penetration ability in rice (*Oryza sativa* L.). *Theoretical and Applied Genetics* 92, 627–636.

Redoña, E.D. and Mackill, D.J. (1996) Mapping quantitative trait loci for seedling vigor in rice using RFLPs. *Theoretical and Applied Genetics* 92, 395–402.

Risch, N.J. (2000) Searching for genetic determinants in the new millennium. *Nature* 405, 847–856.

Robertson, D.S. (1985) A possible technique for isolating genomic DNA for quantitative traits in plants. *Journal of Theoretical Biology* 117, 1–10.

Sasahara, H., Fukuta, Y. and Fukuyama, T. (1999) Mapping of QTLs for vascular bundle system and spike morphology in rice, *Oryza sativa* L. *Breeding Science* 49, 75–81.

Satagopan, J.M., Yandell, B.S., Newton, M.A. and Osborn, T.G. (1996) A Bayesian approach to detect quantitative trait loci using Markov chain Monte Carlo. *Genetics* 144, 805–816.

Sham, P. (1997) *Statistics in Human Genetics.* John Wiley & Sons, New York, 290 pp.

Somerville, C. and Somerville, S. (1999) Plant functional genomics. *Science* 285, 380–383.

Song, W.-Y., Wang, G.-L., Chen, L.-L., Kim, H.-K., Pi, L.-Y., Holsten, T., Gardner, J., Wang, B., Zhai, W.-X., Zhu, L.-H., Fauquet, C. and Ronald, P. (1995) A receptor kinase-like protein encoded by the rice disease resistance gene, *Xa21. Science* 270, 1084–1086.

Sripongpangkul, K., Posa, G.B.T., Senadhira, D.W., Brar, D., Huang, N., Khush, G.S. and Li, Z.K. (2000) Genes/QTLs affecting flood tolerance in rice. *Theoretical and Applied Genetics* 101, 1074–1081.

Stuber, C.W. (1994) Breeding multigenic traits. In: Phillips, R.L. and Vasil, I.K. (eds) *DNA Based Markers in Plants.* Kluwer Academic Publishers, Dordrecht, The Netherlands, pp. 97–115.

Tabien R.E., Li, Z., Paterson, A.H., Marchetti, M.A., Stansel, J.W., Pinson, S.R.M. and Park, W.D. (2000) Mapping of four major rice blast resistance genes from 'Lemont' and 'Teqing' and evaluation of their combinatorial effect for field resistance. *Theoretical and Applied Genetics* 101, 1215–1225.

Taguchi-Shiobara, F., Lin, S.Y., Tanno, K., Komatsuda, T., Yano, M., Sasaki, T. and Oka, S. (1997) Mapping quantitative trait loci associated with regeneration ability of seed callus in rice, *Oryza sativa* L. *Theoretical and Applied Genetics* 95, 828–833.

Tan, Y.F., Xing, Y.Z., Li, J.X., Yu, S.B., Xu, C.G. and Zhang, Q. (2000) Genetic bases of appearance quality of rice grains in Shanyou 63, an elite rice hybrid. *Theoretical and Applied Genetics* 101, 823–829.

Tang D., Wu, W., Li, W., Lu, H. and Worland, A.J. (2000) Mapping of QTLs conferring resistance to bacterial leaf streak in rice. *Theoretical and Applied Genetics* 101, 286–291.

Tanksley, S.D., Young, N.D., Paterson, A.H. and Bonierbale, M.W. (1989) RFLP mapping in plant breeding: new tools for an old science. *Bio/Technology* 7, 257–263.

Temnykh, S., Park, W.D., Ayres, N., Cartinhour, S., Hauck, N., Lipovich, L., Cho, Y.G., Ishii, T. and McCouch, S.R. (2000) Mapping and genome organization of microsatellite sequences in rice (*Oryza sativa* L.). *Theoretical and Applied Genetics* 100, 697–712.

Temnykh, S., DeClerck, G., Lukashova, A., Lipovich, L., Cartinhour, S. and McCouch, S.R. (2001) Computational and experimental analysis of microsatellites in rice (*Oryza sativa* L.): frequency, length variation, transposon associations, and genetic marker potential. *Genome Research* 11, 1441–1452.

Thiellement, H., Plomion, C. and Zivy, M. (2001) Proteomics as a tool for plant genetics and breeding. In: Pennington, S.R. and Dunn, M.J. (eds) *Proteomics.* BIOS Scientific Publisher, Oxford, pp. 289–309.

Uimari, P., Thaller, G. and Hoeschele, I. (1996) The use of multiple markers in a Bayesian method for mapping quantitative trait loci. *Genetics* 143, 1831–1842.

Verhaegen, D., Plomion, C., Gion, J.-M., Poitel, M., Costa, P. and Kremer, A. (1997) Quantitative trait dissection analysis in Eucalyptus using RAPD markers: 1. Detection of QTL in inter-specific hybrid progeny, stability of QTL expression across different ages. *Theoretical and Applied Genetics* 95, 597–608.

Virk, P.S., Ford-Lloyd, B.V., Jackson, M.T., Pooni, H.S., Clemeno, T.P. and Newbury, H.J. (1996) Predicting quantitative variation within rice germplasm using molecular markers. *Heredity* 76, 296–304.

Walling, G.A., Visscher, P.M., Andersson, L., Rothschild, M.F., Wang, L., Moser, G., Groenen, M.A.M., Bidanel, J.P., Cepica, S., Archibald, A.L., Geldermann, H., de Koning, D.J., Milan, D. and Haley, C.S. (2000) Combined analyses of data from quantitative trait loci mapping studies: chromosome 4 effects on porcine growth and fatness. *Genetics* 155, 1369–1378.

Wang, G.-L., Mackill, D.J., Bonman, J.M., McCouch, S.R., Champoux, M.C. and Nelson, R.J. (1994) RFLP mapping of genes conferring complete and partial resistance to blast in a durably resistant rice cultivar. *Genetics* 136, 1421–1434.

Wang, R.R.-C., Li, C.-M. and Chatterton, N.J. (1999) Loss of heterozygosity and accelerated genotype fixation in rice hybrids. *Genome* 42, 789–796.

Wu, P., Zhang, G. and Huang, N. (1996) Identification of QTLs controlling quantitative characters in rice using RFLP markers. *Euphytica* 89, 349–354.

Wu, P., Ni, J.J. and Luo, A.C. (1998) QTLs underlying rice tolerance to low-potassium stress in rice seedlings. *Crop Science* 38, 1458–1462.

Wu, P., Liao, C.Y., Hu, B., Yi, K.K., Jin, W.Z., Ni, J.J. and He, C. (2000) QTLs and epistasis for aluminum tolerance in rice (*Oryza sativa* L.) at different seedling stages. *Theoretical and Applied Genetics* 100, 1295–1303.

Wu, R. (1999) Reporting of the committee on genetic engineering (functional genomics of plants). *Rice Genetics Newsletter* 16, 10–14.

Wu, W.-R., Li, W.-M., Tang, D.-Z., Lu, H.-R. and Worland, A.J. (1999) Time-related mapping of quantitative trait loci underlying tiller number in rice. *Genetics* 151, 297–303.

Xiao, J., Li, J., Yuan, L. and Tanksley, S.D. (1995) Dominance is the major genetic basis of heterosis in rice as revealed by QTL analysis using molecular markers. *Genetics* 140, 745–754.

Xiao, J., Li, J., Yuan, L. and Tanksley, S.D. (1996) Identification of QTLs affecting traits of agronomic importance in a recombinant inbred population derived from subspecific rice cross. *Theoretical and Applied Genetics* 92, 230–244.

Xu, Y. (1994) Application of molecular markers in genetic improvement of quantitative traits in plants. In: *Proceedings of the Third Young Scientists Symposium on Crop Genetics and Breeding*. Publishing House of Agricultural Science and Technology of China, Beijing, pp. 38–49.

Xu, Y. (1997) Quantitative trait loci: separating, pyramiding, and cloning. *Plant Breeding Reviews* 15, 85–139.

Xu, Y. and Shen, Z. (1991) Diallel analysis of tiller number at different growth stages in rice (*Oryza sativa* L.). *Theoretical and Applied Genetics* 83, 243–249.

Xu, Y. and Zhu, L. (1994) *Molecular Quantitative Genetics* [in Chinese]. China Agriculture Press, Beijing, 291 pp.

Xu, Y., Shen, Z., Xu, J., Chen, Y. and Zhu, L. (1993) Mapping quantitative trait loci via restriction fragment length polymorphism markers in rice. *Rice Genetics Newsletter* 10, 135–138.

Xu Y., Zhu, L., Chen, Y., Lu, C., Shen, L., He, P. and McCouch, S.R. (1997) Tagging genes for photo-thermo sensitivity in rice using RFLP and microsatellite markers. In: *Plant and Animal Genome V*, Plant and Animal Genome Conference Organizing Committee, San Diego, California, Poster 149.

Xu, Y., McCouch, S.R. and Shen, Z. (1998) Transgressive segregation of tiller angle in rice caused by complementary action of genes. *Crop Science* 38, 12–19.

Yadav, R., Courtois, B., Huang, N. and McLaren, G. (1997) Mapping genes controlling root morphology and root distribution in a double-haploid population of rice. *Theoretical and Applied Genetics* 94, 619–632.

Yan, J., Zhu, J., He, C., Benmoussa, M. and Wu, P. (1998a) Molecular dissection of developmental behavior of plant height in rice (*Oryza sativa* L.). *Genetics* 150, 1257–1265.

Yan, J., Zhu, J., He, C., Benmoussa, M. and Wu, P. (1998b) Quantitative trait loci analysis for the developmental behavior of tiller number in rice (*Oryza sativa* L.). *Theoretical and Applied Genetics* 97, 267–274.

Yan, J., Zhu, J., He, C., Benmoussa, M. and Wu, P. (1999) Molecular marker-assisted dissection of genotype × environment interaction for plant type traits in rice (*Oryza sativa* L.). *Crop Science* 39, 538–544.

Yano, M. and Sasaki, T. (1997) Genetic and molecular dissection of quantitative traits in rice. *Plant Molecular Biology* 35, 145–153.

Yano, M., Harushima, Y., Nagamura, Y., Kurata, N., Minobe, Y. and Sasaki, T. (1997) Identification of quantitative trait loci controlling heading date in rice using a high-density linkage map. *Theoretical and Applied Genetics* 95, 1025–1032.

Yano, M., Katayose, Y., Ashikari, M., Yamanouchi, U., Monna, L., Fuse, T., Baba, T., Yamamoto, K., Umehara, Y., Nagamura, Y. and Sasaki, T. (2000) *Hd1*, a major photoperiod sensitivity quantitative trait locus in rice, is closely related to the *Arabidopsis* flowering time gene *CONSTANS*. *Plant Cell* 12, 2473–2484.

Yu, S.B., Li, J.X., Xu, C.G., Tan, Y.F., Gao, Y.J., Li, X.H., Zhang, Q.F. and Saghai Maroof, M.A. (1997) Importance of epistasis as the genetic basis of heterosis in an elite rice hybrid. *Proceedings of the National Academy of Sciences USA* 94, 9226–9231.

Zeng, R., Zhang, Z. and Zhang, G. (2000) Identification of multiple alleles at the *Wx* locus in rice using miscrosatellite class and G–T polymorphism. In: Liu, X. (ed.) *Theory and Application of Crop Research*. China Science and Technology Press, Beijing, pp. 202–205.

Zeng, Z.-B. (1993) Theoretical basis of separation of multiple linked gene effects on mapping quantitative trait loci. *Proceedings of the National Academy of Sciences USA* 90, 10972–10976.

Zeng, Z.-B. (1994) Precision mapping of quantitative trait loci. *Genetics* 136, 1457–1466.

Zhang, J.-S., Xie, C., Li, Z.-Y. and Chen, S.-Y. (1999) Expression of the plasma membrane H+-ATPase gene in response to salt stress in a rice salt-tolerant mutant and its original variety. *Theoretical and Applied Genetics* 99, 1006–1011.

Zheng, H., Babu, R.C., Pathan, M.S., Ali, L., Huang, N., Courtois, B. and Nguyen, H.T. (2000) Quantitative trait loci for root-penetration ability and root thickness in rice: comparison of genetic backgrounds. *Genome* 43, 53–61.

Zhuang, J.-Y., Lin, H.-X., Lu, J., Qian, H.-R., Hittalmani, S., Huang, N. and Zheng, K.-L. (1997) Analysis of QTL × environment interaction for yield components and plant height in rice. *Theoretical and Applied Genetics* 95, 799–808.

Zou, J.H., Pan, X.B., Chen, Z.X., Xu, J.Y., Lu, J.F., Zhai, W.X. and Zhu, L.H. (2000) Mapping quantitative trait loci controlling sheath blight resistance in two rice cultivars (*Oryza sativa* L.). *Theoretical and Applied Genetics* 101, 569–573.

10 Marker-assisted Back-cross Breeding: a Case-study in Genotype-building Theory

Frédéric Hospital

INRA, Station de Génétique Végétale, Ferme du Moulon, 91190 Gif-sur-Yvette, France

Introduction

I wish here to provide an overview of some past and more recent results on marker-assisted back-cross (MAB) breeding theory, and discuss the general consequences for marker-assisted selection (MAS) and genotype building (GB). MAB breeding is a well-known procedure for the introgression of a target gene from a donor line into the genomic background of a recipient line. The objective is to reduce the donor genome content (DGC) of the progenies by repeated back-crosses (BCs) to the recipient line. GB means here the use of markers to design new genotypes combining favourable alleles previously detected at a (possibly large) number of loci in (possibly many) different parental lines. Here, the genomic background in which those alleles are combined cannot, in general, be controlled, because the genes are too numerous. The theory in this domain remains largely unexplored and few results are available. For example, de Koning and Weller (1994) and Dekkers and van Arendonk (1998) have considered the optimization of MAS for identified quantitative trait loci (QTL) plus a possible 'polygenic' background controlling the rest of the genetic variation not explained by the identified QTL. These analyses are restricted to one or two identified QTL. Also, van Berloo and Stam (1998) and

Charmet *et al.* (1999) have considered a larger set of identified QTL, each controlled by flanking markers, and studied selection of recombinant inbred lines or doubled haploids based on flanking markers to produce the best hybrid. This analysis is restricted to selection among inbreds for one or two generations only. Hospital *et al.* (2000) studied selection on marker pairs flanking 50 QTL identified in an F_2 population. With a 'QTL complementation strategy', selection of three to five individuals among a total of 200 for ten generations increases the frequency of favourable alleles at the 50 QTL up to 100% when markers are located exactly on the QTL, but only to 92% when marker–QTL distance is 5 cM. The authors conclude that the efficiency of marker-based selection is bounded by the recombinations taking place between the markers and the QTL. Hence, one has to accelerate the response to selection to fix favourable QTL alleles before marker–QTL linkage disequilibrium vanishes. The main limitation identified is the fact that selected individuals are mated at random: the authors suggest that pairwise matings should increase the efficiency of selection. But the theory in this domain remains unexplored.

MAB is of great practical interest in applied breeding schemes, either to manipulate 'classical' genes between élite lines or from genetic resources or to manipulate transgenic constructions. From a theoretical standpoint, it is a 'simple' example of marker-based selection: in general, only two alleles are segregating, and the gametic phase is known because only one chromosome of each pair is issued from effective recombination (the chromosome from the gamete produced by the back-crossed parent). It is, then, also an appropriate case-study to investigate how selection and recombination work together to make it work better in any type of MAS programme.

In BC breeding, markers can be used to: (i) control the target gene (foreground selection) if needed (Melchinger (1990) discussed the optimal scheme to obtain a minimum number of individuals carrying a target gene of known location; Hospital and Charcosset (1997) discussed the optimal number and positions of markers to control a QTL (target gene of uncertain location)); and/or (ii) control the genetic background (background selection). The objective of background selection is to accelerate the return to the recipient genome outside the target gene, by selection of the recipient allele at markers located either on the carrier chromosome (the chromosome carrying the target gene) and/or on non-carrier chromosomes (the other chromosomes). Background selection has already been shown to be efficient by previous theoretical work (e.g. Hillel *et al.*, 1990; Hospital *et al.*, 1992; Groen and Smith, 1995; Visscher *et al.*, 1996) and experimental work (e.g. Ragot *et al.*, 1995). I wish here to focus on recent theoretical developments achieved by our group on two aspects of background selection: the reduction of linkage drag around the target gene and the estimation of recipient genome content in BC progenies.

In any case, one must keep in mind that selection on markers in BC programmes is considered efficient if it permits a return to the recipient genome outside the target gene faster than the normal return rate when no selection on markers is applied (DGC halves at each generation). Hence, the efficiency of MAS should always be compared with this normal rate as a reference.

The Reduction of Linkage Drag in Marker-assisted Back-cross Programmes

The carrier chromosome deserves special consideration in BC programmes because, due to selection for the donor allele at the target locus in each generation, the rate of return to the recipient genotype on this chromosome is slower than on non-carrier chromosomes. Stam and Zeven (1981) provided an equation to calculate this rate of return when no selection on markers is applied. Based on a numerical comparison of these results with the known rate of return on non-carrier chromosomes (DGC halved each generation), Young and Tanksley (1989b) pointed out that the donor genes on the carrier chromosome were the most difficult to eliminate and could persist in the progenies long after the DGC on non-carrier chromosomes has returned to approximately zero if no selection on markers was applied. They provided impressive experimental proof of this statement, based on the a posteriori genotyping of a collection of tomato varieties previously introgressed for a resistance gene.

Size of intact donor chromosome segments around the target gene

The intact donor segment is in any BC generation the chromosome segment of donor origin containing the target locus, which has remained unaltered by cross-overs since the original cross between the donor and recipient parents. Hanson (1959) first provided the theoretical expression for the expected length of this intact segment. This was later revisited by Naveira and Barbadilla (1992), who also provided the corresponding variance. It is important to note that Stam and Zeven (1981) provided the total proportion of donor alleles on the carrier chromosome either on the intact segment or on other non-contiguous blocks of genes elsewhere on the carrier chromosome, which is a different measure of linkage drag. In fact, comparing numerically the proportion of donor alleles on the intact segment with the total proportion shows that the vast majority of unwanted donor alleles are located on the intact donor segment in advanced BC generations. Hence, I shall focus here only on the intact segment as a measure of linkage drag.

Hospital (2001) computed the mean and variance of the length of the intact donor segment around the target gene, when background selection is applied on two markers flanking the gene, one on each side (i.e. size of segment among ideotypes: individuals that are heterozygous at the target locus and homozygous for the recipient allele at both flanking markers) in any BC generation. The numerical results indicate that the expected length of donor segment on each side of the target gene is approximately half of the distance between the gene and the flanking marker in BC1, but the length at more advanced BC generations depends on the marker distance. For distant markers (more than 30 cM), the expected length of donor segment decreases in advanced BC generations, because recombination events accumulate between the target gene and the marker during successive meioses. This is no longer the case for shorter marker distances: for markers at 20 cM from the target gene or closer, the expected size of the donor segment in an advanced BC generation is approximately the same as the expected size in BC1. In this case, recombination events are rare and do not accumulate: in general, the genotypes selected experienced only one crossover, the one that permitted the flanking marker to return to recipient genotype. The basic conclusion is that selecting for distant markers over several successive BC generations cannot provide a better reduction of linkage drag than using close markers. Using very close markers is the only way to reduce linkage drag substantially.

Optimal population sizes

The above results refer to the length of the donor segment in individual genotypes homozygous for the recipient allele at both flanking markers (double recombinants) but say nothing about the probability of obtaining such genotypes. In a classical situation in plant breeding, where, among a whole population, a single individual can be selected and back-crossed to produce the population at the next generation, such probability obviously depends on population sizes. Obviously, using close markers, as recommended above, probably implies screening large populations, which generates large genotyping costs. It is thus important to optimize population sizes, i.e. determine the minimal population sizes (and genotyping effort) necessary to obtain the desired genotypes. Although it is intuitive that, for close-flanking markers, double recombinant genotypes are highly unlikely to be obtained in one single generation (BC1) so that at least two BC generations should be performed (Young and Tanksley, 1989b), the underlying mathematics has been worked out only recently. A solution was first derived by Hospital and Charcosset (1997). This result was used by Frisch *et al.* (1999) with numerical applications in the context of single-generation optimization (population size is optimized to permit the selection of a double recombinant genotype at generation $t + 1$, given that the genotype selected at generation t is known), whereas Hospital (2001) showed that a better optimization is obtained when considering all the planned generations simultaneously. The best optimization strategy is to: (i) determine the maximal number of BC generations that could be performed in a breeding programme; (ii) optimize simultaneously the population sizes at each of those previously defined generations before the programme is started; and (iii) refine the optimization at each generation, when the genotype of the selected individual is known. This requires some numerical computation. A computer program (POPMIN) that performs the corresponding numerical calculations easily was designed (F. Hospital and G. Decoux, 2002) and is freely available at http://moulon.inra.fr/~fred/programs. The results indicate that optimal population sizes should not be the same at each BC generation (using larger population sizes in advanced generations than in early generations reduces the overall number of individuals genotyped during the breeding scheme), as pointed out by Hospital and Charcosset (1997). More generally, the results indicate that a drastic reduction of linkage drag can be obtained at a reasonable cost by performing more than two BC generations. For example, for flanking markers as close as 2 cM on each side of the target gene, the minimum number of individuals that should be genotyped to obtain a double recombinant in BC1 is about 24,000. The same result can be obtained over two generations (BC2 strategy) by genotyping 290 individuals in BC1 and 500 in BC2. Finally, over three generations (BC3 strategy), the optimal population sizes are 120 individuals in BC1, 170 in BC2 and 370 in BC3. In all three strategies, the probability of obtaining a double recombinant for the flanking markers by the end of the breeding programme is above 99%. In the BC3 strategy, the probability of obtaining a double recombinant in BC2 is about 75%. If this happens, the programme is obviously not pursued until BC3 (unless for other reasons not considered here). Hence, planning to perform a maximum of three BC generations (BC3 strategy) permits one in 75% of the cases to obtain a double recombinant in BC2 by genotyping a total of only 290 individuals, which is much less than the 790 individuals necessary with the BC2 strategy. With the BC3 strategy, only in 25% of the cases should the programme be really conducted until generation BC3. Hence, averaging over all possibilities, the mean number of individuals that need to be genotyped to obtain a double recombinant with the BC3 strategy is only about 380, compared with an average of about 760 with the BC2 strategy. Hence, planning at the beginning of the programme to perform more than two BC generations is always a better strategy for optimizing the costs of genotyping (unless a rapid success is really mandatory). This is equivalent to fixing a not-too-low risk of failure per generation (risk of not obtaining a double recombinant at that generation), in particular in early BC generations, which is

the converse of what was advocated by Frisch et al. (1999). Obviously, the strategy and number of individuals to be genotyped should be reconsidered at each generation, once the genotype of the individual selected is known. This is also possible using our computer program POPMIN. In conclusion, planning to perform three or more BC generations and/or increasing the risk per generation has two main advantages. First, for given and affordable population sizes (genotyping effort), it permits a more drastic reduction of linkage drag. This is particularly useful for introgression of genes from exotic genetic resources, which may contain undesirable genes surrounding the gene of interest, for the manipulation of transgenic constructions (genetically modified organisms (GMOs)) when the introgression of the construction only is desired and close markers or better sequences of flanking regions are available, and/or for the derivation of near-isogenic lines (NIL) or congenic lines for the identification and validation of QTL. Secondly, planning to perform more than two BC generations increases the probability of success (obtaining a double recombinant) in advanced BC generations. The optimal population sizes above are defined such that at least one double recombinant is obtained with a given risk. It is then likely that on average more than one is obtained. Background selection for markers on non-carrier chromosomes is then possible among those double recombinants. This permits a better reduction of DGC on other chromosomes. Moreover, background selection on non-carrier chromosomes is more efficient in advanced BC generations (Hospital et al., 1992).

Background Selection on Non-carrier Chromosomes: Estimation of Donor Genome Content

Computation of multilocus genotype frequencies in complex pedigrees

This section is not just related to MAB breeding, though two applications in this field are given in the following sections. However, I want to mention these results because it can prove useful in various areas of MAS and GB theory, as well as for QTL detection.

Computing expected genotype frequencies at several loci (three or more) and/or in complex breeding schemes (back-crossing, hybrid mating, random mating, selfing, full-sib mating or any combination of these) is sometimes necessary in plant breeding. Actually, it is more and more frequent when using marker information, because many theoretical calculations are based on the probabilities of the different possible genotypes at markers (e.g. in QTL detection) or because one wishes to predict the probability of obtaining a particular genotype at markers or loci of interest (see an example above for the reduction of linkage drag in BC). However, such calculations are tedious and barely possible by hand. Hospital *et al.* (1996) have proposed a general algorithm to derive such probabilities automatically by recursion and have provided the corresponding *Mathematica* notebooks (http://moulon.inra.fr/~fred/programs). These recursions were implemented in a general program (MDM), which performs numerical and more powerful calculations, by Servin *et al.* (2002), also available at the above Web page. The programme MDM has various applications in plant and animal genetics. Two examples are provided below.

Precision graphical genotypes

To estimate the genomic composition of individuals using markers, the most basic estimate of DGC could be to score the genotype at the markers and then estimate DGC from the ratio of markers heterozygous for the donor allele over the total number of marker scores. This is a crude estimate that has the major drawback of being highly dependent upon the placement of markers along the genome. If markers are evenly spread and not too far apart from each other, the estimate is not correct (see below) but could be accepted. However, it is self-evident that, if markers are not evenly

distributed (the real situation), weighting them equally is clearly not the best solution.

A first attempt at providing a better estimate of DGC by taking the marker locations into account was made by Young and Tanksley (1989a), who introduced the concept of graphical genotypes to 'portray the parental origin and allelic composition throughout the genome'. This takes into account distances between markers in the sense that a chromosomal segment flanked by two markers of donor type (DD) is considered as 100% donor type, a chromosomal segment flanked by two markers of recipient type (RR) is considered as 0% donor type, and a chromosomal segment flanked by one marker of donor type and one marker of recipient type (DR) is considered as 50% donor type.

Using the program MDM, it is possible to compute, at any point of a segment flanked by two markers, the probability of being of donor type, given the genotypes at the markers and their locations. Averaging over all possible positions between the two markers provides an estimate of DGC: precision graphical genotype (PGG). This shows that the estimate of Young and Tanksley (1989a) is not always correct. In DD segments, DGC is below 100%, due to possible double crossovers between the markers. This error is minimal in BC1 and increases in more advanced BC generations. In RR segments, DGC is above 0%, due to possible double recombinations between the markers. This error is maximal in BC1 and decreases in more advanced BC generations. However, the errors on either DD or RR segments are numerically not very important. In the DR segment, DGC is exactly 50% in BC1 but decreases to below 50% in advanced BC generations. Paradoxically, although the estimate of Young and Tanksley on the DR segment is correct in BC1, it is for the same segments that the error is quantitatively the most important in advanced BC generations. As the general trend in back-crossing is to have more and more markers of recipient type in advanced BC generations, even with no selection on the markers, it is expected that many segments are of DR type: hence the overall error might be important.

Extending these results using MDM, B. Servin (unpublished data) has shown that, when estimating the DGC in a chromosomal segment flanked by two markers at a given generation, not only the genotypes of the two markers at that generation are informative, but the genotypes of the two markers at previous generations also matter, and so do the genotypes of non-flanking markers ('second' markers on the 'left' or on the 'right' of the segment, 'third' markers, and so on). Taking this additional information into account permits in some cases a gain in precision for the estimate of the most probable genotype at any point in the segment. This is useful for graphical genotypes and DGC estimates, but also for any purpose where this type of calculation is necessary, probably the most important one being QTL detection. Using simulation, it was shown that the correlation between the 'true' DGC and its estimate by MDM is very good (B. Servin, unpublished data). The program MDM can be included as a subroutine in any program performing such calculation (e.g. QTL detection programs) and should permit a gain in the precision of the corresponding estimates. However, the amount of this gain remains to be quantified and deserves more work.

Application to maize data

PGGs derived using MDM were applied to experimental data regarding marker-assisted introgression of three favourable QTL alleles between maize élite lines (Bouchez *et al.*, submitted). Three QTL were detected in a recombinant inbred line population. The favourable quantitative trait alleles (QTAs) at these three loci originating from the first parental lines were introgressed into the genomic background of the second parent through three crosses to the second parent (i.e. one non-segregating cross followed by two BCs), followed by one generation of selfing to fix the QTA in a homozygous state. This experiment shows that marker-assisted back-crossing can be used to manipulate QTAs between élite lines, although the validation of QTL effects in introgressed progenies appears easier for simple traits (e.g. earliness) than for more complex traits (e.g. yield), most probably because of stronger genotype–environment interactions. In any case, the experiment is among the few public experimental demonstrations of the efficiency of marker-based selection in BC programmes. In addition, the complex pedigree corresponding to this experiment was a challenging opportunity to apply the method of PGGs, using MDM to estimate the genome content of the products. The results show that, with only about 200 individuals genotyped per generation and a total of 15 markers on non-carrier chromosomes, the return to the recipient parent is close to 100% after two BC and one selfing generations. Chromosomal segments containing the three QTL were efficiently controlled by three markers per segment. However, the small population sizes did not permit a drastic reduction of linkage drag, which was not especially desired here because of the uncertainty about QTL locations. Comparing PGGs with the approximation of Young and Tanksley (1989b) described above indicates that the difference in the estimates can be important – up to ±8% genome content in some cases. The sign of the difference may vary from one chromosome to another, indicating that the error is probably more important qualitatively than quantitatively. The error is particularly important for chromosomal segments flanked by markers of different genotypes and in advanced BC generations, as expected. In particular, MDM can predict possible residual heterozygosity in the final material, where the other approximation obviously cannot.

One possibility is to use PGGs to provide a better estimate of genome content for a set of known markers. Conversely, since the estimate provided is more accurate, this should help reduce the number of markers genotyped and hence reduce the experimental costs. This remains under development.

References

Bouchez, A., Hospital, F., Gallais, A. and Charcosset, A. (2002) Marker-assisted introgression of favorable alleles at quantitative trait loci between maize elite lines. *Genetics* (submitted).

Charmet, G., Robert, N., Perretant, M.R., Gay, G., Sourdille, P., Groos, C., Bernard, S. and Bernard, M. (1999) Marker-assisted recurrent selection for cumulating additive and interactive QTLs in recombinant inbred lines. *Theoretical and Applied Genetics* 99, 1143–1148.

Dekkers, J.C.M. and van Arendonk, J.A.M. (1998) Optimizing selection for quantitative traits with information on an identified locus in outbred populations. *Genetical Research* 71, 257–275.

de Koning, G.J. and Weller, J.I. (1994) Efficiency of direct selection on quantitative trait loci for a two-trait breeding objective. *Theoretical and Applied Genetics* 88, 669–677.

Frisch, M., Bohn, M. and Melchinger, A.E. (1999) Minimum sample size and optimal positioning of flanking markers in marker-assisted backcrossing for transfer of a target gene. *Crop Science* 39, 967–975.

Groen, A.F. and Smith, C. (1995) A stochastic simulation study on the efficiency of marker-assisted introgression in livestock. *Journal of Animal Breeding and Genetics* 112, 161–170.

Hanson, W.D. (1959) Early generation analysis of lengths of heterozygous chromosome segments around a locus held heterozygous with backcrossing or selfing. *Genetics* 44, 833–837.

Hillel, J., Schaap, T., Haberfeld, A., Jeffreys, A.J., Plotzky, Y., Cahaner, A. and Lavi, U. (1990) DNA fingerprint applied to gene introgression breeding programs. *Genetics* 124, 783–789.

Hospital, F. (2001) Size of donor chromosome segments around introgressed loci and reduction of linkage drag in marker-assisted backcross programs. *Genetics* 158, 1363–1379.

Hospital, F. and Charcosset, A. (1997) Marker-assisted introgression of quantitative trait loci. *Genetics* 147, 1469–1485.

Hospital, F., Chevalet, C. and Mulsant, P. (1992) Using markers in gene introgression breeding programs. *Genetics* 132, 1199–1210.

Hospital, F., Dillmann, C. and Melchinger, A.E. (1996) A general algorithm to compute multi-locus genotype frequencies under various mating systems. *Computer Applications in the Biosciences* 12, 455–462.

Hospital, F., Goldringer, I. and Openshaw, S.J. (2000) Efficient marker-based recurrent selection for multiple quantitative trait loci. *Genetical Research* 75, 357–368.

Melchinger, A.E. (1990) Use of molecular markers in breeding for oligogenic disease resistance. *Plant Breeding* 104, 1–19.

Naveira, H. and Barbadilla, A. (1992) The theoretical distribution of lengths of intact chromosome segments around a locus held heterozygous with backcrossing in a diploid species. *Genetics* 130, 205–209.

Ragot, M., Biasiolli, M., Delbut, M.F., Dell'orco, A., Malgarini, L., Thevenin, P., Vernoy, J., Vivant, J., Zimmermann, R. and Gay, G. (1995) Marker-assisted backcrossing: a practical example. In: INRA (ed.) *Techniques et utilisations des marqueurs moléculaires*. Les Colloques, no. 72, INRA, Paris, pp. 45–56.

Servin, B., Dillman, C., Decoux, G. and Hospital, F. (2002) MDM: a program to compute fully informative genotype frequencies in complex breeding schemes. *Journal of Heredity* (in press).

Stam, P. and Zeven, A.C. (1981) The theoretical proportion of the donor genome in near-isogenic lines of self-fertilizers bred by backcrossing. *Euphytica* 30, 227–238.

van Berloo, R. and Stam, P. (1998) Marker-assisted selection in autogamous RIL populations: a simulation study. *Theoretical and Applied Genetics* 96, 147–154.

Visscher, P.M., Haley, C.S. and Thompson, R. (1996) Marker-assisted introgression in backcross breeding programs. *Genetics* 144, 1923–1932.

Young, N.D. and Tanksley, S.D. (1989a) Restriction fragment length polymorphism maps and the concept of graphical genotypes. *Theoretical and Applied Genetics* 77, 95–101.

Young, N.D. and Tanksley, S.D. (1989b) RFLP analysis of the size of chromosomal segments retained around the *tm-2* locus of tomato during backcross breeding. *Theoretical and Applied Genetics* 77, 353–359.

11 Complexity, Quantitative Traits and Plant Breeding: a Role for Simulation Modelling in the Genetic Improvement of Crops

M. Cooper,[1,2] D.W. Podlich,[1,2] K.P. Micallef,[2] O.S. Smith,[1] N.M. Jensen,[1] S.C. Chapman[3] and N.L. Kruger[2]

[1]Pioneer Hi-Bred International Inc., 7300 NW 62nd Avenue, PO Box 1004, Johnston, IA 50131, USA; [2]School of Land and Food Sciences, University of Queensland, Brisbane, Qld 4072, Australia; [3]CSIRO Plant Industry, 120 Meiers Road, Long Pocket Laboratories, Indooroopilly, Qld 4068, Australia

> . . . no one believes an hypothesis except its originator but everyone believes an experiment except the experimenter.
>
> (W.I.B. Beveridge, 1950)

Modelling: a Historical Perspective

Modelling has a long history in genetics. Arguably, it is as old as the concept of proposing testable hypotheses about the behaviour of genotype–environment systems. Therefore, it can be thought of as extending back to and before the founding works of Darwin, Wallace and Mendel. Perhaps we can think of the 21st century as our third century in this scientific enterprise. It is worth noting that successful application of plant breeding has a much longer history than its modern implementation based on our scientific understanding of genetics. The objective of this chapter is to speculate about some of the directions that will be taken in the application of modelling methods in quantitative genetics and plant breeding in the 21st century.

The modelling process in genetics has taken various forms throughout its history.

From the perspective of the discipline of quantitative genetics today, the 19th century was dominated by a growing qualitative appreciation of the implications of inheritance. This appreciation was gleaned and moulded from a dispersed mixture of observational studies of animal and plant diversity and, by today's standards, a limited body of experimental work. Also the formal methods of statistical analysis that we are familiar with today were in their infancy in the second half of 19th century. In the 20th century, the emergence of quantitative genetics as a discipline within genetics is more clearly observed. The neo-Darwinian synthesis of the evolutionary process grew out of the work of key individuals in the first half of the century. The most obvious contributions came from the work by Fisher, Wright and Haldane. Much has been made of and written about the debate between Fisher and Wright on the relationships between and different emphases in their models of the evolutionary process. However, much less attention has been given to the dominant influence of Fisher's views over those of Wright in the development of the modern theory and practice of plant and animal breeding. As we learn more about the detail of the genetic architecture of quantitative traits, it is interesting to contemplate whether this dominance will continue or whether some form of resolution of the two models will emerge.

From the perspective of plant-breeding research, the last quarter of the 20th century was dominated by the possibilities and promises of the molecular biology revolution, particularly the use of transgenics, molecular-marker methodologies and, more recently, structural and functional genomics. Thus, as we move into the 21st century, the practice of plant breeding is shifting from a foundation that was previously based largely on inferring genetics by measurement of phenotype and selection among individuals and families based on their phenotypes to one that is based on direct measurement and manipulation of the genetic code in association with field-based evaluation of phenotypes (Koornneef and Stam, 2001) and a growing preoccupation with the biology of the gene–phenotype relationships that underpin the traits that are the targets for improvement within breeding programmes. The 2010 proposal for investigating the functional genomics of the model plant *Arabidopsis* demonstrates the high level of optimism among plant geneticists that a comprehensive understanding of gene–phenotype relationships is feasible (Somerville and Dangl, 2000). This changing paradigm in plant genetics and plant breeding has many implications for the modelling processes that will be used in quantitative genetics.

An interesting trend that has been stimulated by the improvements in our potential ability to investigate gene–phenotype relationships for traits is that of questioning many of the common assumptions that were made in the formulation of the classical quantitative-genetic theory for polygenic traits. It is widely recognized that many of these assumptions were made to enable tractable treatments of the quantitative-genetic models. It is also widely understood by those involved in developing these models that they are approximations of the genetic architecture of quantitative traits. The need,

feasibility and mechanisms used to relax these assumptions, in order to make the quantitative-genetic models more flexible for a wider range of specific conditions, is often a topic of debate. One thing is clear, algebraic derivation of genetic models is difficult when the simplifying assumptions about model components are relaxed, e.g. number and distribution of effects of loci, epistasis, gene–environment interactions, linkage and epigenetic effects. The complexity of these extended models generally limits their accessibility as tools for the wider plant-breeding audience. Kempthorne (1988) discussed this trend and its implications for model development in quantitative genetics. He concluded that the classical approach of developing tractable algebraic models would have limited applicability as we uncover and want to include more of the detail of the genetic architecture of the traits in our models. He also made a case for the use of high-speed computing as a practical approach for the analysis of more complex genetic models.

With the availability of appropriate hardware and software computing environments, computer simulation has been increasingly used as a methodology for modelling in quantitative genetics. Fraser and Burnell (1970) gave a synthesis of much of the early work. A search of the last 30 years of literature that is stored in the CAB, BA and AGRICOLA electronic referencing databases suggests an increasing trend in the use of simulation methodology in the plant sciences (Fig. 11.1). While it is difficult to interpret trends over time in these databases, since they are a recent innovation and many of the older articles were not published in a way that takes advantage of these archive and search facilities, it is clear that simulation methodology is now being widely used in plant genetics and plant-breeding research. Many of the software implementations used for these applications of simulation methodology are problem-specific. This makes it difficult to modify the software for other questions or problems. We have developed hardware (QCC: QU-GENE Computing Cluster) and software (QU-GENE: Quantitative Genetics) simulation tools that

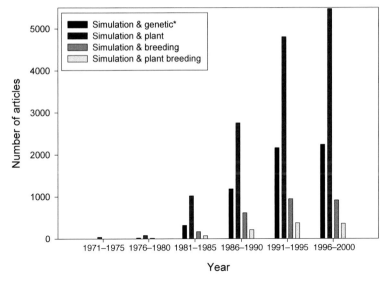

Fig. 11.1. Number of articles published in the last 30 years with the word simulation and either genetic* (* represents all extensions of the word genetic), breeding, plant or plant breeding as a word anywhere in the CAB (1984–2000), BA (1985–2000) and AGRICOLA (1970–2000) databases.

can be applied to a wide range of issues involved in computer modelling of plant-breeding programmes (Podlich and Cooper, 1998; Cooper *et al.*, 1999; Micallef *et al.*, 2001). In this chapter, we shall give an overview of aspects of this work and discuss some applications and research directions.

Quantitative Traits: Simple to Complex

The detail of the genetic architecture of traits contributes to the degree of complexity that is associated with understanding their inheritance and manipulating them in a breeding programme. Many of the important traits that are targeted for manipulation in breeding programmes are quantitative in nature. From a quantitative genetic and plant breeding perspective, the properties of a trait that should be considered in describing its genetic architecture include:

- The number of genes and the distribution of the size of their effects on trait phenotypes.
- The number of alleles for each gene, the distribution of their effects and their frequency within a reference population.

- The linkage arrangements among the genes and the distribution of allele haplotypes within a reference population.
- The extent and form of epistatic interactions among the genes.
- The extent and form of gene–environment interactions.
- The extent and form of any pleiotropic effects of the genes.

Many of these properties of trait architecture are not static within a population or, indeed, within a single plant or crop stand over the duration of a growing season. Therefore, it is necessary to consider their potential to change as plant populations are modified over time in the breeding programme. One property that has the potential to become important when considering population structure over cycles of intermating is the physical distribution of recombination hot spots and cold spots across the genome and the effects of these on the distribution of allele haplotypes and patterns of linkage disequilibrium within the reference populations of a breeding programme.

The availability of molecular-marker technologies and their use to construct high-density genetic maps for model species and the important agricultural plant species have increased our capacity to investigate a number of these properties of trait architecture. Analysis of the phenotypic variation for traits by use of molecular-marker maps in appropriate segregating populations has enabled the resolution of the continuous trait distribution into components associated with regions of the genome, usually referred to as quantitative trait loci (QTL), for a number of traits (e.g. Paterson, 1998). Over the last two decades of the 20th century, QTL mapping became commonplace for many breeding programmes. This work has taught us a lot about the statistical power of this methodology (Beavis, 1998) and given us some clues about the genetic architecture of some quantitative traits (Kearsey and Farquhar, 1998). In most cases, it has been relatively easy to identify large regions of the genetic map (10–30 cM) that are associated with phenotypic variation in specific experiments. A few studies have dissected these large genomic regions into smaller regions, using fine-mapping methods (1–5 cM). Less commonly, targeted QTL regions that have been fine-mapped have been sequenced and a specific gene responsible for the phenotypic variation has been identified and studied (Lukens and Doebley, 1999; Frary *et al.*, 2000).

Together, advances in DNA-sequencing capability and the development of technologies for quantifying gene-expression patterns have catalysed a new range of methods for studying the structural and functional properties of genomes. The complete DNA sequences of several prokaryotic and eukaryotic organisms are now available. Thus, physical maps of the distribution of genes throughout plant genomes are now available for some species. For a number of species, work is under way to align the physical and genetic maps and to investigate genetic diversity at the sequence level for key regions of genomes. Already, this work has extended our thinking about the genetic architecture of many of the traits we have previously studied only at the phenotypic level. Consideration is now being given to the way in which genes operate in networks to influence growth and development, response to environmental signals and the implications of these gene networks for the phenomena of epistasis, pleiotropy and gene–environment interactions. Thus, to our list of the properties that need to be considered in studying the genetic architecture of quantitative traits, we could add:

- Description of the gene networks that underlie traits.
- Definition of the patterns of interconnections among the genes within gene networks.
- Extracellular signal-detection mechanisms and intracellular signal-transduction pathways.
- The importance and influence of *cis*- and *trans*-acting factors in the regulation of gene expression.
- The sensitivity and robustness of these gene networks to perturbations.
- The relationships between the gene networks and the biochemical and physiological processes associated with trait expression at the phenotypic level.

If anything, the developments in molecular genetics and molecular biology have motivated the need for parallel development of modelling capabilities that enable *in silico* evaluation of plant-breeding methodologies for a range of genetic architectures that go beyond many of the assumed models used in classical quantitative genetics.

To understand and quantify the capacity of any breeding strategy to improve the phenotype of a trait, it is necessary to combine consideration of the genetic architecture of the traits with the heritability of the trait in the reference population of the breeding programme. It is daunting to contemplate the complete description of the genetic architecture of a trait for the purposes of directing its genetic improvement in a breeding programme. However, to enable the design of efficient breeding strategies to achieve short-term genetic gain, it is not necessary to consider variation for all of the genes involved in determining the gene–phenotype relationship. For a given reference population, at a given point in time, it can be argued that it is only the segregating genes that need to be considered in evaluating the relative merits of alternative breeding strategies. In contrast, this issue is less clear and the argument of focusing only on the extant allelic variation at segregating loci becomes less compelling when we consider long-term genetic improvement and management of genetic resources.

The above list of considerations, while not intended to be exhaustive, is at least indicative of the magnitude of the task that is involved in establishing a flexible platform for modelling breeding programmes. As a first step in outlining one approach that we have adopted, it is useful to qualitatively sort traits into categories that are based on the factors that contribute to the complexity of their genetic architecture and the heritability of the trait (Table 11.1). Here we have emphasized gene number and the distribution of the size of gene effects, epistasis, gene–environment interactions, linkage and the influence of experimental error on heritability. It is recognized that the concepts of complexity and heritability are continuous and both can be investigated accordingly. Therefore, any preliminary complexity–heritability categorization, such as that in Table 11.1, is somewhat arbitrary and only serves as a first approximation in analysis. However, we have found that this form of categorization is often a useful first step in any investigation.

An objective we have pursued, using the categorization described in Table 11.1, is the analysis of the power of a range of breeding strategies to achieve a response to selection across cycles of a breeding programme for a large number of genetic models representing each of the four broad complexity–heritability categories. To achieve this requires quantification of the properties of the genetic model for each of the four categories described in Table 11.1. For the two categories identified as low-complexity, this is relatively straightforward, and much has already been learnt from the classical quantitative-genetic approaches (e.g. Falconer and Mackay, 1996; Kearsey and Pooni, 1996; Lynch and Walsh, 1998). However, for the two complex categories, this required consideration of how we could incorporate the effects of epistasis and gene–environment interactions into the models. While there is some capacity to consider these properties within the classical linear modelling framework used in quantitative genetics, we sought greater flexibility to consider a wider range of epistatic and gene–environment interaction models. The objective here was to enable us to accommodate some of the non-linear effects expected of gene networks. Some properties of this modelling framework are discussed below.

The E(N:K) model as a framework

In this section, we give an overview of a process that can be used for setting up

Table 11.1. Trait characterization into qualitative categories based on the influences of experimental error on heritability and the number of genes, distribution of gene effects, epistasis, gene–environment interactions and linkage on the complexity of the genetic architecture of the trait.

| Heritability | Genetic architecture complexity | |
	Low	High
Low	High experimental error Few genes with major effects No epistasis No gene–environment interactions No linkage	High experimental error Many genes: major and minor effects Epistasis Gene–environment interactions Linkage
High	Low experimental error Few genes with major effects No epistasis No gene–environment interactions No linkage	Low experimental error Many genes: major and minor effects Epistasis Gene–environment interactions Linkage

simulation investigations to evaluate breeding strategies. The $E(N:K)$ model provides a quantitative framework that allows the construction of relationships between genetic-network models, the quantitative effects of the genes in these networks and gene–phenotype relationships. Kauffman (1993) discussed the foundations of the NK model of gene networks. Within his model, N is the number of genes in the network and K measures the number of genes that have an epistatic influence on the effect of another gene. Thus, genetic background effects, which make the effects of the alleles of the N genes context-specific, can be introduced by the K parameter. To take into consideration the potential for gene–environment interactions, we extended the NK model to allow the effects of the genes in these networks and the form of these networks to change among environments (Podlich and Cooper, 1998). Thus, different NK models can be completely or partially nested within environment types. Here environment types are defined as sets of environmental conditions that contribute to gene–environment interactions. Thus, in the $E(N:K)$ model, E is the number of environment types that contribute to gene–environment interactions within a 'target population of environments' (TPE).

For example, if we consider a simple network based on three genes, where each gene is influenced by one of the other genes, for the NK model, $N = 3$ and $K = 1$ (Fig. 11.2a). If the effects of the genes in the network change between two environment types, then $E = 2$. Thus, this example would be defined as an $E(N:K) = 2(3 : 1)$ model. The colon is used here simply as a mechanism to separate the N and K values when numbers replace the letters. To emphasize the relationship between K and some of the classical quantitative-genetic models used to describe types of epistasis, consider any pair of genes in this network (e.g. genes A and B). The form of their epistatic interaction can be considered to be analogous to what would be classically referred to as digenic epistasis, i.e. there are two genes and one gene ($K = 1$) has an epistatic influence on the other. Similarly, for networks of larger numbers of

genes, if $K = 2$, this would be analogous to trigenic epistasis. Also it is possible to relate different families of $E(N:K)$ models to some of the classical models of quantitative genetics. For example, the family of all possible $1(N:0)$ models would represent the additive genetic models where there is no epistasis or genotype–environment interaction. With $N = 1$, this would be the classical Mendelian locus model. Where $N > 1$ and $K = 0$, we can consider a wide range of additive finite locus models. As we allow N to increase, we can consider the case where N is very large ($N \rightarrow \infty$), approaching a practical approximation of the infinitesimal model. To introduce gene–environment interactions, we let $E > 1$ and similarly, to introduce epistasis, we let $K > 0$. The example described in Fig. 11.2 is only one of many possible ways of defining gene networks. Figure 11.3 shows some forms of gene network, for $N = 6$, that we have considered: (a) interconnected network, (b) disconnected networks, (c) cascade network, and (d) mixtures of interconnected and disconnected networks. The engine of the QU-GENE simulation software platform (Podlich and Cooper, 1998; Fig. 11.4) provides a number of ways for specifying and visualizing the properties of different $E(N:K)$ models.

As Kauffman (1993) discussed for evolutionary scenarios, we currently have little understanding of the detail of most of the gene networks that underlie fitness traits. This is equally true for most of the traits targeted in breeding programmes. However, uncovering this type of detailed information is a research objective of many groups. In the absence of an accurate representation of the genetic architecture of our target traits, it is possible to simulate the properties of many possible $E(N:K)$ models of putative architectures by drawing the effects of the genes from some underlying frequency distribution of effects. Kauffman (1993) considered a number of types of distribution in his analysis of haploid NK models. He found that many of the properties of the NK model were insensitive to the type of distribution from which the effects were drawn. We are currently examining a range of types of distributions for diploid $E(N:K)$ models (uniform, normal,

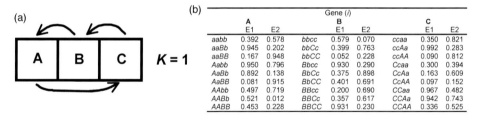

(a)

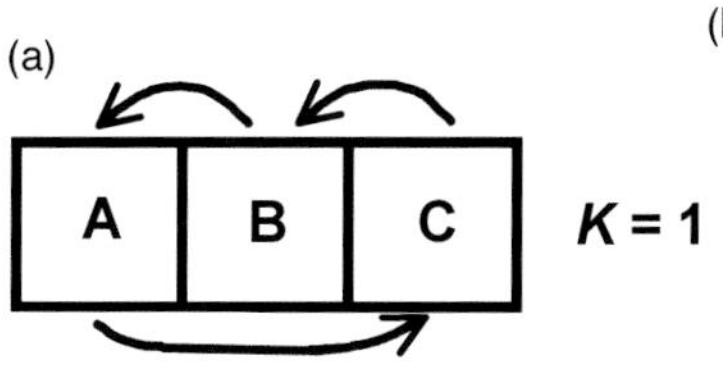

(b)

	Gene (*i*)							
	A			**B**			**C**	
	E1	E2		E1	E2		E1	E2
aabb	0.392	0.578	*bbcc*	0.579	0.070	*ccaa*	0.350	0.821
aaBb	0.945	0.202	*bbCc*	0.399	0.763	*ccAa*	0.992	0.283
aaBB	0.167	0.948	*bbCC*	0.052	0.228	*ccAA*	0.090	0.812
Aabb	0.950	0.796	*Bbcc*	0.930	0.290	*Ccaa*	0.300	0.394
AaBb	0.892	0.138	*BbCc*	0.375	0.898	*CcAa*	0.163	0.609
AaBB	0.081	0.915	*BbCC*	0.401	0.691	*CcAA*	0.097	0.152
AAbb	0.497	0.719	*BBcc*	0.200	0.690	*CCaa*	0.967	0.482
AABb	0.521	0.012	*BBCc*	0.357	0.617	*CCAa*	0.942	0.743
AABB	0.453	0.228	*BBCC*	0.931	0.230	*CCAA*	0.336	0.525

(c)

	Environment Type								TPE
	1				2				
Genotype	w_{11}	w_{21}	w_{31}	W_1	w_{12}	w_{22}	w_{32}	W_2	W_T
aabbcc	0.392	0.579	0.350	*0.440*	0.578	0.070	0.821	*0.490*	**0.465**
aabbCc	0.392	0.399	0.300	*0.364*	0.578	0.763	0.394	*0.578*	**0.471**
aabbCC	0.392	0.052	0.967	*0.470*	0.578	0.228	0.482	*0.429*	**0.450**
aaBbcc	0.945	0.930	0.350	*0.742*	0.202	0.290	0.821	*0.438*	**0.590**
aaBbCc	0.945	0.375	0.300	*0.540*	0.202	0.898	0.394	*0.498*	**0.519**
aaBbCC	0.945	0.401	0.967	*0.771*	0.202	0.691	0.482	*0.458*	**0.615**
aaBBcc	0.167	0.200	0.350	*0.239*	0.948	0.690	0.821	*0.820*	**0.529**
aaBBCc	0.167	0.357	0.300	*0.275*	0.948	0.617	0.394	*0.653*	**0.464**
aaBBCC	0.167	0.931	0.967	*0.688*	0.948	0.230	0.482	*0.553*	**0.621**
Aabbcc	0.950	0.579	0.992	*0.840*	0.796	0.070	0.283	*0.383*	**0.612**
AabbCc	0.950	0.399	0.163	*0.504*	0.796	0.763	0.609	*0.723*	**0.613**
AabbCC	0.950	0.052	0.942	*0.648*	0.796	0.228	0.743	*0.589*	**0.619**
AaBbcc	0.892	0.930	0.992	*0.938*	0.138	0.290	0.283	*0.237*	**0.588**
AaBbCc	0.892	0.375	0.163	*0.477*	0.138	0.898	0.609	*0.548*	**0.513**
AaBbCC	0.892	0.401	0.942	*0.745*	0.138	0.691	0.743	*0.524*	**0.635**
AaBBcc	0.081	0.200	0.992	*0.424*	0.915	0.690	0.283	*0.629*	**0.527**
AaBBCc	0.081	0.357	0.163	*0.200*	0.915	0.617	0.609	*0.714*	**0.457**
AaBBCC	0.081	0.931	0.942	*0.651*	0.915	0.230	0.743	*0.629*	**0.640**
AAbbcc	0.497	0.579	0.090	*0.389*	0.719	0.070	0.812	*0.534*	**0.461**
AAbbCc	0.497	0.399	0.097	*0.331*	0.719	0.763	0.152	*0.545*	**0.438**
AAbbCC	0.497	0.052	0.336	*0.295*	0.719	0.228	0.525	*0.491*	**0.393**
AABbcc	0.521	0.930	0.090	*0.514*	0.012	0.290	0.812	*0.371*	**0.443**
AABbCc	0.521	0.375	0.097	*0.331*	0.012	0.898	0.152	*0.354*	**0.343**
AABbCC	0.521	0.401	0.336	*0.419*	0.012	0.691	0.525	*0.409*	**0.414**
AABBcc	0.453	0.200	0.090	*0.248*	0.228	0.690	0.812	*0.577*	**0.412**
AABBCc	0.453	0.357	0.097	*0.302*	0.228	0.617	0.152	*0.332*	**0.317**
AABBCC	0.453	0.931	0.336	*0.573*	0.228	0.230	0.525	*0.328*	**0.451**

(d)

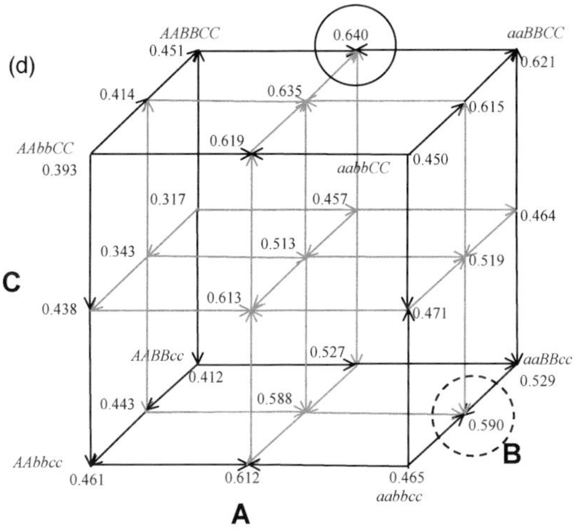

Fig. 11.2. An example of the process of defining genotypic values for an $E(N{:}K) = 2(3{:}1)$ gene network model: (a) network diagram of the epistatic relationships among the three genes; (b) the effects for the nine genotypic combinations for each of the three digenic epistatic sets (A = B → A; B = C → B; and C = A → C) in the two environment-types (E1 and E2) drawn from the uniform distribution; (c) the 27 genotype values for the combinations of three genes in environment type 1 (W_1), environment type 2 (W_2) and the target population of environments (TPE) (W_T); (d) a Boolean hypercube displaying the 27 genotypes in three-dimensional one-mutant neighbour genetic space; W_T genotype values are located beside each genotype, arrows indicate directions of increasing W_T values between one-mutant neighbours, circles indicate adaptive peaks where all one-mutant neighbours have lower W_T values and the solid circle identifies the global peak for the network and the dashed circle a local peak.

(a)   **Interconnected**

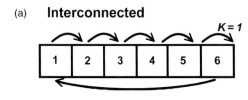

$K = 1$

(b)   **Disconnected**

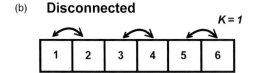

$K = 1$

(c)   **Cascade**

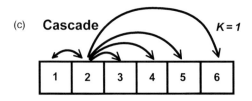

$K = 1$

(d)   **Mixture of connections**

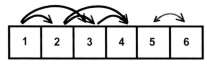

$K = 1$ & $K = 2$
Interconnected and disconnected

Fig. 11.3. Examples of four network diagrams emphasizing different types of network relationships for networks based on six genes: (a) interconnected network; (b) disconnected networks; (c) cascade network; and (d) a mixture of the interconnected and disconnected network types.

exponential). Here we consider some examples where the parameters for the $E(N:K)$ model are drawn from the uniform distribution, i.e. making few assumptions about the effects of the genes. Clearly, as we learn more about the genetic architecture of traits, the random effects that are used in the examples considered here can be replaced by specific gene information from a number of information sources. Sources of information would include results from applying a number of forward and reverse genetic approaches to experimental populations, e.g: (i) experimental estimates of QTL effects from

whole-genome and fine-mapping studies; (ii) experimental estimates of genetic parameters for major genes that behave as Mendelian loci; (iii) estimates of the effects of genes from targeted mutation and gene-expression studies; and (iv) experiments conducted to evaluate the effects of candidate genes. It is expected that an important source of basic data on plant gene–environment systems will come from work on the model species *Arabidopsis* (Somerville and Dangl, 2000).

For the example network shown in Fig. 11.2a, we consider the case where there are two environment types ($E = 2$), and determine the genotypic values for the 27 possible genotypes comprising the three digenic sets, $B \rightarrow A$, $C \rightarrow B$, and $A \rightarrow C$, indicated by the arrows connecting the genes (Fig. 11.2a). For each digenic set, there are nine genotypic combinations (Fig. 11.2b). For each of the nine combinations and within each of the three genotypic sets, a value is drawn at random from the uniform distribution for environment type 1 (E1) and independently for environment type 2 (E2). For any given genotype, the appropriate gene effects are combined to give the value of the genotype in an environment type. There are several ways of combining these gene effects. Here we use the approach adopted by Kauffman (1993) and compute the value of a genotype in environment j (W_j) as the mean of the effects for the digenic combinations that contribute to the genotype (Fig. 11.2c), e.g. for genotype *aabbcc* in E1 combine digenic sets *aabb*, *bbcc* and *aacc* from E1 in Fig. 11.2b. To estimate the genotypic value for the target population of environments (W_T), we compute the weighted mean of the genotypic values from each environment type, where the weights are the frequencies of occurrence of the environment types in the target population of environments (Fig. 11.2c). For the example in Fig. 11.2, both environment types are assumed to occur with equal frequency, i.e. they are both likely to be encountered 50% of the time in the TPE.

Many options exist to graphically represent the genetic space for these gene-network models. For the simple three-gene case in Fig. 11.2, the genotypes can be arranged into a three dimensional Boolean

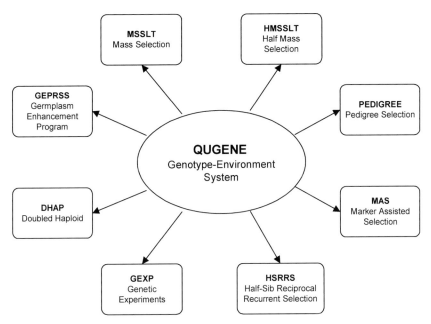

Fig. 11.4. Schematic outline of the structure of the QU-GENE simulation software platform. The central ellipse represents the engine (QUGENE) where the properties of the $E(N{:}K)$ models are specified and the surrounding boxes represent the application modules used to model breeding programmes. (Adapted from Podlich and Cooper, 1998.)

hypercube, where the genotypes are located on the axes of the cube beside their one-mutant neighbours (Fig. 11.2d). Additional information can be added to the hypercube. For example, the genotypic values in the individual environment types or the TPE can be added. It is then possible to identify directions of increasing genotypic value between the one-mutant neighbours and arrows can be added to the axes to highlight these directions. Trajectories across the genetic space, as represented by the cube, can be highlighted by following paths of increasing genotypic value between one-mutant neighbours on the cube. A property of the genetic space that is interesting to observe is the number and the distribution of end-points of these one-mutant neighbour walks across genetic space. These end-points can be thought of as representing performance or adaptive peaks in the genetic space. Further, the regions of the genetic space that lead to these adaptive peaks can be thought of as basins of attraction. In the example (Fig. 11.2d), there are two adaptive

peaks for the TPE. Within this framework, distinctions can be drawn between local and global adaptive peaks in the genetic space and the extent to which there is co-location of these peaks. For our example, the global peak is genotype $AaBBCC$ ($W_T = 0.640$) and there is one local adaptive peak, genotype $aaBbcc$ ($W_T = 0.590$).

A number of statistics can be used to describe the structure of the $E(N{:}K)$ model genetic space using the derived genotypic values (W_j and W_T). One statistic that we have found useful is the correlation of the genotypic values of one-mutant neighbours in the genetic space. Consider three $E(N{:}K)$ $= 2(3{:}K)$ models, where $K = 0$ (Fig. 11.5a), $K = 1$ (Fig. 11.5c) and $K = 2$ (Fig. 11.5e). In all three cases the gene effects were drawn from the uniform distribution following the procedure described in Fig. 11.2. For $K = 0$ (no epistasis; Fig. 11.5a), we have an additive model with one adaptive peak, in this case $aabbcc$ ($W_T = 0.569$). In all cases where $K = 0$, there will be one adaptive peak in the TPE. There will also be only one adaptive peak for

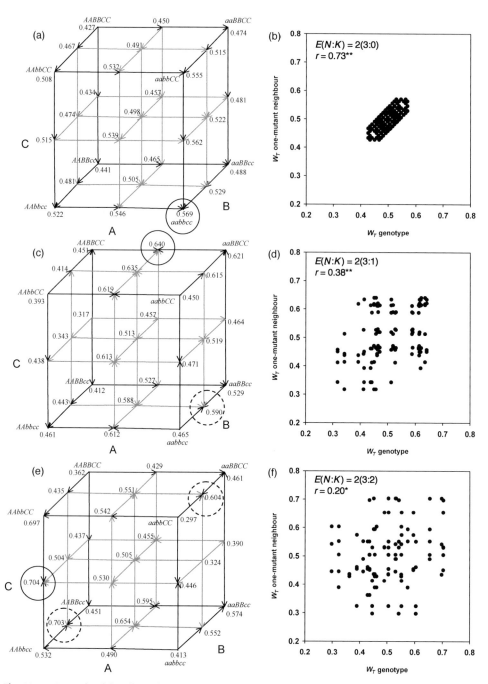

Fig. 11.5. Example of the effect of changing *K* for a three-gene model on the one-mutant genotype value (*W_T*) relationships depicted on the Boolean hypercube, number of adaptive peaks and one-mutant correlation coefficients: (a) Boolean hypercube for *E*(*N*:*K*) = 2(3:0); (b) scatter diagram of the *W_T* genotype values for all 27 genotypes and each of their one-mutant neighbours for *E*(*N*:*K*) = 2(3:0); (c) Boolean hypercube for *E*(*N*:*K*) = 2(3:1); (d) scatter diagram of the *W_T* genotype values for all 27 genotypes and each of their one-mutant neighbours for *E*(*N*:*K*) = 2(3:1); (e) Boolean hypercube for *E*(*N*:*K*) = 2(3:2); (f) scatter diagram of the *W_T* genotype values for all 27 genotypes and each of their one-mutant neighbours for *E*(*N*:*K*) = 2(3:2).

each environment type, but the genotype that possesses the peak can differ among environment types when the effects are drawn independently for each environment. The correlation of the one-mutant neighbours for $K = 0$, in this example, was $r = 0.73**$ (Fig. 11.5b). For both $K = 1$ (Fig. 11.5c) and $K = 2$ (Fig. 11.5e), there was more than one adaptive peak for the TPE. Generally, it is found that, as K increases, the number of local peaks in the genetic space increases. This increase in ruggedness of the genetic space results in a reduction in the correlation of the one-mutant neighbours. In this example, for $K = 1$, the one-mutant correlation $r = 0.38*$ (Fig. 11.5d) and, for $K = 2$, the one-mutant correlation $r = 0.20*$ (Fig. 11.5f).

This quantitative framework can be extended for larger values of E, N and K and many alternative samples of gene effects can be drawn from the selected underlying distributions. While it is feasible to graphically represent two-gene and three-gene networks in the manner of Fig. 11.2, obviously it is more difficult to visualize networks based on larger numbers of genes. However, auto-correlation functions can be applied to quantify the ruggedness of larger neighbourhood

regions in the genetic space that is defined by these larger $E(N{:}K)$ models. Figure 11.6 is a schematic representation of the general trends that we observe across levels of N as both E and K are increased. When $E = 1$ and $K = 0$, there is only a single adaptive peak. When $K > 0$, the ruggedness of the genetic space increases and when $K = N - 1$, the landscape is random and therefore uncorrelated. When $E > 1$, there are shifts in the adaptive peak between the environment types. When $E > 1$ and $K > 0$, there is the potential for multiple peak shifts among the environment types. Thus, by changing the levels of E, N and K, we can use the $E(N{:}K)$ framework to examine a range of trait architectures that range from simple to complex. The important point here is not the detail of any one genetic architecture or set of genetic effects specified in the models, but rather the flexibility that the framework gives for considering many families of genetic models and many parameterizations. This enables investigation of the properties of many putative genetic models within the QU-GENE simulation platform.

We refer to the response surface for the genetic space as defined by the $E(N{:}K)$ model

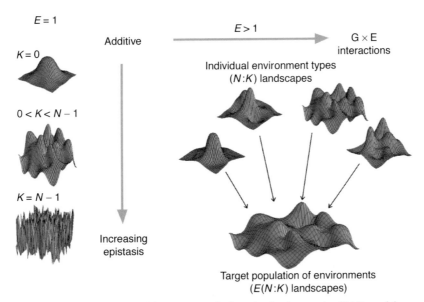

Fig. 11.6. Schematic representation of the structure of adaptation landscapes for $E(N{:}K)$ models as epistasis is introduced by increasing K from $K = 0$ (additive) to $K = N - 1$ (uncorrelated) and genotype–environment interactions are introduced by increasing E from $E = 1$ to $E > 1$.

as an adaptation landscape. Clearly, these concepts are derived from and related to the fitness landscape concepts defined and popularized by Wright (1932). The shape or ruggedness of the adaptation landscape is one way of quantifying and visualizing the complexity of the gene–phenotype relationship for a given genetic architecture. This quantification of the properties of the genetic space enables consideration of both global and local landscape features and their potential for impact on the effectiveness of the plant-breeding process. This has significant advantages when we want to distinguish between modelling quantitative traits in general (global landscape features) and modelling the specific situation that is faced by a particular breeding programme (local landscape features). The recent advances in the range of molecular tools for characterizing the structural and functional properties of the genome, constructing genetic

and physical maps and studying gene-expression patterns provide a sound basis for modelling the genetic architecture of traits at the genomic level and modelling the gene–phenotype relationships for the traits.

In addition to the statistics described above, the properties of the adaptation landscapes that are associated with different $E(N{:}K)$ models can be analysed in terms of the more familiar quantitative genetic concepts of additive and non-additive gene effects and genetic variance. Computing the additive effects of genes for a range of $E(N{:}K)$ models, where $N = 12$ and the structure of the networks is disconnected (Fig. 11.3b), we observe as an emergent property that the typical distribution of effects shows a few genes with large effects and a greater number of genes with smaller effects (Fig. 11.7). These distributions of gene effects have features in common with those of experimentally determined QTL effects (Kearsey

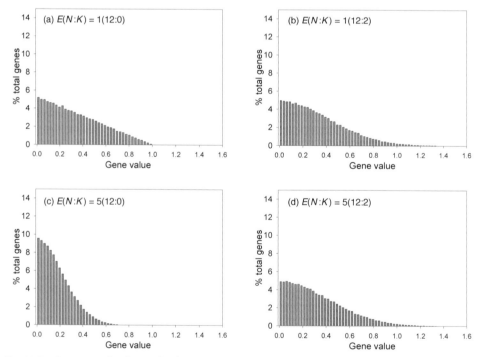

Fig. 11.7. Frequency distribution for the additive effects of genes (Gene value) for (a) an additive model, $E(N{:}K) = 1(12{:}0)$; (b) a model with epistasis but no genotype–environment interactions, $E(N{:}K) = 1(12{:}2)$; (c) a model with genotype–environment interactions but no epistasis, $E(N{:}K) = 5(12{:}0)$; and (d) a model with epistasis and genotype–environment interactions, $E(N{:}K) = 5(12{:}2)$. Distributions based on the simulated effects for 12,000 genes.

and Farquhar, 1998). In this example, based on 12,000 simulated genes within $E(N:K)$ networks, the exact shape of the distribution of additive gene effects depended on the values of E and K. Similarly, the distributions of genotype values tend to be symmetrical and the exact form of the distribution also depends on the values of E and K (Fig. 11.8). Partitioning the total genotypic variation into additive and non-additive components of variance and their interaction with environments indicates increasing complexity, and therefore degree of difficulty for genetic improvement, as both E and K increase (Fig. 11.9). Therefore, using the $E(N:K)$ framework, families of models can be defined to generate genetic scenarios ranging from simple to highly complex. This serves as a basis for using simulation to investigate the relative merits and efficiencies of breeding strategies.

Modelling Breeding Programmes as Search Strategies Exploring Genetic Space

While we are working to advance our capacity to investigate and understand the genetic architecture of traits, we have, at best, a limited understanding of the power of plant-breeding strategies to bring about the desired genetic changes for traits with complex genetic architectures. Computer simulation can be used to gain an understanding of the relationship between the efficiency of a breeding strategy and the genetic architecture of traits. Plant-breeding strategies can be evaluated in combination with the $E(N:K)$ framework for their efficiency as search strategies on the adaptation landscapes associated with the different genetic models. With high-speed computing, it is possible to rapidly evaluate many

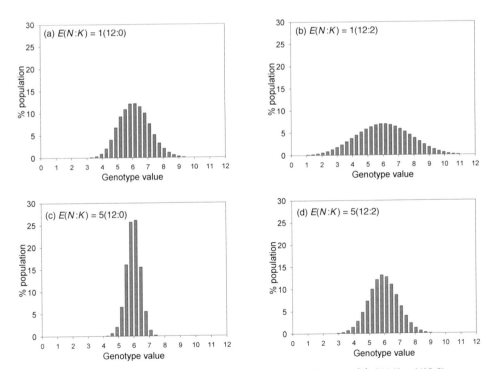

Fig. 11.8. Frequency distribution of genotypic values for: (a) an additive model, $E(N:K) = 1(12:0)$; (b) a model with epistasis but no genotype–environment interactions, $E(N:K) = 1(12:2)$; (c) a model with genotype–environment interactions but no epistasis, $E(N:K) = 5(12:0)$; and (d) a model with epistasis and genotype–environment interactions, $E(N:K) = 5(12:2)$. Distribution for each $E(N:K)$ model is based on the 1000 sets of gene effects drawn from the uniform distribution.

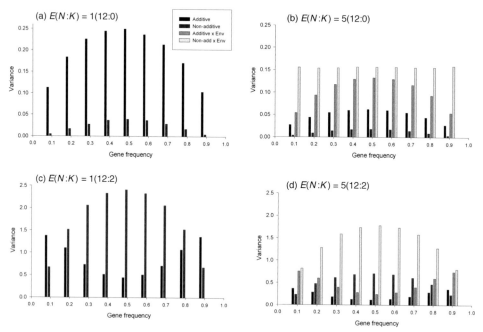

Fig. 11.9. Estimates of additive, non-additive, additive–environment interaction and non-additive–environment interaction components of variance for: (a) an additive model, $E(N:K) = 1(12:0)$; (b) a model with genotype–environment interactions but no epistasis, $E(N:K) = 5(12:0)$; (c) a model with epistasis but no genotype–environment interactions, $E(N:K) = 1(12:2)$; and (d) a model with epistasis and genotype–environment interactions, $E(N:K) = 5(12:2)$. Estimates for each $E(N:K)$ model are based on the 1000 sets of gene effects drawn from the uniform distribution.

breeding strategies across many models of trait architecture (Micallef *et al.*, 2001).

Many crop-breeding programmes can be thought of as open or closed recurrent-selection programmes (Fig. 11.10). Schematically, the breeding programmes involve working with samples of the possible genotypes, evaluating their phenotypic and/or genotypic performance across a sample of environments (ideally, the sample being taken to represent the TPE) (Comstock, 1977; Podlich *et al.*, 1998; Chapman *et al.*, 2000), selecting the superior genotypes based on the available genetic and phenotypic information collected from multienvironment trials and trait-screening experiments, releasing improved genotypes where appropriate and using some of the improved genotypes to initiate a new cycle of the breeding programme.

As part of a larger simulation study, a QU-GENE application module (Fig. 11.3; Podlich and Cooper, 1998) was developed

to compare mass selection and S_1 family selection (Hallauer and Miranda Fo, 1988; Comstock, 1996). Mass selection is based on individual phenotypic performance in a single environment. S_1 family selection is based on selection on the performance of replicated S_1 progeny families derived from individual S_0 plants. The evaluation of S_1 families enabled replication within and across environments, whereas for mass selection there was no replication. For the purposes of this study, it was assumed that both mass and S_1 selection operated as closed recurrent-selection programmes, with the mass selection completing a cycle of selection in 1 year and the S_1 selection completing a cycle in 3 years. Progress from selection over time was monitored in the simulation experiment for the numbers of cycles of recurrent selection equivalent to a period of 30 years. Thus, 30 cycles of mass selection were completed and 10 cycles of S_1 selection were completed. The two breeding strategies

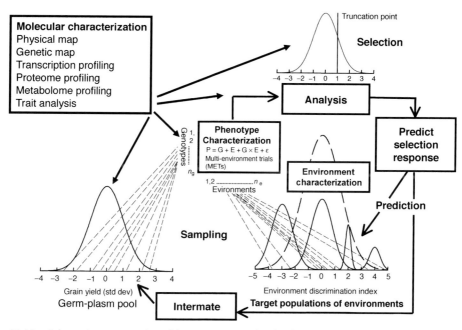

Fig. 11.10. Schematic representation of the processes involved in the conduct of a cycle of a breeding programme (adapted from Cooper and Hammer, 1996).

were examined for their capacity to achieve a response to selection for a range of $E(N:K)$ models. A selection intensity of 10% was applied for both mass (50 individuals selected from 500) and S_1 selection (50 S_1 families selected from 500 S_1 families evaluated).

The $E(N:K)$ models were constructed by changing E (1, 2, 5, 10), N (12, 24, 36) and K (0, 1, 2, 3) and evaluating all possible combinations of these levels of the model factors. For each $E(N:K)$ model combination, 500 sets of gene effects were examined by drawing samples from the uniform distribution and, for each of the 500 parameterizations of each $E(N:K)$ model, ten different starting populations were considered. Each of these model scenarios was examined for three levels of single-plant broad-sense heritability (H = 0.05, 0.50, 1.0). Thus, for the results reported here, the mass and S_1 selection strategies were compared across a total of 720,000 genetic model scenarios. To summarize the results, each $E(N:K)$ model–heritability combination was allocated to one of the four complexity–heritability model categories defined in Table 11.1. The simple-model

low-heritability combination was considered to be those cases where $E = 1$, $N = 12$, $K = 0$ and H = 0.05. The simple-model high-heritability combination included those cases where $E = 1$, $N = 12$, $K = 0$ and H = 0.50 and 1.0. There was only a small effect of increasing H from 0.50 to 1.0; therefore, these results were combined. The complex-model low-heritability combination included all cases where $E > 1$, $N > 12$, $K > 0$ and H = 0.05. The complex-model high-heritability combination included all cases where $E > 1$, $N > 12$, $K > 0$ and H = 0.50 and 1.0.

On average, over all model scenarios, the mass and S_1 strategies made similar selection progress for the first 12 years, with mass selection having a slight advantage initially, whereas the S_1 strategy outperformed the mass selection strategy from year 12 to 30 (Fig. 11.11a). The effect of increasing the complexity of the $E(N:K)$ genetic models by increasing either E (Fig. 11.11b), N (Fig. 11.11c) or K (Fig. 11.11d) was to decrease the rate of progress from selection. Contrasting the estimates of the genetic components of variance between the simple ($E = 1$, $K = 0$) and complex ($E > 1$, $K > 0$) model categories

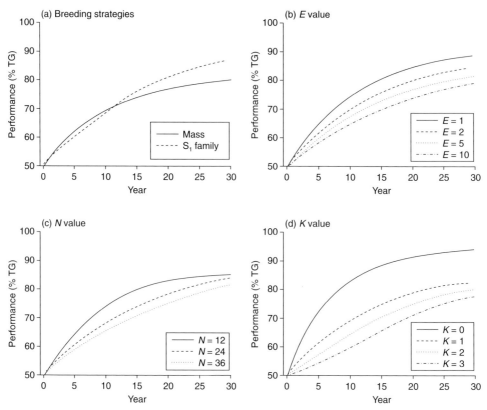

Fig. 11.11. Change in population mean performance, expressed as a percentage of the performance of the target genotype (% TG), over 30 years of simulated breeding for higher trait value by mass and S_1 family selection for 720,000 $E(N:K)$ model scenarios: (a) mean for mass selection and S_1 family selection over all $E(N:K)$ models; (b) mean over all other factors and breeding strategies for different numbers of environment types (E); (c) mean over all other factors and breeding strategies for different numbers of genes (N); (d) mean over all other factors and breeding strategies for different levels of epistasis (K).

(Table 11.1), on average, the additive genetic variance was the dominant component for the simple models and there were no genotype–environment interactions (Fig. 11.12a). However, for the complex models, while additive genetic variance was present, the non-additive component and the genotype–environment interaction components of variance were frequently larger than the additive component (Fig. 11.12b).

Further considerations examined how the rates of genetic progress for mass and S_1 selection changed with the four model complexity–heritability categories described in Table 11.1 (Fig. 11.13). For the simple models (Fig. 11.13a: low heritability, and Fig. 11.13c: high heritability), the rates of

improvement in population mean, expressed on a per year basis, were consistent with expectations in the absence of epistasis and gene–environment interactions. With a high heritability (Fig. 11.13c), population performance was increased to the upper limit by both mass and S_1 selection. However, S_1 selection took approximately three times the number of years to reach this limit, since one cycle of S_1 selection took three times longer than one cycle of mass selection. With a low heritability (Fig. 11.13a), replication was important to accommodate the effects of a higher experimental error and S_1 selection had a greater rate of improvement of performance than mass selection. The S_1 selection strategy managed to improve the population

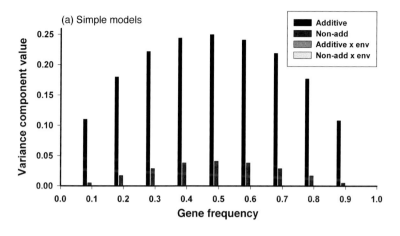

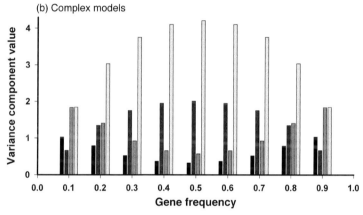

Fig. 11.12. Average estimates of additive, non-additive, additive–environment interaction and non-additive–environment interaction components of variance for all $E(N{:}K)$ model scenarios categorized according to Table 11.1 as: (a) simple genetic models ($E = 1$, $N = 12$ and $K = 0$), and (b) complex genetic models ($E = 2, 5, 10$, $N = 24, 36$, $K = 1, 2, 3$).

to the upper limit, whereas mass selection had not reached the upper limit after 30 years. For the complex models where $E > 1$ and $K > 0$ (Fig. 11.13b: low heritability, and Fig. 11.13d: high heritability), the rate of improvement was much lower than for the simple models and tended to be more linear over the duration of the experiment. Given a high heritability (Fig. 11.13d), the rate of improvement was similar for mass and S_1 selection for the first 18 years, with S_1 selection achieving a higher population mean by year 30 than mass selection. When the heritability was low (Fig. 11.13b), S_1 selection outperformed mass selection across all years, as it had done for the simple model (Fig. 11.13a).

The results of this simulation study serve to emphasize the following points:

- A general computer-simulation platform enables evaluation of any proposed breeding strategy over many alternative genetic-model scenarios. An associated point is, simulation (and theoretical) studies that consider only one or a few putative genetic models should be treated with caution, unless there are strong experimental grounds for accepting the putative model as representative of the local breeding situation.
- Computer simulation enables investigation of genetic properties and expectations of a wide range of complex

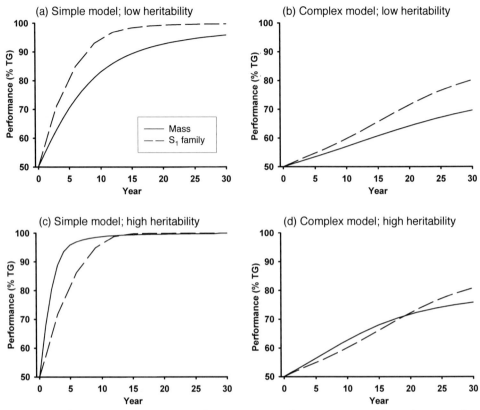

Fig. 11.13. Change in population mean performance, expressed as a percentage of the performance of the target genotype (% TG), over 30 years of simulated breeding for higher trait value by mass and S_1 family selection for the categorization of 720,000 $E(N:K)$ model scenarios according to Table 11.1 as: (a) simple-model low-heritability ($E = 1$, $N = 12$, $K = 0$ and $H = 0.05$); (b) complex-model low-heritability ($E > 1$, $N > 12$, $K > 0$ and $H = 0.05$); (c) simple-model high-heritability ($E = 1$, $N = 12$, $K = 0$ and $H = 0.50$ and 1.00); and (d) complex-model high-heritability ($E > 1$, $N > 12$, $K > 0$ and $H = 0.50$ and 1.00).

genetic models that are extremely difficult to examine through more classical algebraic approaches. Therefore, simulation is one approach for evaluating the implications of relaxing many of the simplifying assumptions that are often used in quantitative-genetic theory.

- It is feasible to evaluate the efficiency of any breeding strategy across a continuum of levels of trait complexity and heritability.
- The availability of high-speed computing platforms enables an extensive evaluation of breeding strategies within a relatively short time frame (Micallef *et al.*, 2001).

From this analysis, it should be clear that it is possible to parameterize any simulation experiment to investigate specific scenarios when appropriate data on the genetic architecture of a trait are available. In fact, one can establish a database of scenarios to be investigated in a systematic fashion, from which conclusions can be drawn as traits become better understood.

Understanding Gene–Phenotype Relationships

The properties of the $E(N:K)$ model examples discussed above were based on analysis of random genetic networks. These investigations can be broadened to concentrate on specific types of gene networks and experimentally derived gene–phenotype relationships for a trait or combinations of traits. A number of models of gene–phenotype relationships have been proposed for quantitative traits (e.g. Doebley and Lukens, 1998). The implications of these models for the power and design of plant breeding strategies can be quantitatively investigated by simulation using the $E(N:K)$ framework. Chapman *et al.* (Chapter 12, this volume) used the $E(N:K)$ framework to evaluate breeding strategies for the genetic improvement of grain sorghum in drought-prone environments in Australia. In their study, experimental data were used to parameterize the $E(N:K)$ models they considered. The experimental data were used to: (i) propose QTL models for four traits (phenology, stay green, osmotic adjustment and transpiration efficiency) that are considered to be important for the adaptation and yield of grain sorghum in the different types of drought-prone environments; (ii) quantify the physiological relationships between the traits, environmental variables and grain yield; and (iii) determine the relative occurrences and influences on yield of the drought-environment types that make up the TPE for grain sorghum in Australia.

The novel aspect of their study was the quantification of QTL or specific gene effects for component 'traits' and their interactions with each other and the environment by using a biophysical crop-simulation model. Potentially, this is a powerful approach for determining the value of a genotype in an environment in terms of the component traits and the integrative trait 'grain yield'. Chapman *et al.* (Chapter 12, this volume) utilized simple additive gene effects to specify the genetic variation for the four component traits that contributed to adaptation under drought. When the effects were measured at the level of grain yield, there were substantial pleiotropic and epistatic effects associated with the genes for the component traits. The genes associated with favourable expression of the different traits varied with selection environment

type. With selection for grain yield, the traits were fixed at different rates depending on their value in the selection environments encountered. Hence, the resultant adaptation landscape for yield was akin to a complex $E(N:K)$ model when compared with an additive trait landscape that was not modulated by gene interactions through biophysical simulation (Chapman *et al.*, 2001).

The issues raised in this preliminary research into a specific gene network identify three layers of complexity in gene–phenotype models in addition to those sources considered in Table 11.1. First, there is the complexity introduced by the way an experiment is measured and analysed. If the genes for a trait are segregating, are the context-dependent gene effects that we observe at the yield level simply indicative of the fact that we are not observing the segregation of the underlying yield-determining physiological components? Secondly, following on from the first point above, we need to consider the level of complexity involved in understanding the hierarchical nature of physiological interactions among traits within an organism and between the organism and the environment during plant growth and development. Physiologists frequently refer to traits being constitutive (i.e. expressed in all environments) or adaptive (i.e. expressed under a specific environmental influence). Adaptive traits are particularly vulnerable to being misinterpreted where it is difficult to quantify the effect of the environment on a particular trait. Thirdly, when analysing trait contributions to adaptation and yield, it must be recognized that component traits interact in a time sequence of expression over the duration of the crop growing season.

With these layers of complexity, the form of the gene–phenotype relationships is likely to be highly variable among target traits. For example, if the trait being considered was resistance to a pathogen with a simple genetic basis, then, in a set of segregating populations, there is likely to be a high correlation between the occurrences of the alternative alleles of the gene and trait phenotypes, i.e. symptoms of susceptibility and expression of resistance in a particular tissue following infection. Obviously, the gene–phenotype relationship is relatively simple in this case and QTL identified for such a trait are likely to be quite well correlated with the gene presence and therefore provide useful input data for genetic simulations. In contrast to the above biotic-stress example, crop adaptation to the biophysical environment and many abiotic stresses is often far more complex. For example, in sorghum, the trait 'stay green' (the ability of plants to keep leaves green at the end of a drought period) is a target of several molecular-mapping programmes (e.g. Subudhi *et al.*, 2000; Tao *et al.*, 2000). Several putative QTL have been identified for the trait. However, while some forms of 'stay green' are known to be 'cosmetic', the physiological mechanism for this form of the trait is unknown. Borrell *et al.* (2000) noted that genotypes of sorghum that differed for their degree of 'stay green' near the end of the season were different in nitrogen (N) content per unit leaf area. This suggests that, given similar requirements for N in the grain of different hybrids, 'stay green' near maturity could in part be simply an effect of differences in uptake of N during vegetative growth in the first part of the season. Such a finding in the 'stay green' mapping populations would have substantial implications for marker programmes determining whether (if any) QTL identified in the mapping studies are even close to the physical site for 'stay green' or are of real value in the target environments. Providing that the biophysical models of gene–phenotype relationships are backed by effective field experimentation to validate the models, they should begin to help determine when an adaptive trait becomes a suitable target for selection.

Synopsis and Speculation

Schrage (2000) in his book *Serious Play* discussed the many roles of simulation as an innovation tool and how it had been used in a range of industrial applications. We would argue that many of the points he makes are also relevant for the application of simulation in the design and evaluation

of plant-breeding strategies. Kempthorne (1988) argued that many of the limitations that are faced in the classical application of algebraic approaches in the development of theoretical models for complex traits in quantitative genetics can be overcome through combining high-performance computing capability with a flexible simulation platform. We have found that this is indeed the case (Podlich and Cooper, 1998; Cooper *et al.*, 1999; Micallef *et al.*, 2001). With the growing body of experimental data on the genetic architecture of traits for our important agricultural species, there is now a significant opportunity to examine the global and local features of the adaptation landscapes that we face in our breeding programmes. To exploit these data sources efficiently requires information-management systems and bioinformatics tools that link these information sources to the simulation platform.

There is a growing body of research into the multifaceted roles of epistasis in the evolutionary process (Wolf *et al.*, 2000). Much of the work on the molecular structure of gene networks and gene regulation suggests that epistasis is likely to play an important role in determining genetic variation in quantitative traits. However, in the construction of prediction equations for plant-breeding purposes, the assumption of the absence of epistasis is virtually ubiquitous. The analyses we have conducted based on random and structured gene networks using the $E(N:K)$ model indicate that the consequences of making this assumption when epistasis is indeed present are in general likely to result in overprediction of genetic gain (e.g. Figs 11.11d and 11.13). The qualitative feature of this result has been appreciated for a long time. It is also appreciated that the extent of the overestimation of expected genetic gain is likely to be context-specific and will depend on the detail of the form of epistasis that exists, the allele frequencies for the genes involved in the gene networks and the linkage relationships among the alleles within the reference population of the breeding programme. These factors, together with the considerable difficulty associated with experimentally demonstrating strong epistatic influences for many quantitative traits, have contributed to the limited attention given to developing prediction equations that explicitly accommodate epistatic effects, other than treating them as a source of error in the prediction equation. Many of these limitations can be overcome by applying a comprehensive computer-simulation approach to the problem.

While it is often argued that there is limited experimental evidence for epistasis in studies of agronomic traits, several empirical studies have shown that epistasis is significant for quantitative traits, such as yield in small-grain crops (as reviewed by Goldringer *et al.*, 1997). Work focused on both describing the genetic architecture of quantitative traits at the level of gene networks and understanding gene–phenotype relationships has emphasized the need to give greater attention to the potential effects of epistatic interactions among genes and the interaction of the genes in these networks with environmental factors (e.g. Doebley and Lukens, 1998; Schlichting and Pigliucci, 1998). Analysis of the implications of these more complex trait architectures is not a trivial task. However, the hardware and software to apply computer-simulation methodology to this problem have advanced greatly, even in the short time since Kempthorne (1988) speculated about the feasibility of a computer-simulation-based approach to advancing quantitative-genetic modelling methodology. Thus, it is feasible to undertake *in silico* investigations of trait architectures that range from the more classical models assumed in quantitative genetics to highly complex gene-network structures that have the potential to interact with environments. Using computer simulation we have found that breeding strategies can indeed differ in their capacity to accommodate some of the effects of complexities, such as genotype–environment interactions (Podlich *et al.*, 1998) and epistasis (Podlich and Cooper, 1999).

Earlier, we commented on the dominant position of the Fisher modelling methodology in the development of plant-breeding theory. In contrast, the details of the gene networks that are in the process of being uncovered, using the tools of molecular genetics, appear to be more compatible with some of the models and concepts discussed by Wright. We do not see these two modelling frameworks as incompatible. Simulation analysis of quantitative-trait models using the $E(N{:}K)$ framework provides scope for exploring several of the important theoretical issues, some of which were debated by Fisher and Wright. The capacity to jointly interpret gene-network models in terms of the structure of adaptation landscapes and the additive and non-additive effects of genes demonstrates some of the common ground that can be found between these two views of the properties of quantitative traits.

Much of the theoretical treatment of breeding strategies considers them as single-search programmes seeking improved genotypes. In practice, larger breeding programmes do not operate as a textbook single-search programme but are better considered as connected multiple-search strategies. By multiple-search strategies here we mean that there are multiple breeding activities semi-independently searching the genetic space and actively comparing and exchanging improved germ-plasm. Podlich and Cooper (1999) used the $E(N{:}K)$ model to examine the efficiency of single-search and multiple-search breeding strategies. In their preliminary study they found that, as the complexity of the genetic architecture of the trait increased, due to the effects of epistasis and gene–environment interactions, there were significant advantages of the multiple-search strategy over a single-search breeding strategy. Clearly, there is much work to be done if we are to understand the ways in which the three key components, breeding strategy, trait architecture and germ-plasm dynamics, interplay to determine the efficiency of the breeding process.

Summary and Conclusions

Plant-breeding programmes are a practical use of financial, human, scientific and genetic resources. These are applied in concert to search genetic space in ways that seek new genotypes that are superior to those genotypes already in hand. Some of the features of the genetic space in which these searches are conducted are being uncovered with the use of genomic technologies. The traits we attempt to manipulate can range in complexity because of differences in the numbers of genes involved, interactions among genes within networks (epistasis) and interactions between genes and environmental conditions, thus influencing the degree of difficulty encountered in manipulating plant development, trait expression, adaptation and ultimately yield and quality. For the resource base of a breeding programme, it would be useful to quantify the capacity to manipulate traits of differing genetic architecture and complexity. At present, we have at best a limited understanding of the relationships between the power of plant-breeding strategies to seek new genotypes and the complexity of the genetic space in which that search is being undertaken. With advances in computer software and hardware and our growing genomic databases, computer modelling of plant-breeding strategies provides a powerful tool for *in silico* innovation and optimization of breeding programmes.

We conclude with the following speculation. A fundamental paradigm in the application of the tools of molecular biology to reveal gene–phenotype relationships is that, through understanding the function of every gene, we shall be able to elucidate, describe and predict the gene–phenotype relationship. The combinatorial complexity of this task is analogous to the concept in physics of describing the properties of every molecule in an ideal gas as a basis for predicting the behaviour of the gas. In fact, due to the potential for high-level interactions between the components of the genotype–environment systems, the combinatorial complexity is likely to be greater in the biological systems than in many of the physical systems. Nevertheless, in physics, if we wish to describe the properties of the gas, we do not describe each gas molecule and accumulate their independent effects; instead, we describe the integrated statistical properties of the gas. We do not need to follow the behaviour of every gas molecule to know

that the gas will tend to expand in space when it is heated. It is probably commonplace that genes function in networks and therefore many of the effects of genes may be context-dependent. Networks have basic properties that determine their behaviour. There are two basic behaviour types for networks: ordered or chaotic (Kauffman, 1993). A key network property is the way in which the genes are interconnected. This is emphasized by the E and K parameters in the $E(N:K)$ model. It seems likely that specific types of network connections have been preferred as evolutionary trajectories by the combined effects of network properties and selection. Thus, the robustness of many of the extant gene networks may have been a target of selection for quantitative traits in evolutionary history. If this is the case, applying genomic tools to identify the key regulatory genes within gene networks will be critical to understanding the genetic architecture of quantitative traits and to achieving their directed manipulation in breeding programmes. With the above arguments in mind, we consider that, in the 21st century, computer simulation will be a central tool in the evaluation of breeding strategies for their power to improve quantitative traits.

Acknowledgements

We thank Oxford University Press, Oxford, UK, for permission to reproduce Fig. 11.4, Research Trends, Trivandrum, India, for permission to reproduce components of Fig. 11.6, and CAB International, Wallingford, UK, for permission to reproduce part of Fig. 11.10.

References

Beavis, W.D. (1998) QTL analyses: power, precision and accuracy. In: Paterson, A.H. (ed.) *Molecular Dissection of Complex Traits*. CRC Press, Boca Raton, pp. 145–162.

Beveridge, W.I.B. (1950) *The Art of Scientific Investigation*. William Heinemann, Melbourne.

Borrell, A.K., Hammer, G.L. and Henzell, R.G. (2000) Does maintaining green leaf area in sorghum improve yield under drought? II. Dry matter production and yield. *Crop Science* 40, 1037–1048.

Chapman, S.C., Hammer, G.L., Butler, D.G. and Cooper, M. (2000) Genotype by environment interactions affecting grain sorghum. III. Temporal sequences and spatial patterns in the target population of environments. *Australian Journal of Agricultural Science* 51, 223–233.

Chapman, S.C., Cooper, M., Podlich, D.W. and Hammer, G.L. (2001) Evaluating plant breeding strategies by simulating gene action and environment effects to predict phenotypes for dryland adaptation. *Agronomy Journal* (in review, Proceedings of Workshop on Crop Modelling and Genomics at the ASA/CSSA/SSSA Meetings, November, 2000, Minneapolis) (in press).

Comstock, R.E. (1977) Quantitative genetics and the design of breeding programs. In: Pollack, E., Kempthorne, O. and Bailey, B. (eds) *Proceedings of the International Conference on Quantitative Genetics*. Iowa State University Press, Ames, pp. 705–718.

Comstock, R.E. (1996) *Quantitative Genetics with Special Reference to Plant and Animal Breeding*. Iowa State University Press, Ames, 421 pp.

Cooper, M. and Hammer, G.L. (1996) Synthesis of strategies for crop improvement. In: Cooper, M. and Hammer, G.L. (eds) *Plant Adaptation and Crop Improvement*. CAB International, Wallingford, pp. 591–623.

Cooper, M., Podlich, D.W., Jensen, N.M., Chapman, S.C. and Hammer, G.L. (1999) Modelling plant breeding programs. *Trends in Agronomy* 2, 33–64.

Doebley, J. and Lukens, L. (1998) Transcriptional regulators and the evolution of plant form. *The Plant Cell* 10, 1075–1082.

Falconer, D.S. and Mackay, T.F.C. (1996) *Introduction to Quantitative Genetics*, 4th edn. Longman, Harlow, Essex, 464 pp.

Frary, A., Nesbitt, T.C., Frary, A., Grandillo, S., van der Knaap, E., Cong, B., Liu, J., Meller, J., Elber, R., Alpert, K.B. and Tanksley, S.D. (2000) *fw2.2*: a quantitative trait locus key to the evolution of tomato fruit size. *Science* 289, 85–88.

Fraser, A. and Burnell, D. (1970) *Computer Models in Genetics*. McGraw-Hill Book Company, New York, 206 pp.

Goldringer, I., Brabant, P. and Gallais, A. (1997) Estimation of additive and epistatic genetic variances for agronomic traits in a wheat

population of doubled-haploid lines of wheat. *Heredity* 79, 60–71.

Hallauer, A.R. and Miranda Fo, J.B. (1988) *Quantitative Genetics in Maize Breeding*, 2nd edn. Iowa State University Press, Ames, 468 pp.

Hospital, F. and Decoux, G. (2002) POPMIN: a program for the numerical optimization of population sizes in marker-assisted backcross programs. *Journal of Heredity* (in press)

Kauffman, S.A. (1993) *The Origins of Order: Self-organization and Selection in Evolution*. Oxford University Press, New York, 709 pp.

Kearsey, M.J. and Farquhar, G.L. (1998) QTL analysis in plants; where are we now? *Heredity* 80, 137–142.

Kearsey, M.J. and Pooni, H.S. (1996) *The Genetical Analysis of Quantitative Traits*. Chapman & Hall, London, 381 pp.

Kempthorne, O. (1988) An overview of the field of quantitative genetics. In: Weir, B.S., Eisen, E.J., Goodman, M.M. and Namkoong, G. (eds) *Proceedings of the Second International Conference on Quantitative Genetics*. Sinauer Associates, Sunderland, pp. 47–56.

Koornneef, M. and Stam, P. (2001) Changing paradigms in plant breeding. *Plant Physiology* 125, 156–159.

Lukens, L.N. and Doebley, J. (1999) Epistatic and environmental interactions for quantitative trait loci involved in maize evolution. *Genetical Research, Cambridge* 74, 291–302.

Lynch, M. and Walsh, B. (1998) *Genetics and Analysis of Quantitative Traits*. Sinauer Associates, Sunderland, 980 pp.

Micallef, K.P., Cooper, M. and Podlich, D.W. (2001) Using clusters of computers for large QU-GENE simulation experiments. *Bioinformatics* 17, 194–195.

Paterson, A.H. (ed.) (1998) *Molecular Dissection of Complex Traits*. CRC Press, Boca Raton, 305 pp.

Podlich, D.W. and Cooper, M. (1998) QU-GENE: a platform for quantitative analysis of genetic models. *Bioinformatics* 14, 632–653.

Podlich, D.W. and Cooper, M. (1999) Modelling plant breeding programs as search strategies on a complex response surface. *Lecture Notes in Computer Science* 1585, 171–178.

Podlich, D.W., Cooper, M. and Basford, K.E. (1998) Computer simulation of a selection strategy to accommodate genotype-by-environment interaction in a wheat recurrent selection programme. *Plant Breeding* 118, 17–28.

Schlichting, C.D. and Pigliucci, M. (1998) *Phenotypic Evolution: a Reaction Norm Perspective*.

Sinauer Associates, Sunderland, Massachusetts, 387 pp.

Schrage, M. (2000) *Serious Play: How the World's Best Companies Simulate to Innovate.* Harvard Business School Press, Boston, 244 pp.

Somerville, C. and Dangl, J. (2000) Plant biology in 2010. *Science* 290, 2077–2078.

Subudhi, P.K., Rosenow, D.T. and Nguyen, H.T. (2000) Quantitative trait loci for the stay green trait in sorghum (*Sorghum bicolor* L. Moench): consistency across genetic backgrounds and environments. *Theoretical and Applied Genetics* 101, 733–741.

Tao, Y.Z., Henzell, R.G., Jordan, D.R., Butler, D.G., Kelly, A.M. and McIntyre, C.L. (2000) Identification of genomic regions associated with stay green in sorghum by testing RILs in multiple environments. *Theoretical and Applied Genetics* 100, 1225–1232.

Wolf, J.B., Brodie, E.D., III and Wade, M.J. (eds) (2000) *Epistasis and the Evolutionary Process.* Oxford University Press, New York, 330 pp.

Wright, S. (1932) The roles of mutation, inbreeding, crossbreeding and selection in evolution. In: Jones, D.F. (ed.) *Proceedings of the Sixth International Congress of Genetics.* Ithaca, New York, pp. 356–366.

12 Linking Biophysical and Genetic Models to Integrate Physiology, Molecular Biology and Plant Breeding

S.C. Chapman,[1] G.L. Hammer,[2] D.W. Podlich[3,4] and M. Cooper[3,4]

[1]CSIRO Plant Industry, 120 Meiers Road, Long Pocket Laboratories, Indooroopilly, Qld 4068, Australia; [2]Agricultural and Production Systems Research Unit (APSRU), Queensland Department of Primary Industries, PO Box 102, Toowoomba, Qld 4350, Australia; [3]School of Land and Food Sciences, University of Queensland, Brisbane, Qld 4072, Australia; [4]Pioneer Hi-Bred International Inc., 7300 NW 62nd Avenue, PO Box 1004, Johnston, IA 50131, USA

Crop Adaptation as Determined by Gene Networks

Syngenta recently announced the complete sequencing of the first crop genome (rice). While the public sequencing of *Arabidopsis thaliana* provides a useful model tool, extrapolation from *Arabidopsis* to other dicots (e.g. Lan *et al.*, 2000) and from rice to other cereals (Moore, 2000) will generate substantial information about the physical make-up and location of genes in the chromosomes of our major food crops. The rice map claims to contain the DNA sequence of every gene, their regulatory sequences and the correspondence between the genome map and the plant breeders' map of inherited traits. While this provides exciting targets for the manipulation of traits associated with tolerance/susceptibility to chemicals, grain quality and pest resistance (Somerville and Somerville, 1999), what opportunities will this avalanche of molecular knowledge reveal to improve biological yields and crop adaptation to physical environments?

The principal objective of a plant-breeding programme is the generation and selection of new gene combinations to create genotypes with trait performance that is superior to current genotypes, within the target population of environments (TPE) (Comstock, 1977). This objective applies equally to conventional, molecular and combined approaches. For relatively simple adaptation targets, such as single-gene pest resistance, access to the genome sequence will allow molecular biologists and breeders to look for the better alleles for pest resistance and even to modify gene sequences or mutate genes to create new resistances (e.g. Collins *et al.*, 1999). Sequencing, though, is only a part of the story. The 99% perspiration that is now to be done is to determine how to create genotypes by combining together the superior combinations of genes as they interact within a crop plant as it grows and responds to the physical environment that it encounters.

Networks of genes coordinate the process of gathering nutrients, photosynthesizing and metabolizing the various compounds that make up a plant and direct its systems for growth, development, resource acquisition and defence against pathogens. Many defence-gene networks can be thought of as having a specific environmental challenge to detect and respond to, often providing a straightforward model for the screening of gene expression for upstream regulation genes (e.g. Schenk et al., 2000). In this way, these defence-gene networks are similar to the gene networks that apparently respond to extremes of abiotic stress and are a major subject of research in molecular biology (e.g. Ingram and Bartels, 1996; proline accumulation pathways: Hare et al., 1999). However, the accumulation of solutes is one of the last processes to occur in stressed plant tissues (Hsiao, 1973; Ludlow and Muchow, 1990) and, as such, extreme abiotic stress can frequently be classified as a catastrophic hazard for crop yield, even if the gene responses help in plant survival (Bidinger et al., 1996). In this respect, the classification of genes as 'stress' genes is misleading, i.e. the responses of the major genes controlling normal growth and development are also stress genes. For example, cell expansion, rather than photosynthesis or solute accumulation, is the first process to be affected by decreased tissue water potential (Hsiao, 1973). So coping with the seasonal variation in radiation, temperature and water supply is largely a matter of 'improved' regulation of the genes that are already expressed under 'optimal' growing conditions. It is this 'intermediate' condition of adaptation to enable the plant to continue to capture resources (i.e. continuing canopy and root expansion) for improved crop yield that will generate economic benefits in the many environments where plants experience substantial, but not catastrophic, levels of stress (Bidinger et al., 1996). Unfortunately, it is also a difficult issue to address, due to the need to control genes that are operational under both optimal and stress conditions.

To document this complexity of interactions, we shall need to link the tools and databases that are developing in all of the research areas (genetics, molecular biology, plant breeding, plant and crop physiology) in order to understand the effects of gene networks associated with crop traits and how these are mediated by genetic variation and the environment. In turn, this requires an analysis framework to interpret the rules that govern this network of gene interactions, rather than trying to define every component. The study of networks now pervades all of science and their description, using mathematical models, is frequently described as science's major challenge (Strogatz, 2001). Cooper et al. (1999) and Cooper et al. (Chapter 11, this volume) have presented a simulation framework that allows us to describe the processes involved in a plant-breeding programme. In this chapter, we expand on the characteristics of that simulation framework to one that captures gene interactions through modelling a biologically complex system.

Plant breeders attempt to manipulate gene networks to improve crop yield and quality and tolerance to external stresses. In the simplest terms, plant breeders try to combine together, in a single cultivar, the best alleles of each gene acting in combination with each other and the environment. This chapter aims to outline how these processes can be simulated (often with incomplete knowledge) to discover optimal methods of plant breeding while considering the entire spectrum of influences from the gene action to the cropping system. Our first objective here is to describe how simple and complex gene networks can be linked to plant response and variation of the phenotype and show how these gene–phenotype relationships can be characterized and improved by marker selection. We further aim to illustrate three points made by Cooper et al. (1999) regarding the use of in silico crop modelling to improve the efficiency of plant breeding: (i) characterizing environments to define the TPE; (ii) assessing the value of specific putative traits in improved plant types; and (iii) enhancing integration of molecular genetic technologies with an example utilizing molecular markers.

Linking Gene to Phenotype – Integration and Signalling across Scales

The genetic-simulation framework underlying the Quantitative Genetics (QU-GENE) model (Podlich and Cooper, 1998; Cooper *et al.*, Chapter 11, this volume), was designed around the commonly used model in quantitative genetics whereby an allelic variant of a gene is defined as a stochastic effect without an explicit requirement to specify the biophysical process controlling the effect. Cooper *et al.* (Chapter 11, this volume) have given examples of how low and high levels of both heritability and complexity of trait behaviour affect the behaviour of stochastically described genes. This behaviour can be biologically specified and perhaps will, in the future, be defined at a biochemical level for limited numbers of segregating genes, using information gathered with genomics approaches. However, a statistical interpretation of all of the allelic values for genes in all pathways, across a season, is unlikely ever to be feasible or desirable, if it does not help us discriminate among breeding methods to improve crop performance. Our challenge is to incorporate the biological basis for gene networks and gene–phenotype relationships into this stochastic framework to capture and interpret how patterns of interconnection among gene effects generate the emergent properties that influence the performance of genotypes across environments.

In a simple example of a gene–phenotype relationship, an enzyme might utilize substrates (brought to the cell by the actions of other genes) to make a new product that can be described as a 'cellular trait' or molecular phenotype (Fig. 12.1). In a cereal, an example of such a trait might be the hormonal switches that cause new cells in the growing meristem to become

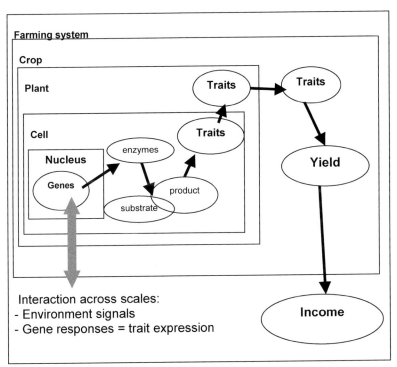

Fig. 12.1. Genes expressing traits that integrate with the farming system over scale and time to generate yield.

'reproductive' rather than 'vegetative' (see Koornneef et al. (1998) for a discussion of the molecular basis for this). These switches are associated with the external signals of temperature and photoperiod and internal development cues. At the plant level, this trait would be physically observed, first, as floral initiation (under a microscope) and, after several weeks, head emergence and flowering of that growing point. Expression of this trait as observed at the crop level could be the 'average flowering date of the population of plants' in the field. That trait then has an eventual effect on yield (e.g. earliness helps to avoid late-season drought, but reduces potential yield) and the viability of the cropping system (earliness uses less nitrogen (N) and water from the soil, but, at the same time, provides less stover to protect against soil erosion).

Scales of integration within a cropping system influence the assembly of gene variants that make up a genotype (Fig. 12.1). At the cellular level, it may be that a single gene controls a part of a pathway. Allelic variants of this gene may cause it to be transcribed in response to a slightly different time or tissue location or set of local (cell?) environmental conditions. It is important to recognize that while the actions of genes propagate upward from the cellular level, the environmental signals also propagate from the cropping system back to the cell. Figure 12.1 implies that the timing and intensity of the ultimate 'signal' for gene action at some point during the growing season (e.g. increased the activity of nitrate transporters in the roots or drought-induced slow-down in leaf growth) may have been a function of the effects of the density of the planted crop or even the effect of a previous crop on the soil resource. In dryland environments, such as Australia, this appreciation of long lag-time effects on crop growth is a key part of a cropping system that 'farms' soil water. For example, excess N application to a wheat crop in a soil containing low stored soil water at planting can greatly lower yields if the season happens to be one of low rainfall, i.e. the excess N encourages excessive early-season growth relative to the water supply available for grain filling later (Cantero-Martinez et al., 1999).

The process of gene action at the cellular level has been extensively studied and modelled in cell culture and simple systems, e.g. GEPASI (Mendes and Kell, 1998; Giersch, 2000). Genes are programmed to be 'switched' on or off in response to internal cues, such as developmental age, and external cues, such as photoperiod or temperature. In reality, the system of 'switching' involves complex networks of genes, with some designed to 'detect' cues, others to initiate a response ('signalling genes') and others to actually produce the outcome ('action genes') that can be measured. Much current molecular research is aimed at determining what the signalling genes for processes are, with the eventual intention that up-regulating these genes will propagate an increase in the activity of the genes controlled by the signal (e.g. Hare et al., 1999). The eventual response to a signal can be due to an up- or down-regulation of the amount of 'action gene' transcribed and/or modification of its activity. Many products in plants are known to participate in 'futile' cycles, which could have the purpose of allowing a plant to respond quickly to a signal to change the activity and/or amount of gene product. Each gene product in a biochemical pathway also has a particular level of efficiency related to its structure, substrate availability and the multitude of other pathways operating in the same cellular compartment. Allelic variation can act at each of these points of signal detection, transcription and actual activation of the product, as well as in other genes that affect these steps. Any observed trait at the tissue or plant level, then, has already been filtered through these steps and its action is substantially masked by this complexity, as well as by the compartmentalization of many reactions and the time required for effects to be observable.

Modelling a simple gene–phenotype relationship for a trait

Models of crop yield in response to environment include both regression-type models (e.g. regression of yield against in-season rainfall) and dynamic biophysical crop-simulation models, e.g. CERES (Ritchie, 1991), SUCROS (Bouman *et al.*, 1996) and APSIM (McCown *et al.*, 1996). Crop-simulation models typically utilize a daily time step, with inputs of soil descriptions and weather, to integrate the growth of the crop across the day and accumulate the value of growth components (weights and structure of plant parts, size of leaf area, root system) and their exchanges of carbon and N and water status over the season. Thus, when accommodating the effects of traits within a crop model, modellers have to consider how the traits propagate across the scales of cell to plant to crop in parallel with the passage of time.

Consider the trait of 'flowering date' as an example. Cropping-system models aim to use daily inputs of weather to represent the genetic effects of development and environmental signals on flowering date. Thornley (1972) proposed that we could build a model of the biochemical switch by accumulation of hormonal signals over time-scales of days and weeks. Certainly this is possible, and yet crop-simulation models have functioned well by specifying temperature and photoperiod models that predict the date of floral initiation and maximum leaf number, such that flowering date (in cereals) becomes a linear function of the constant (thermal) time taken for each leaf to emerge (e.g. Ritchie and Nesmith, 1991). Genotypic differences are represented by different parameter values in these (usually) non-linear functions associated with predicting floral initiation. The APSIM model also attempts to represent the feedback effects on phenology of stresses due to water and N deficits (Hammer *et al.*, 1996b). Apart from capturing genetic variation, crop modelling offers the potential to interpret and predict performance given different management situations, including both long-term decisions (e.g. to maintain system sustainability) and seasonal decisions, such as those associated with planting time and fertility management.

One can see how qualitative traits, such as some seed colour, which depend on allelic variation (presence/absence) that blocks a pathway, can be easily observed in plants. It is perhaps remarkable that there are any quantitative plant- or crop-level traits that appear to have several major genes under simple gene action. However, this may be an emergent property of gene networks (see Cooper *et al.*, Chapter 11, this volume). Plant height (e.g. Fujioka *et al.*, 1988) and flowering time (e.g. Koornneef *et al.*, 1998) are two such traits for which the pathways and action are being discerned. The control of grain number in cereals in different environments at first appears to be much more complex, but agronomists and physiologists have developed some 'short cuts' to describe this control.

Complexity from simplicity – pleiotropic effects of height genes

Pleiotropy can arise directly from a single gene affecting multiple traits (e.g. where the same dwarfing gene reduces the size of all plant parts) or as an indirect effect where a dwarfing gene might reduce the lengths of internodes and thereby affect the canopy development and long-term biomass accumulation by the crop. Differences in patterns of the segregation for traits that appear to be quantitative at the plant level can arise due to variation in the action of multiple genes affecting a single major gene pathway or because the trait is actually an integrated result of several pathways acting over time. A simple example of genes with known effects at the cellular level that can be readily observed at the plant level is the association of gibberellin-synthesis genes with plant height (Phinney, 1956). Quinby and Karper (1954) determined that in grain sorghum, for example, there are four major unlinked genes associated only with internode length, creating five recognized classes of height from zero-dwarf to four-dwarf (see review by Morgan and Finlayson (2000) for some of the comments that follow below).

These height genes can operate independently of the flowering-time genes, which also have a major influence on height. For a given set of height genes, a later-flowering genotype produces longer nodes and reaches a greater height than an early-flowering genotype, i.e. in this case, the height difference would be a pleiotropic effect of a flowering (photoperiod-responsive) gene. While most grain-sorghum-breeding programmes work with 'mid-size' lines (three-gene dwarf), height genes *per se* have been recognized in sorghum as being positively correlated with yield, particularly in environments where plant lodging is not frequent (Morgan and Finlayson, 2000). The reason for this correlation is not clear. Breeders have noted that some single-gene, recessive mutants of maize suffer easily observed pleiotropic effects on leaf and tassel size, whereas in sorghum the mutants tend to be brachytic (confined to internode length). Physiologists are (or should be) more concerned with the effects of this change in stem-elongation pattern following floral initiation on the ability of the crop to utilize radiation and water resources. While there is little evidence that height genes affect root extension, the higher yields indicate that there are apparently other indirect effects of these height genes on canopy development and light-use efficiency and possibly on the accumulation of stem reserves for later use during grain-filling. These effects are likely to be exacerbated (both positively and negatively) by different patterns of drought stress during the season, although there has been little crop physiological research to investigate this. For example, if the taller genotype grows more quickly, and yet uses soil water at the same efficiency as its dwarf mutant, the taller genotype would be likely to yield poorly in a season when soil water is not replenished during grain-filling. This type of season is quite common in north-eastern Australia (see later discussion) and serves to emphasize the flow-through effects of cellular traits at the plant, crop and cropping-system levels (Fig. 12.1).

One summary of these considerations is that an apparently simple trait in terms of gene action (directly affecting a gibberellic acid pathway and influencing potential cell size and elongation rate) causes complex pleiotropic effects on the crop yield in different environments. Most crop models do not simulate height at all and they rarely attempt to capture any feedback mechanisms that result. In contrast, virtually all crop models do capture the effects of flowering time (Ritchie and Nesmith, 1991), which can be broadly considered as having a similar number of major genes and biochemical pathways that are understood to a similar or slightly greater level (Koornneef *et al.*, 1998). It seems that height as a crop trait is largely avoided by crop modellers, perhaps due to its substantial interaction with the notoriously difficult problem of 'interpreting the rules' of carbon partitioning. This issue cannot be avoided once we attempt to simulate grain yield.

Simplicity from complexity – control of grain number under drought stress

A key attribute of most cereal simulation models is prediction of the size of the yield 'sink', usually in terms of grain number. The size of this component is critical to the later prediction of yield, since grain number is almost universally correlated with yield within sets of related genotypes in a given environment. In maize, plant breeders have long recognized (e.g. Lonnquist and Jugenheimer, 1943; Stringfield and Thatcher, 1947) and exploited genetic variation for the ability to set greater numbers of grains per plant under difficult conditions, such as drought (Fischer *et al.*, 1989; Bolaños and Edmeades, 1993) and high density (Troyer and Larkins, 1985). It is one of the few crop physiological traits to benefit from research that encompasses all system levels, from gene expression to crop agronomy and active plant breeding.

Grain number in maize is particularly sensitive to reduced carbon supply (through drought, density or shading) from the time of ear initiation until the linear period of grain filling begins. The major effects under drought have been well described by Edmeades *et al.* (2000), while the general physiology of the kernel setting in maize is covered well by other papers in that monograph. In summary, prior to silking, one can describe the consequence of drought as ovary abortion and, while some grains may then be lost due to effects on pollination, abortion of fertilized embryos also continues during early grain fill. The occurrence of ovary abortion is correlated with a reduction in the growth rate of ears prior to silking and a consequent delay in the date of silking relative to anthesis (i.e. a long anthesis–silking interval (ASI)). Under controlled drought and among lines of similar anthesis date (that are all experiencing the same degree of stress on ear growth), breeders have been able to select superior lines as those with shorter ASI and higher ear number per plant, grain number per ear and grain yield.

The control of ovary abortion has been investigated in some novel experimentation over the last 10 years or so (summarized by Zinselmeier *et al.*, 1999), including the finding that feeding sucrose into the stalks of drought-stressed maize could reduce, but not completely recover, the number of ovaries aborted. Zinselmeier *et al.* (1999) investigated the biochemistry of starch biosynthesis and came to the following conclusions:

1. Under reduced carbohydrate supply, when ovary starch was sufficiently depleted, the ovary was aborted (at ovary water potentials more negative than −1 MPa).
2. When fed sucrose, sucrose was unloaded into ovary pedicel cells (i.e. membrane transport was unlikely to be the limiting step).
3. Even when sufficient sucrose was fed, starch biosynthesis from sucrose in the ovary was still reduced.
4. Blockage in the sucrose–starch biosynthesis pathway appeared to occur at the first step, mediated by acid invertase.

The reduction in acid invertase was associated with down-regulation of genes associated with expression of this enzyme, as well as others in the starch biosynthesis pathway (C. Zinselmeier, personal communication). In young ears under drought stress, transcription of other genes, including cell-cycle, stress-responsive and abscisic-acid (ABA)-related genes, was affected (Sun *et al.*, 1999). In summary, there is apparently some signal that is causing this down-regulation and consequent embryo abortion, despite artificial replenishment of the sucrose supply. At present, this signal is unknown (Zinselmeier *et al.*, 1999) and genetic variation for the effect is not documented.

Crop physiologists and modellers have been undeterred by this lack of knowledge about the cellular-level control of grain number. This notion of deriving crop responses rather than describing them is not new and is akin to the concept of modelling plant-hormone action without modelling the hormones (de Wit and Penning de Vries, 1983). The approach is to ascertain and model the rules and processes governing crop responses to environmental conditions. These rules enable a complicated array of crop responses to emerge as properties, given differing combinations of conditions. Further, robust empirical relations of these emergent properties can be derived for use in crop models, as, for example, with RUE and leaf N status (Sinclair and Horie, 1989).

This effect has been captured at the crop level by a correlation between plant growth rate during these sensitive stages (derived from different treatments or environments) and the number of grains per plant that were set (Fig. 12.2, idealized from Tollenaar *et al.* (2000)). In maize, researchers have recognized that this relationship varies with cultivar (Tollenaar *et al.*, 1992). Edmeades *et al.* (2000) showed that selection, as described above, effectively increased the number of grains produced per unit growth rate, particularly at low growth rates experienced under drought conditions. Though integrated across several weeks, this crop physiological relationship captures the essence of the 'mystery signal' to which Zinselmeir *et al.* (1999) alluded.

What use can we make of this 'reverse physiology', i.e. the process of observing the system to discern control, rather than the cellular components? First, Edmeades *et al.* (2000) and Fig. 12.2 indicate the importance of selection environment in determining the basis for genotypic differences. While there

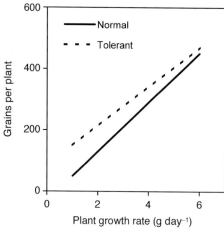

Fig. 12.2. Idealized description of grain number per plant for different rates of plant growth rate between floral initiation and flowering in a 'normal' and 'drought-tolerant' cultivar of maize (derived from Tollenar *et al.*, 1992, and Edmeades *et al.*, 2000).

may be cases where the genotypic differences are consistent across all environments, in maize, stress environments have to be used to reveal the genotypic variation in this trait. Secondly, the slope of the grain-number/growth-rate relationship in a segregating population under a series of different environments could generate a more environmentally stable quantitative trait locus (QTL) than would grain yield. It should describe an inherent self-correcting character of the response to drought. In maize, breeders have found that the attribute ASI is highly correlated with grain number per plant under drought conditions and provides useful QTL *per se* (Ribaut *et al.*, 1996). However, a grain-number/growth-rate QTL would be useful in other cereals and preferable to attempts to explain grain yield in terms of yield components, such as grain number, when the environment effects have confounded such data through genotype–environment interactions (Ribaut *et al.*, 1997). While this is just an example of taking a crop-physiology view, such a QTL or one based on a similar derivation may even be closer to the physical chromosomal location of control of grain set.

The process described above emphasizes a major paradigm of modern approaches to crop modelling – attempting to interpret complexity by modelling from 'the top, down' and aiming to identify conservative, 'genetic' parameters that affect the processes of plant growth and development (Hammer, 1998).

A flexible biophysical crop-simulation platform

In recent years, cropping-system models have adopted an object-orientated modular system of modelling the various processes of soil water and N cycling and species models for plant uptake and utilization of nutrients, plant growth, development and partitioning of carbon and N (McCown *et al.*, 1996). This allows the system components to be tested independently, which is possible only in a truly modular design. Since 1996, the APSIM model has developed a similar approach to modelling crop-growth processes, e.g. one can incorporate multiple approaches of modelling leaf development and compare their performance, independent of changes in the other crop or system processes, such as radiation interception or soil-water balance (Hammer, 1998; www.apsru.gov.au). The open-ended structure of the crop template within the crop model (Fig. 12.3) also allows additional complexity within and among growth processes to be added and tested as knowledge increases. All of the crop/cultivar parameter descriptions are stored external to the code so that the system is easily interfaced with genetic models, such as QU-GENE, which can act as suppliers of genetic parameters. Chapman *et al.* (2002b) detailed an implementation of such linkages between APSIM and QU-GENE.

The submodules in the crop template vary from rational and derived empiricisms (e.g. see the description of modelling flowering date above) to process-based approaches founded on a fairly complete understanding of the dynamics of crop physiology. As outlined in the previous section, important in the template design is the notion of emergent properties. A purely descriptive model may

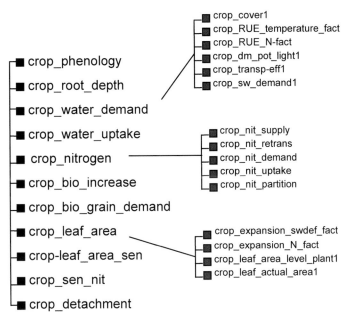

Fig. 12.3.   The structure tree of routines forming part of the generic crop template in APSIM (adapted from Hammer, 1998).

be more accurate in a given situation, but its behaviour is less likely to be suitable when it is forced to the extremes of genotypic and environmental variation. The template approach allows us to evaluate these emergent properties when trying to capture the behaviour of complex biological systems. With the physiological framework in hand, we attend to the questions posed by Cooper *et al.* (1999).

Characterizing Environments in a Cropping System

Knowledge of the structure of the target population of environments is an essential part of designing an efficient plant-breeding programme, particularly when it is impossible to undertake saturated testing of genotype performance in the TPE. Biotic challenges can be substantial barriers to production and must often be overcome through novel breeding and screening approaches. Abiotic challenges, such as drought, are more insidious in their effect. The sequence of environments sampled in

breeding trials is extremely variable for most dryland crops in Australia. Consequently, the selection pressures on breeding lines are quite different from those experienced when the samples of environments are constant across years. The problem for plant-breeding programmes arises when their testing or sampling regimes do not align well with the long-term expectations, i.e. if the Australian sorghum-breeding programme experiences 2–3 years of high rainfall, selection for drought tolerance will have been limited.

The occurrence of drought has been addressed for sorghum in Australia by simulating the drought stress experienced by a crop, given the weather record and soil information (Cooper and Chapman, 1996; Chapman *et al.*, 2000). The simulation model generated a drought-stress index for each day (1.0 = no stress) and, after the season indices were grouped across years and locations, three major patterns were identified. Figure 12.4 shows the patterns of three types of drought simulated from 108 years of data at six locations that represent the major sorghum-production regions. The frequency of occurrence (Fig. 12.5) of these drought patterns over all seasons and locations can be thought of as approximating the boundaries of the TPE. Figure 12.6, averaged for the entire TPE, shows how two different maturity genotypes yield under these different types of stresses. The later-maturing genotype, with a mean flowering date of 71 days after emergence, has an overall yield of 4.08 t ha^{-1} – i.e. 0.64 t ha^{-1} greater than that of the early-maturing genotype (63 days). However, the late-maturing genotype yields less in severe terminal-stress environments, as it cannot 'escape' the effects of terminal drought.

What is the consequence of differences in the sampling of the TPE in plant-breeding trials for the interpretation of genotype performance? In Fig. 12.5, it can also be seen that a sample of 3 years and six locations in 1993–1995 would have sampled these stress types in about the same proportions as they occur in the TPE. However, in 1981–1983, the sample was greatly biased against the

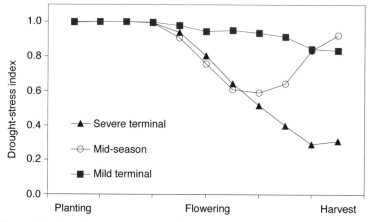

Fig. 12.4. Three major drought-stress patterns in six locations over 108 years of sorghum simulations (adapted from Chapman *et al.*, 2002b).

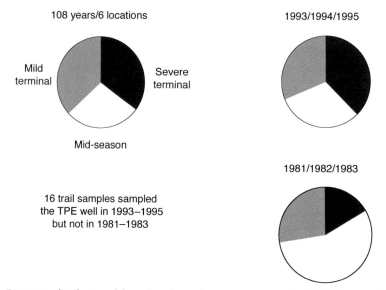

Fig. 12.5. Frequency distribution of three drought-environment types over the TPE (108 years/six locations) or in two samples of the TPE in 1993–1995 or 1981–1983.

severe terminal stress (Fig. 12.4) that causes reranking of the different maturity types (Fig. 12.6). In fact, in 1981–1983, the mean yield advantage of the late (4.52 t ha^{-1}) over the early (3.52 t ha^{-1})-maturity type was 1 t ha^{-1}, whereas in 1993–1995, the advantage was only 0.5 t ha^{-1} and was closer to the long-term difference of 0.64 t ha^{-1}. Hence, the estimates of genotype performance from the years 1993–1995 are, therefore, more representative of the long-term record than the set of years 1981–1983, as the latter sample was biased against the severe terminal stress.

The classification of environments in this way can be utilized directly. Chapman *et al.* (2000) showed that the long-term frequency of severe-stress environments at different locations was associated with differences in their ability to discriminate among cultivars in the sorghum hybrid-testing programme. Hence, the patterns of genotype–environment interactions for grain yield were also affected. In real time, the environments sampled in a hybrid trial can now be simulated and classified as being of a particular drought type to discern whether a set of trials has adequately sampled the TPE. Using the framework described by Cooper *et al.* (Chapter 11, this volume), Podlich and Cooper (1998) have demonstrated that this information can be used to weight the sample environments by their expected frequency in the TPE and improve the rate of selection gain for a variety of genetic models and breeding methods.

Information from these simulations of the environment accommodates cropping-system parameters (rotation, fertilizer, management, etc.) and allows us to define the abiotic TPE for drought. It is difficult to accurately simulate the biotic-stress components of the TPE, even where we can quantify the effect of a pest or disease on crop growth. This is because the long-term population dynamics and epidemics of pests are extremely difficult to parameterize. Hence, biotic effects are not routinely accounted for in cropping-system models. Alternative methods, such as the use of probe genotypes that respond as a bioassay of the presence of a biotic stress, are useful in this instance (e.g. Cooper and Fox, 1996). The simulation of environment and cropping-system effects defines the 'boundary conditions' for the geographical mandate of a breeding programme. In variable environments, it also defines the likelihood that plant breeders will realize a representative sample of environments from the TPE. Assuming that we can define these, the next step is to be able to exploit the available genes controlling the traits that enable adaptation to the TPE.

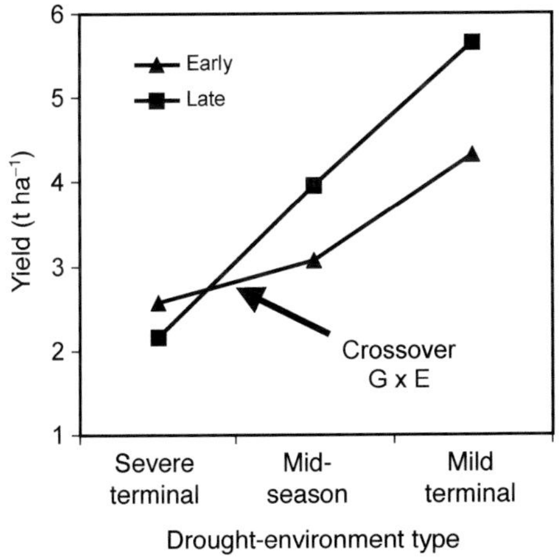

Fig. 12.6. Genotype–environment (G × E) cross-over interaction for simulated grain yield, given genotype maturity and drought-environment type.

Assessing the Value of Putative Traits – Simulating the Gene and Genotype Effects

Defining gene effects on crop growth

There are several major considerations in simulating gene effects on crop growth. What are the traits of interest and is there genetic variation for them? Can we model the gene effect at the scale at which we are developing and operating the model? What is the integrity of the model in the way that its processes interact with each other? Ideally, it will be a 'self-correcting' design that allows the properties of the model to be 'emergent' rather than programmed explicitly.

Muchow *et al.* (1991) and Hammer *et al.* (1996a,b) experimented with a sorghum model to try to establish the contribution to yield of different traits, given the environments of the sorghum region in Australia. They changed the model parameters associated with tillering, stay-green, maturity and transpiration efficiency. While the model could accommodate these effects, at the time this was done, there was little information about how much genetic variation existed for these traits. Furthermore, they did not know how many genes might be associated with such traits. More recently, Chapman *et al.* (2002a,b) have undertaken a similar study, but they incorporated knowledge of genes, QTL and trait variation based on the results of experiments and molecular-marker research conducted through the 1990s. The more recent work has also been able to work with model parameters that the authors believe are more closely aligned with the traits of interest. In a similar approach, White and Hoogenboom (1996) included gene effects as part of the actual parameterization of a model of bean phenology. This took the form of the presence or absence of a gene effect in regression equations that describe the contributions of the genes to the rate of phenological development stages. As the fitting equations in their model were linear, this approach is essentially the same as that used by Chapman *et al.* (2002a,b).

Another way of defining gene effects at the appropriate level within a model is to conduct experiments to measure genetic variation. For example, Yin *et al.* (1999b) observed variation for yield-determining physiological characters in spring barley. These characters were input parameters in their crop model, some of which were later associated with QTL (Yin *et al.*, 1999a). They are at present attempting to predict the variation in yield that arises from the QTL associated with these different yield parameters, i.e. if this is successful, they will have presumably incorporated the effects at an appropriate level. The key to this approach is to find parameters that are stable across environments – the so-called genetic constant parameters, around which

the integrity of a model can be maintained. Where the appropriate molecular-marker mapping experiments are conducted, biologists are beginning to obtain useful information on yield-determining components of plant growth.

It is generally considered unwise for crop-simulation models to attempt to model across more than one or two orders of scale away from the basic scale at which they operate (Sinclair and Seligman, 2000). For example, in a model that operates on a daily time step, with inputs of daily weather (max./ min. temperature, rainfall and radiation), and models light interception at the canopy level, one would not normally attempt to incorporate a photosynthesis model of how the chloroplasts work on a time-scale of seconds. Such a 'cellular' level of modelling, with data inputs every second, would be entirely appropriate to examine how sun flecks affect photosynthesis of small areas of leaf, for example. However, it would not be appropriate to then try to partition the predicted sucrose production to various parts of the crop as simple dry matter, as this produces an internal conflict in the scales used in the model. Hence, to examine gene effects in a crop-simulation model, we have to interpret the perceived or observed trait expression at the cellular level in terms of its expression at the crop level. Careful field experiments involving an appropriate crop-physiology framework are essential to achieve this.

The example of gene effects on flowering date (e.g. Koornneef *et al.*, 1998) and ideas from Thornley (1972) can be expanded here. Assume that we have a gene that causes the growing point to switch from vegetative to reproductive after a period of time has passed. As plants are poorly insulated from the environment, passage of time is normally better described in terms of thermal time, rather than calendar time. This is because biochemical rates of reaction tend to increase within the range of ambient temperature, causing a parallel increase in development rate. The flowering gene may switch on once sufficient thermal time has passed for the accumulation of some 'response' compound or hormone to reach a level that initiates the switch. However, we do not need to model the process at this or the more complex gene-transcription level. Based on field observations, gene action can be registered as a difference in the 'minimum thermal time' that must pass before floral initiation can begin. In a later-flowering genotype, this minimum is simply a larger thermal time. The mechanism may well be that the late-flowering genotype accumulates the 'response' compound more slowly or that the switch requires a greater amount of 'response' compound before it cuts in. To capture this mechanism, it is only necessary to model the induction of the 'gene cascade' that signals the change in state of the meristem so that the simulation model can then account for the changes in other crop-level processes that arise from this. Where major gene effects can be represented as integrated biochemical effects at a daily time step, it is appropriate to include them.

To reiterate an earlier point, in designing models of crop growth and development, modellers should be attempting to capture rules that define the boundary conditions for simulation processes, rather than applying a descriptive structure. This philosophy of parameterization and modelling of the principles of response and feedbacks (cf. description of response) infers that models should be able to express complex behaviour of the type observed in the field, even given simple operational rules at a functional crop physiological level. The sorghum-crop module (APSIM-Sorg) within the APSIM cropping-system model (McCown *et al.*, 1996) contains several deliberate parameterizations to address genetic variation using a 'boundary-conditions' approach (Hammer *et al.*, 1999). The central design of the model aims to simulate genetic variation in 'signalling-type' genes through appropriate modification of the model parameters.

In our work so far, we have focused on four traits for which we have a relatively good understanding of their crop physiology and know that genetic variation exists. These traits are:

- transpiration efficiency (TE), whereby higher TE increases biomass production potential when water is limiting (Mortlock and Hammer, 1999);
- flowering date (phenology (PH)) as influenced by thermal time requirement for floral initiation (Morgan and Finlayson, 2000);
- osmotic adjustment (OA) and its positive effect on increasing grain number and retranslocatable assimilate under drought conditions (Hammer *et al.*, 1999);
- stay green (SG) as a function of the minimum target leaf N (Borrell *et al.*, 2000).

For each trait, the observed genetic variation was associated with a postulated number of genes (n) (suggested from prior experiments) and was divided into equal size effects that equated with the number of expression states ($2n + 1$) associated with the trait, i.e. increased level of expression for a particular trait was associated with a greater number of positive (+ve) alleles for that trait. The number of genes (expression states) assigned to the traits, TE, PH, OA and SG were 5 (11), 3 (7), 2 (5) and 5 (11), respectively (Chapman *et al.*, 2002a). With the exception of OA, which only acted under drought stress, all of the traits were constitutive in action, although their expression was dependent on the environment encountered in any particular situation (Table 12.1). The traits of high TE and high OA were most advantageous in severe terminal-stress environments, whereas high values for PH and SG were of greater importance in the mild terminal-stress environment. As was apparent in Fig. 12.6, late flowering was clearly a disadvantage in the severe-stress environment.

Integrating Molecular Markers via Biophysical Simulation

As related in this volume by Cooper *et al.* (Chapter 11), the QU-GENE simulation platform (Podlich and Cooper, 1998) simulates the stochastic properties of genes,

Table 12.1. From the data set of 4235 genotypes, the mean yield (t ha⁻¹) within each of the three drought-environment types and, for each drought-environment type and trait, the difference between the mean of all genotypes containing the highest expression state for the trait and the mean of all genotypes containing the lowest expression state for the trait (adapted from Chapman et al., 2002b).

Trait	Drought-environment type		
	Severe terminal	Mid-season	Mild terminal
TE	0.82	0.61	0.46
PH	−0.32	0.89	1.36
OA	0.42	0.22	0.18
SG	0.10	0.49	1.03
Mean yield	2.58	3.55	4.99

TE, transpiration efficiency; PH, phenology (flowering time); OA, osmotic adjustment; SG, stay green.

genotypes and environments in the operation of plant-breeding programmes. QU-GENE has two components: the first defines the gene-network–environment structure of the population whereas the second consists of a module (customized) to describe the selection strategy. QU-GENE can model breeding programmes as 'search-strategies that seek higher peaks on the adaptation landscape (genetic space) for a given genotype–environment system'. The rate at which a population improves with selection is monitored by the change in grain yield of successive cycles and in the changes in the fixation (gene frequency) of both positive and negative alleles related to this yield improvement. Statistical analyses determine the effectiveness of searches in 'creating and finding' superior combinations of alleles in the simulated populations. These superior methods of recombination and searching 'genetic space' can then be considered for application in conventional plant-breeding programmes (e.g. Podlich and Cooper, 1998).

The interaction of gene effects with each other and with the environment to determine yield can be handled in two ways. The method used in the original conception of QU-GENE incorporates these effects based on statistically derived or postulated relationships from experiments and published knowledge about gene action. Another way to account for effects is to try to simulate the processes in Fig. 12.1, given some knowledge of the variation in the effects of each gene on several traits. This has been

attempted by linking the definition of gene effects on traits (from QU-GENE) to APSIM. For a single genotype, the effect of the genes on traits and that of traits on grain yield then become the result of their 'integrated effects' via the APSIM biophysical model, i.e. the continuous interaction of soil and weather on growth and yield (Fig. 12.1) are accommodated. This can be repeated for many genotypes across many environments to produce the adaptation landscape for the sorghum TPE of a region (Chapman et al., 2002a,b). In practice, the data set was expanded to include 'all genotypes' in a theoretical population of genotypes to which QU-GENE can be applied to simulate the breeding process.

An example of the APSIM-Sorg model being used to accommodate genotypic variation (minimum, average and maximum expression states) for four traits is given in Fig. 12.7. The thick line ranks the mean yield of each of 54 genotypes when grown at six locations (between central Queensland (CQ) and northern New South Wales (NSW) in Australia) and in 108 years. The lower line shows their respective yields at six locations for 2 years (1937 and 1938) when conditions were relatively dry with the ratio of 4.5 : 5.5 : 0 of 'severe terminal' : 'mid-season' : 'mild terminal' drought environments (as defined in Fig. 12.4). In the following 2 years (1939 and 1940), the weather was more favourable (ratio of 0 : 2 : 8 of the drought environment types) and the yields were almost doubled, though with greater variation among genotypes and a

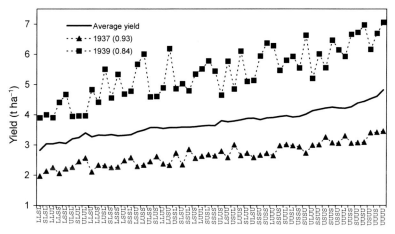

Fig. 12.7. Mean grain yield of 54 simulated genotypes varying (upper, average and lower values) for four traits (TE, PH, OA, SG) and simulated in six locations in 1937 and 1938, 1939 and 1940 or across 108 years.

poorer correlation with the long-term performance. This data set is being used to interpret how different traits interact with different drought-environment types to influence final yield (Chapman *et al.*, 2002b).

The 54 genotypes were expanded to 4235 genotypes by incorporating all of the trait-expression classes for the four traits, utilizing the additive gene effects for each expression state as described above (Chapman *et al.*, 2002b). Each genotype was simulated in all six locations and 108 years to produce an adaptation landscape of gene–gene–environment interactions. We averaged the responses within each of the three drought-environment types described in Fig. 12.4 to produce a table of yields for every genotype in each environment. This created an adaptation landscape for each environment type on which to test different plant-breeding strategies, although here we consider selection only within the mild terminal-stress environment.

Many different breeding strategies can be tested on these landscapes. Briefly, Chapman *et al.* (2002a) set up QU-GENE to run an S_1 recurrent-selection programme with 4 years per cycle. It began with a sample of ten parents from the 4235 genotype classes available, such that the frequencies of each of the 15 controlling genes (one of the alleles for each gene) in the ten parents was 0.2. The first 2 years were used to randomly

intermate the parents, select among 5000 spaced plants and produce 1000 S_1 offspring families. The S_1s were 'evaluated' across 2 years in different drought environments – in practice, by 'looking up' their yields in the relevant table. The best 100 S_1s (10% selection pressure) were then chosen to begin the mating cycle again. To accommodate the effects of variation in the process of choosing parents, each breeding scenario was repeated 200 times to give the estimates of changes in the yield of the S_1 populations.

Chapman *et al.* (2002a) compared phenotypic selection within each of the environment types, as well as using a random sample of environment types from a population of environments structured at the long-term TPE frequency. In this chapter, we have compared, for the mild terminal stress only, the phenotypic strategy with one based on marker selection for all traits. Flanking markers were associated with each trait and examined, given different recombination frequencies (RF) between markers and genes (0, 0.05, 0.2, 0.5) in separate runs. During selection, each S_1 family was assigned a marker score that was the sum of the number of +ve alleles for trait expression present across the 15 genes. Genes were equally weighted, and the top 100 S_1 families were chosen for intermating. In comparing the results, it was assumed that a cycle of completely marker-assisted selection could

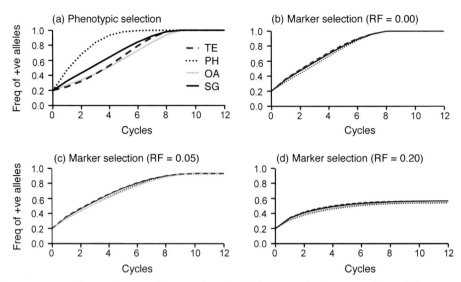

Fig. 12.8. For mild terminal stress, change with cycle of selection in the frequency of +ve alleles (see text) in the selected fraction for four traits using: (a) phenotypic selection for yield or marker selection for three recombination frequencies; (b) 0 (perfect markers); (c) 0.05 (closely linked markers); or (d) 0.20 (poorly linked markers).

be undertaken in 2 years rather than the 4 years taken for phenotype selection.

Figure 12.8 illustrates the change, with cycle of selection, in gene frequency for each trait in the selected fraction of S_1 families. Under phenotypic selection, the PH trait was highly favoured, as later maturity was most beneficial in this environment type, though not in a severe terminal-stress environment (Table 12.1). While positive alleles for all traits were being fixed from the first cycle, PH was fixed more quickly than SG, followed by OA and TE, i.e. when their effects were modulated by the gene–phenotype relationships within a biophysical model, the trait values for yield were quite divergent. This is frequently the case in the practical world, but it is difficult to discern which genetic components of the phenotype are or are not contributing to yield improvement. Under perfect marker selection (i.e. RF = 0) (Fig. 12.8b), the traits were all fixed at the same rate, and yield progress per cycle was similar to, though a little slower than, phenotypic selection (Fig. 12.9a). While lower recombination frequencies limited final yield (Fig. 12.9a), due to inability to combine all of the favourable alleles

(Fig. 12.8c,d), progress was rapid in the first two to three cycles of selection, i.e. even relatively poor markers can be of assistance when the frequencies of positive alleles are low in a broad genetic base. In terms of yield per year, even an RF of 0.2 allowed more rapid progress than phenotypic selection in the first 8 years or two cycles of phenotypic selection.

This scenario compared only phenotypic and complete marker selection. In future experiments, we shall examine marker-assisted selection for one or more traits where molecular-marker and phenotypic data are combined in a selection index.

Summary and Conclusions

In this chapter, we have discussed many of the key aspects of a modelling framework for linking gene effects to their effects on a trait phenotype. Given the complexity of plants as growing systems and while we might one day understand the 'function of every plant gene', it is unlikely that we can build a model of sufficient detail for it to be parameterized to capture the level of

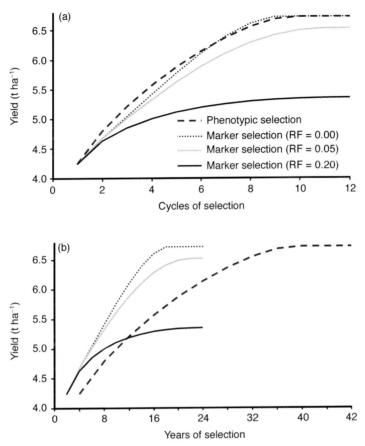

Fig. 12.9. For mild terminal stress, change in yield of the selected fraction of S_1 families for phenotypic or three types of marker selection differing in linkage against (a) cycle or (b) year of selection.

expression and effect of every gene. The combinatorial nature of this problem sets limits on our capacity to evaluate or even simulate all possible genotypes. Instead, by combining good experimentation on the right sets of germ-plasm, it does seem possible that we can capture many of the components that matter at the level of crop growth. While increasing attention is being paid to the gene technologies, the fact that we can measure gene expression and product does not necessarily mean that we can utilize that information. The limits come about due to limits in our current ability to model, as well as limits in the input information to run simulation models, e.g. technologies are simply not precise enough to extrapolate

from satellite data to leaf temperatures at the base of the canopy. As other sciences are beginning to discover though, our major limitation may well be the complexity of the system with which we are working. Utilizing a modular, structured modelling approach allows the integration of new effects and input information as they become known. Though it will not contain all of the possible genes that affect crop growth, the modelling framework will allow us to study complex biological networks and to begin to determine the emergent properties of the network that influence its behaviour. The objective we have focused on here is to be able to apply these interpretations to improving the efficiency of plant breeding. Certainly, it will be exciting to begin to understand the paradigm of genes (rather than genotypes) interacting with environments by utilizing these integrated genetic and biophysical models.

The sciences of molecular biology are beginning to reveal the role of the 10,000–40,000 genes in the biochemical pathways of plants. While it may be possible to eventually determine the 'function' of each gene in single pathways, the complex coordination of pathways will mask the ultimate controls of crop yield for some time. However, we can begin to integrate several types of simulation models to gain insights into how yield is determined by selection among gene networks that comprise both known physiological-trait effects and postulated stochastic effects.

The definition of 'a quantitative heritable trait or character' is still a subject of much debate in genetics. Ultimately, gene networks control traits, and so knowledge of single gene action in a biochemical pathway is likely to be of limited use in understanding crop adaptation to environments. When applied to suitable genetic populations, crop physiology ('forward physiology' when compared with the reductionist 'gene-to-pathway' molecular physiology) can reveal conservative, yield-determining parameters (traits) to incorporate in biophysical simulation models. These models have a framework of realistic data description and environment-input requirements, e.g. time steps of days rather than seconds or minutes. A genetic model can be used to specify the actions and interactions of genes that determine the value of individual traits. In turn, when these combinations of trait values are input to the crop model, the effects interact with the environment to generate an 'adaptation landscape' of fitness or performance phenotype (e.g. yield) for all possible gene/trait networks (genotypes). Thus, the landscape contains the pleiotropic and gene–environment effects associated with the specified gene networks. This complex landscape of biological gene–environment networks can be characterized for its emergent properties and then sampled and searched to identify superior strategies

(breeding methods) to locate genotypes containing superior allelic combinations.

Three areas in which biophysical models can assess *in silico* the multitude of options to improve the efficiency of plant breeding are: (i) characterizing environments to define the TPE; (ii) assessing the value of specific putative traits in improved plant types; and (iii) enhancing the integration of molecular genetic technologies. In this chapter, we demonstrate examples of each of these, following an initial consideration of how to integrate the gene–phenotype effects of traits that are associated with adaptation to the physical environment. It points to research areas where molecular biology can contribute information on the genetic variation for traits that control the physiology of yield. As knowledge improves, there is the opportunity to connect these traits more directly to gene action via models of biochemical pathways.

Acknowledgements

The valuable assistance of technical staff in the Queensland Department of Primary Industries (QDPI) (Greg McLean and Peter de Voil) and the University of Queensland (UQ) (Kevin Micallef) is gratefully acknowledged in the completion of the APSIM and QU-GENE simulations on the computer cluster and processing and summarizing of output.

References

Bidinger, F.R., Hammer, G.L. and Muchow, R.C. (1996) The physiological basis of genotype by environment interaction in crop adaptation. In: Cooper, M. and Hammer, G.L. (eds) *Plant Adaptation and Crop Improvement.* CAB International, Wallingford, UK, pp. 329–348.

Bolaños, J. and Edmeades, G.O. (1993) Eight cycles of selection for drought tolerance in lowland tropical maize. I. Responses in grain yield, biomass, and radiation utilization. *Field Crops Research* 31, 233–252.

Borrell, A.K., Hammer, G.L. and Henzell, R.G. (2000) Does maintaining green leaf area in sorghum improve yield under drought? II.

Dry matter production and yield. *Crop Science* 40, 1037–1048.

Bouman, B.A.M., van Keulen, H., van Laar, H.H. and Rabbinge, R. (1996) The 'School of de Wit' crop growth simulation models: a pedigree and historical overview. *Agricultural Systems* 52, 171–198.

Cantero-Martinez, C., Oleary, G.J. and Connor, D.J. (1999) Soil water and nitrogen interaction in wheat in a dry season under a fallow-wheat cropping system. *Australian Journal of Experimental Agriculture* 39, 29–37.

Chapman, S.C., Cooper, M., Hammer, G.L. and Butler, D. (2000) Genotype by environment interactions affecting yield of grain sorghum. II. Frequencies of different seasonal patterns of drought stress are related to location effects on hybrid yields. *Australian Journal of Agricultural Research* 51, 209–222.

Chapman, S.C., Cooper, M., Podlich, D.W. and Hammer, G.L. (2002a) Evaluating plant breeding strategies by simulating gene action and environment effects to predict phenotypes for dryland adaptation. *Agronomy Journal* (in press).

Chapman, S.C., Cooper, M. and Hammer, G.L. (2002b) Using crop simulation to generate genotype by environment interaction effects for sorghum in water-limited environments. *Australian Journal of Agricultural Research* 53, 379–389.

Collins, N., Drake, J., Ayliffe, M., Sun, Q., Ellis, J., Hulbert, S. and Pryor, T. (1999) Molecular characterization of the maize Rp 1-D rust resistance haplotype and its mutant. *Plant Cell* 11, 1365–1376.

Comstock, R.E. (1977) Quantitative genetics and the design of breeding programs. In: *Proceedings of the International Conference on Quantitative Genetics, 16–21 August 1976.* Iowa State University Press, Ames, USA, pp. 705–718.

Cooper, M. and Chapman, S.C. (1996) Breeding sorghum for target environments in Australia. In: Foale, M.A., Henzell, R.G. and Kneipp, J.F. (eds) *Proceedings of the Third Australian Sorghum Conference, Tamworth, 20–22 February 1996.* Occasional Publication No. 93, Australian Institute of Agricultural Science, Melbourne, pp. 173–187.

Cooper, M. and Fox, P.N. (1996) Environmental characterization based on probe and reference genotypes. In: Cooper, M. and Hammer, G.L. (eds) *Plant Adaptation and Crop Improvement.* CAB International, Wallingford, UK, pp. 529–547.

Cooper, M., Podlich, D.W., Jensen, N.M., Chapman, S.C. and Hammer, G.L. (1999) Modelling plant breeding programs. *Trends in Agronomy* 2, 33–64.

de Wit, C.T. and Penning de Vries, F.W.T. (1983) Crop growth models without hormones. *Netherlands Journal of Agricultural Science* 31, 313–323.

Edmeades, G.O., Bolaños, J., Elings, A., Ribaut, J.-M., Bänziger, M. and Westgate, M.E. (2000) The role and regulation of the anthesis-silking interval in maize. In: Westgate, M.E. and Boote, K.J. (eds) *Physiology and Modeling Kernel Set in Maize.* CSSA Special Publication No. 29, Crop Science Society of America and American Society of Agronomy, Madison, Wisconsin, USA, pp. 43–73.

Fischer, K.S., Edmeades, G.O. and Johnson, E.C. (1989) Selection for the improvement of maize yield under moisture deficits. *Field Crops Research* 22, 227–243.

Fujioka, S., Yamane, H., Spray, C.R., Gaskin, P., MacMillan, J., Phinney, B.O. and Takahashi, N. (1988) Qualitative and quantitative analyses of gibberellins in vegetative shoots of nornali, *dwarf-1*, *dwarf-2*, *dwarf-3*, and *dwarf-5* seedlings of *Zea mays* L. *Plant Physiology* 88, 1367–1372.

Giersch, C. (2000) Mathematical modelling of metabolism. *Current Opinion in Plant Biology* 3, 249–253.

Hammer, G.L. (1998) Crop modelling: current status and opportunities to advance. *Acta Horticulturae* 456, 27–36.

Hammer G.L., Butler, D., Muchow, R.C. and Meinke, H. (1996a) Integrating physiological understanding and plant breeding via crop modelling and optimization. In: Cooper, M. and Hammer, G.L. (eds) *Plant Adaptation and Crop Improvement.* CAB International, Wallingford, UK, pp. 419–441.

Hammer, G.L., Chapman, S.C. and Muchow, R.C. (1996b) Modelling sorghum in Australia: The state of the science and its role in the pursuit of improved practices. In: Foale, M.A., Henzell, R.G. and Kneipp, J.F. (eds) *Proceedings of the Third Australian Sorghum Conference, Tamworth, 20–22 February 1996.* Occasional Publication No. 93, Australian Institute of Agricultural Science, Melbourne, pp. 43–60.

Hammer, G.L., Chapman, S.C. and Snell, P. (1999) Crop simulation modelling to improve selection efficiency in plant breeding programs. In: Williamson, P., Haak, I., Thompson, J. and Campbell, A. (eds) *Proceedings of the Ninth Assembly of the Wheat Breeding Society of*

Australia, Toowoomba, September 1999. Wheat Breeding Society of Australia, Toowoomba, Australia, pp. 79–85.

Hare, P.D., Cress, W.A. and van Staden, J. (1999) Proline synthesis and degradation: a model system for elucidating stress-related signal transduction. *Journal of Experimental Botany* 50, 413–434.

Hsiao, T.C. (1973) Plant responses to water stress. *Annual Reviews of Plant Physiology* 24, 519–570.

Ingram, J. and Bartels, D. (1996) The molecular basis of dehydration tolerance in plants. *Annual Reviews of Plant Physiology and Plant Molecular Biology* 47, 377–403.

Koornneef, M., Alonso-Blanco, C., Peeters, A.J.M. and Soppe, W. (1998) Genetic control of flowering time in *Arabidopsis*. *Annual Reviews of Plant Physiology and Plant Molecular Biology* 49, 345–370.

Lan, T.H., Delmonte, T.A., Reischmann, K.P., Hyman, J., Kowalski, S.P., Mcferson, J., Kresovich, S. and Paterson, A.H. (2000) An EST-enriched comparative map of *Brassica oleracea* and *Arabidopsis thaliana*. *Genome Research* 10, 776–788.

Lonnquist, J.H. and Jugenheimer, R.W. (1943) Factors affecting the success of pollination in corn. *Journal of the American Society of Agronomy* 35, 923–933.

Ludlow, M.M. and Muchow, R.C. (1990) A critical evaluation of traits for improved crop yields in water-limited environments. *Advances in Agronomy* 43, 107–153.

McCown, R.L., Hammer, G.L., Hargreaves, J.N.G., Holzworth, D.P. and Freebairn, D.M. (1996) APSIM: a novel software system for model development, model testing, and simulation in agricultural systems research. *Agricultural Systems* 50, 255–271.

Mendes, P. and Kell, D.B. (1998) Non-linear optimization of biochemical pathways: applications to metabolic engineering and parameter estimation. *Bioinformatics* 14, 869–883.

Moore, G. (2000) Cereal chromosome structure, evolution, and pairing. *Annual Review of Plant Physiology and Plant Molecular Biology* 51, 195–222.

Morgan, P.W. and Finlayson, S.A. (2000) Physiology and genetics of maturity and height. In: Smith, C.W. (ed.) *Sorghum: Origin, History, Technology, and Production.* John Wiley & Sons, Melboune, pp. 227–259.

Mortlock, M.Y. and Hammer, G.L. (1999) Genotype and water limitation effects on transpiration efficiency in sorghum. *Journal of Crop Production* 2, 265–286.

Muchow, R.C., Hammer, G.L. and Carberry, P.S. (1991) Optimising crop and cultivar selection in response to climatic risk. In: Muchow, R.C. and Bellamy, J.A. (eds) *Climatic Risk in Crop Production: Models and Management for the Semiarid Tropics and Subtropics.* CAB International, Wallingford, UK, pp. 235–262.

Phinney, B.O. (1956) Growth responses of single gene mutants of maize to gibberellic acid. *Proceedings of the National Academy of Sciences USA* 42, 185–189.

Podlich, D.W. and Cooper, M. (1998) QU-GENE: a simulation platform for quantitative analysis of genetic models. *Bioinformatics* 14, 632–653.

Quinby, J.R. and Karper, R.E. (1954) Inheritance of height in sorghum. *Agronomy Journal* 46, 211–216.

Ribaut, J.M., Hoisington, D.A., Deutsch, J.A., Jiang, C. and Gonzalez de Leon, D. (1996) Identification of quantitative trait loci under drought conditions in tropical maize. 1. Flowering parameters and the anthesis–silking interval. *Theoretical and Applied Genetics* 92, 905–914.

Ribaut, J.M., Jiang, C., Gonzalez de Leon, D., Edmeades, G.O. and Hoisington, D.A. (1997) Identification of quantitative trait loci under drought conditions in tropical maize. 2. Yield components and marker-assisted selection strategies. *Theoretical and Applied Genetics* 94, 887–896.

Ritchie, J.T. (1991) Specifications of the ideal model for predicting crop yields. In: Muchow, R.C. and Bellamy, J.A. (eds) *Climatic Risk in Crop Production: Models and Management for the Semiarid Tropics and Subtropics.* CAB International, Wallingford, UK, pp. 97–122.

Ritchie, J.T. and Nesmith, D.S. (1991) Temperature and crop development. In: Hanks, J. and Ritchie, J.T. (eds) *Modeling Plant and Soil Systems.* American Society of Agronomy, Madison, Wisconsin, USA, pp. 5–30.

Schenk, P.M., Kazan, K., Wilson, I., Anderson, J.P., Richmond, T., Somerville, S.C. and Manners, J.M. (2000) Coordinated plant defense responses in *Arabidopsis* revealed by microarray analysis. *Proceedings of the National Academy of Sciences USA* 97, 11655–11660.

Sinclair, T.R. and Horie, T. (1989) Leaf nitrogen, photosynthesis, and crop radiation use efficiency: a review. *Crop Science* 29, 90–98.

Sinclair, T.R. and Seligman, N. (2000) Criteria for publishing papers on crop modelling. *Field Crops Research* 68, 165–172.

Somerville, C. and Somerville, S. (1999) Plant functional genomics. *Science* 285, 380–383.

Stringfield, G.H. and Thatcher, L.E. (1947) Stands and methods of planting for corn hybrids. *Journal of the American Society of Agronomy* 39, 995–1010.

Strogatz, S.H. (2001) Exploring complex networks. *Nature* 410, 268–276.

Sun, Y., Helentjaris, T., Zinselmeier, C. and Habben, J.E. (1999) Utilizing gene expression profiles to investigate maize response to drought stress. In: *Proceedings of the 54th Annual Corn and Sorghum Research Conference*. American Seed Trade Association, Washington, DC, pp. 140–153.

Thornley, J.H.M. (1972) A model of a biochemical switch, and its application to flower initiation. *Annals of Botany* 36, 861–871.

Tollenaar, M., Dwyer, L.M. and Stewart, D.W. (1992) Ear and kernel formation in maize hybrids representing three decades of grain yield improvement in Ontario. *Crop Science* 32, 432–438.

Tollenaar, M., Dwyer, L.M., Stewart, D.W. and Ma, B.L. (2000) Physiological parameters associated with differences in kernel set among maize hybrids. In: Westgate, M.E. and Boote, K.J. (eds) *Physiology and Modeling Kernel Set in Maize*. CSSA Special Publication No. 29, Crop Science Society of America and American Society of Agronomy, Madison, Wisconsin, USA, pp. 115–130.

Troyer, A.F. and Larkins, J.R. (1985) Selection for early flowering in corn: 10 late synthetics. *Crop Science* 25, 695–697.

White, J.W. and Hoogenboom, G.H. (1996) Integrating effects of genes for physiological traits into crop growth models. *Agronomy Journal* 88, 416–422.

Yin, X., Kropff, M.J. and Stam, P. (1999a) The role of ecophysiological models in QTL analysis: the example of specific leaf area in barley. *Heredity* 82, 415–421.

Yin, X., Stam, P., Dourleijn, C.J. and Kropff, M.J. (1999b) AFLP mapping of quantitative trait loci for yield-determining physiological characters in spring barley. *Theoretical and Applied Genetics* 99, 244–253.

Zinselmeier, C., Jeong, B.-R. and Boyer, J.S. (1999) Starch and the control of kernel number in maize at low water potentials. *Plant Physiology* 121, 25–35.

13   Tissue Culture for Crop Improvement

Roberta H. Smith and Sung Hun Park

Vegetable and Fruit Improvement Center, 1500 Research Parkway, Suite 120A, Texas A&M University Research Park, College Station, TX 77845, USA

The use of plant cell-, tissue- or organ-culture techniques for crop improvement has many facets. Some techniques are quite simple, requiring minimal laboratory facilities and expertise, whereas others are more complex and require extensive training and equipment and a multidisciplinary approach. The variety of tissue-culture techniques for crop improvement include the following:

- Micropropagation.
- Basic studies of plant morphogenesis.
- Virus-free plants from shoot-apex culture.
- Germ-plasm preservation and transport.
- Androgenesis/gynogenesis for haploid and dihaploid plants.
- Embryo rescue for unique hybrids.
- Cell selection and somaclonal variation for unique germ-plasm.
- Protoplast fusion for somatic hybrids.
- Plant genetic engineering to add unique foreign genes to enhance agronomic characteristics.
- Molecular farming.

The terms plant genetic engineering and plant biotechnology are used interchangeably, and a component of plant biotechnology uses plant-tissue culture and molecular genetics to produce plants with foreign or new genes. Each of these areas provides the plant breeder with unique tools to enhance the conventional process of crop improvement.

In Vitro Clonal Propagation

Micropropagation or clonal propagation is the most widely used tool of plant cell culture because one can clonally propagate a very large number of uniform copies of a selected parent plant in a short period of time (Morel, 1972; Murashige, 1974). This technique is widely used in the ornamental and floriculture industry. Additionally, unique forest/plantation trees, vegetable crops, etc. can also be mass-produced by this method, if it is cost-effective for that particular plant. The technique generally involves isolating a shoot tip or lateral bud that contains a shoot meristem and putting it on a nutrient medium with plant growth regulators to encourage multiple shoots or enhanced axillary branching. Other explants, such as fern runner tips, petiole, stem, floral scapes, leaf sections, etc., also work as long as the cell-culture process involves the direct production of adventitious shoots or somatic embryos from these tissues without a callus intermediate. Any explant that forms callus before adventitious shoot or embryos can result in plants that can be different in phenotype from the parent plant, and this is usually undesirable in propagation.

Basic Studies of Plant Morphogenesis

The basic principles of recovering whole rooted plants from cell culture are central to nearly all the other techniques using plant tissue culture for crop improvement. The extensive commercial use of micropropagation has relied on many very fundamental studies of plant morphogenesis. The basic studies on plant growth regulators, nutrient media, cell growth conditions, explants, etc., which had their origins in the early studies of Haberlandt, Hannig, Kotte, White, Loo, Ball, Gautheret, Nobecourt, Van Overbeek, Skoog, Miller, Morel, Murashige and so many others (for reviews, see Krikorian and Berquam, 1969; Thorpe, 2000), established the foundation for micropropagation. These studies on basic aspects of cell culture and morphogenesis set the stage for the extensive application of these findings to micropropagation and using cell culture to recover virus-free plants. They also laid the foundation for all the useful techniques in plant cell culture as applied to crop improvement.

Virus-free Plants

Many vegetatively propagated plants, such as potato, garlic, pineapple, orchids, carnation, banana, citrus and strawberry, are viral-infected and the virus is further spread by vegetative propagation (Quak, 1977). Many of these plants have multiple viral infections, such as strawberry, with up to 62 viruses and mycoplasms that can infect it (Thorpe and Harry, 1997). Viral contamination by one or more viruses causes many symptoms, including the following:

- Reduced vigor and growth.
- Low yield.
- Necrosis.
- Off colours in the flowers.
- Curling and variegated colour of the leaf blade.
- Decreased rooting of cuttings.
- Plant death.

There are no effective chemical treatments to free plants from viruses. A significant milestone in the use of plant cell culture for ornamental, floriculture and crop plant improvement was the discovery by Morel and Martin (1952) that virus-free tissue in the shoot apex could be isolated and cultured in vitro into plants that were virus-free. This was quite a powerful tool for plant improvement programmes that was not fully utilized on a wide number of plants until the mid-1960s (Smith and Murashige, 1970). Many crop plants now routinely go through in vitro virus clean-up programmes usually combined with thermotherapy. An additional benefit of this process is the elimination of fungal and bacterial contaminants as well. This clean plant material is ideal for germ-plasm storage, as well as international transport.

Germ-plasm Preservation

The use of cell culture for germ-plasm preservation and transport is a very convenient technique. It is used in varying degrees by many germ-plasm repositories, biotechnology companies, commercial propagation laboratories and academic research laboratories. The preservation of wild or diverse population germ-plasm or cultivated crops, especially those that are vegetatively propagated, is a valuable tool for crop improvement (Pauls, 1995). Preservation of germ-plasm is even more critical today with the rapid destruction of habitat and the need to preserve the genetic pool. This technique reduces the labour involved in maintaining plant lines and also minimizes losing the plant material to disease. Plant material can be cold-stored to allow only minimal growth, cryopreserved in liquid nitrogen in the form of shoot tips, somatic embryos and callus and then used as needed (Withers, 1987), or grown with growth-retarding chemicals in the medium to decrease transfer frequency (Thorpe and Harry, 1997). This can minimize maintaining stock plants in the greenhouse or under normal in vitro conditions that require frequent maintenance. Some germ-plasm collection agencies are able to put shoot-tip or lateral-bud material from vegetatively propagated plants out in the field into cell culture and transport the cultures back to the main storage facility, eliminating the transport of large whole-plant material. Additionally, since most *in vitro* plant material is free of microbe or insect contamination and can be, if necessary, certified virus-free, the international transport and exchange of germ-plasm through quarantine facilities is facilitated.

Somaclonal Variation

An important aspect of conventional breeding is creating genetically diverse populations via crossing. Plant tissue-culture techniques, such as somaclonal variation, cell selection, protoplast fusion, embryo rescue and dihaploids via anther culture, can enhance the ability of the plant breeder to obtain unique diversity.

Somaclonal variation was a term coined by Larkin and Scowcroft (1981) to designate the plants from cell culture that were different from the original starting material. In most cases, this was a result of explants producing callus prior to undergoing regeneration via organogenesis (shoot production) or somatic embryogenesis (bipolar embryo production from somatic tissues). It was believed that somaclonal variation would be a unique tool for the plant breeder that would reduce the time it took for conventional variety development and open up access to new sources of genetic variation (Evans *et al.*, 1984). The variation obtained from tissue culture increased with increased length in culture of the callus before regeneration occurred. It is possible to generate new varieties using this approach; however, in most cases the usefulness of this approach has been limited. Very few of these variants were stable and usable in plant-improvement programmes, or the plants had other undesirable agronomic traits. The overall use of somaclonal variants has not played a major role in most plant-improvement programmes (Smith *et al.*, 1993; Karp, 1995; Veilleux and Johnson, 1998). Intense interest in and publication of studies on somaclonal variation and its potential appear to have peaked in 1991 and 1992. The generation of somaclonal variation, however, is very easy to accomplish with minimal facilities and is often a component of plant-breeding programmes, initially getting started using plant cell-culture approaches in crop improvement.

Cell Selection

Similar observations to those for somaclonal variation were made for using cell selection to generate unique germ-plasm. Cells in culture were exposed to selective stresses, such as temperature, salts (e.g. sodium or aluminium), plant toxins, amino acid analogues, etc. (Chaleff, 1983). Cells that survived the selection *in vitro* were then regenerated into plants, and the plants and their progeny were evaluated to see if the cellular level selection would be expressed at the whole-plant level. McHughen (1987) generated salt tolerance in flax (*Linum usitatissimum* L.) from a cell line selected *in vitro* for salt tolerance, and a cultivar, 'Andro', was released and commercially used. Other successes have been reported (Smith *et al.*, 1993); however, in general, considering the level of funding and research effort, little has come out of these strategies.

Androgenesis/Gynogenesis

A useful cell-culture method for obtaining homozygous diploid or haploid plants is anther or microspore culture (Reinert and Bajaj, 1977; Chu, 1982). The ability of cultured anthers to produce haploid plants was first demonstrated by Guha and Maheshwari (1964a,b). They cultured *Datura innoxia* anthers and obtained haploid plants. Since this early discovery, plant-breeding programmes have found this technique effective for obtaining homozygous plants. Plant-breeding programmes have traditionally used back-crossing to obtain homozygosity in parent lines. Anther culture can reduce the time required to generate homozygosity. Culturing the immature pollen, either within the anther or isolated from the anther, can produce haploid or homozygous diploid plants. Haploid plants can be doubled either spontaneously or using colchicine treatments. This tool has been effectively used in barley (*Hordeum vulgare* L.), wheat (*Triticum aestivum* L.), tobacco (*Nicotiana tabaccum* L.), rice (*Oryza sativa* L.) and rapeseed (*Brassica* sp.) breeding programmes (Morrison and Evans, 1988).

Dr Clint MaGill at Texas A&M University released a new rice cultivar using anther culture to reduce the selection time (unpublished). Two rice parent lines were crossed, one that matured very early and another that was a semidwarf with high yields. The anthers of the F_1 progeny were cultured and doubled. This progeny was selected for early maturity, high yield, grain quality and semidwarfness. A new cultivar, 'Texmont', came out of this programme in only 5 years, clearly demonstrating the usefulness of this approach.

Gynogenesis, achieved by culturing immature unpollinated flowers or isolated ovaries or ovules, will also result in haploid plant material (Muren, 1989; Geoffriau *et al.*, 1997). This approach has been successfully used as a practical approach for accelerated inbred line development in onion (*Allium cepa* L.). Some of these haploids spontaneously double (Geoffriau *et al.*, 1997).

Protoplasts

A novel cell-culture technique that generated tremendous excitement in the 1960s was that of fusing protoplasts to create unique somatic hybrid cells. Protoplasts are plant cells that have been enzymatically digested in cellulase and pectinase to remove the rigid cell wall, leaving a fragile, spherical cell surrounded by the plasma membrane (Cocking, 1960). Kao and Michayluk (1974) showed that polyethylene glycol (PEG) could fuse protoplasts, creating hybrid cells. Later, Power *et al.* (1980) reported fusion of sexually incompatible petunia species. Melchers *et al.* (1978) demonstrated the fusion of potato and tomato protoplasts, producing hybrid potato/tomato plants. Zimmerman and

Scheurich (1981) reported a new technique to fuse protoplasts using electrical pulses. Many other reports followed and created an intense interest in this novel way of creating unique hybrid plants.

A further refinement was cybrid production, where the intact protoplast of one parent was fused with a protoplast of another parent whose nucleus was removed by centrifugation or inactivated by X-ray irradiation, so that the resulting fusion product had the nuclear genome of one parent and a cytoplasm with a mixture of both parents. This was mainly focused on developing new cytoplasmic male-sterility systems (Zelcer *et al.*, 1978).

Although numerous fusion hybrids have been produced, mainly in the *Solanaceae* family, protoplast-fusion technology has had limited application. It is difficult to regenerate fusion products and the plants are generally abnormal. Resulting plants have genetic instability, cytoplasmic segregation and variation. With the development of more efficient techniques for generating germ-plasm diversity, such as genetic engineering, protoplast-fusion approaches do not have widespread application. The ability to transfer identified genetic traits and not vary other desirable agronomic traits has received more attention.

Plant Transformation

A major revolution in crop plants and agriculture has been the recent increasing commercialization of transgenic crop plants. Of all the tools of plant cell culture for crop improvement, this has had by far the greatest impact for agricultural crops. This revolution, however, has been developed on the tremendous foundation of information established from the development and use of all the above technologies that have been discussed.

In 1997, 48 transgenic crops were approved for commercialization (Hansen and Wright, 1999). The increase in the use of transgenic crops has been significant – from 4 million acres in 1995 to 100 million acres in 1999 (Krattinger, 2000). The growth has been due to the economic advantage to the farmer in reduced pesticide application, higher yields and lower consumer prices (Krattinger, 2000). These genetically modified organisms (GMOs), as they are often referred to, contain genes for herbicide, insect and disease resistance, enhanced nutritional value and environmental-stress resistance. Although there are several techniques used to insert the foreign genes into the plant cell, all methodologies rely on plant-tissue culture to recover the modified plant. Having a rapid, genotype-independent regeneration protocol for plant regeneration is important for the application of transgene technology. There are several approaches to inserting foreign DNA into plant cells. The decision about which approach to take is determined by the target tissue appropriate for regeneration, an effective vector for DNA delivery, an efficient *in vitro* selection system to identify transgenic cells, transgenic plant recovery at a reasonable frequency and a short *in vitro* stage to avoid somaclonal variation (Hansen and Wright, 1999).

The earliest-developed transformation technique involved using protoplasts and inducing foreign DNA or gene uptake by PEG or electroporation (Cocking, 1977; Fromm *et al.*, 1986; Rhodes *et al.*, 1988; Datta *et al.*, 1990) or *Agrobacterium*-mediated uptake (Horsch *et al.*, 1984). However, again, the major obstacle was regenerating plants from the engineered protoplast. The efficiency and frequencies were very low and required major efforts to obtain normal plants with stable, single-gene-copy insertions.

Microprojectile bombardment or biolistics was a unique approach to inserting foreign genes into plant tissues without using protoplasts, and the added advantage was that any cultivar, monocotyledon or dicotyledon tissue could serve as the target tissue, as long as the target cell could regenerate into a fertile plant (Klein *et al.*, 1988). This method involves making the foreign DNA adhere to the surface of small microprojectiles – gold or tungsten particles – and shooting these into plant cells. The cells can take up the DNA and incorporate it into their genomic DNA at a low frequency. Multiple gene copies and fragments of genes are randomly inserted.

Probably the most widely used method for inserting foreign genes into plants involves using *Agrobacterium tumefaciens* to vector the foreign DNA into the plant cell (Chilton *et al.*, 1977; Smith and Hood, 1995). *Agrobacterium tumefaciens* is attracted to and attaches to wounded plant cells and transfers a piece of its Ti plasmid or T-DNA into the plant cell (Zambryski, 1992). Early work by Fraley *et al.* (1984) and Horsch *et al.* (1985) established an elegant system using *Agrobacterium* to transform plant tissues. Soon, there were numerous reports of transgenic plants. However, transformation of monocotyledons using *Agrobacterium* was not believed to be possible, as *Agrobacterium* did not appear to infect monocotyledons and cause tumours (Smith and Hood, 1995). Until it was unequivocally demonstrated by Chan *et al.* (1993), Hiei *et al.* (1994) and Park *et al.* (1996) that *A. tumefaciens* could transform monocotyledon plants, protoplast genetic engineering and biolistics were the main avenue for inserting agronomically important genes into cereal crops. Overall, the *Agrobacterium* system is probably the most widely used. It is the most efficient system of delivering foreign genes, gives a high frequency for single-gene-copy insertion with minimal gene rearrangement, and requires minimal equipment and facilities.

Improvements in transformation technology that could have a significant impact have been recently publicized. A press release by Delta and Pine Land Company described an agreement with the US Department of Agriculture (USDA)'s Agricultural Research Service (ARS) regarding the ARS pollen transformation system (PTS) (USDA patent issued in July 1999). The PTS involves inserting a gene into cultured pollen and using the transformed pollen to pollinate a flower. Transgenic seed is obtained from the plant carrying the new genetic trait. This system may make transformation technology much easier and more economical.

Molecular Farming

One of the newest and most exciting applications of genetic engineering is molecular farming or using crop plants as a factory to produce unique proteins, ranging from antibiotics, pharmaceuticals and speciality proteins for industrial use to edible vaccines (May *et al.*, 1995; Yusibov *et al.*, 1997). The concept is attractive: for example, in the case of maize (*Zea mays* L.), a protein accumulated in seed can be stored in the maize seed and processed using conventional methods already established for maize, making scale-up easy and the process cost-effective. ProdiGene™ is a company in Texas using this approach to produce speciality proteins in transgenic maize.

There are many concerns regarding the use of genetically engineered crops. These concerns include the following:

- The possibility that the foreign genes will escape into the environment. The concern is that superweeds may develop or, with the transfer of virus resistance, this could cause more virulent virus strains. Also gene transfer could result in contamination of non-genetically modified crops, threatening crop diversity.
- The use of antibiotic genes for selectable markers might increase the number of microbes with antibiotic resistance.
- Food-safety issues, such as the foreign protein produced in plants causing food allergies.
- Ownership and proprietary rights of the germ-plasm are a contentious issue. Some feel that there is an element of bio-piracy by more developed countries and biotechnology companies. Farmers in some countries feel that the big biotechnology companies may take advantage of them.
- Environmental damage, such as insect-resistant crops harming non-target species. The monarch butterfly issue ignited concern that the *Bacillus thuringiensis* (Bt) maize pollen could kill the butterflies. Also, insects can

develop resistance to Bt if it is out in the farmers' fields on a large scale, and this could lead to the use of harsher chemicals.
- Ethical issues of patenting life forms concern many individuals.

All of these concerns are currently being debated. The use of this technology to enhance food production could make it possible to produce more food on the current agricultural acreage. The debate over the widespread use of this technology is still under discussion.

Summary

None of these techniques of cell culture/plant biotechnology can attain their potential unless tightly linked to conventional plant-breeding and crop-production programmes. It is only with the cooperation and input of conventional breeding programmes that appropriate problems and goals can be identified and undertaken for a particular crop. Once a plant product is derived from the cell culture, it must move into field-testing and evaluation for appropriate genotype selection and stabilization, testing, increase, proprietary protection and crop-production stages (Smith *et al.*, 1993; Pauls, 1995). Working together, the plant physiologist, plant virologist, plant molecular biologist and plant breeder have the potential to produce high-quality food on limited acreage for the expanding human population. However, this alone is not the solution to world hunger. Political stability, reduction in poverty, increases in literacy and education, population stabilization and agricultural sustainability are all major components in human survival in a civilized world.

References

Chaleff, R.S. (1983) Isolation of agronomically useful mutants from plant cell cultures. *Science* 219, 676–679.

Chan, M.T., Chang, H.H., Ho, S.L., Tong, W.F. and Yu, S.M. (1993) *Agrobacterium*-mediated production of transgenic rice plants expressing a chimeric α-amylase promoter/β-glucuronidase gene. *Plant Molecular Biology* 22, 491–506.

Chilton, M.-D., Drummond, M.H., Merlo, D.J., Sciaky, D., Montoya, A.L., Gordon, M.P. and Nester, E.W. (1977) Stable incorporation of plasmid DNA into higher plant cells: the molecular basis of crown gall-tumorigenesis. *Cell* 11, 263–271.

Chu, C.-C. (1982) Haploids in plant improvement. In: Vasil, I.K., Scowcroft, W.R. and Frey, K.J. (eds) *Plant Improvement and Somatic Cell Genetics*. Academic Press, New York, pp. 129–158.

Cocking, E.C. (1960) A method for the isolation of plant protoplasts and vacuoles. *Nature* 187, 927–929.

Cocking, E.C. (1977) Uptake of foreign genetic material by plant protoplasts. *International Review of Cytology* 48, 323–343.

Datta, S.K., Peterhans, A., Datta, K. and Potrykus, I. (1990) Genetically engineered fertile indica-rice recovered from protoplasts. *Bio/Technology* 8, 736–740.

Evans, D.A., Sharp, W.R. and Medina-Filho, H.P. (1984) Somaclonal and gametoclonal variation. *American Journal of Botany* 71(6), 759–774.

Fraley, R.T., Horsch, R.B., Matzke, A., Chilton, M.D., Chilton, W.S. and Sanders, P.R. (1984) *In vitro* transformation of *Petunia* cells by an improved method of co-cultivating with *A. tumefaciens* strains. *Plant Molecular Biology* 3, 371–378.

Fromm, M.E., Taylor, L.P. and Walbot, V. (1986) Stable transformation of maize after gene transfer by electroporation. *Nature* 319, 791–793.

Geoffriau, E., Kahane, R. and Rancillac, M. (1997) Variation of gynogenesis ability in onion (*Allium cepa* L.). *Euphytica* 94, 37–44.

Guha, S. and Maheshwari, S.C. (1964a) *In vitro* production of embryos from anthers of *Datura*. *Nature* 204, 497.

Guha, S. and Maheshwari, S.C. (1964b) Cell division and differentiation of embryos in the pollen grains of *Datura in vitro*. *Nature* 212, 97–98.

Hansen, G. and Wright, M.S. (1999) Recent advances in the transformation of plants. *Trends in Plant Science* 4(6), 226–231.

Hiei, Y., Ohta, S., Komari, T. and Kumashire, T. (1994) Efficient transformation of rice (*Oryza sativa* L.) mediated by *Agrobacterium* and sequence analysis of the boundaries of the T-DNA. *Plant Journal* 6, 271–282.

Horsch, R.B., Fraley, R.T., Rogers, S.G., Sanders, P.R., Lloyd, A.M. and Hoffmann, N.L. (1984) Inheritance of functional foreign genes in plants. *Science* 223, 496–498.

Horsch, R.B., Fry, J.E., Hoffmann, N.L., Eichholtz, D., Rogers, S.G. and Fraley, R.T. (1985) A simple and general method for transferring genes into plants. *Science* 227, 1229–1231.

Kao, K.N. and Michayluk, M.R. (1974) A method for high frequency intergeneric fusion of plant protoplasts. *Planta* 115, 355–367.

Karp, A. (1995) Somaclonal variation as a tool for crop improvement. *Euphytica* 85, 295–302.

Klein, T.M., Fromm, M., Weissinger, A., Tomes, D., Schaaf, S., Sletten, M. and Sanford, J.C. (1988) Transfer of foreign genes into intact maize cells with high-velocity micro-projectiles. *Proceedings of the National Academy of Sciences* 85, 4305–4309.

Krattinger, A.F. (2000) Food biotechnology: promising havoc or hope for the poor? *Proteus* 17, 3–8.

Krikorian, A.D. and Berquam, D.L. (1969) Plant cell and tissue cultures: the role of Haberlandt. *The Botanical Review* 35(1), 59–88.

Larkin, P.J. and Scowcroft, W.R. (1981) Somaclonal variation – a novel source of variability from cell cultures for plant improvement. *Theoretical and Applied Genetics* 60, 197–214.

McHughen, A. (1987) Salt tolerance through increased vigor in a flax line (STS-11) selected for salt tolerance *in vitro*. *Theoretical and Applied Genetics* 74, 727–732.

May, G.D., Afza, R., Mason, H.S., Wiecko, A., Novak, F.J. and Arntzen, C.J. (1995) Generation of transgenic banana (*Musca acuminata*) plants via *Agrobacterium*-mediated transformation. *Biotechnology* 13, 486–492.

Melchers, G., Sacristan, M.D. and Holder, A.A. (1978) Somatic hybrid plants of potato and tomato regenerated from fused protoplasts. *Carlsberg Research Communications* 43, 203–218.

Morel, G. (1972) Morphogenesis of stem apical meristem cultivated *in vitro*: applications to clonal propagation. *Phytomorphology* 22, 265–277.

Morel, G. and Martin, C. (1952) Guérison de dahlias atteints d'une maladie à virus. *Comptes Rendus Hébdomadaires des Séances de l'Académie des Sciences (Paris)* 234, 1324–1325.

Morrison, R.A. and Evans, D.A. (1988) Haploid plants from tissue culture: new plant varieties in a shortened time frame. *Bio/ Technology* 6, 684–690.

Murashige, T. (1974) Plant propagation through tissue cultures. *Annual Review of Plant Physiology* 25, 135–166.

Muren, R. (1989) Haploid plant induction from unpollinated ovaries in onion. *HortScience* 24(5), 833–834.

Park, S.H., Pinson, S.R.M. and Smith, R.H. (1996) T-DNA integration into genomic DNA of rice following *Agrobacterium* inoculation of isolated shoot apices. *Plant Molecular Biology* 32, 1135–1148.

Pauls, K.P. (1995) Plant biotechnology for crop improvement. *Biotechnology Advances* 13(4), 673–693.

Power, J.B., Berry, S.F., Chapman, J.V. and Cocking, E.C. (1980) Somatic hybridization of sexually incompatible petunias: *Petunia parodii, Petunia parviflora. Theoretical and Applied Genetics* 57, 1–4.

Quak, F. (1977) Meristem culture and virus-free plants. In: Reinert, J. and Bajaj, Y.P.S. (eds) *Applied and Fundamental Aspects of Plant Cell, Tissue, and Organ Culture.* Springer-Verlag, New York, pp. 598–646.

Reinert, J. and Bajaj, Y.P.S. (1977) Anther culture: haploid production and its significance. In: Reinert, J. and Bajaj, Y.P.S. (eds) *Applied and Fundamental Aspects of Plant Cell, Tissue, and Organ Culture.* Springer-Verlag, New York, pp. 251–264.

Rhodes, C.A., Pierce, D.A., Mettler, D.M., Mascarenhas, D. and Detmer, J.J. (1988) Genetically transformed maize plants from protoplasts. *Science* 240, 204–207.

Smith, R.H. and Hood, E.E. (1995) *Agrobacterium tumefaciens* transformation of monocots. *Crop Science* 35, 301–309.

Smith, R.H. and Murashige, T. (1970) *In vitro* development of the isolated shoot apical meristem of angiosperms. *American Journal of Botany* 57(5), 562–568.

Smith, R.H., Duncan, R.R. and Bhaskaran, S. (1993) *In vitro* selection and somaclonal variation for crop improvement. *International Crop Science Proceedings* 629–632.

Thorpe, T.A. (2000) History of plant cell culture. In: Smith, R.H. (ed.) *Plant Tissue Culture Techniques and Experiments.* Academic Press, San Diego, pp. 1–32.

Thorpe, T.A. and Harry, I.S. (1997) Application of tissue culture to horticulture. In: Altman, A. and Ziv, M. (eds) *Proceedings of the Third International ISHS Symposium on In Vitro Culture and Horticulture Breeding.* Ministry of Flemish Community, Israel, pp. 39–49.

Veilleux, R.E. and Johnson, A.A.T. (1998) Somaclonal variation: molecular analysis, transformation interaction, and utilization. *Plant Breeding Reviews* 16, 229–267.

Withers, L.A. (1987) Long-term preservation of plant cells, tissues and organs. *Oxford Survey, Plant Molecular and Cell Biology* 4, 221.

Yusibov, V., Modelska, A., Steplewski, K., Agadjanyan, M., Weiner, D., Hooper, D.C. and Koprowski, H. (1997) Antigens produced in plants by infection with chimeric plant viruses immunize against rabies virus and HIV-1. *Proceedings of the National Academy of Sciences USA* 94, 5784–5788.

Zambryski, P.C. (1992) Chronicles from the *Agrobacterium*–plant cell DNA transfer story. *Annual Review of Plant Physiology and Plant Molecular Biology* 43, 465–490.

Zelcer, A., Aviv, D. and Galun, E. (1978) Interspecific transfer of cytoplasmic male sterility by fusion between protoplasts of normal *Nicotiana sylvestris* and X-ray irradiated protoplasts of male-sterile *N. tabacum. Zeitschrift für Pflanzenphysiology* 90, 397–407.

Zimmerman, U. and Scheurich, P. (1981) High frequency fusion of plant protoplasts by electric fields. *Planta* 151, 26–32.

14   Transferring Genes from Wild Species into Rice

D.S. Brar and G.S. Khush

International Rice Research Institute, DAPO 7777, Metro Manila, The Philippines

Three cereal crops – rice, wheat and maize – feed the world. These crops supply 49% of the calories consumed by the world population; 23% come from rice, 17% from wheat and 9% from maize. Rice is the primary food source for more than a third of the world's population. It is planted on almost 150 million ha annually or 11% of the world's cultivated land. More than 90% of rice is produced and consumed in Asia. It is also an important staple in Latin America, Africa and the Middle East. Rice is grown under a wide range of agroclimatic conditions. Four major ecosystems are generally recognized: irrigated (55%), rain-fed lowland (25%), upland (12%) and flood-prone (8%).

World rice production has more than doubled, from 257 million t in 1966 to 600 million t in 2000. To meet the growing needs of the ever-increasing human population, however, rice production must increase by 40% during the next 25 years. To achieve this, several biotic and abiotic stresses that adversely affect rice productivity must be overcome. Some of the major diseases and pests affecting rice production include bacterial blight, blast, sheath blight, tungro virus diseases and rice yellow mottle virus (RYMV) and insects, such as the brown planthopper (BPH), stemborer and Asian and African gall midge. Similarly, abiotic stresses, such as drought, cold, salinity,

acidity, iron toxicity and submergence under water (flooding tolerance), adversely affect rice production. Changes in insect biotypes and disease races are a continuing threat to rice production. The genetic variability for some traits, such as resistance to sheath blight, tungro, RYMV and yellow stemborer and tolerance to salinity and acid sulphate conditions, is limited in the cultivated germplasm. There is thus an urgent need to broaden the rice gene pool by introgression of new genes from diverse sources to meet various challenges affecting rice production. Wild species of *Oryza* are an important reservoir of useful genes for rice improvement (Table 14.1).

Wild Relatives of Rice

Rice belongs to the family *Poaceae* (*Gramineae*), tribe *Oryzeae*. This tribe has 11 genera, of which *Oryza* is the only one with cultivated species. *Oryza* has two cultivated and 22 wild species. Of the two cultivated species, *Oryza sativa* ($2n = 24$ *AA*), the Asian rice, is grown worldwide, but *Oryza glaberrima* ($2n = 24$ *AA*), the African rice, is cultivated on a limited scale in West Africa. The wild species have either $2n = 24$ or 48 chromosomes, representing *AA*, *BB*, *CC*, *BBCC*, *CCDD*, *EE*, *FF*, *GG*, and *HHJJ* genomes (Table 14.1). Another

Table 14.1. Chromosome number, genomic composition and potential useful traits of *Oryza* species.

Species	2n	Genome	Number of accessions*	Distribution	Useful or potentially useful traits
O. sativa complex					
O. sativa L.	24	*AA*	84,186	Worldwide	Cultigen
O. glaberrima Steud.	24	$A^g A^g$	1,299	West Africa	Cultigen, tolerance to drought, acidity and iron toxicity, resistance to RYMV, African gall midge, nematodes and weed competitiveness
O. nivara Sharma et Shastry	24	*AA*	1,130	Tropical and subtropical Asia	Resistance to grassy stunt virus
O. rufipogon Griff.	24	*AA*	858	Tropical and subtropical Asia, tropical Australia	Resistance to BB, tungro virus, tolerance to aluminium and soil acidity, source of CMS
O. breviligulata A. Chev. et Roehr. (*O. barthii*)	24	$A^g A^g$	214	Africa	Resistance to GLH, BB, drought avoidance
O. longistaminata A. Chev. et Roehr.	24	$A^l A^l$	203	Africa	Resistance to BB, nematodes, drought avoidance
O. meridionalis Ng	24	$A^m A^m$	46	Tropical Australia	Elongation ability, drought avoidance
O. glumaepatula Steud.	24	$A^{gp} A^{gp}$	54	South and Central America	Elongation ability, source of CMS
O. officinalis complex					
O. punctata Kotschy ex Steud.	24, 48	*BB, BBCC*	59	Africa	Resistance to BPH, zigzag leafhopper
O. minuta J.S. Presl. ex C.B. Presl.	48	*BBCC*	63	Philippines and Papua New Guinea	Resistance to BB, blast, BPH, GLH, tolerant to Shb
O. officinalis Wall ex Watt	24	*CC*	265	Tropical and subtropical Asia, tropical Australia	Resistance to thrips, BPH, GLH, WBPH, BB, stem rot
O. rhizomatis Vaughan	24	*CC*	19	Sri Lanka	Drought avoidance, rhizomatous
O. eichingeri A. Peter	24	*CC*	29	South Asia and East Africa	Resistance to BPH, WBPH, GLH
O. latifolia Desv.	48	*CCDD*	40	South and Central America	Resistance to BPH, high biomass production
O. alta Swallen	48	*CCDD*	6	South and Central America	Resistance to striped stemborer, high biomass production
O. grandiglumis (Doell) Prod.	48	*CCDD*	10	South and Central America	High biomass production
O. australiensis Domin.	24	*EE*	36	Tropical Australia	Resistance to BPH, BB, drought avoidance

	2n	Genome	Distribution	Accessions*	Useful traits
O. meyeriana complex					
O. granulata Nees et Arn. ex Watt	24	*GG*	South and South-East Asia	24	Shade tolerance, adaptation to aerobic soil
O. meyeriana (Zoll. et (Mor. ex Steud.) Baill.	24	*GG*	South-East Asia	11	Shade tolerance, adaptation to aerobic soil
O. ridleyi complex					
O. longiglumis Jansen	48	*HHJJ*	Irian Jaya, Indonesia and Papua New Guinea	6	Resistance to blast, BB
O. ridleyi Hook. F.	48	*HHJJ*	South Asia	15	Resistance to BB, blast, stemborer, whorl maggot
Unclassified					
O. brachyantha A. Chev. et Roehr.	24	*FF*	Africa	19	Resistance to BB, yellow stemborer, leaf-folder, whorl maggot, tolerance to laterite soil
O. schlechteri Pilger	48	*HHKK*	Papua New Guinea	1	Stoloniferous
Related genera			—	15	—

*Accessions maintained at IRRI, The Philippines.
BPH, brown planthopper; GLH, green leafhopper; WBPH, white-backed planthopper; BB, bacterial blight; Shb, sheath blight; CMS, cytoplasmic male sterility; RYMV, rice yellow mottle virus.

tetraploid species, *Porteresia coarctata*, now classified as *Oryza coarctata*, has *HHKK* genome.

The genus *Oryza* has been divided into four species complexes: (i) the *Sativa* complex; (ii) the *Officinalis* complex; (iii) the *Meyeriana* complex; and (iv) the *Ridleyi* complex (Table 14.1). Two species, *Oryza brachyantha* and *Oryza schlechteri*, cannot be placed in any of these groups (Vaughan, 1989).

Sativa *complex*

This complex consists of two cultivated species and six wild taxa. All of them have the *AA* genome and form the primary gene pool for rice improvement. Wild species closely related to *O. sativa* have been variously named. The weedy types of rice have been given various names, such as '*fatua*' and '*spontanea*' in Asia and *Oryza stapfii* in Africa. These weedy forms usually have red endosperm – hence the common name 'red rice'. These weedy species may be more closely related to *Oryza rufipogon* and *Oryza nivara* in Asia and to *Oryza longistaminata* or *Oryza breviligulata* in Africa. One of the species, *Oryza meridionalis*, is distributed across tropical Australia. This species is often sympatric with *Oryza australiensis* in Australia.

Officinalis *complex*

The *Officinalis* complex consists of nine species and is also called the *Oryza latifolia* complex (Tateoka, 1962). This complex has related species groups in Asia, Africa and Latin America. The tetraploid species *Oryza minuta* is sympatric with *Oryza officinalis* in the central islands of Bohol and Leyte in The Philippines. *Oryza rhizomatis* is a new species from Sri Lanka. Another species, *Oryza eichingeri*, grows in forest shade in Uganda. It was found distributed in Sri Lanka (Vaughan, 1989). Two species of this complex, *Oryza punctata* and *O. eichingeri*, are distributed in Africa. Three American species of this complex, *O. latifolia*, *Oryza*

alta and *Oryza grandiglumis*, are tetraploid. *Oryza latifolia* is widely distributed in Central and South America, as well as in the Caribbean islands. A diploid species, *O. australiensis*, occurs in northern Australia in isolated populations.

Meyeriana *complex*

This complex has two diploid species, *Oryza granulata* and *Oryza meyeriana*. *Oryza granulata* grows in South Asia, South-East Asia and south-west China. *Oryza meyeriana* is found in South-East Asia. Another species, *Oryza indandamanica* from the Andaman Islands (India), is a subspecies of *O. granulata*. The species of this complex have unbranched panicles with small spikelets.

Ridleyi *complex*

This complex has two tetraploid species, *Oryza ridleyi* and *Oryza longiglumis*. Both species usually grow in shaded habitats, near rivers, streams or pools. *Oryza longiglumis* is found along the Koembe River, Irian Jaya, Indonesia, and in Papua New Guinea. *Oryza ridleyi* grows across South-East Asia and as far as Papua New Guinea.

Oryza brachyantha

This species is distributed in the African continent. It grows in the Sahel zone and in East Africa, often in laterite soils. It is often sympatric with *O. longistaminata*.

Oryza schlechteri

This is the least studied species of the genus. It was collected from north-east New Guinea. It is a tufted perennial, with 4–5 cm panicles and small, unawned spikelets. It is tetraploid, but its relationship to other species is unknown. Besides *Oryza*, the tribe *Oryzeae* has ten other genera: *Chikusiochloa*, *Hygroryza*, *Leersia*, *Luziola*, *Prosphytochloa*, *Rhynchoryza*, *Zizania*, *Zizaniopsis*, *Porteresia* and *Potamophila*.

Genomic relationships

Various approaches involving morphological differentiation, meiotic chromosome pairing in F_1 hybrids, molecular divergence analysis and fraction 1 protein have been used in determining genome relationships in *Oryza*. On the basis of chromosome pairing in F_1 hybrids, various authors have assigned the genome symbol *AA* for the *Sativa* complex, *BB*, *CC*, *BBCC*, *CCDD* and *EE* for the *officinalis* complex and *FF* for *O. brachyantha* (Table 14.1). Based on molecular divergence analysis, two new genomes *GG* and *HHJJ* have been assigned to the *O. meyeriana* and *O. ridleyi* complexes, respectively (Aggarwal *et al.*, 1997, 1999). Similarly, total genomic hybridization analysis also showed that *O. schlechteri* had a distinct genome (Aggarwal *et al.*, 1996). Hybridization of the total genomic DNA of *P. coarctata* when used as a probe with the DNA of the F_1 hybrid (*O. sativa* × *P. coarctata*) shows strong hybridization with *P. coarctata* and limited cross-hybridization with *O. sativa*, indicating strong divergence between the genomes of *O. sativa* and *P. coarctata* (Brar *et al.*, 1997). Ge *et al.* (1999), on the basis of sequence analysis of nuclear genes (*Adh1*, *Adh2*) and a chloroplast gene (*matK*), proposed the *HHKK* genome for *O. schlechteri* and *P. coarctata*, which further suggested that *P. coarctata* should be treated as an *Oryza* species.

Results of restriction fragment length polymorphism (RFLP), amplified fragment length polymorphism (AFLP), sequence analysis of genes and seed-protein analysis support the genome classification based on morphological and cytological data (Table 14.1). Jena *et al.* (1994) constructed a comparative RFLP map of *O. sativa* (*AA*) and *O. officinalis* (*CC*). The linkage order of mapped RFLP loci on different chromosomes of *O. officinalis* was mostly conserved relative to cultivated rice, but some rearrangements were detected. Nine of the 12 chromosomes of *O. officinalis* were homosequential to those of *O. sativa*. Kennard *et al.* (1999) developed comparative maps of wild rice (*Zizania palustris* $2n = 2x = 30$) and *O. sativa* ($2n = 2x = 24$). Although, the genomes of the two species differ in total DNA content, collinear markers were found for all the rice linkage groups except number 12.

Useful traits of Oryza species

Wild species are an important reservoir of useful genes for resistance to major diseases and insects and tolerance to abiotic stresses and are also a good source of cytoplasmic male sterility (Table 14.1). Many *AA* genome species that are closely related to cultivated rice species possess genes for resistance to diseases, such as bacterial blight, blast, tungro, RYMV and stem rot, and to insects, such as BPH and gall midge, and tolerance to abiotic stresses, such as drought, iron and aluminium toxicity, acidity, etc. Similarly, many distantly related species also possess useful genes for rice improvement. Several incompatibility barriers, such as low crossability and limited recombination between chromosomes of wild and cultivated species, however, limit the transfer of useful genes (Brar and Khush, 1986, 1997; Sitch, 1990; Khush and Brar, 1992). Recent advances in tissue culture, such as embryo rescue, *in vitro* pollination and protoplast fusion, and anther culture have enabled the production of wide hybrids among distantly related species. In addition, molecular marker technology and *in situ* hybridization techniques have made it possible to precisely detect the introgression of chromosome segments from wild into cultivated species.

Alien Introgression in Rice

The main objectives are: (i) to widen the gene pool of rice by transferring useful genes for resistance to major diseases and insects and tolerance to abiotic stresses; (ii) to enhance the grain yield of rice through introgression of useful alleles of wild relatives; and (iii) to precisely determine the mechanism of alien gene transfer, with the possibility of enhancing introgression from distant genomes.

Some of the steps followed for transferring genes from wild species into rice are given below:

1. Identification of useful genetic variability in the wild-species germplasm for target agronomic traits.
2. Production of hybrids between élite breeding lines of rice and wild species through direct crosses and/or through embryo rescue.
3. Continued back-crossing with the recurrent rice parent followed by embryo rescue to produce fertile introgression lines (2n = 24).
4. Evaluation of advanced fertile back-cross progenies (introgression lines) for transfer of target traits from wild species under screenhouse and field conditions.
5. Characterization of alien introgression using isozyme and molecular markers.
6. Chromosomal location of introgressed gene(s) using monosomic alien addition lines (MAALs).

7. Tagging of introgressed alien genes with molecular markers for use in marker-assisted selection.
8. Location of introgressed alien segments on rice chromosomes through fluorescence *in situ* hybridization (FISH).

Production of interspecific hybrids and advanced back-cross (introgression) progenies

Hybrids between rice and wild species of *Oryza* have been produced through direct crosses and/or through embryo rescue (Table 14.2). Cultivated rice and its closely related wild species, *Oryza perennis*, *O. nivara*, *O. glaberrima* and *O. longistaminata*, share the *AA* genome. These wild species can be easily crossed with *O. sativa* and genes from them can be transferred to cultivated rice by conventional crossing and back-crossing procedures. Wild species with genomes other than *AA* are, however, difficult to cross with *O. sativa* and produce completely male-sterile hybrids. Embryo rescue is used to overcome hybrid

Table 14.2. Wide-cross progenies between rice and wild species of *Oryza* produced at IRRI through direct crosses and/or through embryo rescue.

Cross-combination	F₁	Method	Advanced introgression lines (2n = 24)	MAALs (2n = 25)
O. sativa (AA) × *O. rufipogon* (AA)	+	Direct cross	+	–*
O. sativa (AA) × *O. glumaepatula* (AA)	+	Direct cross	+	–
O. sativa (AA) × *O. longistaminata* (AA)	+	Direct cross	+	–
O. sativa (AA) × *O. glaberrima* (AA)	+	Direct cross	+	–
O. sativa (AA) × *O. officinalis* (CC)	+	Embryo rescue	+	+
O. sativa (AA) × *O. rhizomatis* (CC)	+	Embryo rescue	n/a	n/a
O. sativa (AA) × *O. eichingeri* (CC)	+	Embryo rescue	n/a	n/a
O. sativa (AA) × *O. minuta* (BBCC)	+	Embryo rescue	+	+
O. sativa (AA) × *O. latifolia* (CCDD)	+	Embryo rescue	+	+
O. sativa (AA) × *O. alta* (CCDD)	+	Embryo rescue	n/a	n/a
O. sativa (AA) × *O. glandiglumis* (CCDD)	+	Embryo rescue	n/a	n/a
O. sativa (AA) × *O. australiensis* (EE)	+	Embryo rescue	+	+
O. sativa (AA) × *O. brachyantha* (FF)	+	Embryo rescue	+	+
O. sativa (AA) × *O. granulata* (GG)	+	Embryo rescue	+	+
O. sativa (AA) × *O. ridleyi* (HHJJ)	+	Embryo rescue	BC₂F₁	n/a
O. sativa (AA) × *Porteresia coarctata* (HHKK)	+	Embryo rescue	n/a	n/a

*A minus sign (–) indicates that, because of homologous genomes, MAALs are not recovered.
+, available; n/a, not available; MAALs, monosomic alien addition lines.

inviability and to produce interspecific hybrids. Several workers have produced hybrids among *O. sativa* and wild species containing *BB, BBCC, CC, CCDD, EE, FF* and *HHJJ* genomes. These studies investigated genomic homologies and species relationships, but did not attempt to transfer useful traits from wild species into cultivated rice. At the International Rice Research Institute (IRRI), a series of hybrids, MAALs representing seven to 12 chromosomes of six wild species and introgression lines have been produced through direct crosses, as well as through embryo rescue following hybridization between élite breeding lines of rice and several distantly related species

of *Oryza* (Table 14.2, Fig. 14.1). Most of these wide crosses have been used to transfer useful genes into rice (Table 14.3).

An intergeneric hybrid between *O. sativa* and *P. coarctata* has been produced through sexual crosses following embryo rescue (Brar *et al.*, 1997). The hybrid is completely male-sterile. Meiotic chromosome analysis of the hybrid showed 36 univalents at metaphase I, indicating lack of pairing between chromosomes of *O. sativa* and *P. coarctata*. Continued efforts are being made to produce back-cross progenies for the transfer of genes for salinity tolerance from *P. coarctata* to rice cultivars. So far, no back-cross progeny has been produced.

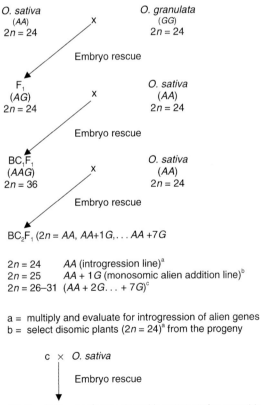

Fig. 14.1. Scheme showing production of monosomic alien addition lines ($2n = 25$) and introgression lines ($2n = 24$) from crosses of rice and distantly related wild species of *Oryza* (modified from Brar and Khush, 1997).

Table 14.3.   Introgression of genes of wild *Oryza* species into rice (modified from Brar and Khush, 1997).

Trait	Wild species	Genome	Accession number
			Donor Oryza species
Transferred to *Oryza sativa*			
Grassy stunt resistance	*O. nivara*	AA	101508
Bacterial blight resistance	*O. longistaminata*	AA	–
	O. officinalis	CC	100896
	O. minuta	BBCC	101141
	O. latifolia	CCDD	100914
	O. australiensis	EE	100882
	O. brachyantha	FF	101232
Blast resistance	*O. minuta*	BBCC	101141
Brown planthopper resistance	*O. officinalis*	CC	100896
	O. minuta	BBCC	101141
	O. latifolia	CCDD	100914
	O. australiensis	EE	100882
White-backed planthopper resistance	*O. officinalis*	CC	100896
Cytoplasmic male sterility	*O. sativa* f. *spontanea*	AA	–
	O. perennis	AA	104823
	O. glumaepatula	AA	100969
Tungro tolerance	*O. rufipogon*	AA	105908
	O. rufipogon	AA	105909
Progenies under evaluation			
Yellow stemborer	*O. longistaminata*	AA	–
Sheath blight resistance	*O. minuta*	BBCC	101141
Increased elongation ability	*O. rufipogon*	AA	CB751
Tolerance to acidity, iron and	*O. glaberrima*	AA	Many
aluminium toxicity	*O. rufipogon*	AA	106412
	O. rufipogon	AA	106423
Tolerance to nematodes	*O. glaberrima*	AA	Many

Production of somatic hybrids through protoplast fusion

Somatic hybrids through protoplast fusion between rice and related species and genera have been produced. Hayashi *et al.* (1988) produced 250 somatic hybrid plants through electrofusion of protoplasts of rice with four wild species, *O. officinalis, O. eichingeri, O. brachyantha* and *Oryza perrieri*. The hybrid nature of these plants was confirmed through morphological, isozyme and karyotype analyses.

Somatic hybrid plants have also been produced through electrofusion of protoplasts of *O. sativa* and *P. coarctata*, a salt-tolerant species (Jelodar *et al.*, 1999). A total of 119 regenerated plants were micropropagated. Putative somatic hybrid plants were identified by randomly amplified polymorphic DNA (RAPD) analysis and eight plant lines were characterized for ploidy level by flow cytometry. One line was allohexaploid ($2n = 72$). The genomic *in situ* hybridization (GISH) analysis showed that the somatic hybrid had complete chromosome complements of both *O. sativa* and *P. coarctata*. Liu *et al.* (1999) produced a highly asymmetric and fertile somatic hybrid through protoplast fusion of *O. sativa* and *Zizania latifolia*. Gamma ray-irradiated mesophyll protoplasts of *Zizania* were electrofused with iodoacetamine-inactivated cell-suspension protoplasts of rice. The two hybrid plants showed $2n = 24$ chromosomes. Southern blot analysis, using total genomic DNA and moderate-copy *Z. latifolia*-abundant DNA sequences, showed intergenomic exchange between rice and *Zizania*.

A new protocol has been used to introduce *Zizania* DNA into rice cells. Abedinia *et al.* (2000) used microprojectile bombardment to introduce DNA from a tertiary gene pool, *Z. palustris*, into rice. High-molecular-weight unfractionated genomic DNA of *Zizania* was bombarded into the callus cells of rice cultivar 'Jarrah'. Selection was performed throughout callus proliferation, regeneration and plant growth. More than 250 transgenic plants were obtained. AFLP analysis revealed many *Zizania*-specific markers in transgenic plants, indicating introgressions of the *Zizania* genome into rice. Further studies are needed to confirm the transmission of introgression in the progeny of transgenic plants.

Introgression from AA genome wild species

Crosses between cultivated rice (*O. sativa*, $2n = 24$ *AA*) and the *AA* genome wild species can be made easily and the genes can be transferred through conventional crossing and back-crossing procedures. The hybrids between *O. sativa* and *O. rufipogon* are partially fertile; however, *O. sativa* × *O. glaberrima* and *O. sativa* × *O. longistaminata* F$_1$s are highly sterile. The first examples of transfer of a useful gene from wild species are the introgression of a gene for grassy stunt virus resistance from *O. nivara* to cultivated rice varieties (Khush, 1977) and the transfer of a cytoplasmic male-sterile (CMS) source from wild rice, *O. sativa* f. *spontanea*, to develop CMS lines for commercial hybrid-rice production (Lin and Yuan, 1980). Other useful genes, such as *Xa21*, for bacterial blight resistance, were transferred to rice from *O. longistaminata* and new CMS sources from *O. perennis* and *Oryza glumaepatula* into rice. More recently, genes for tungro tolerance and tolerance to a moderate level of acidity have been transferred from *O. rufipogon* into indica rice cultivar 'IR64'. A summary of genes transferred from wild species to cultivated rice is given in Table 14.3.

Introgression of gene(s) for resistance to grassy stunt virus

The grassy stunt virus is a serious disease transmitted by BPH. The diseased rice plants either produce no panicles or produce only small panicles with deformed grains. The disease can cause heavy losses in yield, particularly under epidemic conditions. Ling *et al.* (1970) screened more than 6000 accessions of cultivated rice and several wild species of *Oryza* for resistance. Among these materials, only one accession of *O. nivara* (Acc. 101508) was found to be resistant. Crosses were made between improved rice varieties, IR8, IR20, IR24 and *O. nivara*. Following three backcrosses with improved varieties, the gene for grassy stunt resistance was transferred into cultivated germplasm (Khush, 1977). The first grassy stunt-resistant varieties, IR28, IR29 and IR30, were released for cultivation in 1974. Subsequently, many grassy stunt-resistant varieties, such as IR34 and IR36, were released. Since then, grassy stunt resistance gene has been incorporated into many rice varieties developed at IRRI as well as by national rice improvement programmes.

Introgression of gene(s) for resistance to tungro disease

Rice tungro disease (RTD) is the most serious virus disease in South and South-East Asia. The disease is caused either by a single infection or by a double infection with two virus particles, the rice tungro bacilliform virus (RTBV) − a double-stranded DNA virus − and the rice tungro spherical virus (RTSV) − a single-stranded RNA virus. The green leafhopper (GLH) *Nephotettix virescens* is the principal vector of tungro virus. Typical symptoms include yellowing of leaves, stunted growth, delayed heading, shortening of panicles and reduced or no seed set.

There is a limited variability in the cultivated rice germplasm for resistance to RTBV, which is the main cause of tungro symptoms. Kobayashi *et al.* (1993) found 15 accessions of eight wild species resistant to RTBV. Three accessions of *O. rufipogon*

(IRRC Nos 105908, 105909, 105910) showed a low or moderate level of antibiosis to the vector GLH. Crosses were made between the high-yielding indica cultivar IR64 and the above-mentioned three accessions of *O. rufipogon*. Back-crosses were made with the recurrent rice parents, IR64. The back-cross progenies were screened under hot spots in the field. After four back-crosses, uniform progenies resembling the recurrent parent and resistant to tungro disease were selected (G.S. Khush and D.S. Brar, unpublished). Some of the breeding lines are under multilocation testing in The Philippines. One of the élite breeding lines (IR73885-1-4-3-2-1-6) resistant to tungro has been recommended for prerelease in The Philippines. We have also identified progenies from another cross of IR64 × *O. rufipogon* (Acc. no. 106424). These progenies are being back-crossed to IR64 to obtain agronomically desirable and tungro-resistant lines.

Introgression of gene(s) for resistance to bacterial blight

The bacterial blight caused by *Xanthomonas oryzae* pv. *oryzae* is one of the most destructive diseases of rice in Asia. One of the wild species, *O. longistaminata*, was found to be resistant to all races of bacterial blight in The Philippines. Crosses were made between IR24 and *O. longistaminata*. Following four back-crosses with the recurrent rice parent, a gene for resistance to bacterial blight was transferred and designated as *Xa21* (Khush *et al.*, 1990). This gene has a wide spectrum of resistance to bacterial blight. The *Xa21* has been transferred through marker-assisted selection into several other indica lines, such as IR64 and PR106, including élite breeding lines of new-plant-type (NPT) rice (Sanchez *et al.*, 2000; Singh *et al.*, 2001).

Zhang *et al.* (1998) made a cross between the rice cultivar 'Jiagang 30' and *O. rufipogon* (RBB16). Doubled haploids (DHs) from the F$_1$ were produced and lines resistant to bacterial blight were selected. Polymerase chain reaction (PCR) analysis of *Xa21*-specific primers was carried out. The resistance gene was mapped to chromosome 11 and designated as *Xa23*(t). This gene also conferred a very wide spectrum of resistance and showed a highly resistant reaction to all nine races of bacterial blight of The Philippines. At IRRI, we have produced a large number of advanced back-cross progenies (introgression lines) from the cross of an élite breeding line of NPT rice (IR65600-81-5-3-2) and *O. longistaminata*. The NPT line is highly susceptible to all six races of bacterial blight of The Philippines. Introgression for bacterial blight resistance has been achieved. Some of the derived introgression lines show longer panicles, more seeds per panicle and increased grain yield over the recurrent parent.

Incorporation of CMS sources from wild species

A number of CMS sources have been developed in rice. The most commonly used CMS source in hybrid-rice breeding, however, is derived from the wild species *O. sativa* f. *spontanea* (Lin and Yuan, 1980). The cytoplasmic source has been designated as wild abortive (WA), which refers to a male-sterile wild rice plant having abortive pollen. About 95% of the male-sterile lines used in commercial rice hybrids grown in China and other countries have the WA type of cytoplasm.

A new CMS source from *O. perennis* was transferred into indica rice (Dalmacio *et al.*, 1995). This newly developed CMS line has the nuclear genome of IR64 and the cytoplasm of *O. perennis* and is designated as IR66707A. The male-sterility source of IR66707A is different from that of WA. Southern hybridization using mitochondrial DNA (mtDNA)-specific probes showed an identical banding pattern between IR66707A (recipient) and *O. perennis* (donor). It appears that CMS may not be caused by any major rearrangement or modification of mtDNA. Another CMS line (IR69700A) having the cytoplasm of *O. glumaepatula* (*A* genome species) and the nuclear genome of IR64 has been developed (Dalmacio *et al.*, 1996). No good restorer could be identified for these two CMS lines. Many laboratories have transferred CMS from other *AA*-genome wild species; however, this effort did not yield any good restorers.

Introgression for tolerance to abiotic stresses

Little or no work has been done on the transfer of genes for tolerance to abiotic stresses from wild species into rice. We have evaluated several introgression lines derived from the crosses of *O. sativa* cv. IR64 × *O. rufipogon* at hot spots under field conditions for tolerance to abiotic stresses at Iloilo, The Philippines. In a collaborative project with Texas A&M University and Cuu Long Delta Rice Research Institute (CLRRI), Vietnam, quantitative trait loci (QTL) for aluminium tolerance have been identified from the cross of *O. sativa × O. rufipogon*. Élite breeding lines with good agronomic traits and moderately tolerant to iron toxicity, aluminium toxicity and acid sulphate conditions have been identified. Similarly, élite breeding lines from the cross of *O. sativa × O. rufipogon* that possesses increased elongation ability under deepwater conditions are under multilocation testing.

Some of the wild species, such as *O. rufipogon*, grow under natural conditions in acid sulphate soils of Vietnam. Three accessions of *O. rufipogon* (Acc. 106412, 106423, 106424), collected from the Dong Thap Muoi area, Mekong Delta, Vietnam, were used in crosses with IR64. In a collaborative project, 460 lines derived from the crosses of IR64 and *O. rufipogon* were sent to CLRRI, Vietnam, in 1995 for evaluation for tolerance to abiotic stresses. Two sets of nurseries were tested in both target and non-target areas. Selections were made in subsequent generations for increased seed fertility and improved plant type. Three promising lines were selected and tested through the yield-testing network of CLRRI. Of the three breeding lines, IR73678-6-9-B (AS996) has been released as a variety for commercial cultivation in the Mekong Delta, Vietnam. This variety occupies 40,000 ha

(Bui Chi Buu, personal communication). It is a short-duration (95–100 days), semidwarf variety with good plant type suitable for moderately acid sulphate soils and is tolerant to BPH and blast.

Introgression from African rice (*O. glaberrima*) into Asian rice (*O. sativa*)

The cultivars of Asian rice, *O. sativa*, are high-yielding, whereas African rice, *O. glaberrima*, is low-yielding and is susceptible to lodging and grain shattering. *Oryza glaberrima* has several desirable traits, such as resistance to RYMV, African gall midge and nematodes and tolerance to drought, acidity and iron toxicity. Soriano *et al.* (1999) identified one accession of *O. longistaminata* (WL-02) and three accessions of *O. glaberrima* (TOG7235, TOG5674, TOG5675) that were highly resistant to root-knot nematode (*Meloidogyne graminicola*). One of the important features of *O. glaberrima* is its strong weed-competitive ability. Thus, the interspecific hybridization among Asian and African species offers tremendous potential for combining the high productivity of *O. sativa* with the tolerance to biotic and abiotic stresses of *O. glaberrima*. The F$_1$ hybrids between *O. sativa* and *O. glaberrima*, in spite of complete chromosome pairing, are highly sterile. Back-crossing is used to restore fertility and derive agronomically desirable lines.

Major efforts have been made at the West Africa Rice Development Association (WARDA), Africa, to introgress useful genes from *O. glaberrima* into élite breeding lines of *O. sativa* (Jones *et al.*, 1997). More than 1100 *O. glaberrima* accessions were evaluated for various morpho-agronomic characteristics. Most of the *O. glaberrima* accessions have profuse vegetative vigour, with droopy lower leaves, and possess high specific leaf area (SLA). These features contribute to the weed-competitive ability of *O. glaberrima*. Seedling vigour of the progenies derived from *O. sativa × O. glaberrima* varied between 1 (extra vigour) and 5 (normal). Ten progenies scored 1 or 2. Under high-input conditions, several progenies outyielded the *O. glaberrima* parent (CG14) and were equivalent in yield to *O. sativa*. Under low-input management, however, a few progenies outyielded *O. sativa* and some had grain yields similar to or higher than *O. glaberrima*, indicating their adaptation to low-input conditions.

Johnson *et al.* (1998) conducted studies on the weed-competitive ability of *O. glaberrima* (IG10). *Oryza glaberrima* suffered less from competition with weeds and suppressed weeds better than *O. sativa*. IG10 accumulated more biomass, produced more tillers, had higher SLA and, in the earlier stages of growth, partitioned more of its biomass to leaves than *O. sativa*, but IG10 produced a higher yield in competition with weeds.

Dingkuhn *et al.* (1998) compared the growth and yield potential of CG14 (*O. glaberrima*), WAB56-104 (*O. sativa*) and their progeny under different management conditions. CG14 produced two to three times the leaf area index (LAI) and tiller numbers more than WAB56-104. The progenies had intermediate LAI, SLA and leaf chlorophyll content. A conclusion is that, to combine high weed competitiveness and high yield in a single line, we need to develop genotypes with high SLA during vegetative growth (for weed competitiveness) to accelerate leaf area development, followed by a rapid decrease in SLA during the reproductive growth phase (for high yield potential) to ensure high leaf photosynthesis.

At IRRI, a large number of advanced introgression lines have been produced from the crosses of *O. sativa* cvs IR64 and BG90-2 and several accessions of *O. glaberrima*. These progenies are being evaluated in a collaborative project with WARDA for introgression for tolerance to RYMV, African gall midge and abiotic stresses. Some lines for tolerance to iron toxicity have been identified. Molecular analyses indicate frequent recombination between *O. sativa* and *O. glaberrima*.

Development of doubled haploids (DHs) from O. sativa × O. glaberrima

Many laboratories, including WARDA and IRRI and others, are engaged in combining the high productivity of *O. sativa* with the tolerance to major biotic and abiotic stresses and weed-competitive ability of *O. glaberrima*. The two species show strong reproductive barriers, however, and their F_1s exhibit a high level of sterility. Several sterility genes differentiate the two species (Oka, 1988). Anther culture is being explored to develop DH lines from the crosses of *O. sativa × O. glaberrima* in order to overcome sterility and produce homozygous lines, fix recombinants and use the DH lines as mapping populations. Jones *et al.* (1997) reported the production of DH lines from such crosses and their utilization in rice improvement. Among the plants regenerated from anther culture, 52% were haploid, 41% DHs and 7% polyploids. Fifteen per cent of the spontaneously generated DH lines had partial sterility.

At IRRI, we have cultured anthers of 75 F_1s involving ten varieties of *O. sativa* and 18 accessions of *O. glaberrima*. Upon culture of 45,400 anthers from 75 F_1s, no calluses were produced from 34 crosses (Enriquez *et al.*, 2000). The other 41 F_1s showed, on average, 1.3% callus formation from 144,160 cultured anthers. The plant regeneration rate ranged from 0.0 to 77.0%. The anther-derived calluses from 16 F_1s did not show any plant regeneration. Among *O. sativa* lines, an élite breeding line of the NPT IR68552-55-3-2 responded better in producing green plants in crosses with *O. glaberrima*. Similarly, *O. glaberrima* (CG14) in crosses with some of *O. sativa* parents responded favourably to anther culture. Although 562 DH lines could be produced from different crosses, a majority of them showed very high seed sterility (56.2 to 100.0%). The high sterility of DH lines is indicative of the presence of several loci for sterility that differentiate between the Asian and African rice species. Genotyping of DH lines using microsatellite markers indicated a frequent exchange of chromosome segments between *O. sativa* and *O. glaberrima*.

Construction of chromosome substitution lines of O. glaberrima *in the* background of O. sativa

Many laboratories are engaged in constructing chromosome substitution lines (CSLs) of *O. glaberrima*. Such lines would be important for understanding the genetics of useful traits, such as the tolerance to biotic and abiotic stresses and the weed competitiveness of *O. glaberrima*. Doi *et al.* (1997) constructed a series of *O. glaberrima* (Acc. 104038) CSLs in the background of japonica rice, Taichung 65. From 45 BC_3F_2, 907 plants were genotyped and a subset of 39 CSLs that carried homozygous chromosome segments from the donor, *O. glaberrima*, were selected. These lines cover most parts of the *O. glaberrima* genome. Ghesquiere *et al.* (1997) initiated a molecular marker-aided back-cross programme to develop a set of 100 'contig lines', each carrying an *O. glaberrima* chromosome fragment of around 20 cM in the *O. sativa* background.

At IRRI, we are using microsatellite markers to identify BC_2F_3/BC_3F_3 lines from the crosses of *O. sativa* × *O. glaberrima* carrying different segments of *O. glaberrima* in the background of the high-yielding indica cv. IR64 and an élite breeding line of NPT rice. Such CSLs would be an important resource for the functional genomics of rice.

Identification and introgression of yield-enhancing loci/QTL from wild species

Wild species are phenotypically inferior to the cultivated species. As discussed in the preceding sections, these wild species are a rich reservoir of useful genes for resistance to major diseases and insects and for tolerance to abiotic stresses. Recent findings have shown they contain genes capable of improving yield as well. Deleterious genes often mask these favourable genes. Transgressive segregation for yield in crosses of cultivated and wild species suggests that, despite inferior phenotypes, wild species contain genes that can improve quantitative traits, such as yield. Molecular markers have made it possible to identify and

introgress desirable QTL from wild species into élite breeding lines. Tanksley and Nelson (1996) proposed advanced back-cross QTL analysis to discover and transfer valuable QTL alleles from unadapted germplasm, such as wild species, into élite breeding lines of species.

Transgressive segregation for yield and yield components has been obtained in crosses of rice and *AA*-genome species. QTL from wild species of rice for increased yield have been identified. Xiao *et al.* (1996) analysed 300 BC_2 testcross families produced from the cross of *O. sativa* × *O. rufipogon*. On average, each BC_2 test-cross line contained 5% *O. rufipogon* genome. In most cases, introgression of *O. rufipogon* alleles either had no significant effect on yield or was inferior to the alleles of cultivated rice. *Oryza rufipogon* alleles at two marker loci, RM5 (*yld1-1*) on chromosome 1 and RG256 on chromosome 2 (*yld2-1*), were, however, associated with enhanced yield. The alleles *yld1-1* and *yld2-1* were both associated with a significant increase in grains per plant. In another experiment, Xiao *et al.* (1998) identified 68 QTL. Of these, 35 (51%) had trait-improving alleles derived from the phenotypically inferior wild species. Nineteen (56%) of these beneficial QTL alleles had no deleterious effects on other characters.

Recently, Moncada *et al.* (2001) followed the advanced-back-cross breeding strategy and analysed BC_2F_2 populations derived from the cross involving an upland japonica rice cultivar, 'Caiapo', from Brazil and an accession of *O. rufipogon* from Malaysia. The populations were tested under drought-prone, acid soil conditions. Based on analyses of 125 simple sequence repeat (SSR) and RFLP markers, using single-point, interval and composite-interval mapping, two putative *O. rufipogon*-derived QTL were detected for yield, 13 for yield components, four for maturity and six for plant height. Advanced-back-cross QTL analysis showed that certain regions of the rice genome harbour genes that are useful across a range of environments.

Our preliminary results, at IRRI, of advanced-back-cross progenies derived from the crosses of IR65600-81-5-3-2, an élite breeding line of NPT rice, with *O. longi-staminata* and IR64 × *O. rufipogon*, also support transgressive segregation for yield and yield components. These findings show that genes from wild species can increase the yield of élite rice lines, even though wild species are phenotypically inferior to cultivated rice.

Introgression from distantly related genomes

Hybrids between cultivated rice and *AA*-genome wild species can be produced through normal procedures. Hybrids between rice and distantly related wild species are usually difficult to produce. Low crossability and abortion of hybrid embryos are the common features of such crosses. These hybrids are completely male-sterile. Subsequent back-crosses are made with the recurrent rice parent to produce disomic progenies ($2n = 24$). Embryo rescue is used to produce F_1 and back-cross progenies until fertile plants with the normal diploid chromosome complement ($2n = 24$) or $2n = 25$ (MAALs) become available (Fig. 14.1). The fertile progenies are selfed to produce advanced introgression lines and evaluated for transfer of useful traits. We have produced MAALs representing seven to 12 chromosomes from six species (Table 14.2). These MAALs are important cyto-genetic stocks carrying an individual extra chromosome of wild species and serve as an important source of alien genetic variation and for mapping genes on chromosomes.

Introgression from the CC genome

Interspecific hybrids have been produced through embryo rescue between rice and wild species with the *CC* genome. Jena and Khush (1990) produced several introgression lines from the cross of *O. sativa* × *O. officinalis*. Useful genes for resistance to BPH, the white-backed planthopper (WBPH) and bacterial blight have been transferred from *O. officinalis* into an élite breeding line of rice. Several introgression

lines resistant to three BPH biotypes of The Philippines were evaluated for resistance to BPH populations in India and Bangladesh. Many progenies were found to be resistant to BPH in the three countries. A few of the BPH-resistant lines were also resistant to BPH populations in Vietnam. Three breeding lines have been released as varieties for commercial cultivation in the Mekong Delta of Vietnam. IR54751-2-44-15-24-3 was designated as MTL98, IR54751-2-34-10-6-2 as MTL 103 and IR54751-2-41-10-5-1 as MTL 105.

Hirabayashi and Ogawa (1999) analysed recombinant inbred lines (RILs) from the cross between 'Hinohikari' (susceptible japonica) with the IR54742-1-11-17 indica introgression line derived from crosses of *O. sativa* × *O. officinalis*. Two genes for BPH resistance, *bph11*(t) and *bph12*(t), were identified and mapped to chromosomes 3 and 4 of rice. Huang *et al.* (2001) made crosses between the rice cultivar 'Zhensheng 97B' and *O. officinalis*. One of the introgression lines, B5, showed a strong resistance to BPH biotypes 1 and 2 and to field populations collected from Zhejiang Province, China.

We have produced several introgression lines from the crosses of *O. sativa* cvs M202 (japonica) and IR74 (indica) with *O. officinalis* (Acc. 101399). These progenies are under evaluation for the possible introgression of stem-rot resistance into rice.

Introgression from BBCC-genome parents

Interspecific hybrids have been produced between *O. sativa* and the tetraploid wild species *O. minuta* (*BBCC*) (Sitch, 1990). Following back-crossing and embryo rescue, advanced introgression lines have been produced from the cross of *O. sativa* (IR31917-45-3-2) and *O. minuta* (Acc. 101141). Amante-Bordeos *et al.* (1992) evaluated advanced progenies for resistance to bacterial blight and blast. Two introgression lines were resistant – one to race 6 of bacterial blight and another to race PO6-6 of blast. The introgressed blast gene has been designated as *Pi9*(t). It has resistance to several isolates of blast. Brar *et al.* (1996)

evaluated introgression lines derived from *O. sativa* × *O. minuta*. Of the 96 back-cross progenies screened, ten showed introgression for resistance to BPH biotype 1 of The Philippines.

Resistance to sheath blight, another serious disease, is limited or is of a moderate level, in both the cultivated and the wild species. *Oryza minuta*, however, is comparatively tolerant to sheath blight. We are evaluating advanced introgression lines derived from the cross of *O. sativa* × *O. minuta* for tolerance to sheath blight.

Introgression from CCDD-genome parents

A number of workers have produced hybrids between rice and *CCDD*-genome species (Sitch, 1990; Brar *et al.*, 1991). Of the three *CCDD* species, advanced lines derived from the cross of *O. sativa* × *O. latifolia* have been investigated. Introgression from *O. latifolia* for resistance to BPH, WBPH and bacterial blight and for other traits, such as growth duration and purple pigmentation, has been obtained (unpublished).

Introgression from the EE genome

Multani *et al.* (1994) produced hybrids between colchicine-induced autotetraploids of rice and *O. australiensis* ($2n = 24$ *EE*). Introgression was detected for morphological traits, such as long awns and earliness, and for *Amp-3* and *Est-2* allozymes. Of the 600 BC_2F_4 progenies, four were resistant to BPH and one to race 6 of bacterial blight. BPH resistance was found to be controlled by a recessive gene in two of the four lines but was controlled by a dominant gene in the other two lines. One of the lines (IR65782-4-136-2-2) carried the *Bph10* gene.

Introgression from the FF genome

A series of introgression lines has been derived from the cross of *O. sativa* cv. IR56 and the wild species, *O. brachyantha* ($2n = 24$ *FF*). IR56 is susceptible to bacterial blight races 1–4 and 6 from The Philippines, whereas *O. brachyantha* is resistant.

Of the 149 back-cross progenies analysed, 27 showed introgression for resistance to bacterial blight races 1–4 and 6 (Brar *et al.*, 1996).

Introgression from GG-genome

Hybrids have been produced from the cross of *O. sativa* and *O. granulata* (Brar *et al.*, 1991). Advanced progenies have also been produced; however, none of the lines tested has shown introgression of traits from *O. granulata* into rice.

BC_2 progenies derived from the crosses of *O. sativa* with *O. officinalis* (*CC*), *O. australiensis* (*EE*), *O. brachyantha* (*FF*) and *O. granulata* (*GG*) resembled the recurrent rice parent in most morphological traits. This suggested that only a limited amount of recombination between the *A* genome of *O. sativa* and the *C*, *E*, *F* and *G* genomes of wild species occurred.

Introgression from HHJJ-genome parents

Hybrids between rice cv. IR56 and *O. ridleyi* (Acc. 100821) have been produced. The tetraploid *Ridleyi* complex includes two species; *O. ridleyi* and *O. longiglumis*. A few introgression lines (BC_2F_1) from this cross have been produced; however, no introgression could be detected.

Introgression from the HHKK genome

Intergeneric hybrids between *O. sativa* and *P. coarctata* have been produced both through sexual crosses following embryo rescue (Brar *et al.*, 1997) and through protoplast fusion (Jelodar *et al.*, 1999). Due to strong incompatibility barriers, no back-cross progenies could be obtained.

Molecular mapping of introgressed alien genes

Traits introgressed from different wild species into rice are listed in Table 14.3. Some of the introgressed genes have been mapped via linkage to molecular markers.

Mapping of the Xa21 gene for bacterial blight resistance

The *Xa21* gene, introgressed from *O. longistaminata*, has been mapped to chromosome 11, close to the RG103 marker (Ronald *et al.*, 1992). This gene has been transferred through marker-assisted selection into several rice cultivars – IR64, PR106 – and NPT rice (Sanchez *et al.*, 2000; Singh *et al.*, 2001). Wang *et al.* (1995) used a bacterial artificial chromosome (BAC) library and isolated 12 BAC clones that hybridized with the three DNA markers linked to the *Xa21* locus. Jiang *et al.* (1995) used BAC clones and FISH and physically mapped *Xa21* locus to chromosome 11 of rice. The *Xa21* gene has been isolated (Song *et al.*, 1995) via positional cloning. The transgenic plants carrying the cloned *Xa21* show a high level of resistance to bacterial blight pathogen.

Mapping of Bph10(t) for BPH resistance

A gene conferring resistance to three BPH biotypes from The Philippines was introgressed from *O. australiensis* into rice (Multani *et al.*, 1994). MAAL analyses showed that the gene for BPH resistance is located on chromosome 12 of *O. australiensis*. Hence, probes of chromosome 12 were used for an RFLP survey with the recurrent parent, the wild species and the introgression line. All 14 probes were polymorphic in the recurrent parent and the wild species; however, only RG457 detected introgression from *O. australiensis* into the introgression line. Cosegregation for BPH reaction and RG457 was determined from the F$_2$ data. The gene for BPH resistance is linked to RG457, with a distance of 3.68 ± 1.29 cM (Ishii *et al.*, 1994).

Mapping of QTL for BPH resistance

BPH resistance was introgressed from *O. officinalis* into rice cv. 'Zhenshan 97B'. Huang *et al.* (2001) carried out bulk segregant analysis of F$_3$ populations produced from the cross between the BPH-resistant introgression line 'B5', derived from *O.*

officinalis and 'Mingui 63', a rice cultivar susceptible to BPH. QTL analyses revealed that *qbp1* was located in the 14.3 cM interval between R2443 and R1925 on the long arm of chromosome 3. This QTL explained a 26.4% phenotypic variation for BPH resistance. The second QTL, *qbp2*, was mapped to chromosome 4, with a 0.4 cM interval between C820 and R288.

Mapping of a gene for earliness

The introgression line (IR65482-4-136-2-2), derived from a cross of *O. sativa* and *O. australiensis*, was crossed with the recurrent parent (IR31917-45-3-2). The F$_2$ segregation indicated that the introgressed gene for earliness is recessive. Since the gene for earliness is located on chromosome 10 (Sato *et al.*, 1985), probes from chromosome 10 were hybridized with the DNA of the recurrent parent, the wild species and the introgression line. All five probes were polymorphic between recurrent parent and wild species. Only CDO98 detected introgression from *O. australiensis*, however. Cosegregation between CDO98 and days to flowering in F$_2$ showed that the gene for earliness is situated at a distance of 9.96 ± 3.28 cM from CDO98, thus indicating that this recessive gene for earliness is also located on chromosome 10 of *O. australiensis*.

Mapping of the gene (Pi9t) for blast resistance

A gene for blast resistance (*Pi9t*) was introgressed from *O. minuta* (BBCC) into rice (Amante-Bordeos *et al.*, 1992). The introgression line (IR71033-4-127-B) was surveyed using 103 polymorphic RFLP markers, located at an average distance of 20 cM intervals in the rice genome. No linkage, however, could be established between any markers and *Pi9t*. In another experiment, a back-cross population, produced by crossing the introgression line and the susceptible parent IR31917-45-3-2, was analysed. Three RAPD markers have been found to be linked to *Pi9t* (Reimer and Nelson, unpublished).

Molecular characterization of alien introgression

Molecular markers and *in situ* hybridization techniques provide a unique opportunity to determine the extent and process of alien introgression. Jena *et al.* (1992) analysed 52 introgression lines (BC_2F_8) derived from crosses of *O. sativa* × *O. officinalis*. Of the 177 RFLP markers, 174 were polymorphic between the two parents, with one or more enzymes. Most markers were polymorphic with multiple enzymes, but *Hind* digests showed highest polymorphism (85.8%). Of the 174 informative RFLP markers, 28 (16.1%) identified putative *O. officinalis* introgressed segments in one or more of the introgression lines. Individual introgression lines contained 1.1–6.8% introgressed *O. officinalis* segments. Introgressed segments were found on 11 of the 12 rice chromosomes. In a majority of cases, *O. sativa* alleles were replaced by *O. officinalis* alleles. Introgressed segments were smaller in size and similar in plants derived from early and later generations. Single RFLP markers detected most introgressed segments, and the flanking markers were negative for introgression. Brar *et al.* (1996) analysed 29 derivatives of *O. sativa* × *O. brachyantha* and 40 derivatives of *O. sativa* × *O. granulata*. Extensive polymorphism between rice and wild species was observed. Of the six chromosomes surveyed, no introgression was detected from chromosomes 7, 9, 10 or 12 of *O. granulata* and chromosome 10 or 12 of *O. brachyantha*. For each of the remaining chromosomes, one or two RFLP markers showed introgression in some of the derived lines. Although the level of introgression was low, the results showed possibilities of introgressing chromosome segments even from distantly related genomes into cultivated rice and thus the feasibility of transferring useful genes from distant *Oryza* species.

Microsatellite markers show extensive polymorphism between *O. sativa* and *O. glaberrima*. Analysis of DH populations derived through anther culture of *O. sativa* × *O. glaberrima* showed frequent exchange of chromosome segments between *O. sativa* and *O. glaberrima*. In some of the alien introgression lines, non-parental bands were detected. This could result from genomic interactions of cultivated and wild species or an activation of some transposable elements producing novel bands.

In situ hybridization is a powerful technique for characterizing parental genomes in wide hybrids and detecting introgressed alien segments. The protocols for FISH of rice chromosomes have been refined (Fukui *et al.*, 1994). Variability in rDNA loci has been detected through FISH in *Oryza* species (Fukui *et al.*, 1994). Jiang *et al.* (1995) mapped the *Xa21* gene derived from *O. longistaminata* to rice chromosomes, using FISH and BAC clones. Shishido *et al.* (1998) used multicolour FISH to characterize *A*, *B* and *C* genomes in somatic hybrids of rice.

We have used GISH and characterized parental chromosomes in wide hybrids (F_1, BC_1) involving *O. sativa* × *O. officinalis*, *O. sativa* × *O. brachyantha*, *O. sativa* × *O. australiensis* and *O. sativa* × *O. ridleyi*. The alien extra chromosome in MAALs and introgressed segments could also be identified. Abbasi *et al.* (1999) used total genomic DNA of *O. australiensis* as a probe and hybridized it with the meiotic chromosomes of the F_1 hybrid (*O. sativa* × *O. australiensis*). Both autosyndetic and allosyndetic pairing among *A* and *E* genomes could be detected. GISH is now being extended to detect pairing among *A* genome and other distantly related genomes of *Oryza* at pachytene in order to precisely understand the process of alien introgression, particularly of small chromosome segments. Asghar *et al.* (1998) applied FISH for characterizing the chromosomes of *O. sativa* and *O. officinalis* and located rDNA loci on somatic chromosomes of both *O. sativa* and *O. officinalis*. Yan *et al.* (1999) used FISH to characterize *A*- and *C*-genome chromosomes in F_1 and BC_1 of *O. sativa* × *O. eichingeri*.

Mechanism of alien introgression

Molecular analysis of introgression lines derived from crosses of *O. sativa* with *AA*-genome wild species, such as *O. glaberrima* and *O. rufipogon*, revealed recombination resulting from classical crossing over. No hot spots for chromosome recombination could be detected. Cytogenetic and molecular-marker analysis of introgression lines derived from *O. sativa* and distantly related *Oryza* species did not show any evidence of alien chromosome substitution. The results indicate genetic recombination between chromosomes of cultivated and wild species as the cause of alien gene transfer. RFLP analyses of introgression lines showing reciprocal replacement of alleles of *O. officinalis*, *O. australiensis* and *O. brachyantha* with the alleles of *O. sativa* further support alien gene transfer through crossing over, rather than the substitution of a complete chromosome or an arm of a chromosome of wild species (Jena *et al.*, 1992; Ishii *et al.*, 1994; Brar *et al.*, 1996). The rapid recovery of recurrent parent phenotypes in BC$_2$ and BC$_3$ of *O. sativa* × *O. officinalis*, *O. sativa* × *O. australiensis*, *O. sativa* × *O. brachyantha* and *O. sativa* × *O. granulata* is an indication of limited recombination between the *A* genome, on the one hand, and the *C, E, F* and *G* genomes, on the other. Progenies recovered in BC$_2$ of *O. sativa* × *O. officinalis* were so similar to *O. sativa* that they were evaluated in field trials and released as varieties for commercial cultivation in Vietnam.

Most introgressed segments were detected via single RFLP and SSR markers and the flanking markers were negative for introgression. This also supports the conclusion regarding limited recombination and the possible cause for the rapid recovery of the recurrent parent phenotype. Rapid recovery of the recurrent parent phenotypes in the back-cross progenies of wide crosses has been reported in *Gossypium* by Stephens (1949) and *Lycopersicon* by Rick (1969), although a higher number of back-crosses was required to reconstitute the recurrent phenotypes.

Future Outlook on Gene Transfer from Wild Species to Rice

Rice productivity is adversely affected by various biotic and abiotic stresses. There is thus an urgent need to widen the rice gene pool by incorporating genes for such traits from diverse sources. Wild species are an important reservoir of useful genes and offer great potential to incorporate such genes into commercial rice cultivars for resistance to major diseases and insects and tolerance to various abiotic stresses. Moreover, many of the useful alien genes are different from those of the cultivated species and are thus useful in expanding the sources of resistance to various stresses. The introgressed alien genes have become a valuable addition for molecular marker-assisted selection. Several studies have demonstrated transgressive segregation for yield and yield components from crosses of rice and closely related *AA*-genome wild species. Molecular markers have made it possible to identify and introgress such yield-enhancing locus/QTL 'wild alleles' into rice cultivars and thus offer new opportunities for enhancing the grain yield of rice. Future research should focus on overcoming one of the key barriers in the transfer of useful genes from distantly related wild species by enhancing recombination among the homologous chromosomes. One strategy should aim at identifying gene(s) controlling homologous chromosome pairing in *Oryza*. Alien introgression could also be enhanced through tissue culture of wide hybrids resulting from chromosomal exchanges between genomes of cultivated and wild species. With the advances in tissue culture, molecular markers, *in situ* hybridization and genomics, the future outlook for broadening the gene pool of rice through the precise transfer of useful genes from wild species into rice cultivars seems more promising than ever before.

References

Abbasi, F.M., Brar, D.S., Carpena, A.L., Fukui, K. and Khush, G.S. (1999) Detection of auto-syndetic and allosyndetic pairing among *A* and *E* genomes of *Oryza* through genomic *in situ* hybridization. *Rice Genetics Newsletter* 16, 24–25.

Abedinia, M., Henry, R.J., Blakeney, A.B. and Levin, L.G. (2000) Accessing genes in the tertiary gene pool of rice by direct introduction of total DNA from *Zizania palustris* (wild rice). *Plant Molecular Biology Reporter* 18, 133–138.

Aggarwal, R.K., Brar, D.S., Khush, G.S. and Jackson, M.T. (1996) *Oryza schlechteri* Pilger has a distinct genome based on molecular analysis. *Rice Genetics Newsletter* 13, 58–59.

Aggarwal, R.K., Brar, D.S. and Khush, G.S. (1997) Two new genomes in the *Oryza* complex identified on the basis of molecular divergence analysis using total genomic DNA hybridization. *Molecular and General Genetics* 254, 1–12.

Aggarwal, R.K., Brar, D.S., Nandi, S., Huang, N. and Khush, G.S. (1999) Phylogenetic relationships among *Oryza* species revealed by AFLP markers. *Theoretical and Applied Genetics* 98, 1320–1328.

Amante-Bordeos, A., Sitch, L.A., Nelson, R., Dalmacio, R.D., Oliva, N.P., Aswidinnoor, H. and Leung, H. (1992) Transfer of bacterial blight and blast resistance from the tetraploid wild rice *Oryza minuta* to cultivated rice. *Theoretical and Applied Genetics* 84, 345–354.

Asghar, M., Brar, D.S., Hernandez, J.E., Ohmido, N. and Khush, G.S. (1998) Characterization of parental genomes in a hybrid between *Oryza sativa* L. and *O. officinalis* Wall ex Watt. through fluorescence *in situ* hybridization. *Rice Genetics Newsletter* 15, 83–84.

Brar, D.S. and Khush, G.S. (1986) Wide hybridization and chromosome manipulation in cereals. In: Evans, D.H., Sharp, W.R. and Ammirato, P.V. (eds) *Handbook of Plant Cell Culture*, Vol. 4: *Techniques and Applications*. Macmillan, New York, pp. 221–263.

Brar, D.S. and Khush, G.S. (1997) Alien introgression in rice. *Plant Molecular Biology* 35, 35–47.

Brar, D.S., Elloran, R. and Khush, G.S. (1991) Interspecific hybrids produced through embryo rescue between cultivated and eight wild species of rice. *Rice Genetics Newsletter* 8, 91–93.

Brar, D.S., Dalmacio, R., Elloran, R., Aggarwal, R., Angeles, R. and Khush, G.S. (1996) Gene transfer and molecular characterization of introgression from wild *Oryza* species into rice. In: *Rice Genetics III*. International Rice Research Institute, Manila, The Philippines, pp. 477–486.

Brar, D.S., Elloran, R.M., Talag, J.D., Abbasi, F. and Khush, G.S. (1997) Cytogenetic and molecular characterization of an intergeneric hybrid between *Oryza sativa* L. and *Porteresia coarctata* (Roxb.) Tateoka. *Rice Genetics Newsletter* 14, 43–44.

Dalmacio, R., Brar, D.S., Ishii, T., Sitch, L.A., Virmani, S.S. and Khush, G.S. (1995) Identification and transfer of a new cytoplasmic male sterility source from *Oryza perennis* into indica rice (*O. sativa*). *Euphytica* 82, 221–225.

Dalmacio, R.D., Brar, D.S., Virmani, S.S. and Khush, G.S. (1996) Male sterile line in rice (*Oryza sativa*) developed with *O. glumaepatula* cytoplasm. *International Rice Research Notes* 21(1), 22–23.

Dingkuhn, M., Jones, M.P., Johnson, D.E. and Sow, A. (1998) Growth and yield potential of *Oryza sativa* and *O. glaberrima* upland rice cultivars and their interspecific progenies. *Field Crops Research* 57, 57–69.

Doi, K., Iwata, N. and Yoshimura, Y. (1997) The construction of chromosome substitution lines of African rice (*Oryza glaberrima* Steud.) in the background of japonica rice (*O. sativa* L.). *Rice Genetics Newsletter* 14, 39–40.

Enriquez, E.C., Brar, D.S., Rosario, T.L., Jones, M. and Khush, G.S. (2000) Production and characterization of doubled haploids from anther culture of the F_1s of *Oryza sativa* L. $\times$ *O. glaberrima* Steud. *Rice Genetics Newsletter* 17, 67–69.

Fukui, K., Ohmido, N. and Khush, G.S. (1994) Variability in the rDNA loci in genus *Oryza* detected through fluorescence *in situ* hybridization. *Theoretical and Applied Genetics* 87, 893–899.

Ge, S., Sang, T., Lu, B.R. and Hong, D.Y. (1999) Phylogeny of rice genomes with emphasis on origins of allotetraploid species. *Proceedings of the National Academy of Sciences USA* 96(25), 14400–14405.

Ghesquiere, A., Sequier, J., Second, G. and Lorieux, M. (1997) First steps towards a rational use of African rice, *Oryza glaberrima*, in rice breeding through a 'contig line' concept. *Euphytica* 96, 31–39.

Hayashi, Y., Kozuka, J. and Shimamoto, K. (1988) Hybrids of rice (*Oryza sativa* L.) and wild *Oryza* species obtained by cell fusion. *Molecular and General Genetics* 214, 6–10.

Hirabayashi, H. and Ogawa, T. (1999) Identification and utilization of DNA markers linked to genes for resistance to brown planthopper (BPH) in rice [in Japanese]. *Recent Advances in Breeding* 41 (Suppl.), 71–74.

Huang, Z., He, G., Shu, L., Li, X. and Zhang, Q. (2001) Identification and mapping of two brown planthopper resistance genes in rice. *Theoretical and Applied Genetics* 102, 929–934.

Ishii, T., Brar, D.S., Multani, D.S. and Khush, G.S. (1994) Molecular tagging of genes for brown planthopper resistance and earliness introgressed from *Oryza australiensis* into cultivated rice, *O. sativa. Genome* 37, 217–221.

Jelodar, N.B., Blackhall, N.W., Hartman, T.P.V., Brar, D.S., Khush, G.S., Davey, M.R., Cocking, E.C. and Power, J.B. (1999) Intergeneric somatic hybrids of rice (*Oryza sativa* L. (+) *Porteresia coarctata* (Roxb.) Tateoka). *Theoretical and Applied Genetics* 99, 570–577.

Jena, K.K. and Khush, G.S. (1990) Introgression of genes from *Oryza officinalis* Well ex Watt to cultivated rice, *O. sativa* L. *Theoretical and Applied Genetics* 80, 737–745.

Jena, K.K., Khush, G.S. and Kochert, G. (1992) RFLP analysis of rice (*Oryza sativa* L.) introgression lines. *Theoretical and Applied Genetics* 84, 608–616.

Jena, K.K., Khush, G.S. and Kochert, G. (1994) Comparative RFLP mapping of a wild rice, *Oryza officinalis* and cultivated rice, *O. sativa. Genome* 37, 382–389.

Jiang, J., Gill, B.S., Wang, G.L., Ronald, P.C. and Ward, D.C. (1995) Metaphase and interphase fluorescence *in situ* hybridization mapping of the rice genome with bacterial artificial chromosomes. *Proceedings of the National Academy Sciences USA* 92, 4487–4491.

Johnson, D.E., Dingkuhn, M., Jones, M.P. and Mahamane, M.C. (1998) The influence of rice plant type on the effect of weed competition on *Oryza sativa* and *Oryza glaberrima. Weed Research* 38, 207–216.

Jones, M.P., Dingkuhn, M., Aluko, G.K. and Semon, M. (1997) Interspecific *Oryza sativa* L. × *O. glaberrima* Steud. progenies in upland rice improvement. *Euphytica* 92, 237–246.

Kennard, W., Phillips, R., Porter, R. and Grambacher, A. (1999) A comparative map

of wild rice (*Zizania palustris* $2n = 2x = 30$). *Theoretical and Applied Genetics* 99, 793–799.

Khush, G.S. (1977) Disease and insect resistance in rice. *Advances in Agronomy* 29, 265–341.

Khush, G.S. and Brar, D.S. (1992) Overcoming the barriers in hybridization. In: Kallo, G. and Chowdhury, J.B. (eds) *Theoretical and Applied Genetics* 16, 47–61.

Khush, G.S., Bacalangco, E. and Ogawa, T. (1990) A new gene for resistance to bacterial blight from *O. longistaminata. Rice Genetics Newsletter* 7, 121–122.

Kobayashi, N., Ikeda, R., Domingo, I.T. and Vaughan, D.A. (1993) Resistance to infection of rice tungro viruses and vector resistance in wild species of rice (*Oryza* spp.). *Japanese Journal of Breeding* 43, 377–387.

Lin, S.C. and Yuan, L.P. (1980) Hybrid rice breeding in China. In: *Innovative Approaches to Rice Breeding*. International Rice Research Institute, Manila, The Philippines, pp. 35–51.

Ling, K.C., Aguiero, V.M. and Lee, S.H. (1970) A mass screening method for testing resistance to grassy stunt disease of rice. *Plant Disease Reporter* 56, 565–569.

Liu, B., Liu, Z. and Li, X.W. (1999) Production of a highly asymmetric somatic hybrid between rice and *Zizania latifolia* (Griseb): evidence for intergenome exchange. *Theoretical and Applied Genetics* 98, 1099–1103.

Moncada, P., Martinez, C.P., Borrero, J., Chatel, M., Gauch, H., Jr, Guimaraes, E., Tohme, J. and McCouch, S.R. (2001) Quantitative trait loci for yield and yield components in an *Oryza sativa* × *Oryza rufipogon* BC$_2$F$_2$ population evaluated in an upland environment. *Theoretical and Applied Genetics* 102, 41–52.

Multani, D.S., Jena, K.K., Brar, D.S., delos Reyes, B.C., Angeles, E.R. and Khush, G.S. (1994) Development of monosomic alien addition lines and introgression of genes from *Oryza australiensis* Domin. to cultivated rice *O. sativa* L. *Theoretical and Applied Genetics* 88, 102–109.

Oka, H.I. (1988) *Origin of Cultivated Rice. Developments in Crop Science*, Vol. 14. Japan Scientific Society Press, Tokyo, and Elsevier, Amsterdam.

Rick, C.M. (1969) Controlled introgression of chromosomes of *Solanum pennelli* into *Lycopersicon esculentum*: segregation and recombination. *Genetics* 26, 753–768.

Ronald, P.C., Albano, B., Tabien, R., Abenes, L., Wu, K., McCouch, S. and Tanksley, S.D. (1992) Genetic and physical analysis of rice bacterial blight resistance locus, *Xa21*. *Molecular and General Genetics* 236, 113–120.

Sanchez, Z.C., Brar, D.S., Huang, N., Li, Z. and Khush, G.S. (2000) Sequence tagged sited marker-assisted selection for three bacterial blight resistance gene in rice. *Crop Science* 40, 792–797.

Sato, S., Sakamoto, I. and Nakasone, S. (1985) Location of *Ef1* for earliness on Nishimura's seventh chromosome. *Rice Genetics Newsletter* 2, 59–60.

Shishido, R., Apisitwanich, S., Ohmido, N., Okinaka, Y., Mori, K. and Fukui, K. (1998) Detection of specific chromosome reduction in rice somatic hybrids with the *A*, *B*, and *C* genomes by multicolor genomic *in situ* hybridization. *Theoretical and Applied Genetics* 97, 1013–1018.

Singh, S., Sidhu, J.S., Huang, N., Vikal, Y., Li, Z., Brar, D.S., Dhaliwal, H.S. and Khush, G.S. (2001) Pyramiding three bacterial blight resistance genes (*xa5*, *xa13* and *Xa21*) using marker assisted selection into indica rice cultivar PR106. *Theoretical and Applied Genetics* 102, 1011–1015.

Sitch, L.A. (1990) Incompatibility barriers operating in crosses of *Oryza sativa* with related species and genera. In: Gustafson, J.P. (ed.) *Genetic Manipulation in Plant Improvement II*. Plenum Press, New York, pp. 77–94.

Song, W.Y., Wang, G.L., Chen, L.L., Kim, H.S., Pi, Y.L., Holsten, T., Gardner, J., Wang, B., Zhai, W.X., Zhu, L.H., Fauquet, C. and Ronald, P.A. (1995) A receptor kinase like protein encoded by the rice disease resistance gene, *Xa21*. *Science* 270, 1804–1806.

Soriano, I.R., Schmit, V., Brar, D.S., Prot, J.C. and Reversat, G. (1999) Resistance to rice root-knot nematode *Meloidogyne graminicola* identified in *Oryza longistaminata* and *O. glaberrima*. *Nematology* 1, 395–398.

Stephens, S.G. (1949) The cytogenetics of speciation in *Gossypium* I. Selective elimination of the donor parent genotype in interspecific backcrosses. *Genetics* 34, 627–637.

Tanksley, S.D. and Nelson, J.C. (1996) Advanced backcross QTL analysis: a method for the simultaneous discovery and transfer of valuable QTLs from unadapted germplasm into elite breeding lines. *Theoretical and Applied Genetics* 92, 191–203.

Tateoka, T. (1962) Taxonomic studies of *Oryza* L. *O. latifolia* complex. *Botanical Magazine* 75, 418–427.

Vaughan, D.A. (1989) *The Genus* Oryza *L.: Current Status of Taxonomy*. IRRI Research Paper Series No. 138. IRRI, Manila, The Philippines, 21 pp.

Wang, G.L., Holsten, T.E., Song, W.Y. and Ronald, P.C. (1995) Construction of a rice bacterial artificial chromosome library and identification of clones linked to the *Xa21* disease resistant locus. *The Plant Journal* 7, 525–533.

Xiao, J., Grandillo, S., Ahn, S.N., McCouch, S.R., Tanksley, S.D., Li, J. and Yuan, L. (1996) Genes from wild rice improve yield. *Nature* 384, 223–224.

Xiao, J., Li, J., Grandillo, S., Ahn, S.N., Yuan, L., Tanksley, S.D. and McCouch, S.R. (1998) Identification of trait-improving quantitative trait loci alleles from a wild rice relative, *Oryza rufipogon*. *Genetics* 150, 899–909.

Yan, H., Min, S. and Zhu, L. (1999) Visualization of *Oryza eichingeri* chromosomes in intergenomic hybrid plants from *O. sativa* × *O. eichingeri* via fluorescent *in situ* hybridization. *Genome* 42, 48–51.

Zhang, Q., Lin, S.C., Zhao, B.Y., Wang, C.L., Wang, W.C., Zhou, Y.L., Li, D.Y., Chen, C.B. and Zhu, L.H. (1998) Identification and tagging of a new gene for resistance to bacterial blight (*Xanthomonas oryzae* pv. *oryzae*) from *O. rufipogon*. *Rice Genetics Newsletter* 15, 138–142.

Section II

Genotype–Environment Interaction and Stability Analysis

15 Genotype–Environment Interaction: Progress and Prospects

Manjit S. Kang

Department of Agronomy, Louisiana State University, Baton Rouge, LA 70803-2110, USA

Introduction

Genotype–environment interaction (GEI) is an age-old, universal issue that relates to all living organisms. Genotypes and environments interact to produce an array of phenotypes. GEI can be defined as the difference between the phenotypic value and the value expected from the corresponding genotypic and environmental values (Baker, 1988). When responses of two genotypes to different levels of environmental stress are compared, an interaction is described statistically as the failure of the two response curves to be parallel (Baker, 1988). GEI is the variation caused by the joint effects of genotypes and environments (Dickerson, 1962).

Breeders/agronomists usually test a diverse array of genotypes in diverse environments, which implies that GEI is to be expected. According to Haldane (1947), GEI is important only if genotypes switch ranks from one environment to another. GEIs can be grouped into two broad categories: crossover and non-crossover interactions; a brief discussion of each follows.

Crossover and non-crossover interaction

Crossover (qualitative) interaction

The differential response of cultivars to diverse environments is referred to as a crossover interaction when cultivar ranks change from one environment to another. A main feature of crossover interaction is intersecting lines in a graphical representation. If the lines do not intersect, there is no crossover interaction (Kang, 1998).

In crop breeding, the crossover interaction is more important than non-crossover interaction (Baker, 1990). Since the presence of a crossover interaction has strong implications for breeding for specific adaptation, it is important to assess the frequency of crossover interactions (Singh *et al.*, 1999). According to Gregorius and Namkoong (1986), crossover interaction is not only non-additive in nature but also non-separable. Lack of crossover interaction for quantitative trait loci (QTL) even in the presence of significant GEI has been reported (Lee, 1995; Beavis and Keim, 1996). The reader may refer to Beavis and Keim (1996), Cornelius *et al.* (1996), Crossa *et al.* (1996) and Singh *et al.* (1999) for further discussion on crossover and non-crossover interactions.

Variation among genotypes in phenotypic sensitivity to the environment (GEI) may necessitate the development of locally adapted varieties (Falconer, 1952). If no one genotype has superiority in all situations, GEI indicates the potential for genetic differentiation of populations under prolonged selection in different environments (Via, 1984).

Non-crossover (quantitative) interaction

These interactions represent changes in magnitude of genotype performance (quantitative), but rank order of genotypes across environments remains unchanged, i.e. genotypes that are superior in one environment maintain their superiority in other environments. Non-crossover interactions may mean that genotypes are genetically heterogeneous but test environments are more or less homogeneous or that genotypes are genetically homogeneous but environments are heterogeneous. All identical genotypes grown in constant (ideal) environments should perform consistently. Any departure from the ideal environment leads to GEI.

Importance of GEI

Thus far, agricultural production has kept pace with the world's population growth mainly because of the innovative ideas and efforts of agricultural researchers. The world population, currently 6 thousand million, is expected to almost double – to 10 thousand million – by the middle of the 21st century (Kang, 2002). The key to doubling agricultural production is increased efficiency in the utilization of resources (increased productivity per hectare and per dollar) and this includes a better understanding of GEI and ways of exploiting it.

The importance of GEI is highlighted by Gauch and Zobel (1996):

> Were there no interaction, a single variety of wheat (*Triticum aestivum* L.) or corn (*Zea mays* L.) or any other crop would yield the most the world over, and furthermore the variety trial need be conducted at only one location to provide universal results. And were there no noise, experimental results would be exact, identifying the best variety without error, and there would be no need for replication. So, one replicate at one location would identify that one best wheat variety that flourishes worldwide.

The importance of GEI can be seen from the relative contributions of new cultivars and improved management to yield increases from direct comparisons of yields of old and new varieties in a single trial (Silvey, 1981). Genetic improvements have been estimated to account for about 50% of the total gains in yield per unit area for major crops during the past 50–60 years (Silvey, 1981; Simmonds, 1981; Duvick, 1992, 1996). The remainder of the yield gain is attributable to improved management and cultural practices. Barley yield data from the UK (1946–1977: mean yield for 1946 = 2.3 t ha^{-1} and for 1977 = 3.9 t ha^{-1}) indicated that the environmental contribution was 10–30% and the genetic contribution 30–60%; the remainder 25–45% of the yield gain was attributed to GEI (Simmonds, 1981). For wheat for the same period (1946–1977: mean yield for 1946 = 2.4 t ha^{-1} and for 1977 = 4.7 t ha^{-1}), yield gain was attributed as follows: 40–60% to the environment (E), 25–40% to the genotype (G) and 15–25% to GEI (Simmonds, 1981). The GEI confounds precise partitioning of the contributions of improved cultivars and improved environment/technology to yield (Silvey, 1981). Thus, the combined contributions of G and genotype–environment (GE) effects can be substantial (40–60% wheat and 70–90% in barley).

GEI occurs during and has an impact on all stages of a breeding programme and has enormous implications for the allocation of resources. A large GEI could mean the establishment of two full-fledged breeding stations in a region, instead of one, thus requiring increased input of resources (manpower, land and money).

Heritability of a trait plays a key role in determining genetic advance from selection. As a component of the total phenotypic variance (the denominator in any heritability equation), GE interaction affects heritability negatively. The larger the GEI component, the smaller the heritability estimate; thus, progress from selection would be limited.

A large GEI reflects the need for testing cultivars in numerous environments (locations and/or years) to obtain reliable results. If the weather patterns and/or management practices differ in target areas, testing must be done at several sites representative of the target areas.

Kang (1993a) discussed the disadvantages of discarding genotypes evaluated in only one environment in early stages of a breeding programme. The discarded genotypes might have the potential to do well at another location or in another year. Thus, some potentially useful genes could be 'lost' due to limited testing. An example from six-row barley illustrates this point well. A total of 288 barley lines were evaluated in the Magreb countries and in International Center for Agricultural Research in Dry Areas (ICARDA)'s yield trials at three locations (Ceccarelli *et al.*, 1994). Of the 103 lines selected at ICARDA and 154 lines at the Magreb, only 49 were selected at both locations.

Performance evaluation is the second component of a breeding programme. Testing done in one environment provides only limited information. Multienvironment testing provides additional useful information, e.g. a GEI component can be estimated. In addition, multienvironment testing yields better estimates of variance components and heritability. Therefore, GEI need not be perceived only as a problem.

As the magnitude of a significant interaction between two factors increases, the usefulness and reliability of the main effects are correspondingly decreased. Since GEI reduces the correlation between phenotypic and genotypic values, the difficulty in identifying truly superior genotypes across environments is magnified.

Obviously, the cost of cultivar evaluation increases as additional testing is carried out. However, with additional test environments, a breeder/agronomist can identify cultivars with specific adaptation as well as those with broad adaptation, which will not be possible from testing in a single environment. Broad adaptation provides stability against the variability inherent in an ecosystem, but specific adaptations may provide a significant yield advantage in particular environments (Wade *et al.*, 1999). Multienvironment testing makes it possible to identify cultivars that perform consistently from year to year (small temporal variability) and those that perform consistently from location to location (small spatial variability). Temporal stability is desired by and beneficial to growers, whereas spatial stability is beneficial to seed companies and

breeders. Stability of performance can be ascertained via stability statistics (Lin *et al.*, 1986; Kang, 1990; Kang and Gauch, 1996).

Achievements

The GEI issue received focused attention in 1990 when an international symposium on 'Genotype-by-Environment Interaction and Plant Breeding' was held on 12 and 13 February at the Louisiana State University campus in Baton Rouge (Kang, 1990). The various GEI issues have come to the forefront in many breeding programmes throughout the world. Reviews and extensive bibliographies (Aastveit and Mejza, 1992; Annicchiarico and Perenzin, 1994; Denis and Gower, 1996; Denis *et al.*, 1996a; Kang, 1998; Piepho, 1998), conference/symposia proceedings (Rao *et al.*, 1988, 1993; Cooper and Hammer, 1996; Zavala-Garcia and Treviño-Hernández, 2000) and books (Gauch, 1992; Prabhakaran and Jain, 1994; Hildebrand and Russell, 1996; Kang and Gauch, 1996; Hoffmann and Parsons, 1997; Basford and Tukey, 1999; Hall, 2001) have since been published. The issue is not only important in plant-breeding programmes but also in animal-breeding programmes (Lin and Lin, 1994; Montaldo, 2001).

GEI presents many challenges for breeders and has significant implications in both applied plant- and animal-breeding programmes. The breeder is faced with developing separate populations for each site type where genotypic rankings drastically change and/or is faced with selecting genotypes that generally perform well across many sites (McKeand *et al.*, 1990). Gains are expected to be greater with the first approach, but costs would also likely be higher; the second approach, while less expensive, yields smaller gains. Denis and Gower (1996) suggested that plant breeders should consider GEI to avoid missing a variety that performed, on average, poorly but did well when grown in specific environments or selecting a variety that, on average, performed well but did poorly when grown in a particular environment.

Table 15.1.   Univariate parametric, univariate non-parametric and multivariate methods compared in Flores *et al.* (1998).

Type	Statistics included
Univariate parametric methods	S2X1: environmental variance (S_{xi}^2) Attributed to Roemer (1917) in Becker and Leon (1988). Measures deviations from the genotypic mean. Minimum variance in different environments = stable genotype
	EBRAS: Eberhart and Russell (1966): optimal yield stability measured via regression represented by high mean yield, moderate to high b_i or responsive to favourable environments, and low deviations from regression (S_{di}^2)
	TAI: regression approach employed by Tai (1971). Uses two statistics, as is the case with Eberhart and Russell (1966): linear response of a genotype to environmental effects (α_i) and deviation from linear response (λ_i) Stable genotype: (α_i, λ_i) = (−1, 1)
	SHUKLA: unbiased estimate of GEI variance attributed to each genotype (σ_i^2) proposed by Shukla (1972) Comparison of σ_i^2 with pooled error (σ_0^2) from ANOVA: a significant F test indicates instability
	CV: Francis and Kannenberg (1978) proposed combined use of yield and CV_i. CV is plotted against mean yield across environments. Low CV and high mean yield = stable, desirable genotype
	PI: superiority measure (P_i) proposed by Lin and Binns (1988) = distance mean square between genotype's response and maximum response in each environment (averaged across environments). Low P_i = high stability
Univariate non-parametric methods	S1O, S2O, S3O and S6O: Hühn (1979) proposed non-parametric stability statistics ($S_i^1, S_i^2, S_i^3, S_i^6$). The lowest value for each statistic represents highest stability S_i^1: the mean of the absolute rank differences of a genotype across environments S_i^2: represents variance of ranks across environments S_i^3 and S_i^6: represent mean rank of each genotype (stability) (formulas differ)
	KANG: Kang (1988) developed a rank-sum method, with genotypes with the highest yield receiving the rank of 1 and the lowest estimated value of σ_i^2 (Shukla's stability variance) receiving the rank of 1. The sum of the two ranks determines the final ranking of genotypes. The genotype with the lowest sum is regarded as most desirable
	KETRANK: proposed by Ketata *et al.* (1989). Plots mean rank across environments against standard deviation of ranks for all genotypes KETYIELD: also proposed by Ketata *et al.* (1989). Plots mean yield across environments against standard deviation of yields for all genotypes A genotype is regarded as stable if its KETRANK or KETYIELD is relatively consistent across environments (low mean rank, i.e. high yield, and a low standard deviation)
	FOXRANK: proposed by Fox *et al.* (1990). This stratified ranking technique was applied to unadjusted means Procedure scores the number of environments in which each genotype ranked in the top, middle and bottom third. Genotypes found in the top third are regarded as well adapted and stable
	STAR: Flores (1993) proposed this method, in which a star is drawn for each genotype; the length of each star represents mean yield in an environment The largest and most regular polygon represents the highest-yielding and most stable genotype

Table 15.1. *Continued.*

Type	Statistics included
Multivariate methods	UPGMA: unweighted pair-group arithmetic mean (Sokal and Michener, 1958). Genotypes are clustered via a dissimilarity matrix of squared Euclidean distance and an UPGMA method fusion strategy If a check cultivar is included in a trial, it can be used as a benchmark for other genotypes in the same trial (Lin and Binns, 1985) Gadheri *et al.* (1980) proposed to conduct an ANOVA at various truncation points and to determine the level at which mean squares within groups are not significantly greater than the estimated error. This method is laborious
	LIN: cluster analysis proposed by Lin (1982). It is a dissimilarity measure for a pair of genotypes (estimated as the squared distance between them adjusted for the average effects of genotypes) Involves calculation of genotype by environment interaction at each fusion cycle
	FOXROS: proposed by Fox and Rosielle (1982). Genotypes are clustered via a dissimilarity matrix of squared Euclidean distance and an incremental sum-of-squares fusion strategy
	AMMI: Zobel *et al.* (1988) proposed the additive main effects and multiplicative interaction (AMMI) model to partition GEI. Provides a biplot and gives information on main effects and interactions of genotypes and environments Genotypes with first principal-component axis value close to zero indicate general adaptation to environments
	PPCC: the principal coordinate analysis proposed by Westcott (1987) Analysis is based on genotype means for each environment, starting with the lowest-yielding environment. Environments are added to the data set in ascending order of yield. Stable and high-yielding genotypes consistently show above-average performance
	CA: correspondence analysis simultaneously represents genotypes and environments. It involves choosing of scores that maximize correlation of rows and columns. A variant of principal-component analysis A genotype is regarded as stable if its first and second correspondence-analysis scores are near zero (Lopez, 1990)
	YIELD: used as a reference

ANOVA, analysis of variance.

Denis *et al.* (1996b) presented a number of models that can account for heteroscedasticity in GE tables. They presented a general scheme for describing heteroscedasticity with a reduced number of parameters using the mixed-model framework, which allows new parsimonious models.

Since the 1970s, various attempts have been made to jointly capture the effects of G and GE interaction. Simultaneous selection for yield and stability of performance is an important consideration in breeding programmes. No methods developed so far have been universally adopted. Flores *et al.*

(1998) compared 22 univariate and multivariate methods to analyse GEI. Additional information on each method is provided in Table 15.1. These 22 methods were classified into three main groups (Flores *et al.*, 1998): in group 1, statistics are mostly associated with yield level and show little or no correlation with stability parameters; in group 2, both yield and stability of performance are considered simultaneously to reduce the effect of GEI; and group 3 emphasizes only stability. Group 1 includes YIELD, PI, UPGMA, FOXRANK and FOXROS; group 2 includes S6O, PPCC, STAR, AMMI and

KANG; and group 3 includes TAI, LIN, CA, SHUKLA and EBRAS.

Hussein *et al.* (2000) provided a comprehensive statistical analysis system (SAS) program for computing univariate and multivariate stability statistics for balanced data. Their program provides estimates of more than 15 stability-related statistics.

Path coefficient analysis has been effectively used to investigate GEI in potato by Tai and Coleman (1999). The path analysis has not found much favour with most researchers. Nevertheless, Tai has expounded on the merits of this method (Tai, 1990).

Piepho (2000b) proposed a mixed-model method to detect QTL with significant mean effect across environments and to characterize the stability of effects across multiple environments. He treated environment main effects as random, which meant that both environmental main effects and QTL–environment interaction effects could be regarded as random.

Biadditive factorial regression models, which encompass both factorial regression and biadditive (additive main effect and multiplicative interaction (AMMI)) models, have also been evaluated (Brancourt-Hulmel *et al.*, 2000). The biadditive factorial regression models involved environmental covariates related to each deviation and included environmental main effect, sum of water deficits, an indicator of nitrogen stress, sum of daily radiation, high temperature, pressure of powdery mildew and lodging (Brancourt-Hulmel *et al.*, 2000). The models explained about 75% of the interaction sum of squares. The biadditive factorial biplot provided relevant information about the interaction of the genotypes with respect to environmental covariates.

The biplot method originated with Gabriel (1971). Others have used this method in describing GEI. The versatility of the GGE (G = genotype effect and GE = genotype–environment effect) biplot has only recently been elucidated (Yan *et al.*, 2000). The GGE biplot approach has captured the imagination of plant breeders and production agronomists like no other approach ever has.

In addition to dissecting genotype-by-environment interactions, GGE Biplot helps analyse genotype-by-trait data, genotype-by-marker data, and diallel cross data (Yan *et al.*, 2000, 2001; Yan, 2001; Yan and Hunt, 2001, 2002; Yan and Rajcan, 2002). These aspects make the GGE biplot a most comprehensive tool in quantitative genetics and plant breeding (see Yan and Hunt, Chapter 19, this volume).

Causes of Genotype–Environment Interaction

To be able to understand GEI and utilize it effectively in breeding programmes, information is needed on the factors responsible for the differential response of genotypes to variable environments. A factor may be present at optimal, suboptimal or superoptimal levels. When present at a level other than optimal, it represents a stress. According to Baker (1988), differences in the rate of increase in response of genotypes at suboptimal levels would reflect differences in efficiency, and differences in the rate of decrease at superoptimal levels would reflect differences in tolerance. Without the presence of stresses, genotype attributes, such as efficiency and tolerance, cannot be identified and investigated. In this section, the effects of environmental stress on the plant genome in general and biotic and abiotic factors that may be responsible for GEI are considered.

Environmental effect on genome

An understanding of plant stress responses is essential because of predicted global environmental changes and their impact on the production of food and fibre. Stress is a physiological response to an adverse environmental factor(s). Plants respond to a variety of environmental cues: nutrients, toxic elements and salts in the soil solution, gases in the atmosphere, light of different wavelengths, mechanical stimuli, gravity, wounding, pests, pathogens and symbionts

(Crispeels, 1994). Plants have incorporated a variety of environmental signals into their developmental pathways that have provided for their wide range of adaptive capacities over time (Scandalios, 1990).

Environmental stresses have been shown to elicit specific responses at the DNA level in a number of organisms. A differentiated cell expresses an array of genes required for its stable functioning and metabolic roles (Scandalios, 1990). In response to severe environmental changes, a genome can respond by selectively regulating (increasing or decreasing) the expression of specific genes.

Interspecific variation in DNA amounts is correlated with various quantitative properties of cells, and these may secondarily affect the quantitative characters of the whole plant (Bachmann *et al.*, 1985; Cavalier-Smith, 1985a,b; Bennett, 1987). Highly significant differences of up to 32% in DNA content were found in meristems of seedlings from 35 natural populations of hexaploid *Festuca arundinacea* (Ceccarelli *et al.*, 1992). In cultivated maize, variation in genome size has been reported to be as high as 38.8% (Laurie and Bennett, 1985; Rayburn *et al.*, 1985). Maize lines from higher latitudes of North America had significantly lower nuclear DNA amounts than those from lower latitudes (Rayburn *et al.*, 1985). Rayburn and Auger (1990) determined the nuclear DNA content of 12 southwestern US maize populations collected at various altitudes and observed a significant positive correlation between genome size and altitude. Higher amounts of DNA at higher elevation have also been found in teosinte (Laurie and Bennett, 1985).

Herrera-Estrella and Simpson (1990) investigated the influences of environmental factors on genes involved in photosynthesis. The mechanism of regulation may vary from one species to another (Herrera-Estrella and Simpson, 1990).

Biotic stresses

Biotic stress factors are a major limitation to plant productivity and a dominant element in plant ecology and evolution (Higley *et al.*, 1993). Biotic stresses and interactions among them and/or with abiotic factors remain poorly understood; however, they have significant relevance to GEI in plants.

Plants may respond to pathogen infection by inducing a long-lasting, broad-spectrum, systemic resistance to subsequent infections (Ryals *et al.*, 1994). Induced disease resistance has been referred to as physiological acquired immunity, induced resistance or systemic acquired resistance (SAR). Differences in insect and disease resistance among genotypes can be associated with stable or unstable performance (Baker, 1990).

Abiotic stresses

The major abiotic stresses are atmospheric pollutants, soil stresses (salinity, acidity and mineral toxicity and deficiency), temperature (heat and cold), water (drought and flooding) and tillage operations (Blum, 1988; Clark and Duncan, 1993; Specht and Laing, 1993; Unsworth and Fuhrer, 1993). Genetic variation exists for plant responses to the above stress factors. Breeding for tolerance to air pollutants has considerable potential (Unsworth and Fuhrer, 1993).

With stress caused by suboptimal levels of water, nutrients and solar radiation, it should be possible to identify genotypes that are efficient or inefficient in using the respective resource. Woodend and Glass (1993) demonstrated the presence of GEI for potassium-use efficiency in wheat.

Responses to temperature

Rapid temperature changes, particularly those toward the upper end of the adaptation range for individual plant species, can produce dramatic changes in the pattern of gene expression. Heat-shock responses are plants' protective measures against potentially lethal, rapid-rate, upward departures from the optimal temperature (Pollack *et al.*, 1993). Tolerance of protein synthesis and seedling growth to a previously lethal high temperature can be induced by prior short exposure to a sublethal high temperature that triggers the synthesis of a specific set of proteins – the heat-shock proteins (HSPs) – via mRNA that is newly transcribed in response to high temperature. In the meantime, the synthesis of normal cellular proteins is reduced or shut down. This process is detectable within minutes of the onset of stress (Ougham and Howarth, 1988). HSPs are induced at different temperatures in different species. The rule of thumb is that temperature must be ~10°C higher than the optimal temperature for a particular species.

Oxidative stress

A common feature of different stress factors is an increased production of reactive oxygen species in plant tissues, but their mode of action varies depending on whether oxidants are generated outside (e.g. by oxidizing air pollutants) or inside (e.g. high radiation, low temperatures or nutrient deficiency) a plant cell (Polle and Rennenberg, 1993). It is important to understand both the mode of action of different stress factors and the critical physiological properties that limit ameliorative mechanisms at the subcellular level (Polle and Rennenberg, 1993).

Scandalios (1990) summarized plant responses to environmental stress, pointing out that activated oxygen species (endogenous – by-products of normal metabolism – and exogenous – triggered by environmental factors) were highly reactive molecules capable of causing extensive damage to plant cells. The effects of oxidative stress can range from simple inhibition of enzyme function to the production of random lesions in proteins and nucleic acids and the peroxidation of membrane lipids. Loss of membrane integrity can cause decreased mitochondrial and chloroplast functions, which, in turn, can lower the plant's ability to fix carbon and to properly utilize the resulting products (Scandalios, 1990). This decrease in metabolic efficiency results in reduced yield.

How to Deal with GEI

The presence of crossover interactions has important implications for breeding strategies that aim to improve either broad or specific adaptation or some combination of both components of adaptation (Cooper et al., 1999). Eisemann et al. (1990) listed three ways of dealing with GEI in a breeding programme: (i) ignoring them, i.e. using genotypic means across environments even when GEI exists; (ii) avoiding them; or (iii) exploiting them. Interactions should not be ignored when they are significant and of the crossover type.

The second way of dealing with these interactions, i.e. avoiding them, involves minimizing the impact of significant interactions. One approach is to group similar environments (forming mega-environments) via a cluster analysis. With environments being more or less homogeneous, genotypes evaluated in them would not be expected to show crossover interactions. By clustering environments, potentially useful information may be lost. International research centres, such as the International Maize and Wheat Improvement Center (CIMMYT), aim to identify maize and wheat genotypes with broad adaptation (i.e. stable performance across diverse environments) at many international sites. Such an objective cannot be achieved by restricting (clustering) test environments.

The third approach encompasses stability of performance across diverse environments by analysing and interpreting genotypic and environmental differences. This approach allows researchers to select genotypes with consistent performance, identify the causes of GEI and provide the opportunity to correct the problem. When the cause for the unstable performance of a genotype is known, either the genotype can be improved by genetic means or a proper environment (inputs and management) can be provided to enhance its productivity.

A genotype that performs consistently (high-yielding) across many environments would possibly possess broad-based, durable resistances/tolerances to the biotic and abiotic environmental factors that it encountered during development. The more the breeders know about the crop environment, the better job they can do of judiciously targeting appropriate cultivars to production environments.

In the next section, the concepts of stability are presented. A methodology for identifying stable genotypes and environmental factors that may be responsible for stable or unstable performance is also given.

Concepts of stability

Stability is a central keyword for plant breeders analysing GE data. A simple corresponding statistical term is 'dispersion around a central value' (Denis *et al.*, 1996a). There are two concepts of stability: static and dynamic. The static concept means that a genotype has a stable performance across environments and there is no among-environment variance. This would mean that a genotype would not respond to high levels of inputs, such as fertilizer. This type of stability would not be beneficial for the farmer, and it has been referred to as the biological concept of stability (Becker, 1981), which is equivalent to Lin *et al.*'s (1986) type 1 stability. In type 1 stability, a genotype is regarded as stable if its among-environment variance is small.

The dynamic concept means that a genotype has a stable performance, but, for each environment, its performance corresponds to the estimated level or predicted level. There would be agreement between the estimated or predicted level and the level of actual performance (Becker and Leon, 1988). This concept has been referred to as the agronomic concept (Becker, 1981), which is equivalent to Lin *et al.*'s (1986) type 2 stability. In type 2, a genotype is regarded as stable if its response to environments is parallel to the mean response of all genotypes in a test.

Lin *et al.* (1986) defined four groups of stability statistics. Group A is based on deviation from the average genotype effect (DG), group B on the GEI term (GEI) and groups C

and D on either DG or GEI. The formulae of groups A and B represent sums of squares and those of groups C and D represent a regression coefficient or deviation from regression. They integrated type 1, type 2 and type 3 stabilities with the four groups: group A was regarded as type 1, groups B and C as type 2, and group D as type 3 stability. In type 3 stability, a genotype is regarded as stable if the residual mean square from the regression model on the environmental index is small (Lin *et al.*, 1986). Lin and Binns (1988) proposed the type 4 stability concept on the basis of predictable and unpredictable non-genetic variation: the predictable component is related to locations and the unpredictable component is related to years. Lin and Binns (1988) suggested the use of a regression approach for the predictable portion and the mean square for years within locations for each genotype as a measure of the unpredictable variation. The latter was called the type 4 stability statistic.

Stability statistics

Plant breeding can exploit wide adaptation by selecting genotypes that yield well across large geographical areas or mega-environments (Witcombe, 2001). Mega-environments are broad (frequently discontinuous transcontinental) areas that are characterized by similar biotic and abiotic stresses, cropping-system requirements and consumer preferences (Witcombe, 2001). Several methods have been developed to analyse GEI and to select genotypes that perform consistently across many environments (Lin *et al.*, 1986; Becker and Leon, 1988; Kang, 1990; Kang and Gauch, 1996; Weber *et al.*, 1996). The earliest approach was the linear regression analysis (Mooers, 1921; Yates and Cochran, 1938). Finlay and Wilkinson (1963), Eberhart and Russell (1966) and Tai (1971) popularized variations of the regression approach, assuming an expected linear response of yield to environments. The merits and demerits of several methods were discussed by Kang and Miller (1984). Kang *et al.* (1987b) concluded that Shukla's (1972) stability variance and Wricke's (1962) ecovalence were equivalent methods and they ranked genotypes identically for stability (rank correlation coefficient = 1.00). These types of measures are useful to breeders and agronomists, as they provide the contribution of each genotype to total GEI. They can also be used to evaluate testing locations by identifying those locations with a similar GEI pattern (Glaz *et al.*, 1985). Other statistical methods that have received significant attention are pattern analysis (DeLacy *et al.*, 1996), the AMMI model (Gauch and Zobel, 1996), the shifted multiplicative model (SHMM) (Cornelius *et al.*, 1996; Crossa *et al.*, 1996), the non-parametric methods of Hühn (1996), which are based on cultivar ranks, the probability of outperforming a check (Eskridge, 1996) and Kang's rank-sum method (Kang, 1988, 1993b). The methods of Hühn (1996) and Kang (1988, 1993b) integrate yield and stability into one statistic that can be used as a selection criterion.

Dashiell *et al.* (1994) evaluated the usefulness of several stability statistics for simultaneously selecting for high yield and stability of performance in soybean. Fernandez (1991) also evaluated stability statistics for similar purposes. Recently, Flores *et al.* (1998) and Hussein *et al.* (2000) conducted comparative evaluations of 22 and 15 stability statistics/methods, respectively.

Simultaneous selection for yield and stability

Growers would prefer to use a high-yielding cultivar that performs consistently from year to year (temporal adaptation) and might be willing to sacrifice some yield if they are guaranteed, to some extent, that a cultivar would produce consistently from year to year (Kang *et al.*, 1991).

Kang (1993b) discussed the motivation for emphasizing stability in the selection process. He enumerated the consequences to growers of researchers' committing type I (rejecting the null hypothesis when it is true) and type II errors (accepting the null hypothesis when it is false) relative to selection on the basis of yield alone (conventional method (CM)) and that on the basis of yield and stability. Simultaneous selection for yield and stability reduces the probability of committing type II errors (probability = β). Generally, type II errors constitute the most serious risk for growers (Glaz and Dean, 1988; Johnson *et al.*, 1992). The combined rate of committing a type II error for simultaneous selection for yield and stability will be the product of β for comparisons of overall yield mean and β for comparisons of GEI means.

Several methods of simultaneous selection for yield and stability and relationships among them were discussed by Kang and Pham (1991). The development and use of the yield–stability statistic (YS_i) demonstrated the significance and rationale of incorporating stability in selecting genotypes tested across a range of environments (Kang, 1993b). A QBASIC computer program (STABLE) for calculating this statistic was developed and is available free of charge (Kang and Magari, 1995).

The stability component in YS_i is based on Shukla's (1972) stability-variance statistic (σ_i^2). Shukla (1972) partitioned GEI into components, one corresponding to each genotype, and referred to it as stability variance. Lin *et al.* (1986) classified σ_i^2 as type 2 stability, meaning that it was a relative measure dependent on genotypes included in a particular test. Pazdernik *et al.* (1997) analysed soybean seed yield, protein and oil concentrations and stability statistics. They concluded that Hühn's rank-based S_i^1 and S_i^2 statistics and Kang's YS_i statistic could be used by breeders to select parents to improve protein concentration and stability by combining stable high-yielding lines with stable high-protein lines. They further suggested that the same statistics could be used by consultants and variety-testing personnel to aid in making recommendations to soybean producers.

Covariates and stability

Yield stability or GEI for yield is a complex issue. Yield stability depends on plant characteristics, such as resistance to pests and tolerance to environmental stress factors. By determining factors responsible for GEI or stability/instability, breeders can improve cultivar stability. If instability was caused by susceptibility to a disease, breeding for resistance to that disease should reduce losses in disease-inducing environments and increase genotype stability.

It is important to know the underlying causes of GEI (Kang, 1998; Haji and Hunt, 1999). An observational description of GEI is not very useful unless one knows the elements that cause the environmental differentiation (Federer and Scully, 1993). The use of environmental variables as covariates was suggested and/or employed by several researchers (Freeman and Perkins, 1971; Hardwick and Wood, 1972; Shukla, 1972; Wood, 1976; Kang and Gorman, 1989; van Eeuwijk *et al.*, 1996; Piepho *et al.*, 1998). Individual components of the environment (rainfall, temperature, fertility, etc.), used as covariates in explaining GEI, can greatly increase the reliability of predictions relative to cultivar performance. Environmental characterization can be achieved directly, by measuring environmental variables, which can be physical, biological or nutritional, or indirectly, by measuring plant responses to capture the influence of environmental conditions on plant performance (Brancourt-Hulmel *et al.*, 2000). Winter-wheat data from Ontario revealed that January temperatures, together with moisture supply before anthesis, were associated with some of the GEIs (Haji and Hunt, 1999).

A fertility score was used as an environmental covariate in Germany (Piepho, 2000a). This score, ranging between 0 and 100, incorporates several variables, including soil type and the geological age of parent material. Piepho (2000a), who provided confidence limits for estimated risks, argued that, if yield depends on environmental covariates, risk for a specific environment can be estimated on the basis of covariate information, thus yielding a more specific risk assessment.

Methods of assessing the contributions of weather variables and other factors (covariates) that contribute to GEI are available (Shukla, 1972; Denis, 1988; van Eeuwijk *et al.*, 1996; Magari *et al.*, 1997). Contributions of different environmental variables to GEI have been reported by several researchers (Saeed and Francis, 1984; Gorman *et al.*, 1989; Kang and Gorman, 1989; Kang *et al.*, 1989; Rameau and Denis, 1992). Additional reports have appeared on the use of covariates in the past 5 years (Magari *et al.*, 1997; Vargas *et al.*, 2001; Yan and Hunt, 2001).

In the following linear model, GEI is explained in terms of the covariate used, as shown by Shukla (1972):

$$Y_{ijk} = \mu + \alpha_i + \theta_{ij} + \beta_k + b_k z_i + \varepsilon_{ijk} \quad (15.1)$$

where Y_{ijk} = observed trait value, μ = grand mean, α_i = environmental effect, θ_{ij} = blocks within environments effect, β_k = cultivar effect, b_k = regression coefficient of the kth genotype's yield in different environments, z_i = an environmental covariate and ε_{ijk} = experimental error.

When a number of environmental variables are considered, the combination of two or more variables would remove more heterogeneity from GEI than individual variables do. Methods developed by van Eeuwijk *et al.* (1996) may be helpful for this purpose. Magari *et al.* (1997) identified precipitation as the single most important environmental factor that contributed to GEI for ear-moisture loss rate in maize. They identified precipitation + growing degree-days from planting to black-layer maturity (GDD-BL) and relative humidity + GDD-BL as the two-factor combinations that explained larger amounts of GEI.

Vargas *et al.* (2001) found the most important variables that explained nitrogen (N)–year interaction to be minimum temperature in January–March and maximum temperature in April. Evaporation rates for December and April were important covariates for describing tillage–year and summer crop–year interactions, whereas precipitation in December and sun hours in February explained year–manure interaction (Vargas *et al.*, 2001).

Stability variance for unbalanced data

Plant breeders often deal with unbalanced data. Searle (1987) classified unbalancedness as planned unbalanced data and missing observations. Both categories of unbalancedness may occur, but planned unbalancedness (a situation when, for different reasons, one does not have data for all genotypes in all environments) is more difficult to handle. Researchers have used different approaches for studying GEI in unbalanced data (Freeman, 1975; Pedersen *et al.*, 1978; Zhang and Geng, 1986; Gauch and Zobel, 1990; Rameau and Denis, 1992; Piepho, 1994). Usually environmental effects are considered as random and cultivar effects as fixed. Inference on random effects using least squares, in the case of unbalanced data, is not appropriate because information on variation among random effects is not incorporated (Searle, 1987). For this reason, mixed model equations (MMEs) are recommended (Henderson, 1975).

The values of Shukla's (1972) σ_i^2 can be negative because they are calculated as the differences of two statistically dependent sums of squares, which is a negative feature of this approach. Computation of σ_i^2 is impossible from unbalanced data, but genotype$_k$–environment variance components ($\sigma^2_{g(k)e}$) can be estimated using the maximum likelihood approach. The general linear model for randomized complete-block-design experiments conducted in different environments is:

$$Y_{ijk} = \mu + \alpha_i + \theta_{ij} + \beta_k + \gamma_{ik} + \varepsilon_{ijk} \quad (15.2)$$

Using matrix notation, Equation (15.2) can be written as:

$$y = 1\mu + X\beta + W\alpha + U\theta + \Sigma_k Z_k a_k + \varepsilon \quad (15.3)$$

where y = vector of observed yield data, 1 = vector of ones, X = design matrix for fixed effects (genotypes), β = vector of genotype effects, W and α are, respectively, a design matrix for and a vector of environmental effects, U and θ are, respectively, a design matrix for and a vector of replications within environment effects, Z_k and a_k are, respectively, a design matrix for and a vector of GEI effects and ε is the vector of residuals. Equation (15.3) can be solved using Henderson's (1975) MME. The levels of random factors are generally assumed to be independent.

The restricted maximum-likelihood (REML) methodology is generally preferred to maximum-likelihood estimates because it considers the degrees of freedom for fixed effects for calculating error. The calculation of REML stability variances for unbalanced data allows one to obtain a reliable estimate of stability parameters, and overcomes the difficulties of manipulating unbalanced data (Kang and Magari, 1996).

Testing and Breeding Strategies

The best approach for breeders and geneticists would be to understand the nature and causes of GEI and to try to minimize its deleterious implications and exploit its beneficial potential through appropriate breeding, genetic and statistical methodologies (Kang and Gauch, 1996). Appropriate analyses of data can provide an opportunity for exploiting GEI using applied analytical methods, such as AMMI, the use of climatic factors in explaining GEI and the evaluation of risk of production and the optimal allocation of land resources to various genotypes for selection in heterogeneous environments (Singh *et al.*, 1999). Some of the important strategies for accomplishing this are outlined below.

Breeding for resistance/tolerance to stresses

Resistance or tolerance to any type of stress, biotic or abiotic, is essential for stable performance (Khush, 1993; Duvick, 1996). Sources of increased crop productivity include enhanced yield potential, heterosis, modified plant types, improved yield stability, gene pyramiding and exotic and transgenic germ-plasm (Khush, 1993). It is important to identify the factor(s) that are responsible for GEI. If interaction is caused by European corn-borer (ECB) damage, a gene conferring resistance to ECB (e.g. the *Bt* gene) could be inserted into one of the two inbred parents of the susceptible genotype.

Brancourt-Hulmel (1999) used crop diagnosis with the analysis of interaction by factorial regression in wheat. She provided an agronomic explanation of GEI and defined responses or parameters for each genotype and each environment. Earliness at heading, susceptibility to powdery mildew, and susceptibility to lodging were the major factors responsible for GEI. In the same study (Brancourt-Hulmel, 1999), factorial regression revealed that water deficits during the formation of grain number and N level also were associated with GEI.

To alleviate GEI concerns caused by stresses, breeders need to know as much about the various characteristics of genotypes as possible. They also need to characterize environments as fully as possible. Knowledge of soil characteristics and ranges of weather variables and stresses that plant materials will be exposed to is a prerequisite to exploiting the beneficial potentials of the genotypes and environments and to targeting appropriate cultivars to specific environments.

Economically important characters in crop species are generally quantitative in nature. For improving quantitative traits, breeders need to know what genetic factors are involved, where they are located on chromosomes and what type of inheritance they exhibit. Recent advances in molecular genetics have provided some of the best tools for obtaining insights into the molecular mechanisms associated with GEI. Molecular markers, such as restriction fragment length polymorphisms (RFLPs), can be employed to find genomic regions with stable responses. Molecular markers have paved the way for investigating the QTL–environment interaction (QEI) (Beavis and Keim, 1996), which will ultimately provide a better genetic understanding and possible regulation of this phenomenon. Regions of plant genomes that provide stable responses across diverse environments can be identified by determining the linkage of QTL to RFLPs, which should make it possible for breeders to manipulate QTL in the same fashion as single genes that control qualitative traits. Wang *et al.* (1999) reported a new methodology based on mixed linear models to map QTL with digenic epistasis and QEIs. Reliable estimates of QTL main effects (additive and epistatic effects) can be obtained with the maximum-likelihood estimation method, and QEI effects (additive–environment interaction and epistatic effects–environment interaction) can be obtained with the best linear unbiased prediction (BLUP) method (Wang *et al.*, 1999).

It is highly desirable to identify QTL for a complex trait (say, high yield) that is expressed in a number of environments. Crossa *et al.* (1999) found that higher maximum temperature in low- and intermediate-altitude sites affected the expression of some QTL, whereas minimum temperature affected the expression of other QTL, in tropical maize. Jiang *et al.* (1999) used molecular markers to investigate adaptation differences between highland and lowland tropical maize. They concluded that breeding for broad thermal adaptation should be possible by pooling genes showing adaptation to specific thermal regimes, albeit at the expense of reduced progress for specific adaptation. Molecular marker-assisted selection would be an ideal tool for this task because it could reduce linkage drag caused by the unintentional transfer of undesirable traits (Jiang *et al.*, 1999).

Breeding for stability/reliability of performance

Evans (1993) pointed to the need for developing new cultivars with broad adaptation to a number of diverse environments (selection for adaptability) and to the need of farmers to use new cultivars with reliable or consistent performance from year to year (reliability). Smith *et al.* (1990) pointed out that genetic improvement for low-input conditions would require capitalizing on GEI and that slower or limited gains in low-input or stress environments suggested that conventional high-input management of breeding nurseries and evaluation trials might not effectively select genotypes with improved performance at low-input levels. This viewpoint is also highlighted by Ceccarelli *et al.* (2001). Because of the successes in favourable environments, plant breeders have tried to solve the problems of poor farmers living in unfavourable environments by simply extending the same methodologies and philosophies applied to favourable, high-potential environments, without considering the possible limitations associated with the presence of a large GEI (Ceccarelli *et al.*, 2001). Selection in good environments is favoured because it is believed that heritabilities are higher there than in poor environments (Blum, 1988). Singh and Ceccarelli (1995) suggested, however, that there was no relationship between yield level and magnitude of heritability. Rosielle and Hamblin (1981) examined theoretical aspects of selection for yield in stress and non-stress environments. They showed that selection for tolerance to stress generally reduced mean yield in non-stress environments and that selection for mean productivity generally increased mean yields in both stress and non-stress environments. Bramel-Cox (1996) reviewed relevant literature on breeding for reliability of performance in unpredictable environments.

To be reliable, a stability statistic must be based on a large number of environments (more than ten). Information on stability can usually be obtained in the final stages of a breeding programme, when replicated tests are conducted. From the standpoint of individual growers, stability across years (temporal) is most important. A breeder could test cultivars or lines for 10–15 years and identify those that have temporal stability. Crosses could then be made among the most stable cultivars to develop source material (germ-plasm) that would be utilized for developing inbred lines or pure lines.

Therefore, extensive cultivar testing across years is a precursor to cultivar development.

Stability of cultivars would be enhanced if multiple resistances/tolerances to stress factors were incorporated into the germ-plasm used for cultivar development. If every cultivar (different genotypes) possessed equal resistance/tolerance to every major stress encountered in diverse target environments, GEI would be reduced. Conversely, if genotypes possessed differential levels of resistance (a heterogeneous group) and, somehow, we could make all target environments as homogeneous as possible, GEI would again be reduced. Since we do not have any control over unpredictable environments from year to year, the best approach would be the former.

Stability analyses can be used to identify durable resistance to disease pathogens (Jenns *et al.*, 1982). If a cultivar–pathogen-isolate interaction exists, it would be necessary to identify a cultivar that has general resistance instead of specific resistance.

Kang *et al.* (1987a) examined whether stability of one trait was correlated with stability of another trait. If the stability (stability variance, ecovalence or any other stability statistic) of two traits were reasonably well, positively correlated, concurrent selection for stabilities of the two traits might be possible.

Measure interaction at intermediate growth stages

A crop is exposed to variable environmental factors throughout the growing season. Generally, researchers investigate the causes of GEI at the final harvest stage. To critically investigate GEI, one may need to record environmental variables and plant-growth measurements at weekly intervals. This would help determine what effect, if any, the environmental variables from an earlier period had on GEI at intermediary stages and on final yield. This may provide a better understanding of the dynamic process of yield formation.

Early multienvironment testing

Usually, there is a shortage of seed at the earliest stages of breeding, which prevents extensive testing. However, in a clonally propagated crop, such as sugarcane or potato, one stalk of sugarcane or one tuber of potato can be divided into at least two pieces and planted in more than one environment. Similarly, in other crops, if only 20 kernels are available, one could plant ten seeds each in two diverse environments. In the absence of a GEI, one would obtain a better evaluation of the genotypes, but, if GEI was present, one would obtain information about the consistency or inconsistency of performance of genotypes early in the programme. This strategy would prevent gene loss or genetic erosion, which could occur if testing was done in only one environment, and would also result in an increased breeding effort without a corresponding increase in expenditure of resources.

Optimal resource allocation

GEI can be employed to judiciously allocate resources in a breeding programme (Pandey and Gardner, 1992; Magari *et al.*, 1996). Carter *et al.* (1983) estimated that, at a low level of treatment–environment interaction (10% of error variance), testing in at least two environments was necessary to detect treatment differences of 20% and it required at least seven environments to detect smaller (10%) treatment differences for growth-analysis experiments in soybean. With a larger magnitude of interaction, a larger number of environments would be needed for a given level of precision in treatment differences.

Magari *et al.* (1996) used multienvironment (different planting dates) data for ear-moisture loss rate in maize, which exhibited planting-date–genotype interaction. The relative efficiency for the benchmark protocol (11 plants per replication, three replications and three planting dates) was regarded as the reference value (100%). The relative efficiency for five plants per plot in four replications and three planting dates was equivalent to that for the benchmark protocol. A relative efficiency of 100% could also be achieved with a sample of four planting dates, three replications and three to four plants per plot. When the number of replications was increased to four in each of four planting dates, only two plants per plot were needed to achieve a relative efficiency of 100%. The number of planting dates (environments) was found to be a critical factor in determining the precision of an experiment.

Future Prospects

With the increasing impact of molecular biology on breeding and genetics, we are at the dawn of molecular plant breeding. Molecular approaches are being incorporated at various levels in crop-breeding programmes. I expect that molecular biology (including molecular genetics, biochemistry and plant physiology) will play an enhanced role in breeding crop species and overcoming the constraints imposed on genotypes by their interaction with environmental factors. For example, cloning of genes for cold tolerance obtained from cold-tolerant plant species and insertion of these genes into cold-sensitive crop species could overcome stress imposed by a cold climate on the latter. Physiology will also increase our knowledge of signal transduction in plants in response to environmental cues.

The area of QEI has seen vigorous growth in the past 10 years. This issue is expected to continue to expand and further investigations will contribute to our understanding of the complex relationship between crop performance and the environment. Theory for QEI has lagged behind, but progress is being made in this area (Van Eeuwijk *et al.*, Chapter 16, this volume). Applied research, such as that of Moutiq *et al.* (Chapter 17, this volume), will contribute much to the understanding of the QEI.

Statistical models for handling GEI data, especially mixed-model analyses, are being advanced. The work of Smith *et al.* (Chapters

21 and 22, this volume) and Crossa and Cornelius (Chapter 20, this volume) on AMMI and sites regression model (SREG), of Van Eeuwijk *et al.* (Chapter 16, this volume) on regression and of Balzarini (Chapter 23, this volume) on mixed models (best linear unbiased estimate (BLUE) and BLUP) represent important trends for the future.

The GGE biplot technique, which has been popularized by Yan and Hunt (Chapter 19, this volume), is a versatile statistical/ quantitative genetic methodology. GGE biplot methods not only dissect GEI and QEI but also aid in analysing genotype–trait data, genotype–marker data and diallel-cross data. The use of this methodology is expected to expand worldwide in the next decade.

Annicchiarico (Chapter 24, this volume) proposes the use of artificial environments to aid in selecting for adaptability. According to him, assessing the value of a specific adaptation strategy has an obvious interest for the globally orientated breeding programmes of large seed companies or international research centres. Fitting cultivars to an environment, instead of modifying the environment to fit widely adapted cultivars, and safeguarding the biodiversity of cultivated material can contribute to food security and enhance the effect of this strategy by integrating participatory breeding schemes.

To ensure the stability of crop production, the basic crop germ-plasm pools would need to be broadened (Sperling *et al.*, 2001). I expect that there would be a greater emphasis on participatory plant breeding, which involves scientists, farmers, consumers, extension personnel, industry and others, in the future. The role of participatory plant breeding is expected to expand, especially in developing countries. This should help broaden the genetic base of crops and stabilize food production as a result of farmers' developing, identifying and using locally adapted crop varieties that are farmer-acceptable and farmer-accessible.

Participatory plant breeding has not received as much emphasis in Africa as it has in Asia (Cromwell and Van Oosterhout, 2000), but it will receive an impetus in the 21st century because it seeks to deliver planting material that is closely in line with farmers' needs, more quickly than is possible through conventional plant breeding. Participatory plant breeding can make a real contribution to supporting farmers' efforts to maintain a wide range of crop varieties on-farm (Cromwell and Van Oosterhout, 2000). Duvick (2002) envisions that the farmer-breeders (acting either as individuals or in associations, such as communities) and their non-governmental organization (NGO) partners would produce varieties with utility in farming systems that are not well served (or not served at all) by formal plant breeding, either public or private. Decentralization (participatory plant breeding) is essential for exploiting specific adaptation fully and making positive use of GEI (Ceccarelli *et al.*, 2001).

References

Aastveit, A.H. and Mejza, S. (1992) A selected bibliography on statistical methods for the analysis of genotype × environment interaction. *Biuletyn Oceny Odmian* 25, 83–97.

Annicchiarico, P. and Perenzin, M. (1994) Adaptation patterns and definition of macroenvironments for selection and recommendation of common wheat genotypes in Italy. *Plant Breeding* 113, 197–205.

Bachmann, K., Chambers, K.L. and Price, H.J. (1985) Genome size and natural selection: observations and experiments in plants. In: Cavalier-Smith, T. (ed.) *The Evolution of Genome Size.* John Wiley & Sons, Chichester, UK, pp. 267–276.

Baker, R.J. (1988) Differential response to environmental stress. In: Weir, B.S., Eisen, E.J., Goodman, M.M. and Namkoong, G. (eds) *Proceedings of the Second International Conference on Quantitative Genetics.* Sinauer, Sunderland, Massachusetts, pp. 492–504.

Baker, R.J. (1990) Crossover genotype–environmental interaction in spring wheat. In: Kang, M.S. (ed.) *Genotype-by-Environment Interaction and Plant Breeding.* Louisiana State University Agricultural Center, Baton Rouge, Louisiana, pp. 42–51.

Basford, K.E. and Tukey, J.W. (1999) *Graphical Analysis of Multiresponse Data: Illustrated with a Plant Breeding Trial.* Chapman & Hall/ CRC, Boca Raton, Florida.

Beavis, W.D. and Keim, P. (1996) Identification of quantitative trait loci that are affected by environment. In: Kang, M.S. and Gauch, H.G., Jr (eds) *Genotype-by-Environment Interaction*. CRC Press, Boca Raton, Florida, pp. 123–149.

Becker, H.C. (1981) Correlations among some statistical measures of phenotypic stability. *Euphytica* 30, 835–840.

Becker, H.C. and Leon, J. (1988) Stability analysis in plant breeding. *Plant Breeding* 101, 1–23.

Bennett, M.D. (1987) Variation in genomic form in plants and its ecological implications. *New Phytologist* 106 (Suppl.), 177–200.

Blum, A. (1988) *Plant Breeding for Stress Environments*. CRC Press, Boca Raton, Florida.

Bramel-Cox, P.J. (1996) Breeding for reliability of performance across unpredictable environments. In: Kang, M.S. and Gauch, H.G., Jr (eds) *Genotype-by-Environment Interaction*. CRC Press, Boca Raton, Florida, pp. 309–339.

Brancourt-Hulmel, M. (1999) Crop diagnosis and probe genotypes for interpreting genotype environment interaction in winter wheat trials. *Theoretical and Applied Genetics* 99, 1018–1030.

Brancourt-Hulmel, M., Denis, J.-B. and Lecomte, C. (2000) Determining environmental covariates which explain genotype environment interaction in winter wheat through probe genotypes and biadditive factorial regression. *Theoretical and Applied Genetics* 100, 285–298.

Carter, T.E., Jr, Burton, J.W., Cappy, J.J., Israel, D.W. and Boerma, H.R. (1983) Coefficients of variation, error variances, and resource allocation in soybean growth analysis experiments. *Agronomy Journal* 75, 691–696.

Cavalier-Smith, T. (1985a) Eucaryote gene number, non-coding DNA and genome size. In: Cavalier-Smith, T. (ed.) *The Evolution of Genome Size*. John Wiley & Sons, Chichester, UK, pp. 69–103.

Cavalier-Smith, T. (1985b) Cell volume and the evolution of eukaryotic genome size. In: Cavalier-Smith, T. (ed.) *The Evolution of Genome Size*. John Wiley & Sons, Chichester, UK, pp. 105–184.

Ceccarelli, M., Falistocco, E. and Cionini, P.G. (1992) Variation of genome size and organization within hexaploid *Festuca arundinacea*. *Theoretical and Applied Genetics* 83, 273–278.

Ceccarelli, S., Erskine, W., Hamblin, J. and Grando, S. (1994) Genotype by environment interaction and international breeding programmes. *Experimental Agriculture* 30, 177–187.

Ceccarelli, S., Grando, S., Amri, A., Asaad, F.A., Benbelkacem, A., Harrabi, M., Maatougui, M., Mekni, M.S., Mimoun, H., El-Einen, R.A., El-Felah, M., El-Sayed, A.F., Shreidi, A.S. and Yahyaoui, A. (2001) Decentralized and participatory plant breeding for marginal environments. In: Cooper, H.D., Spillane, C. and Hodgkins, T. (eds) *Broadening the Genetic Bases of Crop Production*. CAB International, Wallingford, UK, pp. 115–135.

Clark, R.B. and Duncan, R.R. (1993) Selection of plants to tolerate soil salinity, acidity, and mineral deficiencies. In: Bruxton, D.R., Shibles, R., Forsberg, R.A., Blad, B.A., Asay, K.H., Paulsen, G.M. and Wilson, R.F. (eds) *International Crop Science I*. Crop Science Society of America, Madison, Wisconsin, pp. 371–379.

Cooper, M. and Hammer, G.L. (eds) (1996) *Plant Adaptation and Crop Improvement*. CAB International, Wallingford, UK, ICRISAT, Patancheru, India, and IRRI, Manila, The Philippines.

Cooper, M., Rajatasereekul, S., Immark, S., Fukai, S. and Basnayake, J. (1999) Rainfed lowland rice breeding strategies for Northeast Thailand. I. Genotypic variation and genotype × environment interactions for grain yield. *Field Crops Research* 64, 131–151.

Cornelius, P.L., Crossa, J. and Seyedsadr, M.S. (1996) Statistical tests and estimates of multiplicative models for GE interaction. In: Kang, M.S. and Gauch, H.G., Jr (eds) *Genotype-by-Environment Interaction*. CRC Press, Boca Raton, Florida, pp. 199–234.

Crispeels, M.J. (ed.) (1994) *Introduction to Signal Transduction in Plants: a Collection of Updates*. American Society of Plant Physiologists, Rockville, Maryland.

Cromwell, E. and Van Oosterhout, S. (2000) On-farm conservation of crop diversity: policy and institutional lessons from Zimbabwe. In: Brush, S.B. (ed.) *Genes in the Field*. International Plant Genetic Resources Institute, Rome, Italy, International Development Research Centre, Ottawa, Canada, and Lewis Publishers, Boca Raton, Florida, pp. 217–238.

Crossa, J., Cornelius, P.L. and Seyedsadr, M.S. (1996) Using the shifted multiplicative model cluster methods for crossover GE interaction. In: Kang, M.S. and Gauch, H.G., Jr (eds) *Genotype-by-Environment Interaction*. CRC Press, Boca Raton, Florida, pp. 175–198.

Crossa, J., Vargas, M., van Eeuwijk, F.A., Jiang, C., Edmeades, G.O. and Hoisington, D. (1999)

Interpreting genotype × environment interaction in tropical maize using linked molecular markers and environmental covariables. *Theoretical and Applied Genetics* 99, 611–625.

Dashiell, K.E., Ariyo, O.J. and Bello, L. (1994) Genotype × environment interaction and simultaneous selection for high yield and stability in soybeans (*Glycine max* (L.) Merr.). *Annals of Applied Biology* 124, 133–139.

DeLacy, I.H., Cooper, M. and Basford, K.E. (1996) Relationships among analytical methods used to study genotype-by-environment interactions and evaluation of their impact on response to selection. In: Kang, M.S. and Gauch, H.G., Jr (eds) *Genotype-by-Environment Interaction*. CRC Press, Boca Raton, Florida, pp. 51–84.

Denis, J.-B. (1988) Two-way analysis using covariates. *Statistics* 19, 123–132.

Denis, J.-B. and Gower, J.C. (1996) Asymptotic confidence regions for biadditive models: interpreting genotype–environment interactions. *Applied Statistics* 45, 479–493.

Denis, J.-B., Gauch, H.G., Jr, Kang, M.S., Van Eeuwijk, F.A. and Zobel, R.W. (1996a) Bibliography on genotype-by-environment interaction. In: Kang, M.S. and Gauch, H.G., Jr (eds) *Genotype-by-Environment Interaction*. CRC Press, Boca Raton, Florida, pp. 405–409.

Denis, J.-B., Piepho, H.-P. and Van Eeuwijk, F.A. (1996b) *Mixed Models for Genotype by Environment Tables with an Emphasis on Heteroscedasticity*. Technical Report, Département de Biométrie, Laboratoire de Biométrie, INRA-Versailles, France, 23 pp.

Dickerson, G.E. (1962) Implications of genetic–environmental interaction in animal breeding. *Animal Production* 4, 47–64.

Duvick, D.N. (1992) Genetic contributions to advances in yield of US maize. *Maydica* 37, 69–79.

Duvick, D.N. (1996) Plant breeding, an evolutionary concept. *Crop Science* 36, 539–548.

Duvick, D.N. (2002) Crop breeding in the twenty-first century. In: Kang, M.S. (ed.) *Crop Improvement: Challenges in the Twenty-first Century*. Food Products Press, Binghamton, New York, pp. 1–15.

Eberhart, S.A. and Russell, W.A. (1966) Stability parameters for comparing varieties. *Crop Science* 6, 36–40.

Eisemann, R.L., Cooper, M. and Woodruff, D.R. (1990) Beyond the analytical methodology, better interpretation and exploitation of GE interaction in plant breeding. In: Kang, M.S.

(ed.) *Genotype-by- Environment Interaction and Plant Breeding*. Louisiana State University Agricultural Center, Baton Rouge, Louisiana, pp. 108–117.

Eskridge, K.M. (1996) Analysis of multiple environment trials using the probability of outperforming a check. In: Kang, M.S. and Gauch, H.G., Jr (eds) *Genotype-by-Environment Interaction*. CRC Press, Boca Raton, Florida, pp. 273–307.

Evans, L.T. (1993) *Crop Evolution, Adaptation and Yield*. Cambridge University Press, New York.

Falconer, D.S. (1952) Selection for large and small size in mice. *Journal of Genetics* 51, 470–501.

Federer, W.T. and Scully, B.T. (1993) A parsimonious statistical design and breeding procedure for evaluating and selecting desirable characteristics over environments. *Theoretical and Applied Genetics* 86, 612–620.

Fernandez, G.C.J. (1991) Analysis of genotype × environment interaction by stability estimates. *HortScience* 26, 947–950.

Finlay, K.W. and Wilkinson, G.N. (1963) The analysis of adaptation in a plant breeding programme. *Australian Journal of Agricultural Research* 14, 742–754.

Flores, F. (1993) Interaccion genotipo-ambiente en *Vicia faba* L. Doctoral dissertation, University of Cordoba, Spain.

Flores, F., Moreno, M.T. and Cubero, J.I. (1998) A comparison of univariate and multivariate methods to analyze G × E interaction. *Field Crops Research* 56, 271–286.

Fox, P.N. and Rosielle, A.A. (1982) Reducing the influence of environmental main-effects on pattern analysis of plant breeding environments. *Euphytica* 31, 645–656.

Fox, P.N., Skovmand, B., Thompson, B.K., Braun, H.J. and Cormier, R. (1990) Yield and adaptation of hexaploid spring triticale. *Euphytica* 47, 57–64.

Francis, T.R. and Kannenberg, L.W. (1978) Yield stability studies in short-season maize: I. A descriptive method for grouping genotypes. *Canadian Journal of Plant Science* 58, 1029–1034.

Freeman, G.H. (1975) Analysis of interactions in incomplete two-way tables. *Applied Statistics* 24, 46–55.

Freeman, G.H. and Perkins, J.M. (1971) Environmental and genotype–environmental components of variability. VIII. Relations between genotypes grown in different environments and measures of these environments. *Heredity* 27, 15–23.

Gabriel, K.R. (1971) The biplot graphic display of matrices with application to principal component analysis. *Biometrika* 58, 453–467.

Gadheri, A., Everson, E.H. and Cress, C.E. (1980) Classification of environments and genotypes in wheat. *Crop Science* 20, 707–710.

Gauch, H.G., Jr (1992) *Statistical Analysis of Regional Yield Trials: AMMI Analysis of Factorial Designs.* Elsevier, Amsterdam, The Netherlands.

Gauch, H.G. and Zobel, R.W. (1990) Imputing missing yield trial data. *Theoretical and Applied Genetics* 70, 753–761.

Gauch, H.G., Jr and Zobel, R.W. (1996) AMMI analysis of yield trials. In: Kang, M.S. and Gauch, H.G., Jr (eds) *Genotype-by-Environment Interaction.* CRC Press, Boca Raton, Florida, pp. 85–122.

Glaz, B. and Dean, J.L. (1988) Statistical error rates and their implications in sugarcane clone trials. *Agronomy Journal* 80, 560–562.

Glaz, B., Miller, J.D. and Kang, M.S. (1985) Evaluation of cultivar-testing locations in sugarcane. *Theoretical and Applied Genetics* 71, 22–25.

Gorman, D.P., Kang, M.S. and Milam, M.R. (1989) Contribution of weather variables to genotype × environment interaction in grain sorghum. *Plant Breeding* 103, 299–303.

Gregorius, H.R. and Namkoong, G. (1986) Joint analysis of genotypic and environmental effects. *Theoretical and Applied Genetics* 72, 413–422.

Haji, H.M. and Hunt, L.A. (1999) Genotype × environment interactions and underlying environmental factors for winter wheat in Ontario. *Canadian Journal of Plant Science* 79, 497–505.

Haldane, J.B.S. (1947) The interaction of nature and nurture. *Annals of Eugenics* 13, 197–205.

Hall, A.E. (2001) *Crop Responses to Environment.* CRC Press, Boca Raton, Florida.

Hardwick, R.C. and Wood, J.T. (1972) Regression methods for studying genotype–environment interactions. *Heredity* 28, 209–222.

Henderson, C.R. (1975) Best linear unbiased estimation and prediction under a selection model. *Biometrics* 31, 423–447.

Herrera-Estrella, L. and Simpson, J. (1990) Influence of environmental factors on photosynthetic genes. In: Scandalios, J.G. and Wright, T.R.F. (eds) *Advances in Genetics.* Academic Press, New York, pp. 133–163.

Higley, L.G., Browde, J.A. and Higley, P.M. (1993) Moving toward new understandings of biotic stress and stress interactions. In: Bruxton,

D.R., Shibles, R., Forsberg, R.A., Blad, B.L., Asay, K.H., Paulsen, G.M. and Wilson, R.F. (eds) *International Crop Science I.* Crop Science Society of America, Madison, Wisconsin, pp. 749–754.

Hildebrand, P.E. and Russell, J.T. (1996) *Adaptability Analysis: a Method for the Design, Analysis and Interpretation of On-farm Research-extension.* Iowa State University Press, Ames, Iowa.

Hoffmann, A.A. and Parsons, P.A. (1997) *Extreme Environmental Change and Evolution.* Cambridge University Press, Cambridge, UK.

Hühn, M. (1979) Beiträge zur Erfassung der phänotypischen Stabilität. I. Vorschlag einiger auf Ranginformationen beruhenden Stabilitätsparameter. *EDP in Medicine and Biology* 10, 112–117.

Hühn, M. (1996) Nonparametric analysis of genotype × environment interactions by ranks. In: Kang, M.S. and Gauch, H.G., Jr (eds) *Genotype-by-Environment Interaction.* CRC Press, Boca Raton, Florida, pp. 235–271.

Hussein, M.A., Bjornstad, A. and Aastveit, A.H. (2000) SASG × ESTAB: a SAS program for computing genotype × environment stability statistics. *Agronomy Journal* 92, 454–459.

Jenns, A.E., Leonard, K.J. and Moll, R.H. (1982) Stability analyses for estimating relative durability of quantitative resistance. *Theoretical and Applied Genetics* 63, 183–192.

Jiang, C., Edmeades, G.O., Armstead, I., Lafitte, H.R., Hayward, M.D. and Hoisington, D. (1999) Genetic analysis of adaptation differences between highland and lowland tropical maize using molecular markers. *Theoretical and Applied Genetics* 99, 1106–1119.

Johnson, J.J., Alldredge, J.R., Ullrich, S.E. and Dangi, O. (1992) Replacement of replications with additional locations for grain sorghum cultivar evaluation. *Crop Science* 32, 43–46.

Kang, M.S. (1988) A rank-sum method for selecting high-yielding, stable corn genotypes. *Cereal Research Communications* 16, 113–115.

Kang, M.S. (ed.) (1990) *Genotype-by-Environment Interaction and Plant Breeding.* Louisiana State University Agricultural Center, Baton Rouge, Louisiana.

Kang, M.S. (1993a) Issues in GE interaction. In: Rao, V., Hanson, I.E. and Rajanaidu, N. (eds) *Genotype–Environment Interaction Studies in Perennial Tree Crops.* Palm Oil Research Institute of Malaysia, Kuala Lumpur, pp. 67–73.

Kang, M.S. (1993b) Simultaneous selection for yield and stability in crop performance trials: consequences for growers. *Agronomy Journal* 85, 754–757.

Kang, M.S. (1998) Using genotype-by-environment interaction for crop cultivar development. *Advances in Agronomy* 62, 199–252.

Kang, M.S. (2002) Preface. In: Kang, M.S. (ed.) *Crop Improvement: Challenges in the Twenty-first Century*. Food Products Press, Binghamton, New York, xv–xix.

Kang, M.S. and Gauch, H.G., Jr (eds) (1996) *Genotype-by-Environment Interaction*. CRC Press, Boca Raton, Florida.

Kang, M.S. and Gorman, D.P. (1989) Genotype × environment interaction in maize. *Agronomy Journal* 81, 662–664.

Kang, M.S. and Magari, R. (1995) STABLE: basic program for calculating yield–stability statistic. *Agronomy Journal* 87, 276–277.

Kang, M.S. and Magari, R. (1996) New developments in selecting for phenotypic stability in crop breeding. In: Kang, M.S. and Gauch, H.G., Jr (eds) *Genotype-by-Environment Interaction*. CRC Press, Boca Raton, Florida, pp. 1–14.

Kang, M.S. and Miller, J.D. (1984) Genotype × environment interactions for cane and sugar yield and their implications in sugarcane breeding. *Crop Science* 24, 435–440.

Kang, M.S. and Pham, H.N. (1991) Simultaneous selection for high yielding and stable crop genotypes. *Agronomy Journal* 83, 161–165.

Kang, M.S., Glaz, B. and Miller, J.D. (1987a) Interrelationships among stabilities of important agronomic traits in sugarcane. *Theoretical and Applied Genetics* 74, 310–316.

Kang, M.S., Miller, J.D. and Darrah, L.L. (1987b) A note on relationship between stability variance and ecovalence. *Journal of Heredity* 78, 107.

Kang, M.S., Harville, B.G. and Gorman, D.P. (1989) Contribution of weather variables to genotype × environment interaction in soybean. *Field Crops Research* 21, 297–300.

Kang, M.S., Gorman, D.P. and Pham, H.N. (1991) Application of a stability statistic to international maize yield trials. *Theoretical and Applied Genetics* 81, 162–165.

Ketata, H., Yan, S.K. and Nachit, M. (1989) Relative consistency performance across environments. In: *International Symposium on the Physiology and Breeding of Winter Cereals for Stressed Mediterranean Environments, Montpellier, France, 3–6 July 1989.*

Khush, G.S. (1993) Breeding rice for sustainable agricultural systems. In: Buxton, D.R., Shibles, R., Forsberg, R.A., Blad, B.L., Asay, K.H., Paulsen, G.M. and Wilson, R.F. (eds)

International Crop Science I. Crop Science Society of America, Madison, Wisconsin, pp. 189–199.

Laurie, D.A. and Bennett, M.D. (1985) Nuclear DNA content in the genera *Zea* and *Sorghum*: intergeneric, interspecific and intraspecific variation. *Heredity* 55, 307–313.

Lee, M. (1995) DNA markers and plant breeding programs. *Advances in Agronomy* 55, 265–344.

Lin, C.S. (1982) Grouping genotypes by a cluster method directly related to genotype–environment interaction mean square. *Theoretical and Applied Genetics* 62, 277–280.

Lin, C.S. and Binns, M.R. (1985) Procedural approach for assessing cultivar–location data: pairwise genotype–environment interactions of test cultivars with checks. *Canadian Journal of Plant Science* 65, 1065–1071.

Lin, C.S. and Binns, M.R. (1988) A method of analyzing cultivar × location × year experiments: a new stability parameter. *Theoretical and Applied Genetics* 76, 425–430.

Lin, C.S., Binns, M.R. and Lefkovitch, L.P. (1986) Stability analysis: where do we stand? *Crop Science* 26, 894–900.

Lin, C.Y. and Lin, C.S. (1994) Investigation of genotype–environment interaction by cluster analysis in animal experiments. *Canadian Journal of Animal Science* 74, 607–612.

Lopez, J. (1990) Estudio de la base genetica del contenido en taninos condensados en la semilla de las habas (*Vicia faba* L.). Doctoral dissertation, University of Cordoba, Spain.

McKeand, S.E., Li, B., Hatcher, A.V. and Weir, R.J. (1990) Stability parameter estimates for stem volume for loblolly pine families growing in different regions in the southeastern United States. *Forest Science* 36, 10–17.

Magari, R., Kang, M.S. and Zhang, Y. (1996) Sample size for evaluating field ear moisture loss rate in maize. *Maydica* 41, 19–24.

Magari, R., Kang, M.S. and Zhang, Y. (1997) Genotype by environment interaction for ear moisture loss rate in corn. *Crop Science* 37, 774–779.

Montaldo, H.H. (2001) Genotype by environment interactions in livestock breeding programs: a review. *Interciencia* 26(6), 229–235.

Mooers, C.A. (1921) The agronomic placement of varieties. *Journal of American Society of Agronomy* 13, 337–352.

Ougham, H.J. and Howarth, C.J. (1988) Temperature shock proteins in plants. In: Long, S.P. and Woodward, F.J. (eds) *Plants and Temperature Symposium of the Society* for Experimental Biology. Company of Biologists, Cambridge, pp. 259–280.

Pandey, S. and Gardner, C.O. (1992) Recurrent selection for population, variety, and hybrid improvement in tropical maize. *Advances in Agronomy* 48, 1–87.

Pazdernik, D.L., Hardman, L.L. and Orf, J.H. (1997) Agronomic performance and stability of soybean varieties grown in three maturity zones of Minnesota. *Journal of Production Agriculture* 10, 425–430.

Pedersen, A.R., Everson, E.H. and Grafius, J.E. (1978) The gene pool concept as basis for cultivar selection and recommendation. *Crop Science* 18, 883–886.

Piepho, H.-P. (1994) Missing observations in analysis of stability *Heredity* 72, 141–145. (Correction 73 (1994), 58.)

Piepho, H.-P. (1998) Methods of comparing the yield stability of cropping systems – a review. *Journal of Agronomy and Crop Science* 180, 193–213.

Piepho, H.P. (2000a) Exact confidence limits for covariate-dependent risk in cultivar trials. *Journal of Agricultural Biological Environmental Statistics* 5, 202–213.

Piepho, H.-P. (2000b) A mixed model approach to mapping quantitative trait loci in barley on the basis of multiple environment data. *Genetics* 156, 2043–2050.

Piepho, H.-P., Denis, J.B. and Van Eeuwijk, F.A. (1998) Predicting cultivar differences using covariates. *Journal of Agricultural Biological Environmental Statistics* 3, 151–162.

Pollack, C.J., Eagles, C.F., Howarth, C.J., Schunmann, P.H.D. and Stoddart, J.L. (1993) Temperature stress. In: Fowden, L., Mansfield, T. and Stoddart, J. (eds) *Plant Adaptation to Environmental Stress*. Chapman & Hall, New York, pp. 109–132.

Polle, A. and Rennenberg, H. (1993) Significance of antioxidants in plant adaptation to environmental stress. In: Fowden, L., Mansfield, T. and Stoddart, J. (eds) *Plant Adaptation to Environmental Stress*. Chapman & Hall, New York, pp. 263–273.

Prabhakaran, V.T. and Jain, J.P. (1994) *Statistical Techniques for Studying Genotype–Environment Interactions*. South Asian Publishers, New Delhi, India.

Rameau, C. and Denis, J.-B. (1992) Characterization of environments in long-term multi-site trials in asparagus, through yield of standard varieties and use of environmental covariates. *Plant Breeding* 109, 183–191.

Rao, T.D.P., Rao, D.V.S. and Rai, S.C. (1988) Symposium on statistical aspects of stability

of crop yields. *Journal of Indian Society of Agricultural Statistics* 60, 70–79.

Rao, V., Henson, I.E. and Rajanaidu, N. (eds) (1993) *Genotype × Environment Interaction in Perennial Tree Crops*. International Society of Oil Palm Breeders and Palm Oil Research Institute of Malaysia, Kuala Lumpur.

Rayburn, A.L. and Auger, J.A. (1990) Genome size variation in *Zea mays* ssp. *mays* adapted to different altitudes. *Theoretical and Applied Genetics* 79, 470–474.

Rayburn, A.L., Price, H.J., Smith, J.D. and Gold, J.R. (1985) C-banded heterochromatin and DNA content in *Zea mays*. *American Journal of Botany* 72, 1610–1617.

Rosielle, A.A. and Hamblin, J. (1981) Theoretical aspects of selection for yield in stress and non-stress environments. *Crop Science* 21, 943–946.

Ryals, J., Uknes, S. and Ward, E. (1994) Systemic acquired resistance. *Plant Physiology* 104, 1109–1112.

Saeed, M. and Francis, C.A. (1984) Association of weather variables with genotype × environment interaction in grain sorghum. *Crop Science* 24, 13–16.

Scandalios, J.G. (1990) Response of plant anti-oxidant defense genes to environmental stress. In: Scandalios, J.G. and Wright, T.R.F. (eds) *Advances in Genetics*. Academic Press, New York, pp. 1–41.

Searle, S.R. (1987) *Linear Models for Unbalanced Data*. John Wiley & Sons, New York.

Shukla, G.K. (1972) Some statistical aspects of partitioning genotype–environmental components of variability. *Heredity* 29, 237–245.

Silvey, V. (1981) The contribution of new wheat, barley and oat varieties to increasing yield in England and Wales 1947–78. *Journal of National Institute of Agricultural Botany* 15, 399–412.

Simmonds, N.W. (1981) Genotype (G), environment (E) and GE components of crop yields. *Experimental Agriculture* 17, 355–362.

Singh, M. and Ceccarelli, S. (1995) Estimation of heritability using varietal trials data from incomplete blocks. *Theoretical and Applied Genetics* 90, 142–145.

Singh, M., Ceccarelli, S. and Grando, S. (1999) Genotype × environment interaction of crossover type: detecting its presence and estimating the crossover point. *Theoretical and Applied Genetics* 99, 988–995.

Smith, M.E., Coffman, W.R. and Barker, T.C. (1990) Environmental effects on selection under high and low input conditions. In:

Kang, M.S. (ed.) *Genotype-by-Environment Interaction and Plant Breeding.* Louisiana State University Agricultural Center, Baton Rouge, Louisiana, pp. 261–272.

Sokal, R.R. and Michener, C.D. (1958) A statistical method for evaluating systematic relationships. *University of Kansas Scientific Bulletin* 38, 1409–1438.

Specht, J.E. and Laing, D.R. (1993) Selection for tolerance to abiotic stresses – discussion. In: Bruxton, D.R., Shibles, R., Forsberg, R.A., Blad, B.L., Asay, K.H., Paulsen, G.M. and Wilson, R.F. (eds) *International Crop Science I.* Crop Science Society of America, Madison, Wisconsin, pp. 381–382.

Sperling, L., Ashby, J., Weltzien, E., Smith, M. and McGuire, S. (2001) Base-broadening for client-oriented impact: insights drawn from participatory plant breeding field experience. In: Cooper, H.D., Spillane, C. and Hodgkins, T. (eds) *Broadening the Genetic Bases of Crop Production.* CAB International, Wallingford, UK, pp. 419–435.

Tai, G.C.C. (1971) Genotypic stability analysis and its application to potato regional trials. *Crop Science* 11, 184–190.

Tai, G.C.C. (1990) Path analysis of genotype-environment interactions. In: Kang, M.S. (ed.) *Genotype-by-Environment Interaction and Plant Breeding.* Louisiana State University Agricultural Center, Baton Rouge, Louisiana, pp. 273–286.

Tai, G.C.C. and Coleman, W.K. (1999) Genotype × environment interaction of potato chip colour. *Canadian Journal of Plant Science* 79, 433–438.

Unsworth, M.H. and Fuhrer, J. (1993) Crop tolerance to atmospheric pollutants. In: Bruxton, D.R., Shibles, R., Forsberg, R.A., Blad, B.L., Asay, K.H., Paulsen, G.M. and Wilson, R.F. (eds) *International Crop Science I.* Crop Science Society of America, Madison, Wisconsin, pp. 363–370.

Van Eeuwijk, F.A., Denis, J.-B. and Kang, M.S. (1996) Incorporating additional information on genotypes and environments in models for two-way genotype by environment tables. In: Kang, M.S. and Gauch, H.G., Jr (eds) *Genotype-by-Environment Interaction.* CRC Press, Boca Raton, Florida, pp. 15–49.

Vargas, M., Crossa, J., Van Eeuwijk, F.A., Sayre, K.D. and Reynolds, M.P. (2001) Interpreting treatment × environment interaction in agronomy trials. *Agronomy Journal* 93, 949–960.

Via, S. (1984) The quantitative genetics of polyphagy in an insect herbivore. I. Genotype–environment interaction in larval performance of different host plant species. *Evolution* 38, 881–895.

Wade, L.J., McLaren, C.G., Quintana, L., Harnpichitvitaya, D., Rajatasereekul, S., Sarawgi, A.K., Kumar, A., Ahmed, H.U., Sarwoto, Singh, A.K., Rodriquez, R., Siopongco, J. and Sarkarung, S. (1999) Genotype by environment interactions across diverse rainfed lowland rice environments. *Field Crops Research* 64, 35–50.

Wang, D.L., Zhu, J., Li, Z.K. and Paterson, A.H. (1999) Mapping QTLs with epistatic effects and QTL × environment interactions by mixed linear model approaches. *Theoretical and Applied Genetics* 99, 1255–1264.

Weber, W.E., Wricke, G. and Westermann, T. (1996) Selection of genotypes and prediction of performance by analyzing GE interactions. In: Kang, M.S. and Gauch, H.G., Jr (eds) *Genotype-by-Environment Interaction.* CRC Press, Boca Raton, Florida, pp. 353–371.

Westcott, B. (1987) A method of assessing the yield stability of crop genotypes. *Journal of Agricultural Science* 108, 267–274.

Witcombe, J.R. (2001) The impact of decentralized and participatory plant breeding on the genetic base of crops. In: Cooper, H.D., Spillane, C. and Hodgkins, T. (eds) *Broadening the Genetic Bases of Crop Production.* CAB International, Wallingford, UK, pp. 407–417.

Wood, J.T. (1976) The use of environmental variables in the interpretation of genotype–environment interaction. *Heredity* 37, 1–7.

Woodend, J.J. and Glass, A.D.M. (1993) Genotype–environment interaction and correlation between vegetative and grain production measures of potassium use-efficiency in wheat (*T. aestivum* L.) grown under potassium stress. *Plant and Soil* 151, 39–44.

Wricke, G. (1962) Uber eine Methode zur Erfassung der okologischen Streubreite in Feldversuchen. *Zeitschrift für Pflanzenzucht* 47, 92–96.

Yan, W. (2001) GGEbiplot – a Windows application for graphical analysis of multi-environment trial data and other types of two-way data. *Agronomy Journal* 93(5), 1111–1118.

Yan, W. and Hunt, L.A. (2001) Genetic and environmental causes of genotype by environment interaction for winter wheat yield in Ontario. *Crop Science* 41, 19–25.

Yan, W. and Hunt, L.A. (2002) Biplot analysis of diallel data. *Crop Science* 42, 21–30.

Yan, W. and Rajcan, I. (2002) Biplot analysis of test sites and trait relations of soybean in Ontario. *Crop Science* 42, 11–20.

Yan, W., Hunt, L.A., Sheng, Q. and Szlavnics, Z. (2000) Cultivar evaluation and mega-environment investigation based on the GGE biplot. *Crop Science* 40(3), 597–605.

Yan, W., Cornelius, P.L., Crossa, J. and Hunt, L.A. (2001) Two types of GGE biplots for analyzing multi-environment trial data. *Crop Science* 41(3), 656–663.

Yates, F. and Cochran, W.G. (1938) The analysis of groups of experiments. *Journal of Agricultural Science* 28, 556–580.

Zavala-Garcia, F. and Treviño-Hernández, T.E. (eds) (2000) *Simposium interaccion genotipo × ambiente.* SOMEFI-CSSA-UG, Irapuato, Gto, Mexico.

Zhang, Q. and Geng, S. (1986) A method of estimating varietal stability for long-term trials. *Theoretical and Applied Genetics* 71, 810–814.

Zobel, R.W., Wright, M.J. and Gauch, H.G., Jr (1988) Statistical analysis of a yield trial. *Agronomy Journal* 80, 388–393.

16 Analysing QTL–Environment Interaction by Factorial Regression, with an Application to the CIMMYT Drought and Low-nitrogen Stress Programme in Maize

F.A. Van Eeuwijk,[1] J. Crossa,[2] M. Vargas[2] and J.-M. Ribaut[2]

[1]Wageningen University, Department of Plant Sciences, Laboratory of Plant Breeding, PO Box 386, 6700 AJ Wageningen, The Netherlands; [2]International Maize and Wheat Improvement Center (CIMMYT), Lisboa 27, Apdo. Postal 6-641, 06600 Mexico, DF, Mexico

Introduction

With the increasing omnipresence of marker technology in plant breeding, the classical problem of how to handle genotype–environment interaction (GEI) is gradually being absorbed into more basic questions towards the existence and description of differential gene expression, where the term 'gene' is replaced by 'quantitative trait locus' (QTL). Because of this process, the need has arisen for statistical models that are applicable in the contexts of both GEI and QTL–environment interaction (QEI). This chapter attempts to develop a class of statistical models that are useful for the analysis of GEI as well as QEI.

For the analysis of GEI, the class of factorial regression models has been shown to be a powerful tool (Van Eeuwijk et al., 1996). Taking the two-way analysis of variance (ANOVA) model as a departure point for the description of genotype–environment (GE) data, in factorial-regression models variation due to main effects and interaction is partitioned in parts due to regression on covariables and deviations from regression.

In their most elementary form, factorial-regression models are examples of the application of contrasts in ANOVA. A well-written general treatment of contrasts in ANOVA can be found in Kuehl (2000). For factorial regression, estimation and testing take place within the framework of standard least-squares theory and thus neither present theoretical problems nor require special software (Denis, 1988, 1991). For some recent examples of applications of factorial regression to GEI problems, see Vargas et al. (1999) and Voltas et al. (1999a,b).

For the detection and localization of QTL and the estimation of QTL effects, two major approaches can be distinguished. A computer-intensive, theoretically demanding approach is built on mixture-model theory within a maximum-likelihood framework. An approximate alternative is based on regression theory, where the predictors are derived from the conditional probabilities for specific QTL genotypes, given flanking-marker information and the position of the QTL relative to the flanking markers. A good description of both approaches is given by Lynch and Walsh (1998). Differences in results between the mixture-model approach and the regression approach seem minor, except for the case of closely linked or interacting QTL (Kao, 2000). Therefore, we would, in general, prefer regression-based approaches to QTL analysis because computational requirements for this approach are considerably less demanding. Furthermore, inclusion of the statistical design and additional treatment structure in a QTL analysis seems, a priori, easier within a regression framework.

Theory for QTL detection and estimation has developed strongly during the last decade and a half. Application of this theory has become somewhat of a routine issue in breeding programmes. Still, theory for QEI is scarce and applications of such theory are few. Noteworthy contributions that emphasize mixture-model approaches include, among others, Jansen et al. (1995), Jiang and Zeng (1995) and Korol et al. (1998). Some recent uni- and multivariate regression-based approaches, building on the work by Haley and Knott (1992), were proposed by, among others, Sari-Gorla et al. (1997), Caliński et al. (2000) and Hackett et al. (2001). Using the criteria of transparency and simplicity of application, the methodology developed by Sari-Gorla et al. (1997) merits special attention. For QTL analysis in recombinant inbred lines (RILs), they used weighted least-squares regressions with F tests for determining the presence of QTL main effects and QEI. The problem of multiple testing was elegantly addressed by a sequentially rejective Bonferroni method. Additional block and treatment terms were incorporated without any problems. A forward selection procedure was used for the construction of a cofactor set in their variant of composite interval mapping.

In this chapter, we present a method of QTL analysis based on regression, which is clearly inspired by the work of Sari-Gorla et al. (1997). A first outline of this method appeared in Van Eeuwijk et al. (2001). What distinguishes our approach is, first, the close connection of our models for QEI analysis with factorial-regression models for GEI. We see QEI analysis as a direct elaboration of GEI analysis, and thus think that the

methods of analysis should be, if not the same, at least very similar. As a complementary alternative to the sequentially rejective Bonferroni procedure of testing for QTL and QEI, we propose a randomization test procedure, combining ideas of Manly (1997) with those of Churchill and Doerge (1994). We also propound a new method to capture the effects of other QTL and background genetic variation when searching for QTL in composite interval mapping. The idea is to refrain from choosing a limited number of markers for membership of the cofactor set by including all markers but penalizing the corresponding regression coefficients in order to sidestep collinearity problems. The methodology will be illustrated by an application to data from the drought- and low-nitrogen-stress breeding programme of the International Maize and Wheat Improvement Center (CIMMYT).

Factorial Regression for Analysis of Two-way Genotype–Environment Tables

For the expected response of a genotype i $(i = 1, \ldots I)$ in an environment j $(j = 1, \ldots J)$, we take the two-way ANOVA model as a reference model $\mu_{ij} = \mu + E_j + G_i + GE_{ij}$, where E_j denotes the environmental main effect, G_i the genotypic main effect and GE_{ij} the two-way GEI. Errors are assumed to be independent, of constant variance and normal. Generalization to more complex error structures would require a change from standard least-squares regression to weighted least squares. Following Denis (1988, 1991) and the ANOVA module of Genstat (1993), our preferred choice for identification constraints of parameters is 'sum-to-zero' constraints over the running indices.

In comparison with the reference ANOVA model, the distinguishing characteristic for factorial-regression models is the introduction of one or more genotypic covariables, x_a $(a = 1, \ldots A)$, with values x_{ia} and/or environmental covariables, z_b $(b = 1, \ldots B)$, with values z_{jb} to partition main effects and GEI of the ANOVA model. For example, $G_i = x_{ia}\rho_a + G_i^*$, where the genotypic main effect, G_i, is regressed on the genotypic covariable x_a, leading to a coefficient ρ_a and a residual genotypic effect, G_i^*. By the same token, E_j can be replaced by $\beta_b z_{jb} + E_j^*$, where β_b is a regression coefficient, z_b an environmental covariable and E_j^* a residual environmental effect. Finally, the interaction can be replaced by various cross-products of parameters and covariables. First, the interaction term is split in regressions per environment on the genotypic covariable, x_a: $GE_{ij} = x_{ia}\rho_{ja} + GE_{ij}^*$, where the interaction is described by the differential environmental expression parameters ρ_{ja}. Secondly, an environmental covariable, z_b, is introduced: $GE_{ij} = \beta_{ib} z_{jb} + GE_{ij}^*$, and the interaction resides in the differences in genotypic sensitivity, β_{ib}. Finally, the interaction is described by a one-parameter cross-product term that contains a regressor constructed from a cross-product of a genotypic covariable with an environmental covariable: $GE_{ij} = \kappa_{ab}x_{ia}z_{jb} + GE_{ij}^*$, with κ_{ab} being a scaling constant for the cross-product $x_{ia}z_{jb}$. Useful identification constraints are again sum-to-zero constraints, where covariables must be centred and genotypic parameter vectors must be orthogonal to genotypic covariables, and the same must be true for the environmental parameter vectors and covariables (Denis, 1988, 1991). This parameterization can be obtained by ensuring that every covariable that is introduced in the model is made orthogonal to all covariables already in the model (Genstat, 1993). Of course, there is no reason why more than one covariable cannot be incorporated in the model, for example:

$$GE_{ij} = \sum_{b=1}^{B} \beta_{ib} z_{jb} + GE_{ij}^*$$

Covariables may be quantitative or qualitative. In the case of qualitative covariables, a grouping is imposed on genotypes or environments. Quantitative examples of genotypic covariables, x_a, are disease- and stress-resistance scores, whereas a qualitative example is a classification of genotypes on the basis of origin. For examples of quantitative environmental covariables, think of temperature, humidity, radiation

and number of sun hours, whereas year, location and ecozone classifications are examples of qualitative environmental covariables. A catalogue of factorial regression models can be found in Van Eeuwijk et al. (1996). In the context of this chapter, we mention a special application of factorial regression to the analysis of GEI in maize by Crossa et al. (1999), where molecular-marker scores were used as genotypic covariables.

Factorial-regression Models for QTL Analysis

QTL analysis can be considered as an application of factorial regression by interpreting this analysis as the partitioning of the genotypic main effect, G_i, into a part due to QTL and a residual:

$$G_i = \sum_{q=1}^{Q} x_{iq}\rho_q + G_i^*$$

where ρ_q stands for the QTL effect and x_{iq} is the value of the genetic predictor q for genotype i, which is a function of type of genetic effect (additive, dominance, epistatic), marker information (genotype) and position within the genome. The number of QTL has been set at Q. A similar partitioning for the interaction leads to:

$$GE_{ij} = \sum_{q=1}^{Q} x_{iq}\rho_{jq} + GE_{ij}^*$$

where the QEI effects, ρ_{jq}, represent the deviations from the QTL main effect, ρ_q. An alternative parameterization, which partitions the joint effect of $G_i + GE_{ij}$ in a part due to environment-specific QTL and a residual, is:

$$G_i + GE_{ij} = (G + GE)_{ij} = \sum_{q=1}^{Q} x_{iq}\rho_{jq} + (G + GE)_{ij}^*$$

A further step of analysis could be to regress the QTL effects, ρ_{jq}, on an environmental covariable, z_b. For the environment-specific QTL effects $\rho_{jq'}$ at position q', this leads to:

$$x_{iq'}\rho_{jq'} = x_{iq'}\left(\kappa_{q'b} z_{jb} + \rho_{jq'}^*\right)\kappa_{q'b}x_{iq'}z_{jb} + x_{iq'}\rho_{jq'}^*$$

where $\kappa_{q'b}$ is a scaling constant for the cross-product of genetic predictor $x_{q'}$ and environmental covariable z_b, while the parameters $\rho_{jq'}^*$ represent residual QTL effects. In this way, direct links can be established between phenotypic expression as induced by identified chromosome regions and environmental factors. A special situation occurs when the environmental covariable z_b is taken as equal to the environmental main effect, E_j. The resulting model is the QTL equivalent of the classical regression on the mean model for GEI that was popularized by Finlay and Wilkinson (1963). In the QTL context, this model was first suggested by Korol *et al.* (1998).

The particularity of QTL analysis within a factorial-regression framework resides in the construction of the genetic predictors x_q. Let the marker genotypes at a specific marker locus be MM, Mm and mm. For an additive QTL effect appearing at this locus, the corresponding genetic predictor values, x_{iq}^{add}, would be 1, 0 and –1, respectively, whereas, for a dominance QTL effect, the genetic predictor values, x_{iq}^{dom}, would read 0, 1 and 0, respectively. When these genetic predictors at marker positions have been constructed, one can perform a simple form of QTL analysis, namely QTL mapping by marker regression. More powerful methods of QTL analysis, like simple and composite interval mapping, also require values for the genetic predictors in between marker positions. Here, the additive genetic predictors are constructed as the difference between the conditional probability of the QTL genotype being QQ versus the conditional probability of it being qq. The conditioning is on the flanking marker genotypes and the evaluation position inside the marker bracket. The dominance predictor then follows from the conditional probability for the heterozygote genotype, Qq. Explicit expressions for the construction of these genetic predictors can be found in many papers and books. We mention the seminal paper of Haley and Knott (1992) for the construction of F_2-related predictors and Lynch and Walsh (1998) for a general description of the principles. A most useful paper in this context is Jiang and Zeng (1997), which describes algorithms for the construction of genetic predictors for many

types of populations, with dominant and codominant markers and allowing for missing marker information.

Estimation and Testing for QTL and QEI in Factorial Regression

As remarked, the factorial regression model is a simple linear regression model, for which parameter estimation is typically done by ordinary least-squares procedures. A natural choice to test for the presence of QTL main effects and QEI effects at a particular position would be to use an F test like that used for testing contrasts in two-way ANOVA, where the contrasts for QTL main effect and QEI are defined by the values of the genetic predictors (additive and dominance) at that position. The numerator of these tests will contain the mean square due to regression on the genetic predictor(s). For the denominator, various options exist. One option is to base the denominator mean square on the mean or median of the intrablock error estimates of the individual trials. Another option is to use the deviations from the regressions on the genetic predictors.

Whichever choice is made, for QTL analyses there will always remain the problem of multiple testing. Genetic predictors are constructed at small intervals along the chromosomes and at each position a test for QTL presence will be performed. Thus, the genome-wide type I error should, in some way, be controlled. An attractive, two-step testing procedure was proposed by Sari-Gorla *et al.* (1997). In the first step, a test is carried out for the presence of a QTL at a certain test position, t, by comparing the fit of the model $\mu_{ij} = \mu + E_j + x_{it}\rho_{jt}$ with that of the model $\mu_{ij} = \mu + E_j$, i.e. a model with environment-specific QTL is compared with a model without QTL. The significance of the resulting variance ratio is assessed by a sequentially rejective Bonferroni procedure or Holm's simultaneous testing procedure (Neter *et al.*, 1996). The principle of this procedure is as follows. First sort the P values of the individual F tests at all test positions from small to large. For N

evaluation positions, the null hypothesis of no environment-specific QTL is rejected as long as the sorted P values are smaller than $\alpha/(N - k + 1)$, where α is the genome-wide level of test and k is the position of the P value under test in the sorted vector of P values. Thus the smallest P value is compared with α/N, the second smallest with $\alpha/(N - 1)$, the next with $\alpha/(N - 2)$, etc. In the second step of the procedure, for those positions at which the null hypothesis of no environment-specific QTL was rejected, a further test is performed for the existence of QEI, the fit of the model $\mu_{ij} = \mu + E_j + x_{it}\rho_{jt}$ is compared with that of the model $\mu_{ij} = \mu + E_j + x_{it}\rho_t$. As in the second step, only a limited number of tests will be executed; there is considerably less urgency for type I error control.

An alternative test procedure, based on randomization, is the following. To derive the null distribution of test statistics for QTL main effects and QEI, the vectors of genetic predictors over marker loci per offspring (F$_2$, RIL, etc.) are randomized with respect to the vectors of phenotypic observations across environments per offspring: that is, all genetic information pertaining to a specific offspring is kept together, just like all the phenotypic information, but the connections between the genetic and phenotypic vectors are broken. Next, test statistics can be defined, such as variance ratios, LOD scores or squared correlation coefficients. For each randomization, the maximum value of the test statistic across all evaluation positions on the genome is stored. After a sufficiently large number of randomizations, the quantiles of the null distribution for the test statistic can then be calculated, and the P value of the original test statistic, as calculated on the non-randomized data, can be determined. This randomization procedure extends, on the one hand, ideas of Manly (1997) on randomization tests for two-way ANOVA and multivariate regression and, on the other hand, work by Churchill and Doerge (1994) on controlling type I error in randomization tests for QTL main effects.

The reason for using two test procedures in our QTL analysis alongside each other is that the sequentially rejective Bonferroni procedure seems somewhat liberal, whereas the randomization procedure seems somewhat conservative. Simulation work may lead to clarification of this issue.

A Simple Correction for Genetic Effects Elsewhere on the Genome

By correction for genetic effects elsewhere on the genome, the power to detect QTL at a particular position increases. A convenient way for effecting this correction is by including a number of markers, so-called cofactors, close to putative QTL. The cofactor set is usually assembled by regression subset-selection strategies, such as stepwise forward (e.g. Sari-Gorla *et al.*, 1997) and stepwise backward (e.g. Jansen and Stam, 1994). After the compilation of a cofactor set, i.e. a set of putative QTL, in a second round the whole genome is scanned again for QTL, where an evaluation window is chosen within which selected cofactors are temporally deleted, while retaining all cofactors outside the evaluation window. This procedure is commonly referred to as composite interval mapping. In the factorial regression framework, it can be understood as the comparison of the fit of the model $\mu_{ij} = \mu + E_j + \sum_{m \in C} x_{im}\rho_{jm} + x_{it}\rho_{jt}$ with the fit of $\mu_{ij} = \mu + E_j + \sum_{m \in C} x_{im}\rho_{jm}$, where C designates the set of cofactors.

The choice of the cofactor set depends on the subset selection procedure that was used. The quality of the correction for genetic effects elsewhere on the genome, following from the inclusion of a cofactor set in the model, is dependent on the position of the evaluation window under test. The optimal cofactor set for a specific window need not necessarily be the same as the total of the selected markers across the whole genome minus the selected markers inside the test window. Also, the significance of QTL effects can easily change by dropping or including cofactors. Therefore, it would be preferable to correct for genetic effects outside the evaluation window by some other means than a cofactor set constructed from a regression subset procedure.

The alternative we propose is to inspect the genome chromosome by chromosome (chromosomes are independent units by definition). When testing for QTL at a specific chromosome, the original phenotypic data are replaced by the residuals of a multivariate multiple regression of the phenotypic data (responses across environments) on all markers of the complementary set of chromosomes (all chromosomes minus the one under test). In that way, the genetic effects due to QTL at other chromosomes will have been removed before testing at the present chromosome commences. As the number of markers will be high in relation to the number of offspring groups or the number of predictors will be large in relation to the number of observations, severe collinearity problems can be expected. To solve these collinearity problems, the regression coefficients should be smoothed or penalized in some way. At present, we use a partial least-squares procedure to deal with the collinearity problem (Helland, 1988), although alternative smoothers and shrinkage-estimation procedures are under study. To find an acceptable amount of smoothing of the regression coefficients, a cross-validation procedure is used (Osten, 1988), which is implemented in the partial least-squares module of GENSTAT (1993).

Example

To illustrate the methodology proposed above, we shall take a look at an analysis of

data obtained in the CIMMYT programme for drought and low-nitrogen stress. Part of these data was analysed before. For the results of these analyses and a detailed description of the data material, both phenotypic and molecular, see Ribaut et al. (1996, 1997). In this chapter, we shall discuss analyses of yield and the length of the anthesis–silking interval (ASI), observed on a set of 211 F_2-derived F_3 families that were evaluated in eight trials in the years 1992, 1994 and 1996, under various stress conditions related to water and nitrogen availability. For each trial, environmental information was recorded in the form of radiation, maximum and minimum temperature, precipitation and sun hours during the main developmental stages of preflowering (vegetative), flowering and postflowering (grain filling). Table 16.1 gives a brief description of the trials plus the means for yield and ASI. A short ASI is considered indicative of drought-stress tolerance. The ASI is used as a secondary character for improving yield under stress conditions. In Table 16.1, one can observe that, across the eight trials we analysed, there existed a tendency of lower, more negative, ASIs with higher average yields, i.e. less stress.

We shall report here only on the QTL analyses for the additive genetic effects relative to the first chromosome. The genetic predictors were calculated following the algorithm of Jiang and Zeng (1997). The available set of mapped restriction fragment length polymorphism (RFLP) markers for the whole of the genome consisted of 132 RFLP markers, of which 21 were located on the first chromosome. Factorial regressions

Table 16.1. Description of CIMMYT drought- and low-nitrogen-stress trials.

Trial code	Year	Sowing	Nitrogen	Drought stress	Mean yield in t ha^{-1}	ASI in days
NS92a	1992	Winter	Normal	No	10.5	−1.6
IS92a	1992	Winter	Normal	Intermediate	6.4	−1.0
SS92a	1992	Winter	Normal	Severe	3.7	−0.9
IS94a	1994	Winter	Normal	Intermediate	4.2	1.8
SS94a	1994	Winter	Normal	Severe	4.1	1.9
LN96a	1996	Winter	Low	No	1.8	2.9
LN96b	1996	Summer	Low	No	1.0	3.3
HN96b	1996	Summer	High	No	4.9	−1.1

were calculated every 3.33 cM and at the markers. At each evaluation position, the amount of variation due to the QTL main effect and QEI was calculated. These were expressed as a percentage of the total amount of variation due to the genotypic main effect plus GEI. Thus, the test statistics calculated were the percentage of variation due to a QTL main effect, R^2_{QTL}, the percentage of variation due to QEI, R^2_{QEI}, and the percentage of variation due to the fit of an individual QTL for each environment, $R^2_{\text{QTL + QEI}}$. Note that the total variation was calculated after correction of the data for genetic effects on the other chromosomes, i.e. we applied the form of composite interval mapping described in the preceding section. Figure 16.1 contains, for yield, the profiles for the three test statistics plus critical values ($\alpha = 0.05$) based on 100 randomizations, as described in the section on estimation and testing. The comparison of the QTL + QEI and QEI profiles with the critical values reveals that there are good reasons to assume the existence of environment-specific QTL in the region between roughly 100 and 180 cM. In contrast, inspection of Fig. 16.2,

which contains the R^2 profiles for ASI, shows that, for ASI, there is good evidence for a QTL in the region between 170 and 230 cM, but no indication for environment-specific expression of QTL, as nearly all QTL activity is of a main-effect nature. It is interesting to see that, for chromosome 1, there is no support for the hypothesis of common QTL for yield and ASI.

We continue with the further analysis of yield. As a double check for the significance of the environment-dependent QTL on chromosome 1, we also used the procedure based on the sequentially rejective Bonferroni tests. For a test statistic, we used the variance ratio of the mean square for regression (on the additive genetic predictors per environment) divided by the mean square for deviations from regression. In Fig. 16.3, the part of the chromosome that was found to have a significant test result according to this procedure is boxed. The results of the sequential Bonferroni and randomization procedure were found to correspond well in this case. In general, the randomization tests appeared to be more stringent than the sequential Bonferroni

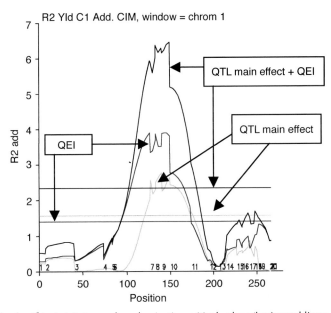

Fig. 16.1. Profiles for R^2 test statistics and randomization critical values (horizontal lines; $\alpha = 0.05$) for additive QTL main effect, QEI and QTL main effect + QEI for yield at chromosome 1. Numbers 1 to 21 indicate marker positions.

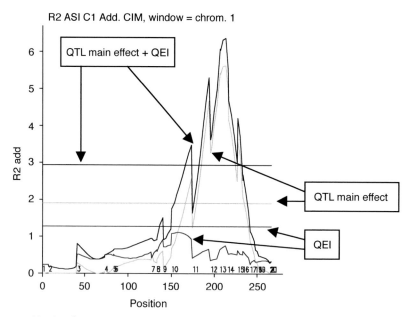

Fig. 16.2. Profiles for R^2 test statistics and randomization critical values (horizontal lines; $\alpha = 0.05$) for additive QTL main effect, QEI and QTL main effect + QEI for ASI on chromosome 1. Numbers 1 to 21 indicate marker positions.

procedure. To estimate the most likely place for the QTL, the R^2 profile was smoothed and the location of the maximum of the smoothed curve was determined to be at 140 cM (139.86 cM).

Results of analyses of variance and factorial regressions at 140 cM are presented in Tables 16.2 and 16.3. The upper segment of Table 16.2 provides the results of the two-way ANOVA of the yield data. Apparently, most of the variation can be attributed to differences between the trials (environments). This is to be expected for a set of trials that differed so much in environmental conditions. The interesting statistics are the sums of squares and mean squares due to F_2 families and GEI. The GE term (for GEI) represents a substantial part of the genotype-related variation (sum of squares for genotypic main effect plus GEI). As the more relevant part of GEI is usually lumped with large amounts of noise in the GE term of the two-way ANOVA, the F test for GE is not very informative. Nevertheless, the GE term in the upper segment of Table 16.2 is clearly significant when tested against the intra-block estimate, 0.75 (with at least 200

degrees of freedom). In the middle segment of Table 16.2, the results of the correction for genetic effects at the other chromosomes (2–10) can be found. About a quarter of the total genotype-related yield variation can be ascribed to QTL (genes) on the other chromosomes. In principle, the degrees of freedom for genotypes and GE should be adjusted for the correction. There are various ways of doing this. However, as the adjustment would have very little influence on further inferences, we left the degrees of freedom unaltered.

The last segment of Table 16.2 is the most interesting one, as now the variation due to additive QTL at chromosome 1 (140 cM) is addressed. First, the eight environment-dependent QTL do not seem to be responsible for a major amount of variation – 6.3% of the adjusted genotype-related variation in yield. QEI dominated the QTL main effect, when comparing the sums of squares. The differences between the environments become clear when the effects are studied (Table 16.3). The additive QTL allele at 140 cM reduced yield by between 0.31 and 0.84 t ha^{-1} in the 1992 and 1994 trials,

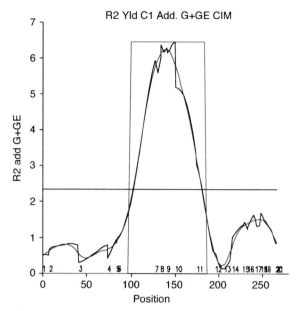

Fig. 16.3. Profiles for R^2 test statistic and randomization critical value (horizontal lines; $\alpha = 0.05$) for additive QTL main effect + QEI for yield on chromosome 1, smoothed R^2 profile, and trajectory of significant test results on the basis of a sequentially rejective Bonferroni procedure (boxed). Numbers 1 to 21 indicate marker positions.

which were mainly drought-stress-related trials, whereas it did not affect yield in the low-nitrogen trials of 1996 and even increased yield by 0.35 t ha^{-1} in the high-nitrogen trial of 1996. The second and third data columns of Table 16.3 demonstrate the partitioning of the QTL effects per environment (first data column) into QTL main effect and QTL interaction effects. The QTL interaction effects were negative for the 1992 and 1994 trials and positive for the 1996 trials. When the QTL interaction effects were regressed on the set of environmental covariables, minimum temperature at flowering exhibited a close relation with these interaction effects (Table 16.2 bottom, Table 16.3 last two columns). The effect of the environment-specific QTL effects (QEI) can be understood as an increase in yield by 0.065 t ha^{-1} for each degree (Celsius) that the minimum temperature at flowering was higher. For genotypes that are homozygous at this QTL, the increase will, of course, double. This model did not do equally well for each environment, as can be seen in the last column of Table 16.3. Note that, because

of the QTL main effect, yield will, at any rate, be decreased by 0.33 t ha^{-1} for each QTL allele present.

Final Comments

In this chapter, we have tried to give an outline of a conceptually simple method for analysing QEI. The model and inference we used do not require special-purpose software or sophisticated programming in one of the major statistical packages. Thus, the methods should be within reach for many practitioners.

As is the case with any method, our method is also open to improvement. One obvious criticism can be that no attention is given to the occurrence of closely linked QTL. To counteract this comment, it can be remarked that the method of analysis that has been described above is rather easily adapted to allow for a few extra cofactors on the chromosome under study. The easiest way to investigate the existence of potential ghost QTL would be to play around with a

Table 16.2. Partitionings of yield variation at position 140 cM on chromosome 1. An estimate for error, derived from the median intra-block error, was 0.75.

Source of variation	Degrees of freedom	Sum of squares	Mean square
Environment (E)	7	12,777.169	1,825.310
G + GE	1,680	3,212.868	1.914
F_2 family (G)	210	1,382.102	6.581
GE	1,470	1,829.700	1.245
Total	1,687	15,988.970	
G + GE	1,680	3,212.868	1.914
QTL + QE Chrom. 2–10	–*	760.990	–
G + GE adjusted	1,680*	2,451.547	
F_2 family (G) adjusted	210	748.879	3.566
GE adjusted	1,470	1,702.668	1.158
G + GE adjusted	1,680	2,451.547	
QTL + QEI 140 cM Chrom. 1	8	154.069	19.259
QTL main effect	1	67.892	67.982
QEI	7	86.177	12.311
Min. temp. flow.	1	65.810	65.810
Residual QEI	6	20.368	3.395
Deviations	1,672	2,297.478	1.374

*For the correction of the yield data due to genetic effects at chromosomes 2–10, degrees of freedom might be discounted (see text).

Table 16.3. QTL effects for yield, corresponding to three models. The fit for the factorial regression model for QEI is expressed as the product* of the regression coefficient and the value for the minimum temperature during flowering in degrees Celsius, the latter given as the deviation from the mean for that temperature (across the eight environments).

Trial code	QTL per environment — QTL effect	QTL + QEI model — QTL main effect	QTL + QEI model — QTL interaction	Factorial regression on min. temp. flow. for QEI — Fit regression on min. temp. flow.	Factorial regression on min. temp. flow. for QEI — Residual QTL interaction
NS92a	−0.62	−0.33	−0.29	0.065*−2.88	−0.10
IS92a	−0.55	−0.33	−0.22	0.065*−3.76	0.03
SS92a	−0.84	−0.33	−0.51	0.065*−4.76	−0.19
IS94a	−0.60	−0.33	−0.27	0.065*−3.08	−0.07
SS94a	−0.31	−0.33	0.02	0.065*−2.98	0.21
LN96a	0.00	−0.33	0.33	0.065* 1.26	0.25
LN96b	−0.08	−0.33	0.25	0.065* 8.12	−0.28
HN96b	0.35	−0.33	0.68	0.065* 8.09	0.15
Standard error	0.132	0.042	0.119	0.0083[a]	0.116

[a]Standard error of slope.

pair of markers at increasing distances from the location of an earlier identified QTL, as suggested by Hackett et al. (2001).

A second point of criticism could be that the factorial-regression model we use does not fully exploit correlations between environments as the variance–covariance structure is the identity structure typical of standard ANOVA and regression models. Again, the factorial-regression model can be extended so as to have a more complicated variance–covariance structure. Once a number of putative QTL have been identified with standard factorial-regression models, more elaborate mixed models, like those described by Piepho (2000), can be fitted to improve estimation of QTL effects and investigate the QTL basis for genetic correlations between environments and traits. This subject is currently under study.

References

Caliński, T., Kaczmarek, Z., Krajewski, P., Frova, C. and Sari-Gorla, M. (2000) A multivariate approach to the problem of QTL localization. *Heredity* 84, 303–310.

Churchill, G.A. and Doerge, R.W. (1994) Empirical threshold values for quantitative trait mapping. *Genetics* 138, 963–971.

Crossa, J., Vargas, M., Van Eeuwijk, F.A., Jiang, C., Edmeades, G.O. and Hoisington, D. (1999) Interpreting genotype × environment interaction in tropical maize using linked molecular markers and environmental covariables. *Theoretical and Applied Genetics* 99, 611–625.

Denis, J.-B. (1988) Two-way analysis using covariates. *Statistics* 19, 123–132.

Denis, J.-B. (1991) Ajustement de modèles linéaires et bilinéaires sous contraintes linéaires avec données manquantes. *Revue de Statistique Appliquée* 39, 5–24.

Finlay, K.W. and Wilkinson, G.N. (1963) The analysis of adaptation in a plant-breeding programme. *Australian Journal of Agricultural Research* 14, 742–754.

Genstat 5 Committee (1993) *Genstat 5 Release 3 Reference Manual*. Numerical Algorithms Group, Oxford, 796 pp.

Hackett, C.A., Meyer, R.C. and Thomas, W.T.B. (2001) Multi-trait QTL mapping in barley using multivariate regression. *Genetical Research* 77, 95–106.

Haley, C.S. and Knott, S.A. (1992) A simple regression method for mapping quantitative trait loci in line crosses using flanking markers. *Heredity* 69, 315–324.

Helland, I.S. (1988) On the structure of partial least squares. *Communications in Statistics, Part B Simulations and Computations* 17, 581–607.

Jansen, R.C. and Stam, P. (1994) High resolution mapping of quantitative traits into multiple loci via interval mapping. *Genetics* 136, 1447–1455.

Jansen, R.C., Van Ooijen, J.W., Stam, P., Lister, C. and Dean, C. (1995) Genotype by environment interaction in genetic mapping of multiple quantitative trait loci. *Theoretical and Applied Genetics* 91, 33–37.

Jiang, C. and Zeng, Z.-B. (1995) Multiple trait analysis of genetic mapping for quantitative trait loci. *Genetics* 140, 1111–1127.

Jiang, C. and Zeng, Z.-B. (1997) Mapping quantitative trait loci with dominant and missing markers in various crosses for two inbred lines. *Genetica* 101, 47–58.

Kao, C.-H. (2000) On the difference between maximum likelihood and regression interval mapping in the analysis of quantitative trait loci. *Genetics* 156, 855–865.

Korol, A.B., Ronin, Y.I. and Nevo, E. (1998) Approximate analysis of QTL–environment interaction with no limits on the number of environments. *Genetics* 148, 2015–2028.

Kuehl, R.O. (2000) *Design of Experiments: Statistical Principles of Research Design and Analysis*, 2nd edn. Duxbury Press, Pacific Grove, California.

Lynch, M. and Walsh, B. (1998) *Genetics and the Analysis of Quantitative Traits*. Sinauer Associates, Sunderland, Massachusetts.

Manly, B.F.J. (1997) *Randomization, Bootstrap and Monte Carlo Methods in Biology*, 2nd edn. Chapman & Hall, London, 397 pp.

Neter, J., Kutner, M.H., Nachtsheim, C.J. and Wasserman, W. (1996) *Applied Linear Statistical Models*, 4th edn. Irwin, Chicago, 1408 pp.

Osten, D.W. (1988) Selection of optimal regression models via cross-validation. *Journal of Chemometrics* 2, 39–88.

Piepho, H.-P. (2000) A mixed model approach to mapping quantitative trait loci in barley on the basis of multiple environment data. *Genetics* 156, 2043–2050.

Ribaut, J.-M., Hoisington, D.A., Deutsch, J.A., Jiang, C. and Gonzalez-de-Leon, D. (1996) Identification of quantitative trait loci under drought conditions in tropical maize. 1. Flowering parameters and the anthesis–silking interval. *Theoretical and Applied Genetics* 92, 905–914.

Ribaut, J.-M., Jiang, C., Gonzalez-de-Leon, D., Edmeades, G.O. and Hoisington D.A. (1997) Identification of quantitative trait loci under drought conditions in tropical maize. 2. Yield components and marker-assisted selection strategies. *Theoretical and Applied Genetics* 94, 887–896.

Sari-Gorla, M., Caliński, T., Kaczmarek, Z. and Krajewski, P. (1997) Detection of QTL × environment interaction in maize by a least squares interval mapping method. *Heredity* 78, 146–157.

Van Eeuwijk, F.A., Denis, J.-B. and Kang, M.S. (1996) Incorporating additional information on genotypes and environments in models for two-way genotype by environment tables. In: Kang, M.S. and Gauch, H.G., Jr (eds) *Genotype-by-Environment Interaction.* CRC Press, Boca Raton, Florida, pp. 15–49.

Van Eeuwijk, F.A., Crossa, J., Vargas, M. and Ribaut, J.M. (2001) Variants of factorial regression for analysing QTL by environment interaction. In: Gallais, A., Dillmann, C. and Goldringer, I. (eds) Eucarpia, *Quantitative genetics and breeding methods: the way ahead.* Les Colloques 96, INRA Editions, Versailles, pp. 107–116.

Vargas, M., Crossa, J., Van Eeuwijk, F.A., Ramírez, M.E. and Sayre, K. (1999) Using AMMI, factorial regression, and partial least squares regression models for interpreting genotype × environment interaction. *Crop Science* 39, 955–967.

Voltas, J., Van Eeuwijk, F.A., Sombrero, A., Lafarga, A., Igartua, E. and Romagosa, I. (1999a) Integrating statistical and ecophysiological analysis of genotype by environment interaction for grain filling of barley in Mediterranean areas. I. Individual grain weight. *Field Crops Research* 62, 63–74.

Voltas, J., Van Eeuwijk, F.A., Araus, J.L. and Romagosa, I. (1999b) Integrating statistical and ecophysiological analysis of genotype by environment interaction for grain filling of barley in Mediterranean areas. II. Grain growth. *Field Crops Research* 62, 75–84.

17 Elements of Genotype–Environment Interaction: Genetic Components of the Photoperiod Response in Maize

R. Moutiq,[1] J.-M. Ribaut,[2] G.O. Edmeades,[3] M.D. Krakowsky[1] and M. Lee[1]

[1]Department of Agronomy, Iowa State University, Ames, IA 50011, USA; [2]International Maize and Wheat Improvement Center (CIMMYT), Lisboa 27, Apdo. Postal 6-641, 06600 Mexico, DF, Mexico; [3]Pioneer Hi-Bred International Inc., PO Box 609, Waimea, HI 96796, USA

Introduction

Genotype–environment interaction remains one of the major sources of vexation and opportunity in plant improvement. Identification of the primary genetic and environmental determinants of the numerous interactions will be a step forward for plant-improvement programmes. Contemporary biological and information sciences are beginning to provide the foundation necessary to conduct studies that link genes with phenotypes and responses to defined signals from the environment.

The centre of origin and diversity of maize (*Zea mays* L.) is in the tropics, but the crop is planted between 58°N and 40°S latitude (Hallauer and Miranda, 1981). Maize has a short-day photoperiod response and has adequate allelic variation to effectively convert the photoperiod response to day neutrality. Photoperiod sensitivity limits the evaluation and exchange of germ-plasm between temperate and tropical breeding programmes. Therefore, maize genetic diversity has not been well exploited. In temperate maize, less than 4% originates from tropical or exotic germ-plasm (Darrah and Zuber, 1985).

In maize, the inheritance of photoperiod response is not well defined. Photoperiod sensitivity is a quantitative trait and its inheritance is mostly additive (Russell and Stuber, 1985). Two to 19 loci were reported to control flowering (Giesbrecht, 1960; Francis, 1972). Koester *et al.* (1993) reported two mechanisms that control the inheritance of flowering: base maturity and photoperiod sensitivity. Kim *et al.* (1991) and Koester *et al.* (1993) detected few quantitative trait loci (QTL) related to flowering and they suggested that chromosome 8 might be involved in the photoperiod response.

In other cereals, three classes of genes controlling flowering time are known: vernalization genes, photoperiod genes and 'earliness *per se*' genes, which control flowering independently from the environment. Other mutations have been reported, but they have not been related to any of these three classes. In barley (*Hordeum vulgare* L.), loci for 'earliness *per se*' – *ea*, *ea*$_{sp}$, *ea*$_c$, *ea*$_k$ and *ea7* – were detected as loci controlling flowering and/or photoperiod sensitivity (Nilan, 1964; Takahashi and Yasuda, 1971; Gallagher *et al.*, 1991; Von Wettstein-Knowles, 1992). The genes ea_{sp}, ea_c and ea_k are activated by short days (Gallagher *et al.*, 1991). Also in barley, *Sh*, *Sh2* and *Sh3* are loci responsible for winter and spring growth habit (Takahashi and Yasuda, 1971). Two genes associated with photoperiod response were detected: *Ppd-H1* regulates flowering only in long days and *Ppd-H2* has significant effect only under short days (Laurie *et al.*, 1994). Epistatic interactions contributing to the control of flowering between *Ppd* and *Sh2* were observed (Karsai *et al.*, 1997). In wheat (*Triticum aestivum* L.), chromosomes 1A, 4B, 6B, 3B and 7D are involved in controlling flowering time (Halloran and Boydell, 1967). Marcellos and Single (1971) classified wheat genotypes in four classes based on their response to photoperiod: least, slightly, moderately and strongly sensitive. Later, *Ppd1*, *Ppd2* and *Ppd3* were identified as major loci for photoperiod sensitivity on chromosomes 2D, 2B and 2A (Law *et al.*, 1978) and seem to be homologous to *Ppd-H1* loci in barley (Laurie *et al.*, 1994). In rice (*Oryza sativa* L.), different photoperiod-sensitivity genes – *E1-E3*, *Se2-Se5*, *Se-1n*, *Se-1u*, *I-Se-1* and *En-Se-1* – were identified (Okumoto and Tanisaka, 1997). In sorghum (*Sorghum bicolor*), four genes were identified that control flowering time (Morgan, 1994). These genes were termed maturity genes, or *Ma1*, *Ma2*, *Ma3* and *Ma4*.

In *Arabidopsis*, multiple pathways control flowering time and currently 80 genes and loci are known to be involved.

1. The photoperiod promotion pathway initiates flowering in response to photoperiod through a number of genes that sense and respond to day length. It groups mutants that have their flowering time delayed in long days but not in short days.
2. The autonomous promotion pathway includes the genes that promote flowering independently from environmental signals.
3. The gibberellic acid (GA) promotion pathway plays a promotive role in flowering based on signals mediated by GA. The application of GA accelerates the flowering time of wild-type plants under short days and of the late-flowering mutants under long days. Under non-inductive photoperiods, the *ga1* mutant does not flower unless provided with GA.
4. The floral transition (FT) subgroup includes mutants that flower late in long days but have properties that distinguish them from the photoperiod pathway (Simpson *et al.*, 1999).

A better understanding of the inheritance of photoperiod sensitivity in maize and the identification of associated QTL might be of interest to breeders in order to have an easy and rapid exchange of germ-plasm across latitudes. Identification of molecular markers closely associated with major QTL controlling photoperiod sensitivity might enable indirect classification of germ-plasm for photoperiod response and provide a basis for the marker-assisted conversion of germ-plasm.

The objectives of the present study are to: (i) identify and locate QTL related to flowering and photoperiod response in a maize population evaluated in long- and short-day environments; (ii) determine their genetic control; (iii) compare these QTL with previously identified QTL in maize and related crops; and (iv) relate QTL with possible candidate genes.

Material and Methods

Plant material

A sensitive tropical inbred line CML9 was crossed to a relatively insensitive temperate line A632Ht. The F_1 generation was self-pollinated and the F_2 generation was grown at Tlaltizapan, Mexico (18°N, 99°W and 940 m above sea level (masl)), and self-pollinated to produce the F_3 lines used for phenotypic evaluations. Leaf samples were harvested from F_2 plants, frozen, ground and stored at -18°C. A selection of 236 F_3 lines was made for evaluation.

Field design

A total of 236 F_3 lines and both parents, A632Ht and CML9, were evaluated in different photoperiods and years. Each parent was repeated twice in each replication. A 24×10 alpha (0, 1) lattice with two replications was used. The F_3 lines and the parents were planted in single-row plots, 2.5 m long with 20 cm between rows. Experiments were planted under two different day lengths and during 3 successive years. There were a total of six environments:

1. Tlaltizapan, Mexico (18°N, 99°W and 940 masl), cycle B (June to November) in 17 h of day length in 1995.
2. Tlaltizapan cycle B in normal day length, 13 h, in 1995.
3. Tlaltizapan, cycle A (November–April) with a day length of 11.5 h in 1996.
4. Tlaltizapan, cycle A with day length of 17 h in 1996.
5. Ames at the Iowa State University (ISU) Agronomy and Agricultural Engineering Research Center, West Ames, with a day length of 15.5 h in 1997.
6. Tlaltizapan, cycle B under normal day length of 13 h in 1997.

Artificial light provided by 150-watt bulbs was used to extend the day length to 17 h in environments 1 and 4.

Phenotypic data

Herein, the primary trait of interest is the number of days from sowing to anthesis (i.e. AD). Anthesis was defined as the date on which 50% of the plants in a plot exerted anthers. Separate analyses of AD in long- and short-day environments were conducted. Subsequently, the data from the long-day environments (1, 4 and 5 above) were combined and used in the analysis of variance and QTL detection. The same was done for data collected in the short-day environments (2, 3 and 6). To estimate the adjusted means used in QTL analyses, entries were considered fixed, while complete and incomplete blocks were random (Cardinal

et al., 2001). To calculate variances, entries, complete and incomplete blocks were all considered random (Cardinal *et al.*, 2001). Broad-sense heritabilities on an entry-mean basis were calculated as previously established (Fehr, 1987). Exact confidence intervals (CIs) for heritability estimates were calculated (Knapp *et al.*, 1985).

Linkage mapping

Restriction fragment length polymorphisms (RFLPs) and simple sequence repeats (SSRs) were used to make the linkage map. DNA was extracted from the parental lines and the 236 F_2 plants. Procedures for RFLP analysis have been described (Hoisington *et al.*, 1994). SSRs were used to detect more polymorphic loci, especially to cover gaps left with RFLP markers according to published procedures (Senior *et al.*, 1996). A total of 129 loci were used to make the genetic map.

MAPMAKER Version 3.0 (Lander *et al.*, 1987) was used to construct the linkage map based on the genetic data of 236 individuals. Markers were assigned to linkage groups based on a minimum LOD score of 3.0 and a maximum Haldane distance of 50 cM. For chromosome 4, the minimum distance was extended to 54 because of one gap (53 cM) therein. The 'three-point' command was used for each linkage group. The 'order' command for multipoint analysis was repeated and the best order from it was used as a starting-point for each linkage group. The command 'try' was used to place the remaining markers in their appropriate linkage group. The chi-square test at $P = 0.05$ detected segregation distortion at 14 (11%) loci. Ten linkage groups were obtained and the total length of the map is 1658.3 cM. The map is mostly in agreement with previous published maps (Davis *et al.*, 1999).

QTL mapping

The composite interval mapping method (Zeng, 1994) facilitated by PLABQTL version 1.1 (Utz and Melchinger, 1996) was used for

QTL mapping. The analysis was conducted for each of the two groups of environments (i.e. the long-day environments and the short-day environments). Cofactors were selected based on the output of the stepwise method using the 'cov select' command. The cofactors close to the putative QTL were then chosen as final cofactors. The threshold LOD score was estimated based on the output of 1000 permutations using PLABQTL. The identified QTL were then included in a model analysed by forward and backward regression facilitated by the 'seq/s' statement of PLABQTL. The best model was chosen based on the Akaiki information criterion (AIC) values (Jansen, 1993). Epistatic interactions between all pairs of marker loci were tested using the Epistacy programme (Holland, 1998). The significant interaction terms with a P value less than 0.00026 (Holland et al., 1997) were included in the final model. Interactions were declared significant and maintained in the final model when $P \leq 0.05$ for the interactions, as well as for the markers close to the QTL.

Results

Field evaluation

In general, AD was higher in long-day than in short-day environments for both parents, as well as for the F_3 lines. The parent CML9 had a stronger response to photoperiod than A632Ht. The difference between the two parents was 7 days in short-day environments and 34 days in long-day environments (Table 17.1). The difference of AD for the same parent in different photoperiods was 6 days for A632Ht and 33 days for CML9 (Table 17.1). These results confirmed that CML9 is more photoperiod-sensitive than A632Ht. The mean AD of the F_3 lines was higher in long-day environments, 71 days in short-day environments and 92 days in long days (Table 17.1). The distribution of AD values in short- and long-day environments was clearly separated. Transgressive segregants were observed only in

short-day environments (Table 17.1). The heritabilities on an entry-mean basis were similar in both photoperiods: 0.85 (0.82–0.88 95% CI) and 0.88 (0.85–0.90 95% CI) in short- and long-day environments, respectively (Table 17.1).

QTL mapping

Different sets of QTL were detected in long-and short-day environments. In long days, five putative QTL were detected on chromosomes 2, 3, 8, 9 and 10. The total phenotypic variation associated with these QTL was 60% (Table 17.2). In short-day environments, QTL were identified on chromosomes 1, 2, 3, 4 (umc353), 4 (npi444), 5 and 9. The total phenotypic variation explained by these QTL was 40% (Table 17.2). Comparing long- and short-day environments, the QTL on chromosome 2 had a similar map position in both

Table 17.1. Means, variances and entry-mean basis heritabilities of AD in the CML9 × A632Ht maize population in long- and short-day environments.

Photoperiod	Long days (days)	Short days (days)
Parents		
CML9	112	79
A632Ht	78	72
LSD	4	5
F_3 lines		
Mean	92	71
Range	82–102	65–77
LSD	5	3
Variances		
Genotypic	18	5
95% CI	15–22	4–6
Error	10	2
95% CI	9–12	1–2
G × E	2	2
95% CI	1–2	2–4
Heritability	0.88	0.85
95% CI	0.85–0.90	0.82–0.88

LSD, least significant difference at $\alpha = 5\%$; CI, confidence interval; G × E, genotype × environment.

photoperiods. Even though the QTL on chromosome 9 had close map positions in both photoperiods (Fig. 17.1), the contrasting parental effects suggested that two different QTL are involved and both are photoperiod-dependent. QTL on chromosomes 3 (npi108a), 8, 9 (umc39d) and 10 were associated with AD only in long days. QTL on chromosomes 1, 3 (umc10), 4 (umc353), 4 (npi444), 5 and 9 (umc81) were associated with AD only in short days (Fig. 17.1 and Table 17.2).

All QTL identified herein in both photoperiods had highly significant additive effects and four QTL had significant dominance effects. QTL in long days had higher additive effects than in short-day environments. CML9 alleles were associated with higher AD (i.e. later flowering) at most QTL in both photoperiods, with additive genetic effects from 1 to 4 days in long-day environments and from 0.5 to 1.0 day in short-day environments (Table 17.2). In long days, the QTL on chromosomes 8 and 10 had the largest additive effects: 2.5 and 4 days (Table 17.2). In short days, QTL on chromosomes 3 (umc10) and 9 had the largest additive values but with contrasting effects: 1 and −1,

respectively (Table 17.2). Dominance effects were negative, indicative of a decrease of AD (i.e. earlier flowering) for most QTL in both photoperiods. However, these effects were statistically significant only for the QTL on chromosome 2 in short days and on chromosomes 2 and 3 in long days. The QTL on chromosome 10 in long-day environments had significant dominance effects that increased AD (positive value; Table 17.2). Overall, dominance effects for AD were towards earlier anthesis in both long- and short-day environments (Table 17.2). A significant epistatic effect between umc96 on chromosome 3 and npi203 on chromosome 4 was detected in long-day environments but not in short-day environments (Table 17.3).

Discussion

Phenotypic and genotypic data

The tropical parent CML9 is highly sensitive to day length, since it had a stronger response to photoperiod than the temperate parent A632Ht. The AD of the population of F_3 lines was significantly higher in long-day

Table 17.2. Chromosomal location, additive and dominance effects and partial R^2 of QTL associated with AD in long-day and short-day environments in the CML9 × A632Ht maize population.

Environment	Chromosome	Adjacent marker	Additive effect † (days)	Partial R^2 additive	Dominance effect	Partial R^2 dominance
Long days	2	Umc38a	1.16**	7.7	−1.66**	7.1
	3	Npi108a	1.57**	11.2	−1.14*	2.7
	8	Umc138b	2.59**	27.9	−0.48	0.6
	9	Umc39d	2.38**	17.7	0.60	0.4
	10	Npi264	4.27**	45.5	0.86*	1.7
			Total adjusted R^2: 60			
Short days	1	Umc23	0.62**	4.9	−0.30	0.5
	2	Umc38a	0.78**	8.9	−0.76**	4.0
	3	Umc10	1.17**	15.0	−0.45	1.3
	4	Umc353	0.95**	9.5	−0.34	0.6
	4	Npi444	0.50**	3.0	0.52	1.5
	5	Npi409	−0.59**	4.8	−0.11	0.1
	9	Umc81	−0.97**	12.6	−0.31	0.7
			Total adjusted R^2: 40			

*Significant at $P < 5\%$; **significant at $P < 1\%$.
†Positive additive effect means that the CML9 alleles increase the value of the trait.

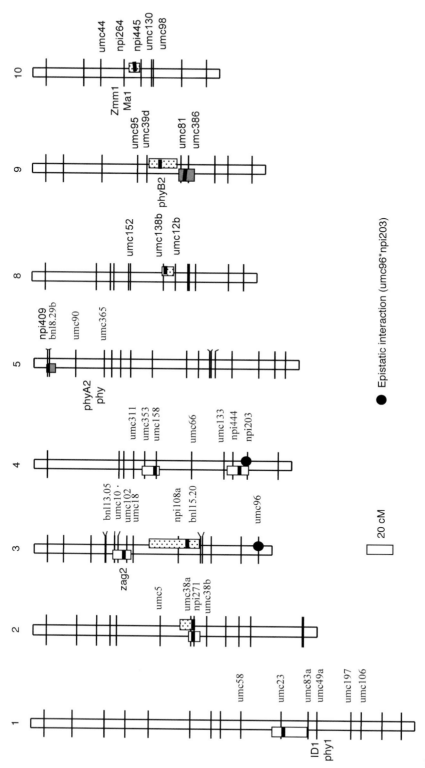

Fig. 17.1. Genetic map of CML9 × A632Ht population, location of QTL and epistatic interaction related to days from sowing to anthesis (AD) in long- and short-day environments and position of candidate genes. The boxes are the support interval for each QTL and the wide black strip in each box is the most probable position of the QTL. White or grey boxes (left side of the chromosome) represent QTL in short days. Alleles from CML9 increase AD for white-box QTL, while alleles from A632Ht increase AD for grey-box QTL. Dotted boxes (right side of the chromosome) represent QTL in long days with alleles from CML9 increasing AD. Candidate genes are in bold on the left side of the chromosome.

Table 17.3. Significant epistatic effects in long-day environments in the CML9 × A632Ht maize population.

Source	Chromosomes	F	Probability
npi264	10	17.44	< 0.0001
umc138b	8	16.70	< 0.0001
umc81	9	5.21	0.0074
npi108a	3	7.02	0.0015
npi271	2	7.73	0.0008
npi203	4	0.10	0.9055
umc96	3	0.61	0.5475
umc96* npi203	3*4	4.51	0.0024

environments. The distribution of the lines was clearly separated in both photoperiods, and transgressive progenies were observed towards earlier flowering in the short-day environments. The presence of transgressive progeny for flowering time has been observed across different plant species, such as rice (Dung *et al.*, 1998; Lin *et al.*, 1998) and barley (Stracke and Borner, 1998). The segregation distortion of some alleles might indicate a non-random transmission of alleles from parents to offspring. This was previously observed in other maize populations (Edwards *et al.*, 1987; Abler *et al.*, 1991; Koester *et al.*, 1993). This distortion should not affect the association between a marker and the phenotype (Koester *et al.*, 1993), but it can affect the precision of the QTL position (Lorieux *et al.*, 1995). Segregation distortion was observed for alleles of the loci close to the QTL on chromosomes 1 (umc23), 3 (umc10), 4 (npi444) and 10 (npi264).

QTL mapping

Different sets of QTL were found in long- and short-day environments. One QTL was photoperiod-independent and ten QTL were photoperiod-dependent. The QTL on chromosome 2 had similar positions in both photoperiods and seems to control AD independently from photoperiod perhaps in a manner similar to that of the 'autonomous promotion pathway', as described in *Arabidopisis* (Simpson *et al.*, 1999), or

'earliness *per se*', as defined in barley (Laurie *et al.*, 1995). The QTL on chromosomes 3, 8, 9 (umc39d) and 10 were detected only in long-day environments, suggesting that these QTL are photoperiod-sensitive and become effective only when day length is above the critical value for maize (i.e. 14.5 h) (Francis, 1972). These QTL might be classified in the photoperiod promotion pathway, as described by Simpson *et al.* (1999) in *Arabidopsis*. The QTL on chromosomes 1, 3 (umc102), 4 (npi444), 4 (umc353), 5 and 9 (umc81) were related to AD only in short-day environments. These QTL are also photoperiod-sensitive but promote flowering when day length is lower than the critical value for maize. Therefore, photoperiod-sensitive genes may be subdivided into two subclasses: a group of genes detected in long days (the QTL on chromosomes 3, 8, 9 (umc39d) and 10) and a second subclass associated with flowering in short days (the QTL on chromosomes 1, 3 (umc102), 4 and 9 (umc81)).

Alleles for earliness were only from the A632Ht parent in long days, but in short days two earliness alleles (chromosomes 5 and 9) were from CML9. Consequently, the combination of earliness alleles from both parents might explain the trangressive segregation for earliness. AD mostly showed earliness dominant to lateness in both photoperiods. Previous studies reported that earliness was dominant to lateness in other maize populations in long and short days (Yang Yun-Kuei, 1949; Giesbrecht, 1960; Bubeck *et al.*, 1993; Cardinal *et al.*, 2001). The same result was observed in other crops, such as wheat, where early heading was found to be partially dominant to late heading (Klaimi and Qualset, 1973). In contrast, dominance towards lateness was prevalent in sorghum (Lin *et al.*, 1995).

One significant digenic interaction was detected in long days, but none was detected in short days. Similar epistatic effects between QTL or flowering genes were previously detected in maize (Rebai *et al.*, 1997), wheat (Klaimi and Qualset, 1973; Pan *et al.*, 1994), barley (Gallagher *et al.*, 1991), soybean (Cober *et al.*, 1996) and *Arabidopsis*

(Kuittinen *et al.*, 1997). In the analyses for epistatic effects, the closest loci and not the exact position of the QTL have been used. Interactive effects might be underestimated with simple analysis of variance because of possible recombination between the QTL and the linked marker used in the analyses.

Comparison with other maize populations and related species

Comparison of QTL position with other maize studies was based on common loci or using the Pioneer Composite 1999 map as a reference when no common loci are available. The QTL detected herein on chromosome 1 was linked to the QTL identified previously in long-day conditions (Veldboom *et al.*, 1994; Cardinal *et al.*, 2001). Also, a QTL closely linked to that region of chromosome 1 has been reported in a previous study conducted in long- and short-day environments (Koester *et al.*, 1993). The QTL on chromosome 2 was linked to the previously identified QTL for AD in long days (Cardinal *et al.*, 2001) and the QTL for time to silking in short days (Bubeck *et al.*, 1993). On chromosome 3, two independent studies conducted in long days (Zehr *et al.*, 1992; Cardinal *et al.*, 2001) found two different QTL each in regions linked to those detected herein (umc10 in long days and npi108 in short days). A QTL linked to umc10 was also previously identified in short days (CIMMYT, 1994). On chromosome 8, a QTL was previously detected only in long days in the same region as detected herein (Stuber *et al.*, 1992; Zehr *et al.*, 1992; Koester *et al.*, 1993; Veldboom *et al.*, 1994). A chromosome segment from early-flowering germ-plasm, introgressed in a back-crossing programme for earliness, was also observed in the same region (Kim *et al.*, 1991; Koester *et al.*, 1993). On chromosome 9 (umc39d), the QTL found herein only in long days was linked to a QTL previously detected in a similar day length (Veldboom *et al.*, 1994). However, in the same chromosomal region, Koester *et al.* (1993) found a QTL in both

short and long days. This might suggest that at least two different QTL are located in that region. The second QTL found herein on chromosome 9 (umc81) only in short days had a similar map position to a QTL previously identified in long days (Abler *et al.*, 1991; Nourse, 2000; Cardinal *et al.*, 2001). An independent study (Nourse, 2000) conducted in both photoperiods found a QTL in that region only in long days, suggesting that more than one QTL occurs in that region – one effective in long days and the other in short days. On chromosome 10, a QTL was found herein only in long days. Previous studies detected a QTL in the same region only in long days (Abler *et al.*, 1991; Nourse, 2000). Koester *et al.* (1993) found a QTL in that region in long- and short-day environments. An introgressed region from early material was identified in that location in a back-crossing programme for earliness (Koester *et al.*, 1993). That location might harbour at least two different QTL – one expressed in long days and the other in short days.

Candidate genes and QTL in different species were identified in similar map positions (same species) or corresponding regions (different species) as QTL detected herein. On chromosome 1, a rice QTL controlling flowering was detected in the region corresponding to the maize region where the QTL is detected herein (Li *et al.*, 1995). Genes such as *Id1* (indeterminate growth 1) and *phy1* (phytochrome 1), which are known for their effect on flowering time, were associated with this region as well. A QTL controlling flowering in sorghum was detected in long days in linkage group B (Lin *et al.*, 1995), which corresponds to maize chromosome 2 at the region where the QTL is detected herein. On chromosome 3 in the region of umc10, a candidate gene *zag2* (*Zea agamous* homolog 2), with 49% identity with an agamous gene (*AG*) of *Arabidopsis*, might be involved in controlling flowering time, as in *Arabidopsis*. *AG* restricts flowering to short-day environments when mutated (*ag*) in *Arabidopsis* and the overexpression of *AG* alleles results in early flowering (Simpson *et al.*, 1999). On chromosome 5, candidate genes *phyA2* and *phy* were linked to the region where QTL were found herein. On chromosome 9, the QTL is linked to a locus defined by *phyB2*, a gene involved in the perception of light signals. On chromosome 10, a homologous gene to *zag2*, *zmm1* in maize, is within the same region as the QTL herein. The *Ma1* gene and a QTL associated with flowering in sorghum (Lin *et al.*, 1995) also mapped to the corresponding region of maize chromosome 10 (linkage group D of sorghum). In rice, flowering genes *Se1* and *Se3* mapped to chromosome 6, a region paralogous to maize chromosome 10 (Paterson *et al.*, 1995). The homologous regions in wheat and barley harboured photoperiod genes *ppd1*, *ppd2* and *ppd3* in wheat (Hart *et al.*, 1993) and *pPD-H1* in barley (Laurie *et al.*, 1994). The paralogous region to maize chromosome 10 in rice (Paterson *et al.*, 1995) also contained a QTL related to flowering time (Li *et al.*, 1995). Moreover, the proportion of the phenotypic variation explained by the QTL found herein on chromosome 10 was the largest ($R^2 = 24$–46%) and that degree of association was observed for the corresponding regions in sorghum (Lin *et al.*, 1995) and barley (Laurie *et al.*, 1994).

References

Abler, B.S.B., Edwards, M.D. and Stuber, C.W. (1991) Isozymic identification of quantitative trait loci in crosses of elite maize inbreds. *Crop Science* 31, 267–274.

Bubeck, D.M., Goodman, M.M., Beavis, W.D. and Grant, D. (1993) Quantitative trait loci controlling resistance to gray leaf spot in maize. *Crop Science* 33, 838–847.

Cardinal, A.J., Lee, M., Sharopova, N., Woodman-Clikeman, W.L. and Long, M.J. (2001) Genetic mapping and analysis of quantitative trait loci for resistance to stalk tunneling by the European corn borer in maize. *Crop Science* 41, 835–845.

CIMMYT (1994) QTL data for populations Ki3 × CML139 and CML131 × CML67. In maize database. http://www.agron.missouri.edu/locus.html

Cober, E.R., Tanner, J.W. and Voldeng, H.D. (1996) Genetic control of photoperiod response in early-maturing, near-isogenic soybean lines. *Crop Science* 36, 601–605.

Darrah, L.L. and Zuber, M.S. (1985) United States farm germplasm base and commercial breeding strategies. *Crop Science* 26, 1109–1113.

Davis, G.L., McMullen, M.D., Baysdorfer, C., Musket, T., Grant, D., Stanebell, M., Xu, G., Polacco, M., Koster, L., Melia-Hancock, S., Houchins, K., Chao, S. and Coe, E.H., Jr (1999) A maize map standard with sequenced core markers, grass genome reference points and 932 expressed sequence tagged sites (ESTs) in a 1736-locus map. *Genetics* 152, 1137–1172.

Dung, L.V., Inukai, T. and Sano, Y. (1998) Dissection of a major QTL for photoperiod sensitivity in rice: its association with a gene expressed in an age-dependent manner. *Theoretical and Applied Genetics* 97, 714–720.

Edwards, M.D., Stuber, C.W. and Wendel, J.F. (1987) Molecular-marker-facilitated investigations of quantitative-trait loci in maize. I. Numbers, genomic distribution and types of gene action. *Genetics* 116, 113–125.

Fehr, W.R. (1987) *Principles of Cultivar Development*. Macmillan, New York.

Francis, C.A. (1972) Photoperiod sensitivity and adaptation in maize. *Proceedings of Annual Corn and Sorghum Research Conference* 27, 119–131.

Gallagher, L.W., Soliman, K.M. and Vivar, H. (1991) Interactions among loci conferring photoperiod insensitivity for heading time in spring barley. *Crop Science* 31, 256–261.

Giesbrecht, J. (1960) The inheritance of maturity in maize. *Canadian Journal of Plant Science* 40, 490–499.

Hallauer, A.R. and Miranda, J.B. (1981) *Quantitative Genetics in Maize Breeding*. Iowa State University Press, Ames, USA.

Halloran, G.M. and Boydell, C.W. (1967) Wheat chromosomes with genes for photoperiodic response. *Canadian Journal of Genetics and Cytology* 9, 394–398.

Hart, G.E., Gale, M.D. and McIntosh, R.A. (1993) Linkage maps of *Triticum aestivum* (hexaploid wheat, $2n = 42$, genomes A, B and D) and *T. tauschii* ($2n = 14$, genome D). In: O'Brien, S.J. (ed.) *Genetic Maps: Locus Maps of Complex Genomes*, 6th edn. Cold Spring Harbor Laboratory Press, Cold Spring Harbor, New York, pp. 6.204–6.219.

Hoisington, D.A., Khairallah, M. and Gonzalez-de-Leon, D. (1994) *Laboratory Protocols: CIMMYT Applied Molecular Genetics Laboratory*. CIMMYT, Mexico.

Holland, J.B. (1998) Epistacy: a SAS program for detecting two-locus epistatic interactions using genetic marker information. *Journal of Heredity* 89, 374–375.

Holland, J.B., Moser, H.S., O'Donoughue, L.S. and Lee, M. (1997) QTLs and epistasis associated with vernalization responses in Oat. *Crop Science* 38, 1306–1316.

Jansen, R.C. (1993) Interval mapping of multiple quantitative trait loci. *Genetics* 135, 205–211.

Karsai, I., Meszaros, K., Hayes, P.M. and Bedo, Z. (1997) Effects of loci on chromosomes 2(2H) and 7(5H) on developmental patterns in barley (*Hordeum vulgare* L.) under different photoperiod regimes. *Theoretical and Applied Genetics* 94, 612–618.

Kim, T.S., Phillips, R.L., Shaver, D.L. and Openshaw, S.J. (1991) Use of backcross derived lines to detect linkages of RFLPs and maturity loci in maize. In: *Agronomy Abstracts*. American Society of Agronomy, Madison, Wisconsin, p. 101.

Klaimi, Y.Y. and Qualset, C.O. (1973) Genetics of heading time in wheat (*Triticum aestivum* L.) I. The inheritance of photoperiodic response. *Genetics* 74, 139–156.

Knapp, S.J., Stroup, W.W. and Ross, W.M. (1985) Exact confidence intervals for heritability on a progeny mean basis. *Crop Science* 25, 192–194.

Koester, R.P., Sisco, P.H. and Stuber, C.W. (1993) Identification of quantitative trait loci controlling days to flowering and plant height in two near isogenic lines of maize. *Crop Science* 33, 1209–1216.

Kuittinen, H., Sillanpaa, M.J. and Savolainen, O. (1997) Genetic basis of adaptation: flowering time in *Arabidopsis thaliana*. *Theoretical and Applied Genetics* 95, 573–583.

Lander, E.S., Green, P., Abrahamson, J., Barlow, A., Daly, M.J., Lincoln, E.S. and Newburg, L. (1987) Mapmaker: an interactive computer package for constructing primary genetic linkage maps of experimental and natural populations. *Genomics* 1, 174–181.

Laurie, D.A., Pratchett, N., Bezant, J.H. and Snape, J.W. (1994) Genetic analysis of a photoperiod response gene on the short arm of chromosome 2(2H) of *Hordeum vulgare* (barley). *Heredity* 72, 619–627.

Laurie, D.A., Pratchett, N., Bezant, J.H. and Snape, J.W. (1995) RFLP mapping of five major genes and eight quantitative trait loci controlling flowering time in a winter × spring barley (*Hordeum vulgar* L.) cross. *Genome* 38, 575–585.

Law, C.N., Sutka, J. and Worland, A.J. (1978) A genetic study of day-length response in wheat. *Heredity* 41, 185–191.

Li, Z.K., Pinson, S.R., Stansel, J.W. and Park, W.D. (1995) Identification of quantitative trait loci (QTL) for heading date and plant height in cultivated rice (*Oryza sativa* L.). *Theoretical and Applied Genetics* 91(2), 374–381.

Lin, S.Y., Sasaki, T. and Yano, M. (1998) Mapping quantitative trait loci controlling seed dormancy an heading date in rice, *Oryza sativa* L., using backcrosses inbred lines. *Theoretical and Applied Genetics* 96(8), 997–1003.

Lin, Y.R., Schertz, K.F. and Paterson, A.H. (1995) Comparative analysis of QTLs affecting plant height and maturity across the *Poaceae*, in reference to an interspecific sorghum population. *Genetics* 141, 391–411.

Lorieux, M., Goffinet, B., Perrier, X., Gonzalez de Leon, D. and Lanaud, C. (1995) Maximum likelihood models for mapping genetic markers showing segregation distortion. 2. F2 populations. *Theoretical and Applied Genetics* 90, 81–89.

Marcellos, H. and Single, M.V. (1971) Quantitative responses of wheat to photoperiod and temperature in the field. *Australian Journal of Agricultural Research* 22(3), 343–357.

Morgan, P.W. (1994) Genetic regulation of flowering in sorghum. In: *Proceedings of the Forty-ninth Annual Corn and Sorghum Industry Research Conference*. American Seed Trade Association, Washington, DC, pp. 217–226.

Nilan, R.A. (1964) *The Cytology and Genetics of Barley*. Washington State University Press, Pullman, Washington, USA.

Nourse, S.M. (2000) Molecular marker analysis of quantitative genetic traits in maize. PhD dissertation, University of Hawaii, Honolulu.

Okumoto, W. and Tanisaka, T. (1997) Trisomic analysis of a strong photoperiod-sensitivity gene E1 in rice (*Oryza sativa* L.). *Euphytica* 95(3), 301–307.

Pan, A., Hayes, P.M., Chen, F., Chen, T.H.H., Blake, T., Wright, S., Karsai, I. and Bedo, Z. (1994) Genetic analysis of the components of winterhardiness in barley (*Hordeum vulgare* L.). *Theoretical and Applied Genetics* 89, 900–910.

Paterson, A.H., Lin, Y.R., Li, Z., Schertz, K.F., Doebley, J.F., Pinson, S.R.M., Liu, S.C., Stansel, J.W. and Irvine, J.E. (1995) Convergent domestication of cereal crops by independent mutations at corresponding genetic loci. *Science* 269, 1712–1718.

Rebai, A., Blanchard, P., Perret, D. and Vincourt, P. (1997) Mapping quantitative trait loci silking date in a diallel cross among four lines of maize. *Theoretical and Applied Genetics* 95, 451–459.

Russell, W.K. and Stuber, C.W. (1985) Genotype × photoperiod and genotype × temperature interactions for maturity in maize. *Crop Science* 25, 152–158.

Senior, M.L., Chin, E.C.L., Lee, M., Smith, J.S.C. and Stuber, C.W. (1996) Simple sequence repeat markers developed from maize sequences found in the GENBANK database: map construction. *Crop Science* 36, 1676–1683.

Simpson, G.G., Gendall, A.R. and Dean, C. (1999) When to switch to flowering. *Annual Reviews of Cell and Developmental Biology* 99, 519–550.

Stracke, S. and Borner, A. (1998) Molecular mapping of the photoperiod response gene *ea7* in barley. *Theoretical and Applied Genetics* 97, 797–800.

Stuber, C.W., Lincoln, S.E., Wolff, D.W., Helentjaris, T. and Lander, E.S. (1992) Identification of genetic factors contributing to heterosis in a hybrid from two elite maize inbreeding lines using molecular markers. *Genetics* 132, 823–839.

Takahashi, R. and Yasuda, S. (1971) Genetics of earliness and growth habit in barley. In: Nilan, R.A. (ed.) *Proceedings of the 2nd International Barley Genetics Symposium.* Washington State University Press, Pullman, Washington, pp. 388–408.

Utz, H.F. and Melchinger, A.E. (1996) PLABQTL: a program for composite interval mapping of QTL. *Journal of Quantitative Trait Loci* 2, 1.

Veldboom, L.R., Lee, M. and Woodman, W.L. (1994) Molecular marker facilitated studies in an elite maize population. I. Linkage analysis and determination of QTL for morphological traits. *Theoretical and Applied Genetics* 88, 7–16.

Von Wettstein-Knowles, P. (1992) Cloned and mapped genes: current status. In: Shewry, P.R. (ed.) *Barley: Genetics, Biochemistry, Molecular Biology and Biotechnology.* CAB International, Wallingford, UK, pp. 73–98.

Yang Yun-Kuei (1949) A study on the nature of gene controlling hybrid vigour, as it affects silking time and plant height in maize. *Agronomy Journal* 41, 309–312.

Zehr, B.E., Dudley, J.W., Chojecki, J., Saghai-Maroof, M.A. and Mowers, R.P. (1992) Use of RFLP markers to search for alleles in a maize population for improvement of an elite hybrid. *Theoretical and Applied Genetics* 83, 903–911.

Zeng, Z. (1994) Precision mapping of quantitative trait loci. *Genetics* 136, 1457–1468.

18 Mechanisms of Improved Nitrogen-use Efficiency in Cereals

Amarjit S. Basra and Sham S. Goyal

Department of Agronomy and Range Science, University of California, Davis, CA 95616, USA

Introduction

Fertilization of intensively managed crop systems is essential in maintaining or increasing world food production, which is heavily dependent on nitrogen (N) input to maximize yield potential. Because N can be limiting for plant growth, approximately 85 million metric tons of nitrogenous fertilizers are now added to soils worldwide annually, up from only 1.3 million t in 1930 (Frink *et al.*, 1999). N fertilizers supplement the natural soil nutrient supply to satisfy the demand of crops, to compensate for N lost by removal of plant products, N leaching and gaseous N loss and to improve or maintain productive soil conditions for agriculture (Tabachow *et al.*, 2001). Humans now use some 160 million t of N year^{-1}, of which 98 million t are industrially fixed by the Haber–Bosch process (83 million t for use as agricultural fertilizer, 15 million t for industry), 22 million tons are made during combustion and the rest is fixed during the cultivation of leguminous crops and fodders (Jenkinson, 2001). This has markedly increased the burden of combined N entering rivers, lakes and shallow seas, as well as increasing the input of ammonia (NH_3), nitrous oxide (N_2O), nitric oxide (NO) and nitrogen dioxide (NO_2) to the atmosphere. Excess N flowing down the Mississippi River each year is estimated

to be worth US$750,000,000 (Malakoff, 1998).

Worldwide, the nitrogen-use efficiency (NUE) of cereal production is approximately 33%, calculated as (total cereal N removed) – (N coming from the soil + N deposited in the rainfall)/fertilizer N applied to cereals (Raun and Johnson, 1999). An increase in NUE of 20% would result in savings in excess of US$750 million year^{-1} worldwide. Also efficient use of N fertilizers decreases negative impacts on the environment and reduces the drain of natural resources.

In the last two decades, world population has increased by 1.6 thousand million, most of this increase being in Asia. It is not surprising that most of the increase in N fertilizer use, from 61 million t in 1980/81 to 80 million t in 1997/98, has also been in Asia (Jenkinson, 2001). The present world population growth rate of 1.8% is estimated to create a shortfall of 20 million t of cereal grains by the year 2020. This implies that additional N fertilizer would need to be produced and used for cereal production in both irrigated and rain-fed agroecosystems. Manufacturing the fertilizer for today's needs already requires 544×10^9 MJ of fossil-fuel energy annually (Mudhar and Hignett, 1987a,b), which is a tremendous depletion of a non-renewable source. To obtain the projected grain yield of 8.0 t ha^{-1} in irrigated rice by the year 2025, it will be necessary to apply 280 kg N ha^{-1} at 33% fertilizer N recovery efficiency (Cassman and Pingali, 1995). This means that N fertilizer use in irrigated rice in Asia would increase from 15.5 to 43.6 million t, nearly a 300% increase in N use for a 63% increase in yield.

Although improved technologies of fertilizer application have been developed (see Cassman *et al.*, 1998), farmer acceptance and adoption have not been very encouraging because of the associated higher labour requirements. On the other hand, the adoption of new N-efficient cultivars by farmers is likely to be rapid, because no additional cost is involved and existing cropping systems and soil/water management practices are generally not affected. It is, therefore, important to select for cereal cultivars and hybrids that absorb and metabolize N in more efficient ways. This has not received due attention in the past and must be accorded high priority for crop improvement.

Unfortunately, resource-poor farmers in marginal areas have not realized the benefits of increased yield potential as in high-input cropping systems. Nitrogen deficiency can also be a problem where N is applied at suboptimal levels because of the potential for low economic return resulting from drought or excessive leaching of nitrate (NO_3^-). In the rice–wheat double-cropping sequence of South Asia, a negative balance of the primary nutrients still exists, even with the recommended rate of fertilization of this system (Singh and Singh, 2001). A rice–wheat sequence that yields 7 t ha^{-1} of rice and 4 t ha^{-1} of wheat removes more than 300 kg N from the soil. The system, in fact, is now showing signs of fatigue and is no longer exhibiting increased production with increases in input use.

The average global grain yield per unit area of the major staple crops, wheat, rice and maize, more than doubled during the period between 1940 and 1980, powered by changes in the genetic potential of the crop, plant-available N and improved agronomic management strategies (Evans, 1993). However, recent analyses suggest a progressive decline in the annual rate of increase in cereal yield, so that, at present, the annual rate of yield increase is below the rate of population increase. To consolidate further gains in yield enhancement, there is a dire need to understand in a more integrated manner the physiological and genetic causes of these improvements, especially in relation to NUE.

The Terminology

Nutrient use efficiency comprises nutrient uptake efficiency and nutrient utilization efficiency (Janssen, 1998). As a consequence, changes in nutrient use efficiency may be brought about by changes in uptake efficiency, in utilization efficiency or in both. Nutrient uptake efficiency is the ratio of actual nutrient uptake to potential nutrient supply. Nutrient utilization efficiency is the ratio of yield to actual nutrient uptake, where yield refers to dry matter in the harvestable plant parts (e.g. grains). Both depend on the availability of the nutrient in relation to other growth factors (e.g. crop production potential, availability of other nutrients, water, irradiation, etc.) in comparison to the availability of the nutrient considered. Nutrient use efficiency is the product of the two: the ratio of yield to potential supply. For practical purposes, potential supply is the maximum quantity of a nutrient that is taken up when all other nutrients and growth factors are at optimum (Janssen *et al.*, 1990). Balanced nutrition is the best guarantee for the simultaneous optimum use of all nutrients. Clark (1990) detailed the physiology of nutrient uptake and utilization efficiency in cereals.

Ladha *et al.* (1998) used the following definitions for evaluation of N-efficient cultivars:

- Efficiency of acquisition or recovery of N = plant N content/available N.
- Physiological efficiency with which N is used to produce biomass = plant biomass/plant N content.
- Physiological efficiency with which N is used to produce grain = grain yield/plant N content.

Because it is difficult to measure the amount of N available from soil and fertilizer, the relative efficiency for N acquisition can be evaluated for cultivars (of similar duration) growing under the same experimental conditions. Total plant N content or uptake alone can be used as a measure of the efficiency for acquisition or recovery. While evaluating NUE of different genotypes, the ratios and absolute contents should be compared at multiple N levels, allowing estimation of both intercept and slope of the response function (Ladha *et al.*, 1998). This comparison allows estimates of genotype–N interactions.

N and Growth of Cereals

During the past 45 years, N fertilization has been a powerful tool in increasing grain yield, particularly in cereals. It is recognized that the optimal combination of new genotypes and higher N rates was the driving force of the Green Revolution. On the other hand, there is increasing concern about the negative effects on the environment and human health because of excessive use of N fertilizer. Thus, the supply of sufficient N fertilizer to achieve optimum grain yield and quality while reducing the risk of pollution caused by inappropriate N applications is a crucial task in cereal production.

Addition of N leads to increased dry-matter accumulation in vegetative plant parts and to increased final yields in cereal crops (Hageman and Lambert, 1988). It is involved in all of the plant's metabolic processes, its rate of uptake and partitioning being largely determined by supply and demand during the various stages of plant development. In cereals, N has a dominant role in dry-matter production and accumulation from germination up to the heading stages (Austin *et al.*, 1977; Heitholt *et al.*, 1990; Delogu *et al.*, 1998). The availability of genotypic variability in crop growth and N uptake rate may be used to assist the improvement of dry-matter yield via selection (Greef *et al.*, 1999).

The dependence of plant growth rate on N concentration in vegetative organs appears to be a general finding. It has been shown both within a given species as a response to external N supply (Agren and Ingestad, 1987; Hirose, 1988; Garnier, 1998) and for different species grown under non-limiting external N supply (Poorter *et al.*, 1990; Van der Werf *et al.*, 1993; Garnier and Vancaeyzeele, 1994). Similarly, there is usually a positive relationship between light-saturated rates of photosynthesis and the N concentration of leaves (Field and Mooney, 1986; Peng *et al.*, 1995). The drop in photosynthesis with the shortage of N is well documented at both leaf and canopy levels (Evans, 1989; Sinclair and Horie, 1989; Sinclair and Muchow, 1999). N deficiency also reduces light interception and radiation-use efficiency (RUE) (Gallagher and Biscoe, 1978; Whitfield and Smith, 1989; Abbate *et al.*, 1995). N deficiency decreases assimilate production at flowering and kernel set by reducing radiation interception and RUE (Gifford *et al.*, 1984; Uhart and Andrade, 1995).

Theoretical studies have suggested that RUE would increase if N is preferentially allocated to the more illuminated leaves in the upper layers of the canopy (Field, 1983; Hirose and Werger, 1987). Whether N distribution can be modified to maximize RUE of cereals has not been methodically addressed (see Dreccer *et al.*, 1998). In addition, N deficiency reduces water-use efficiency (WUE) in wheat (Brown, 1971; Heitholt, 1989; Nielsen and Halvorson, 1991), while fertilizer application dramatically increases WUE (Caviglia and Sadras, 2001). In fact, it is widely accepted that fertilizer management is an important tool for the improvement of WUE (Cooper *et al.*, 1987). Another way to achieve this is through selection of cultivars with enhanced early vigour (Richards *et al.*, 1993).

Pre-anthesis N uptake in winter cereals represents 75–90% of total N in the plant at harvest (Austin *et al.*, 1976; Spiertz and Ellen, 1978; Heitholt *et al.*, 1990). Under conditions of high soil fertility, even post-anthesis N uptake is important because it is positively correlated to kernel protein content and to the N harvest index (Perez et al., 1983).

Although most of the N in the wheat grain originates from N taken up prior to anthesis, the N uptake capacity is retained at anthesis (Van Sanford and MacKown, 1987; Oscarson et al., 1995). The rate of grain N accumulation is linear for most of the grain-filling period (Sofield et al., 1977), which can be met by either the active uptake of N during the grain-filling period and/or the remobilization of N from vegetative tissues. N remobilization is probably the most important source of grain N under most production systems (Simmons, 1987).

This idea is in agreement with previous studies performed on maize hybrids correlating the efficiency of primary N assimilation and N remobilization with yield and its components (Reed et al., 1980; Purcino et al., 1998). Similarly, Teyker et al. (1989) and Plenet and Lemaire (1999) also concluded that, in addition to what is required for its vegetative growth, the plant must absorb and store an excess of N, which is then further metabolized and translocated to the kernels. Studies by Hirel et al. (2001) strengthened the concept that increased productivity in maize genotypes was due to their ability to accumulate NO_3^- in their leaves during vegetative growth and to efficiently remobilize this stored N during grain filling. This NO_3^- pool is usually stored in the vacuole and serves as an osmoticum and as a source of mineral N when the soil supply becomes depleted (Crawford and Glass, 1998). In rice, Ying et al. (1998) have noted a key role of leaf N as a reservoir of remobilizable N that is needed to sustain high grain yield in tropical and subtropical environments. Hence, leaf NO_3^- content at the early stages of plant development may be a good marker for selecting genotypes with enhanced grain yield and grain N content (Hirel et al., 2001).

The N uptake in the grains can be taken as an indicator of N sufficiency (Goos et al., 1982), because most of the N accumulated is translocated to the grains (Sherchand and Paulsen, 1985). Grain filling appears to be largely under the control of N availability (Frederick and Bauer, 1999).

The rate of N demand by the developing grain may, in fact, control the rate of leaf senescence. Frederick (1997) found that the photosynthetic activity of flag leaves decreases rapidly when the developing grain reaches about half of its size. During grain filling, only a limited amount of N is taken from the soil and N remobilization from senescent vegetative tissues is the dominant process. Late applications of N (at booting stage or later) usually increase leaf N concentration and may delay senescence (Banziger et al., 1994; Tindall et al., 1995).

Genetic Improvement for NUE

Although conventional breeding procedures have been successful in enhancing cereal yield, there have been few real attempts to understand in a more integrated manner the physiological and genetic causes of these improvements, especially in relation to NUE (Hirel et al., 2001). A plant's high yield ability as related to N fertilization is usually assessed as agronomic NUE, an indicator of the amount of yield per unit of applied N (Novoa and Loomis, 1981; Craswell and Godwin, 1984). Hence, for a given level of fertilization, differences in grain yield would match differences in NUE. Thus, in selecting improved cultivars, breeders empirically select those that are more efficient in N absorption and utilization. N utilization efficiency reflects the ability of the plant to translate the N taken up into economic yield (grains). This parameter has been extensively used to compare different species or cultivars at different levels of N fertility (Ortiz-Monasterio et al., 1997). Delogu et al. (1998) found that barley out-performed wheat in this respect, suggesting a higher ability of barley to generate yield, particularly at low N input. Ideal cultivars would be those that perform well under low soil fertility conditions but also respond well to applied fertilizer (Ladha et al., 1998).

Various studies on landraces and wild, old, intermediate and new genotypes of cereals showed that the new varieties not only produced higher grain yields than the old varieties but also were more efficient at using nutrients (Vose, 1990; El Bassam, 1998). Some cultivars were identified as being multiple-nutrient use-efficient. These are considered to be low-input cultivars. Once genotypic variation in nutrient uptake and utilization is found, physiological or morphological factors that are responsible for such variation can be further examined (Inthapanya *et al.*, 2000).

Several studies have analysed the relationships between wheat breeding and the N economy (mainly absorption and partitioning), most of which concluded that wheat breeding did not consistently modify the amount of N absorbed by the crop (Fischer and Wall, 1976; Paccaud *et al.*, 1985; Slafer *et al.*, 1990; Feil, 1992; Calderini *et al.*, 1995, 1999). Therefore, as modern cutivars outyield their old counterparts, plant breeding increased the NUE. None the less, it is important to recognize that genetic variation in total N uptake exists in wheat (Austin *et al.*, 1977; Cox *et al.*, 1985; Heitholt *et al.*, 1990), and future breeding programmes could exploit this variability to increase the biomass production in new cultivars (Calderini *et al.*, 1999). Studies recently carried out using wheat lines released by the International Maize and Wheat Improvement Center (CIMMYT) between 1950 and 1985 (Ortiz-Monasterio *et al.*, 1997) have demonstrated that total N uptake has increased, suggesting genetic improvement in N uptake efficiency. Goyal and Huffaker (1986) reported a novel method for studying N transport kinetics using wheat seedlings, which may be exploited for the rapid screening of genotypes for superior N uptake.

Cereal-breeding programmes have generally focused on maximizing yield potential under the conditions where N supply is adequate. High-yielding hybrids and lines typically respond favourably to increased N inputs. Selection for yield in environments with low N should be more effective than selection under high N; however, such environments are not normally favoured by breeders because of increased environmental variability and the reduced heritability of grain yield under low-N conditions (see Blum, 1988). The efficiency of selection

for yield in low-N environments may be improved by selection for correlated secondary traits, such as improved N uptake by seedlings (Teyker *et al.*, 1989), high plant NO_3^- content (Molaretti *et al.*, 1987) and mobilization of accumulated N from vegetative organs during grain production (Eghball and Maranville, 1991; Tollenaar, 1991; Plenet and Lemaire, 1999; Rajcan and Tollenaar, 1999).

There are essential agronomic, economic and ecological reasons that make the search for cultivars that are more efficient and better adapted to less favourable nutrition an important breeding task. However, opportunities for progress may be limited because there is no clear understanding of the manner in which the major components of nutrient efficiency are inherited, and most reports have been concerned with early growth stages, with very little information at the reproductive growth stages (Gorny and Sodkiewicz, 2001). In seedlings, both additive and non-additive gene effects were significant for the accumulation and uptake efficiency of N (Gorz *et al.*, 1987; Ahsan *et al.*, 1996). The contribution of non-additive gene effects for generative efficiency indices appears to be less evident than that found at the vegetative growth stage (Gorny, 1999). However, the observed nutrient shortage-induced decrease of heritabilities and the enhanced expression of non-additive gene effects, as well as the interactions of genotypic effects with fertilization rates and growth stages, suggest that obtaining nutrient-efficient cultivars under low-input conditions would be a difficult task (Gorny and Sodkiewicz, 2001). Experience of breeders targeting specific environments (Sivapalan *et al.*, 2000), however, confirms that selection progress depends on a rather more specific strategy and technology. Further attention needs to be directed towards genetic studies in possibly diverse nutritional regimes on populations originating from a diverse germ-plasm.

In analysing a genotype–environment (G × E) interaction, an index for each environment (the mean performance of all genotypes in an environment) may be used as a suitable index of its environmental productivity (Westcott, 1986). The performance of each genotype can be plotted against this index. When a nutrient is deficient in the natural soil, natural selection probably leads to the development of plants that store a higher concentration of that nutrient in the seed, for the benefit of succeeding generations (Bonfil and Kafkafi, 2000).

The existence of an interaction of genotype × level of N fertilization has been shown in maize (Moll et al., 1987; Bertin and Gallais, 2000). At high N input, variation in NUE has been attributed to variation in N uptake capability, whereas, at low N input, variation in NUE is mainly due to differences in N utilization efficiency (Di Fonzo et al., 1982; Bertin and Gallais, 2000). These differences in the expression of genetic variability have also been confirmed following the detection of specific quantitative trait loci (QTL) for a given level of fertilization (Agrama et al., 1999). This suggests that several sets of genes are differentially expressed according to the amount of N provided for the plant. We need to understand better the G × E interactions in the expression of these QTL, through a collaboration between crop physiologists and molecular biologists, for improving the usefulness of molecular biology in contributing to breeding for complex quantitative traits such as NUE and yield potential.

Nitrogen Supply

Most plant species are able to absorb and assimilate NO_3^-, ammonium (NH_4^+), urea and amino acids as nitrogen sources, but the response to a particular form of N is species-specific. In most soils, NO_3^- and NH_4^+ are the predominant sources of N that are available for plant nutrition. In a typical aerobic agricultural soil, NO_3^- is the predominant form (Marschner, 1995), though high concentrations are generally not maintained in soils due to substantial losses from plants, runoff and microbial denitrification (Raun and Johnson, 1999). Optimal plant growth, however, is usually achieved when

N is supplied in both forms (Bloom et al., 1992, 1993).

The overuse or improper use of nitrogenous fertilizers to maximize biomass accumulation in cereal production is hazardous to the environment. Excess NO_3^- pollutes not only the soil and groundwater but also the produce itself. N_2O released from denitrification of NO_3^- pollutes the air and can damage the ozone layer in the stratosphere. Improved fertilizer N management in cereal production should aim at maximum N absorption during those stages when N is most efficiently translated into grain yield, and in a manner such that applied N is not prone to losses from the soil–plant system (see Singh and Singh, 2001).

Some early studies reported hereditary differences in seedling responses to NO_3^- and NH_4^+ forms of N when grown in nutrient solutions (Harvey, 1939). Although the NO_3^- forms of fertilizer provided the highest average response in maize, certain inbreds responded well to the NH_4^+ form of N (Moll et al., 1982). Genetic studies evaluating N response in breeding lines and maize hybrids suggested polygenic inheritance (Pollmer et al., 1979; Eghball and Maranville, 1991).

Wheat N uptake was increased by 35% when one-quarter of the N was supplied as NH_4^+, compared with all N as NO_3^- (Wang and Below, 1992). High-yielding maize genotypes were unable to absorb NO_3^- during ear development (Pan et al., 1984), and there may be a potential advantage of NH_4^+ nutrition for grain production (Tsai et al., 1992). In regard to energy costs, assimilation of NO_3^- requires the energy equivalent of 20 mol ATP mol^{-1} NO_3^-, whereas NH_4^+ assimilation requires only 5 mol ATP (Salsac et al., 1987), mainly because NO_3^- has to be reduced prior to assimilation. While one might expect NH_4^+ to be preferred by plants, as its assimilation requires less energy than that of NO_3^-, only a few species perform well when NH_4^+ is the only or the predominant form of N. The exclusive supply of N as NH_4^+ is harmful to many plant species and can cause poor root and shoot growth and reduced mineral cation contents relative to those of plants receiving NO_3^- or ammonium nitrate (NH_4NO_3) nutrition (Goyal and Huffaker, 1984; Marschner, 1995; Gerendas *et al.*, 1997).

Several hypotheses have been put forth to explain why NH_4^+ is toxic to plants (Von Wiren *et al.*, 2000). In part, growth depression is directly related to NH_4^+ uptake, as the assimilation of NH_4^+ is accompanied by about equimolar H^+ production. These protons are excreted most probably due to increased H^+-ATPase activity, leading to an acidification of the rhizosphere and thus to repressed cation uptake. Another reason may be the absence of NO_3^-, which is not only an important osmoticum, but also an essential counter-ion for cation translocation in the xylem and a signal for the expression of genes involved in N uptake, N assimilation, organic acid metabolism and starch synthesis (Stitt, 1999). In addition, exclusive nutrition can cause a hormonal imbalance, leading to a stunted growth phenotype (Peuke *et al.*, 1998; Walch-Liu *et al.*, 2000).

Species-specific variation in NH_4^+ toxicity has been observed in cereals, e.g. barley is known to be susceptible to toxicity, while rice is known for its exceptional tolerance even to high levels of NH_4^+. Britto *et al.* (2001) have proposed that the operation of an energy-intensive NH_4^+ extrusion mechanism by barley root cells at high levels of external NH_4^+ appears to be central to the toxicity syndrome. Such a process must carry a substantial energetic burden (40% increase in root respiration) that is independent of N metabolism, and is accompanied by decline in growth. In rice, in contrast, a cellular defence strategy has evolved that is characterized by an energetically neutral equilibration of NH_4^+ at high external NH_4^+ concentration.

Ammonium and Nitrate Uptake

N uptake by plant roots has long been studied, using various methods of chemical analysis, tracer techniques, ion-selective microelectrodes, etc. (Goyal and Huffaker,

1986; Siddiqi *et al.*, 1990; Glass *et al.*, 1992; Henriksen *et al.*, 1992; Colmer and Bloom, 1998; Plassard *et al.*, 1999; Newman, 2001). More recently, attention has been focused on the transporters themselves, located in the plasma membrane of cells (for reviews, see Forde, 2000; Von Wiren *et al.*, 2000). The problems of N uptake are complex and there are a number of transport systems. The kinetic properties of transport systems measured in whole plants are highly variable and mainly dependent on the nutritional status of the plant, which is in turn affected by environmental factors, such as light, temperature and previous external substrate availability.

When NH_4^+ and NO_3^- ions are present together, it appears that the NH_4^+ net influx is greater than the net influx of NO_3^- from equimolar solutions, and the presence of NH_4^+ tends to inhibit the uptake of NO_3^- (Henrikson *et al.*, 1990; Taylor and Bloom, 1998). Colmer and Bloom (1998) found a difference between the mature region and the growing (meristematic and elongation) region of maize and rice roots for NH_4^+ and NO_3^- fluxes, the uptake being more in the latter. Despite much variability between roots, H^+ fluxes were outward in NH_4^+, but in NO_3^- they were inward at the growing region and were mixed in the mature region (Plassard *et al.*, 1999). When N availability is limited, root N demand has a clear priority over shoot N demand, leading to a rapid decrease in N translocation to the shoot (Kronzucker *et al.*, 1998).

The concentration-dependent influx of NH_4^+ into intact plant roots exhibits biphasic kinetics with two distinct components. At < 1 mM external NH_4^+, the influx approaches Michaelis–Menten kinetics, whereas, at higher concentrations, uptake rates seem to increase linearly (Wang *et al.*, 1993). High-affinity transport systems and low-affinity transport systems (LATS) are distinguished by their apparent K_m values (i.e. the Michaelis–Menten constant – the substrate concentration that allows the reaction or transport process to proceed at one-half of its maximum rate). The high affinity of a transporter is responsible for nutrient acquisition at low external concentrations, whereas low affinity often correlates with high capacity for the maintenance of large influxes at high external availability. Thus, a distinction between high-affinity/low-capacity and low-affinity/high-capacity NH_4^+ transport systems reflects their physiological role more precisely than does a distinction based on affinity alone. High-affinity NH_4^+ transporters are induced in N-starved roots, whereas other transporters may be regarded as the 'workhorses' that are active when conditions are conducive to NH_4^+ assimilation (Von Wiren et al., 2000).

The net uptake by plant roots is the difference between the concomitant influx and efflux of the ion. High intracellular concentrations of NH_4^+ build up due to uptake by roots or amino acid breakdown, which leads to a constant leakage of NH_4^+ from roots at a rate of 11–29% that of the NH_4^+ influx (Wang et al., 1993; Feng et al., 1998). In rice roots, NH_4^+ efflux is decreased by up to 50% in the presence of NO_3^-, whereas NO_3^- fluxes and metabolism are strongly repressed by NH_4^+ (Kronzucker et al., 1999). NH_3 losses from the leaves of field-grown plants can account for up to 5% of the shoot N content (Asman et al., 1998), which is mainly released during photorespiration and to a minor extent from amino acid transport and catabolism. Whether NH_4^+ leakage or efflux is mediated by membrane transporters or largely by diffusion of NH_3 is still unclear (Von Wiren et al., 2000).

In soil solution, NO_3^- is carried toward the root by bulk flow and is absorbed into the epidermal and cortical cells of the root. NO_3^- is actively transported through the combined activities of a set of both low- and high-affinity NO_3^- transport systems, with the influx of NO_3^- being driven by the H^+ gradient across the plasma membrane (Crawford and Glass, 1998; Forde, 2000; Tischner, 2000; Fraisier et al., 2001). Once inside, the NO_3^- may be reduced to NH_4^+ and then incorporated into amino acids, undergo transmembrane efflux, be taken up and stored in the vacuoles or translocated to the shoot via xylem for reduction and vacuolar storage (also for osmoregulation) in the leaves. In most plant species, only a small proportion of the absorbed NO_3^- is assimilated in the root, and the remainder is transported to the shoot.

The accumulated evidence from kinetic studies indicates that roots have at least three distinct NO_3^- uptake systems, two of which have a high affinity for NO_3^-, while the third has a low affinity (for a review, see Forde, 2000). One of the high-affinity systems is strongly induced in the presence of an external NO_3^- supply and is known as the inducible high-affinity transport system (iHATS), while the second high-affinity system (the CHATS) is constitutively expressed. The LATS, which appears to be constitutively expressed, is most important at external NO_3^- concentrations of more than 1 mM. It has been suggested that NO_3^- is taken up in both low- and high-affinity transport via 2 : 1 H^+/NO_3^- symports (Glass et al., 1992).

In plants well supplied with N, NO_3^- uptake systems are repressed, evidently to limit amino acid synthesis to a set level corresponding to the N 'demand' for growth (Imsande and Touraine, 1994; Gojon et al., 1998). However, the nature of the key metabolites that are used by the plant to monitor its N status and the molecular pathway by which changes in these key metabolites cause feedback repression of NO_3^- assimilation are not known precisely.

Nutrient uptake is related to the size and efficiency of the root mass and the energy supply. Continuous N uptake during the grain-filling period has been associated with the ability to maintain root growth after silking in maize (Mackay and Barber, 1986). Whereas Rajcan and Tollenaar (1999) found differences in NUE between an older and newer maize hybrid, Duvick (1984) reported no hybrid × N interactions for hybrids representing five decades of yield improvement in the USA when grown at two N levels.

Because NUE in grain production varies under different climatic, soil and management conditions (Keating et al., 1991; Sinclair and Muchow, 1995; Muchow, 1998), there is a need to determine the minimum N requirement for a given yield level to maximize NUE. Maximum NUE of sorghum was smaller than that of maize (48 vs. 61 g per grain g^{-1} N absorbed), and was associated with a higher grain N concentration in sorghum (Muchow, 1994, 1998). An approach to improving grain yield and the efficiency of N utilization might be to genetically decrease the minimum grain N, but this may be at the expense of the nutritive value of the grain, e.g. the lowered protein content may make the grain less desirable as a food or feed (Muchow and Sinclair, 1994; Sinclair and Muchow, 1995). Simulation modelling can be used in defining both the crop N demand and the potential losses from the system under different environmental conditions.

In wheat, grain N concentrations were found to be negatively associated with grain yield (Austin *et al.*, 1980; Paccaud *et al.*, 1985; Feil and Geisler, 1988; Slafer *et al.*, 1990; Canevara *et al.*, 1994; Calderini *et al.*, 1995). This negative association, despite the net increase due to increased yields, highlights the dilution effect of increases in grain yield (Calderini *et al.*, 1999). However, it must be mentioned here that grain N concentration is a major, but not the only, measure of grain quality, which may be increased by selecting genotypes with better protein composition (Canevara *et al.*, 1994).

Nitrate Assimilation

NO_3^- assimilation starts with its reduction to nitrite, catalysed by nitrate reductase (NR), and is followed by nitrite reductase-catalysed reduction of nitrite to NH_4^+ and then the incorporation of NH_4^+ into amino acids, catalysed primarily by glutamine synthetase and glutamate synthase (reviewed in Crawford, 1995; Campbell, 1999; Stitt, 1999; Tischchner, 2000). NO_3^- reduction occurs in the cytosol of cells in both shoots and roots and uses NAD(P)H as the source of reductant. However, there is growing evidence that NR can also be located outside the plasma membrane (Campbell, 1999). Nitrite reduction occurs in chloroplasts of green tissues and in plastids of the roots with reduced ferredoxin as the reductant. In green tissues, reductant originates from photosynthetic electron transport and, in non-green tissues,

primarily from the oxidative pentose phosphate pathway. In addition to reductant, organic acids are needed for NH_4^+ incorporation into amino acids and maintenance of cellular pH because NO_3^- reduction leads to alkalinization.

The reduction of NO_3^- to nitrite catalysed by NR represents the first enzymatic step of primary N assimilation. The regulation of NR is remarkably complex and is subjected to mechanisms controlling both its synthesis and its catalytic activity (Campbell, 1999). The NR gene expression is induced by NO_3^- and other factors, such as light or sugars (Galangau *et al.*, 1988; Pouteau *et al.*, 1989; Gowri *et al.*, 1992; Li and Oaks, 1994; Appenroth *et al.*, 2000; Klein *et al.*, 2000) and is repressed by glutamine or related downstream metabolites that are formed from NO_3^- (Hoff *et al.*, 1994). Regulation of NR expression in leaves by NO_3^- and N metabolism is completely overridden when sugars fall below a critical level (Klein *et al.*, 2000). Genomic analysis revealed that NO_3^- induces not just one but a diverse array of novel metabolic and potential regulatory genes (Wang *et al.*, 2000).

An appraisal of the available literature in the field reveals that, except for a few studies (Pace *et al.*, 1982; Kleinhofs and Warner, 1990; Gojon *et al.*, 1998), physiological characterizations have never included measurements of the *in vivo* capacity for NO_3^- reduction in combination with NO_3^- uptake. In spite of much previous work on the subject (Hageman *et al.*, 1980; Aslam and Huffaker, 1982; Gojon *et al.*, 1991), it is still unknown whether and to what extent the enzyme activities *in vitro* reflect NO_3^- reduction rates *in vivo* (Kaiser *et al.*, 2000).

The efficiency with which N is used varies with plant species and with environmental conditions. Several studies have identified differences between C_3 and C_4 plants in N usage and photosynthetic nitrogen-use efficiency (PNUE) (photosynthesis per unit leaf N), which result from their different modes of carbon fixation (see Taub and Lerdau, 2000, and references therein). The ability of C_4 plants to concentrate carbon dioxide (CO_2) at the site of carboxylation allows them to attain higher photosynthetic rates of CO_2 fixation for a given amount of ribulose bisphosphate carboxylase (RUBISCO) than C_3 plants, resulting in a higher PNUE. Variation in photosynthetic rates among species, at a given level of leaf N, has been attributed to differences in the proportion of leaf N that is used for the synthesis of photosynthetic enzymes rather than for other leaf constituents, e.g. structural proteins, chlorophyll, nucleic acids, etc. (Evans, 1989).

The effect of light on NR gene expression cannot be fully explained, and yet it is possible to replace the light effect by glucose or sucrose (see Tischner, 2000). Both photosynthetically active light and light acting through phytochrome are known to influence the NR expression. Light operating via phytochrome has been implicated in the regulation of NR gene expression, particularly in etiolated tissues (Melzer et al., 1989; Appenroth et al., 2000). In green plants, the light effect is probably mediated via sugars produced by photosynthesis. This is further supported by observations that diminished CO_2 assimilation caused by water stress led to a decrease in NR transcripts (Foyer et al., 1998), and exposure of plants to elevated CO_2 increased transcript levels (Larios et al., 2001).

During the past decade, the post-translational regulation of NR via protein phosphorylation and dephosphorylation has been intensively studied in response to light/dark transitions (for a recent review, see Kaiser et al., 1999). After transfer from light to dark, NR is phosphorylated by an NR kinase, and the subsequent binding of a dimeric 14-3-3 protein converts phospho-NR into a form sensitive to inhibition by magnesium ion (Mg^{2+}). Upon reillumination of the leaves, phospho-NR is dephosphorylated by a protein phosphatase, rendering it active, i.e. a form insensitive to inhibition by free Mg^{2+}. Through this process, NR in leaves is rapidly inactivated in the dark and activated in the light. If ethylenediamine tetra-acetic acid (EDTA) chelates cations, NR becomes fully active. This permits an estimation of the activation state of the NR, which reflects the percentage of non-phosphorylated NR (NR_{act}) in a tissue

extract. Maximum NR activity (NR_{max}) gives the total amount of functional NR present in the extract. However, the activation state of NR is not always correlated with total NR activity in leaves (Man et al., 1999). Only under conditions optimal for photosynthesis (light and high ambient CO_2) are reliable estimates of NO_3^- reduction rates in leaves given by NR_{act} (Kaiser et al., 2000). Under other conditions, NR_{act} considerably overestimates rates of NO_3^- reduction *in vivo*.

Rapid inactivation of leaf NR also occurs in the light when CO_2 fixation is prevented by CO_2 removal from the air (Kaiser and Brendle-Behnisch, 1991; Lejay et al., 1997) or stomatal closure in response to drought (Foyer et al., 1998). Moreover, feeding leaves with sugars in darkness was shown to prevent a decrease in extractable NR activity (De Cires et al., 1993) and to increase the activation state of NR (Provan and Lillo, 1999). Kaiser and Huber (1997) have shown that artificial activation of NR in the dark also stabilizes the steady-state pool of NR. Thus, a high level of NR protein might exist concomitantly with a high activation state. On the other hand, low NR protein levels appear to be compensated by a high activation state, as suggested from experiments with tobacco mutants having a decreased number of functional *nia* genes (Scheible et al., 1997), and the compensation of low leaf NR_{max} in low-NO_3^-/long-day grown barley by a higher activation state of NR in the light and by significantly less dark inactivation (Man et al., 1999).

Several lines of evidence have recently indicated that NR is responsible for NO production in leaves (Yamasaki, 2000). NR also has the ability to convert NO to the extremely toxic peroxynitrate ($ONOO^-$) under aerobic *in vitro* conditions. Like active oxygen species, such as O_2^- and hydrogen peroxide (H_2O_2), active nitrogen species, including NO and $ONOO^-$, induce oxidative damage to DNA, lipids, amino acids and other biomolecules. Because NR activity is highly regulated by transcriptional and post-translational mechanisms in response to many environmental conditions (Kaiser et al., 1999), it is probable that such strict regulation of NR is beyond that needed for N assimilation but it may be necessary for regulating the production of NO and active nitrogenous species. Mounting evidence also suggests that NO is a novel regulator of plant growth and development (Beligni and Lamattina, 2001). Hence, NR-mediated NO production has tremendous implications for plant N metabolism and developmental regulation.

Differential Efficiency of Nitrogen Utilization in C_3 and C_4 Cereals

In C_4 leaves, assimilation of NO_3^- to NH_3 occurs in the mesophyll cells, while CO_2 fixation in the C_3 pathway occurs in the bundle-sheath cells. In C_3, both processes occur in the mesophyll cells. But it is not apparent why this would make a difference in efficiency. This is an outstanding riddle in NO_3^- assimilation that needs to be resolved with innovative approaches.

Since more NO_3^- accumulates in C_3 relative to C_4 cereals (Martin *et al.*, 1983), it is probable that the uptake of NO_3^- is more active or more abundant in C_3 than in C_4 cereals or that NR is less active or less abundant. Differential uptake of NO_3^- by intact seedlings of barley and maize was noted by Sehtiya and Goyal (2000). Oaks *et al.* (1990) grew barley and maize seedlings with three levels of NO_3^- (1, 5 and 20 mM) and measured the loss of NO_3^- from the medium, its accumulation in leaf tissue, NR activity and the levels of total soluble protein. It was clearly shown that the uptake of NO_3^-, tissue NO_3^- levels and NR were higher in barley than in maize. At the same time, levels of soluble protein were lower relative to those found in maize. Thus, although there is ample extractable NR in barley leaves, it is not correlated to NUE. Similarly, an inverse relationship has been observed between NR activity and grain yield in maize (Reed *et al.*, 1980; Hirel *et al.*, 2001).

Studies in our laboratory have confirmed that neither the NR amount nor its activation state appears to limit *in vivo* NO_3^- reduction in a C_3 plant like barley. Instead, (an)other metabolic factor(s) may actually be limiting. Several metabolites could possibly be involved in the process. Total leaf concentrations of sugars, glutamine and glutamate and of glucose-6-phosphate correlated neither with the NR activity nor with its activation state (Man *et al.*, 1999). Cytosolic NO_3^- levels in leaves also appear to be homoeostatically controlled and are hardly suboptimal.

In contrast, cytosolic NADH concentrations are extremely low (0.5 µM) or barely detectable (Heineke *et al.*, 1991) and appear to limit NO_3^- reduction *in vivo* in NO_3^--sufficient plants (Kaiser *et al.*, 2000). *In vivo* NO_3^- reduction rates matched *in vitro* NR activity only when the leaves were exposed to 5% CO_2 during light (Kaiser *et al.*, 2000). Hence, one might assume that NAD(P)H export from chloroplast to cytosol, e.g. via the P_i/triose phosphate translocator, depends on the rate of photosynthetic CO_2 fixation that is relatively much higher in C_4 as compared with C_3 plants. The advantage of C_4 leaf morphology and its associated metabolism would then be twofold: (i) photorespiration is reduced relative to C_3 plants; and (ii) there is an enhanced supply of NADH in the mesophyll cells.

Although there are many differences in the metabolism of C_3 and C_4 plants, a major difference between these two patterns of photosynthesis is the contribution of photorespiration to both carbon and N metabolism. When photorespiration is reduced in C_3 plants either by increasing ambient levels of CO_2 or reducing levels of O_2, the NUE is enhanced, but this effect is not apparent in C_4 plants (Evans, 1989; Hocking and Meyer, 1991). Oaks (1994) postulated that the reduced efficiency of NO_3^- utilization in C_3 leaves was related to a carbon deficiency caused by the inhibition of the mitochondrial pyruvate decarboxylase by photorespiratory NH_4^+. However, this has not been proved.

Yet another possibility might be that photorespiratory metabolites act as inhibitors of NR; however, attempts to prove this have been unsuccessful so far (see Kaiser *et al.*, 2000). On the other hand, it has been proposed that glycine oxidation during photorespiration supports NO_3^- reduction in

C_3 plants but not in C_4 plants (Kumar and Abrol, 1989).

Although not involved in the primary net assimilation of N, the NH_4^+ handled by photorespiratory-N cycle in a C_3 leaf could be up to 20 times that handled by the reduction of NO_3^- (Canvin, 1990). This may be interpreted to mean that the C_3 plants have to assimilate much more NH_3 than C_4 plants, because the former have to reassimilate the NH_3 coming from photorespiration. However, our studies have shown that, even at comparable rates of photosynthetic carbon fixation, the maize leaves maintain the superiority in NO_3^- reduction compared with barley leaves, and maize NO_3^- reduction efficiency is not affected by high NH_4^+ concentrations (A.S. Basra and S.S. Goyal, unpublished). Moreover, Yin *et al.* (1998) have shown that plant leaves can tolerate high levels of NH_3 without a negative influence on photosynthesis and transpiration.

Conclusion

Fertilization of agricultural systems is essential in maintaining or increasing world food production. Worldwide increases in N fertilizer use are expected as global population increases, particularly in developing countries. In view of the poor NUE of cereal production, it is vital to understand its physiological and genetic basis in order to create crop plants with enhanced NUE, achieving maximal growth with minimal fertilizer input. It is most important to develop a successful breeding programme of crop cultivars for low-input conditions.

There is little doubt that the more precisely we can determine the individual components of a complex trait such as NUE, the more rapid genetic progress will be. There exists enough phenotypic and genotypic variability for it to be exploited to identify key components of yield improvement. Not only must the traits be related to yield, but they must also be highly heritable, have low $G \times E$ interactions, and their expression must not be compensated by other related traits that negate their effect on grain yield.

Moreover, we need innovative simple techniques to screen them in a quick, reliable and relatively inexpensive manner. Not many traits fulfil these requirements. In the case of difficult-to-measure traits or traits with low heritability or high $G \times E$ interactions, molecular-marker technologies offer considerable potential. The combination of agronomic and physiological studies with quantitative genetic approaches will allow the use of molecular markers to identify key structural or regulatory loci involved in the expression of a quantitative trait and the selection of genotypes more efficient in terms of N use. Advances in genome sequencing and mapping are allowing the precise location of key genes controlling the expression of desired traits – in other words, the opportunity of translating 'traits to genes'. In turn, this technology will be of great potential for plant breeders in carrying out marker-assisted selection for improved NUE in relation to yield.

Significant progress has been made in dissecting the mechanisms of N uptake and assimilation by plants at the molecular level. As knowledge about the mechanisms advances, cereal NUE may be improved in the future by manipulating the pathways of N uptake and assimilation through genetic engineering and breeding approaches. The new biotechnologies now allow the plant breeder much more scope than has ever been possible before for genotypic, and hence phenotypic, modification. The benefits of such developments would be substantial in terms of income and food for the people, reduced demand for N fertilizers and reducing N losses, all of which would also generate environmental benefits.

References

Abbate, P.E., Andrade, F.H. and Culot, J.F. (1995) The effects of radiation and nitrogen on number of grains in wheat. *Journal of Agricultural Science* 124, 351–360.

Agrama, H.A.S., Zakaria, A.G., Said, F.B. and Tunistra, M. (1999) Identification of quantitative trait loci for nitrogen use efficiency in maize. *Molecular Breeding* 5, 187–195.

Agren, G.I. and Ingestad, T. (1987) Root : shoot ratio as a balance between nitrogen productivity and photosynthesis. *Plant Cell and Environment* 10, 579–586.

Ahsan, M., Wright, D. and Virk, D.S. (1996) Genetic analysis of salt tolerance in spring wheat (*Triticum aestivum* L.). *Cereal Research Communications* 24, 353–360.

Appenroth, K.J., Meco, R., Jourdan, V. and Lillo, C. (2000) Phytochrome and post-translational regulation of nitrate reductase in higher plants. *Plant Science* 159, 51–56.

Aslam, M. and Huffaker, R.C. (1982) *In vivo* nitrate reduction in roots and shoots of barley (*Hordeum vulgare* L.) seedlings in light and darkness. *Plant Physiology* 70, 1009–1013.

Asman, W.A.H., Sutton, M.A. and Schloerring, J.K. (1998) Ammonia: emission, atmospheric transport and deposition. *New Phytologist* 139, 27–48.

Austin, R.B., Edrich, J.A., Ford, M.A. and Blackwell, A.D. (1976) *Report for 1975*. Plant Breeding Institute, Cambridge, pp. 140–141.

Austin, R.B., Ford, M.A., Edrich, J.A. and Blackwell, R.D. (1977) The nitrogen economy of winter wheat. *Journal of Agricultural Science* 88, 159–167.

Austin, R.B., Bingham, J., Blackwell, R.D., Evans, L.T., Ford, M.A., Morgan, C.L. and Taylor, M. (1980) Genetic improvement in winter wheat yield since 1900 and associated physiological changes. *Journal of Agricultural Science* 94, 675–689.

Banziger, M., Feil, B. and Stamp, P. (1994) Competition between nitrogen accumulation and grain growth for carbohydrates during grain filling of wheat. *Crop Science* 34, 440–446.

Beligni, M.V. and Lamattina, L. (2001) Nitric oxide in plants: the history is just beginning. *Plant, Cell and Environment* 24, 267–278.

Bertin, P. and Gallais, A. (2000) Genetic variation for nitrogen use efficiency in a set of recombinant maize inbred lines. I. Agrophysiological results. *Maydica* 45, 53–66.

Bloom, A.J., Sukrapanna, S.S. and Warner, R.L. (1992) Root respiration associated with ammonium and nitrate absorption and assimilation by barley. *Plant Physiology* 99, 1294–1301.

Bloom, A.J., Jackson, L.E. and Smart, D.R. (1993) Root growth as a function of ammonium and nitrate in the root zone. *Plant, Cell and Environment* 16, 199–206.

Blum, A. (1988) *Plant Breeding for Stress Environments.* CRC Press, Boca Raton, Florida, 223 pp.

Bonfil, D.J. and Kafkafi, U. (2000) Wild wheat adaptation in different soil ecosystems as expressed in the mineral concentration of seeds. *Euphytica* 114, 123–134.

Britto, D.T., Siddiqi, M.Y., Glass, A.D.M. and Kronzucker, H.J. (2001) Futile transmembrane NH_4^+ cycling: a cellular hypothesis to explain ammonium toxicity in plants. *Proceedings of the National Academy of Sciences USA* 98, 4255–4258.

Brown, P.L. (1971) Water use and soil water depletion by dryland winter wheat as affected by nitrogen fertilization. *Agronomy Journal* 63, 43–46.

Calderini, D.F., Dreccer, M.F. and Slafer, G.A. (1995) Genetic improvement in wheat yield and associated traits: a re-examination of previous results and the latest trends. *Plant Breeding* 114, 108–112.

Calderini, D.F., Reynolds, M.P. and Slafer, G.A. (1999) Genetic gains in wheat yield and associated physiological changes during the twentieth century. In: Satorre, E.H. and Slafer, G.A. (eds) *Wheat: Ecology and Physiology of Yield Determination.* Food Products Press, Binghamton, New York, pp. 351–377.

Campbell, W.H. (1999) Nitrate reductase structure, function and regulation: bridging the gap between biochemistry and physiology. *Annual Reviews of Plant Physiology and Plant Molecular Biology* 50, 277–303.

Canevara, M.G., Romani, M., Corbellini, M., Perenzin, M. and Borghi, B. (1994) Evolutionary trends in morphological, physiological, agronomical and qualitative traits of *Triticum aestivum* L. cultivars bred in Italy since 1900. *European Journal of Agronomy* 3, 175–185.

Canvin, D.T. (1990) Photorespiration and CO_2-concentrating mechanisms. In: Dennis, D.T. and Turpin, D.H. (eds) *Plant Physiology, Biochemistry and Molecular Biology.* Longman Scientific-Technical, Singapore, pp. 253–273.

Cassman, K.G. and Pingali, P.L. (1995) Intensification of irrigated rice systems: learning from the past to meet future challenges. *Geo Journal* 35, 299–305.

Cassman, K.G., Peng, S., Olk, D.C., Ladha, J.K., Reichardt, W., Dobermann, A. and Singh, U. (1998) Opportunities for increased nitrogen-use efficiency from improved resource management in irrigated rice systems. *Field Crops Research* 56, 7–39.

Caviglia, O.P. and Sadras, V.O. (2001) Effect of nitrogen supply on crop conductance, water- and radiation-use efficiency of wheat. *Field Crops Research* 69, 259–266.

Clark, R.B. (1990) Physiology of cereals for mineral nutrient uptake, use, and efficiency. In: Baligar, V.C. and Duncan, R.R. (eds) *Crops as Enhancers of Nutrient Use*. Academic Press, New York, pp. 131–209.

Colmer, T.D. and Bloom, A.J. (1998) A comparison of NH_4^+ and NO_3^- net fluxes along roots of rice and maize. *Plant, Cell and Environment* 21, 240–246.

Cooper, P.J.M., Gregory, P.J., Tully, D. and Harris, H.C. (1987) Improving water use efficiency of annual crops in rainfed systems of West Asia and North Africa. *Experimental Agriculture* 23, 113–158.

Cox, C.M., Qualset, C.O. and Rains, D.W. (1985) Genetic variation for nitrogen assimilation and translocation in wheat. I. Dry matter and nitrogen accumulation. *Crop Science* 25, 430–435.

Craswell, E.T. and Godwin, D.C. (1984) The efficiency of nitrogen fertilizers applied to cereals in different climates. *Advances in Plant Nutrition* 1, 1–55.

Crawford, N.M. (1995) Nitrate: nutrient and signal for plant growth. *Plant Cell* 7, 859–868.

Crawford, N.M. and Glass, A.D.M. (1998) Molecular and physiological aspects of nitrate uptake in plants. *Trends in Plant Science* 3, 389–395.

De Cires, A., de la Torre, A. and Lara, C. (1993) Involvement of CO_2 fixation products in the light–dark modulation of nitrate reductase activity in barley leaves. *Physiologia Plantarum* 89, 577–581.

Delogu, G., Cattivelli, L., Pecchioni, N., De Falcis, D., Maggiore, T. and Stanca, T. (1998) Uptake and agronomic efficiency of nitrogen in winter barley and winter wheat. *European Journal of Agronomy* 9, 11–20.

Di Fonzo, N., Motto, M., Maggiore, T., Sabatino, R. and Salamini, F. (1982) N uptake, translocation and relationships among N related traits in maize as affected by genotype. *Agronomie* 2, 789–796.

Dreccer, M.F., Slafer, G.A. and Rabbinge, R. (1998) Optimization of vertical distribution of canopy nitrogen: an alternative trait to increase yield potential in wheat. *Journal of Crop Production* 1, 47–77.

Duvick, D.N. (1984) Genetic contributions to yield gains of U.S. hybrid maize, 1930 to 1980. In: Fehr, W.R. (ed.) *Genetic Contributions to Yield Gains of Five Major Crop Plants*. Crop Science Society of America–American Society of Agronomy, Madison, Wisconsin, pp. 1–47.

Eghball, B. and Maranville, J.W. (1991) Interactive effects of water and nitrogen stresses on nitrogen utilization efficiency, leaf water status and yield of corn genotypes. *Communications in Soil Science and Plant Analysis* 22, 1367–1382.

El Bassam, N. (1998) A concept of selection for 'low input' wheat varieties. *Euphytica* 100, 95–100.

Evans, J.R. (1989) Photosynthesis and nitrogen relationships in leaves of C_3 plants. *Oecologia* 78, 9–19.

Evans, L.T. (1993) *Crop Evolution, Adaptation, and Yield*. Cambridge University Press, Cambridge, New York, 500 pp.

Feil, B. (1992) Breeding progress in small grain cereals – a comparison of old and modern cultivars. *Plant Breeding* 108, 1–11.

Feil, B. and Geisler, G. (1988) Untersuchungen zur Bildung und Verteilung der Biomasse bei alten und neuen deutschen Sommerweizensorten. *Journal of Agronomy and Crop Science* 161, 148–156.

Feng, J., Volk, R.J. and Jackson, W.A. (1998) Source and magnitude of ammonium generation in maize roots. *Plant Physiology* 118, 1361–1368.

Field, C. (1983) Allocating leaf nitrogen for the maximization of carbon gain: leaf age as a control on the allocation program. *Oecologia* 56, 341–347.

Field, C. and Mooney, H.A. (1986) The photosynthesis–nitrogen relationship in wild plants. In: Givinish, T.A. (ed.) *On the Economy of Plant Form and Function*. Cambridge University Press, London, pp. 22–25.

Fischer, R.A. and Wall, P.C. (1976) Wheat breeding in Mexico and yield increases. *Journal of Australian Institute of Agricultural Science* 42, 139–148.

Forde, B.G. (2000) Nitrate transporters in plants: structure, function and regulation. *Biochimica et Biophysica Acta* 1465, 219–235.

Foyer, C.H., Valadier, M.H., Migge, A. and Becker, T.W. (1998) Drought-induced effects on nitrate reductase activity and mRNA and on the coordination of nitrogen and carbon metabolism in maize leaves. *Plant Physiology* 117, 283–292.

Fraisier, V., Dorbe, M.F. and Daniel-Vedele, F. (2001) Identification and expression analyses of two genes encoding putative low-affinity nitrate transporters from *Nicotiana plumbaginifolia. Plant Molecular Biology* 45, 181–190.

Frederick, J.R. (1997) Winter wheat leaf photosynthesis, stomatal conductance, and leaf N concentration during reproductive development. *Crop Science* 37, 1819–1826.

Frederick, J.R. and Bauer, P.J. (1999) Physiological and numerical components of wheat yield. In: Satorre, E.H. and Slafer, G.A. (eds) *Wheat: Ecology and Physiology of Yield Determination*. Food Products Press, Binghamton, New York, pp. 45–65.

Frink, C.R., Waggoner, P.E. and Ausubel, J.H. (1999) Nitrogen fertilizer: retrospect and prospect. *Proceedings of the National Academy of Sciences USA* 96, 1175–1180.

Galangau, F., Daniel-Vedele, F., Moureaux, T., Dorbe, M.F., Leydecker, M.T. and Caboche, M. (1988) Expression of leaf nitrate reductase gene from tomato and tobacco in relation to light dark regimes and nitrate supply. *Plant Physiology* 88, 383–388.

Gallagher, J.N. and Biscoe, P.V. (1978) Radiation absorption, growth and yield of cereals. *Journal of Agricultural Science* 91, 47–60.

Garnier, E. (1998) Nitrogen use efficiency from leaf to stand level: clarifying the concept. In: Lambers, H., Poorter, H. and van Vuuren, M.M.I. (eds) *Inherent Variation in Plant Growth. Physiological Mechanisms and Ecological Consequences*. Backhuys Publishers, Leiden, pp. 515–538.

Garnier, E. and Vancaeyzeele, S. (1994) Carbon and nitrogen content of congeneric annual and perennial grass species: relationships with growth. *Plant, Cell and Environment* 17, 399–407.

Gerendas, J., Zhu, Z., Bendixen, R., Ratcliffe, R.G. and Sattelmacher, B. (1997) Physiological and biochemical processes related to ammonium toxicity in higher plants. *Zeitschrift fuer Pflanzenernaehrung und Bodenkunde* 160, 239–251.

Gifford, R.M., Thorne, J.H., Hitz, W.D. and Giaquinta, R.T. (1984) Crop productivity and assimilate partitioning. *Science* 225, 801–808.

Glass, A.D.M., Shaff, J.E. and Kochian, L.V. (1992) Studies of the uptake of nitrate in barley. IV. Electrophysiology. *Plant Physiology* 99, 456–463.

Gojon, A., Wakrim, R., Passama, L. and Robin, P. (1991) Regulation of NO_3^- assimilation by anion availability in excised soybean leaves. *Plant Physiology* 96, 398–405.

Gojon, A., Dapoigny, L., Lejay, L., Tillard, P. and Rufty, T.W. (1998) Effects of genetic modification of nitrate reductase expression on $^{15}NO_3^-$ uptake and reduction in *Nicotiana* plants. *Plant, Cell and Environment* 21, 43–53.

Goos, R.J., Westfall, D.G., Ludwick, A.E. and Goris, J.E. (1982) Grain protein content as an indicator of N sufficiency for winter wheat. *Agronomy Journal* 74, 130–133.

Gorny, A.G. (1999) Inheritance of the nitrogen and phosphorus utilization efficiencies in spring barley at the vegetative growth stages under high and low nutrition. *Plant Breeding* 118, 511–516.

Gorny, A.G. and Sodkiewicz, T. (2001) Genetic analysis of the nitrogen and phosphorus utilization efficiencies in mature spring barley plants. *Plant Breeding* 120, 129–132.

Gorz, H.J., Haskins, F.A., Pedersen, J.F. and Ross, W.M. (1987) Combining ability effects for mineral elements in forage sorghum hybrids. *Crop Science* 27, 216–219.

Gowri, G., Ingemarsson, B., Redinbaugh, M.G. and Campbell, W.H. (1992) Nitrate reductase transcript is expressed in the primary response of maize to environmental nitrate. *Plant Molecular Biology* 18, 55–64.

Goyal, S.S. and Huffaker, R.C. (1984) Nitrogen toxicity in plants. In: Hauck, R.D. (ed.) *Nitrogen in Crop Production*. American Society of Agronomy–Crop Science Society of America–Soil Science Society of America, Madison, Wisconsin, pp. 97–118.

Goyal, S.S. and Huffaker, R.C. (1986) A novel approach and a fully automated microcomputer-based system to study kinetics of NO_3^-, NO_2^-, and NH_4^+ transport simultaneously by intact wheat seedlings. *Plant, Cell and Environment* 9, 209–215.

Greef, J.M., Ott, H., Wulfes, R. and Taube, F. (1999) Growth analysis of dry matter accumulation and N uptake of forage maize cultivars affected by N supply. *Journal of Agricultural Science* 132, 31–43.

Hageman, R.H. and Lambert, R.J. (1988) The use of physiological traits for corn improvement. In: Sprague, G.F. and Dudley, J.W. (eds) *Corn and Corn Improvement*, 3rd edn. Agronomy Monograph No. 18, American Society of Agronomy, Madison, Wisconsin, pp. 431–462.

Hageman, R.H., Reed, A.J., Femmer, R.A., Sherrad, J.H. and Dalling, M.J. (1980) Some new aspects of the *in vivo* assay for nitrate reductase in wheat (*Triticum aestivum* L.) leaves. *Plant Physiology* 65, 27–32.

Harvey, P.H. (1939) Hereditary variation in plant nutrition. *Genetics* 24, 437–461.

Heineke, D., Riens, B., Grosse, H., Hoferichter, P., Peter, U., Fluegge, U.I. and Heldt, H.W. (1991) Redox transfer across the inner chloroplast envelope membrane. *Plant Physiology* 95, 833–837.

Heitholt, J.J. (1989) Water use efficiency and dry matter distribution in nitrogen- and water-stressed winter wheat. *Agronomy Journal* 81, 464–469.

Heitholt, J.J., Croy, L.I., Maness, N.O. and Nguyen, H.T. (1990) Nitrogen partitioning in genotypes of winter wheat differing in grain N concentration. *Field Crops Research* 23, 133–144.

Henriksen, G.H., Bloom, A.J. and Spanswick, R.M. (1990) Measurement of net fluxes of ammonium and nitrate at the surface of barley roots using ion selective microelectrodes. *Plant Physiology* 93, 271–280.

Henriksen, G.H., Raman, D.R., Walker, L.P. and Spanswick, R.M. (1992) Measurement of net fluxes of ammonium and nitrate at the surface of barley roots using ion-selective microelectrodes. II. Patterns of uptake along the root axis and evaluation of the microelectrode flux estimation technique. *Plant Physiology* 99, 734–747.

Hirel, B., Bertin, P., Quiellere, I., Bourdoncle, W., Attagnant, C.D., Gouy, A., Cadiou, S., Retailliau, C., Falque, M. and Gallais, A. (2001) Towards a better understanding of the genetic and physiological basis for nitrogen use efficiency in maize. *Plant Physiology* 125, 1258–1270.

Hirose, T. (1988) Modelling the relative growth rate as a function of plant nitrogen concentration. *Physiologia Plantarum* 72, 185–189.

Hirose, T. and Werger, M.J.A. (1987) Maximizing daily canopy photosynthesis with respect to the leaf nitrogen allocation pattern in the canopy. *Oecologia* 72, 520–526.

Hocking, P.J. and Meyer, C.P. (1991) Effects of CO_2 enrichment and nitrogen stress on growth and partitioning of dry matter and nitrogen in wheat and maize. *Australian Journal of Plant Physiology* 18, 339–396.

Hoff, T., Truong, H.N. and Caboche, M. (1994) The use of mutants and transgenic plants to study nitrate assimilation. *Plant, Cell and Environment* 17, 489–506.

Imsande, J. and Touraine, B. (1994) N demand and the regulation of nitrate uptake. *Plant Physiology* 105, 3–7.

Inthapanya, P., Sipaseuth, Sihavong, P., Sihathep, V., Chanphengsay, M., Fukai, S. and Basnayake, J. (2000) Genotype differences in nutrient uptake and utilization for grain yield production of rainfed lowland rice under fertilized and non-fertilized conditions. *Field Crops Research* 65, 57–68.

Janssen, B.H. (1998) Efficient use of nutrients: an art of balancing. *Field Crops Research* 56, 197–201.

Janssen, B.H., Guiking, F.C.T., van der Eijk, D., Smaling, E.M.A., Wolf, J. and van Reuler, H. (1990) A system for quantitative evaluation of the fertility of tropical soils (QUEFTS). *Geoderma* 46, 299–318.

Jenkinson, D.S. (2001) The impact of humans on the nitrogen cycle, with focus on temperate arable agriculture. *Plant and Soil* 228, 3–15.

Kaiser, W.M. and Brendle-Behnisch, E. (1991) Rapid modulation of spinach leaf nitrate reductase activity by photosynthesis. I. Modulation *in vivo* by CO_2 availability. *Plant Physiology* 96, 363–367.

Kaiser, W.M. and Huber, S.C. (1997) Correlation between apparent activation state of nitrate reductase (NR), NR hysteresis and degradation of Nr protein. *Journal of Experimental Botany* 48, 1367–1374.

Kaiser, W.M., Weiner, H. and Huber, S.C. (1999) Nitrate reductase in higher plants: a case study for transduction of environmental stimuli into control of catalytic activity. *Physiologia Plantarum* 105, 385–390.

Kaiser, W.M., Kandlbinder, A., Stoimenova, M. and Glaab, J. (2000) Discrepancy between nitrate reduction rates in intact leaves and nitrate reductase activity in leaf extracts: what limits nitrate reduction *in situ*? *Planta* 210, 801–807.

Keating, B.A., Godwin, D.C. and Watiki, J.M. (1991) Optimising nitrogen inputs in response to climatic risk. In: Muchow, R.C. and Bellamy, J.A. (eds) *Climatic Risk in Crop Production: Models and Management for the Semiarid Tropics and Subtropics*. CAB International, Wallingford, UK, pp. 329–358.

Klein, D., Morcuende, R., Stitt, M. and Krapp, A. (2000) Regulation of nitrate reductase expression in leaves by nitrate and nitrogen metabolism is completely overridden when sugars fall below a critical level. *Plant, Cell and Environment* 23, 863–871.

Kleinhofs, A. and Warner, R.L. (1990) Advances in nitrate assimilation. In: Miflin, B.J. and Lea, P.J. (eds) *The Biochemistry of Plants*, Vol. 16, *Intermediary Nitrogen Metabolism*. Academic Press, New York, pp. 89–120.

Kronzucker, H.J., Schjoerring, J.K., Erner, Y., Kirk, G.J.D., Siddiqi, M.Y. and Glass, A.D.M. (1998) Dynamic interactions between root

NH$_4^+$ influx and long-distance N translocation in rice: insights into feed-back processes. *Plant and Cell Physiology* 39, 1287–1293.

Kronzucker, H.J., Siddiqi, M.Y., Glass, A.D.M. and Kirk, G.J.D. (1999) Nitrate–ammonium synergism in rice: a subcellular flux analysis. *Plant Physiology* 119, 1041–1045.

Kumar, P.A. and Abrol, Y.P. (1989) Effect of photorespiratory inhibitors on *in vivo* nitrate reduction in the leaves of barley and maize. *Canadian Journal of Botany* 67, 3426–3427.

Ladha, J.K., Kirk, G.J.D., Bennett, J., Peng, S., Reddy, C.K., Reddy, P.M. and Singh, U. (1998) Opportunities for increased nitrogen-use efficiency from improved lowland rice germplasm. *Field Crops Research* 56, 41–71.

Larios, B., Aguera, E., de la Haba, P., Perez-Vicente, R. and Maldonado, J.M. (2001) A short-term exposure of cucumber plants to rising atmospheric CO$_2$ increases leaf carbohydrate content and enhances nitrate reductase expression and activity. *Planta* 212, 305–312.

Lejay, L., Quillere, I., Roux, Y., Tillard, P., Cliquet, J.B., Meyer, C., Morot-Gaudry, J.F. and Gojon, A. (1997) Abolition of posttranscriptional regulation of nitrate reductase partially prevents the decrease in leaf NO$_3^-$ reduction when photosynthesis is inhibited by CO$_2$ deprivation, but not in darkness. *Plant Physiology* 115, 623–630.

Li, X. and Oaks, A. (1994) Induction and turnover of nitrate reductase in *Zea mays*. *Plant Physiology* 106, 1145–1149.

Mackay, A.D. and Barber, S.A. (1986) Effect of nitrogen on root growth of two genotypes in the field. *Agronomy Journal* 78, 699–703.

Malakoff, D. (1998) Death by suffocation in the Gulf of Mexico. *Science* 281, 190–192.

Man, H.M., Abd-El Baki, G.K., Stegmann, P., Weiner, H. and Kaiser, W.M. (1999) The activation state of nitrate reductase is not always correlated with total nitrate reductase activity in leaves. *Planta* 209, 462–468.

Marschner, H. (1995) *Mineral Nutrition of Higher Plants*. Academic Press, London, 674 pp.

Martin, F., Winspear, M.J., MacFarlane, J.D. and Oaks, A. (1983) Effect of methionine sulfoximine on the accumulation of ammonia in C$_3$ and C$_4$ leaves: the relationship between NH$_3$ accumulation and photorespiratory activity. *Plant Physiology* 71, 177–181.

Melzer, J.M., Kleinhofs, A. and Warner, R.L. (1989) Nitrate reductase regulation: effects of nitrate and light on nitrate reductase mRNA accumulation. *Molecular and General Genetics* 217, 341–346.

Molaretti, G., Bosio, M., Gentinetta, E. and Motto, M. (1987) Genotypic variability for N-related traits in maize: identification of inbred lines with high or low levels on NO$_3$-N in the stalks. *Maydica* 32, 309–323.

Moll, R.H., Kamprath, E.J. and Jackson, W.A. (1982) Analysis and interpretation of factors which contribute to efficiency of nitrogen utilization. *Agronomy Journal* 74, 562–564.

Moll, R.H., Kamprath, E.J. and Jackson, W.A. (1987) Development of nitrogen efficient prolific hybrids of maize. *Crop Science* 27, 181–186.

Muchow, R.C. (1994) Effect of nitrogen on yield determination in irrigated maize in tropical and subtropical environments. *Field Crops Research* 38, 1–13.

Muchow, R.C. (1998) Nitrogen utilization efficiency in maize and grain sorghum. *Field Crops Research* 56, 209–216.

Muchow, R.C. and Sinclair, T.R. (1994) Nitrogen response of leaf photosynthesis and canopy radiation-use efficiency in field-grown maize and sorghum. *Crop Science* 34, 721–727.

Mudhar, M.S. and Hignett, T.P. (1987a) Energy requirements, technology and resources in fertilizer sector. In: Helsel, Z.R. (ed.) *Energy in Plant Nutrition and Pest Control*. Elsevier, New York, pp. 25–62.

Mudhar, M.S. and Hignett, T.P. (1987b) Energy Efficiency, economics, and policy in the fertilizer sector. In: Helsel, Z.R. (ed.) *Energy in Plant Nutrition and Pest Control*. Elsevier, New York, pp. 133–163.

Newman, I.A. (2001) Ion transport in roots: measurement of fluxes using ion-selective microelectrodes to characterize transporter function. *Plant, Cell and Environment* 24, 1–14.

Nielsen, D.C. and Halvorson, A.D. (1991) Nitrogen fertility influence on water stress and yield of winter wheat. *Agronomy Journal* 83, 1065–1070.

Novoa, R. and Loomis, R.S. (1981) Nitrogen and plant production. *Plant and Soil* 58, 177–204.

Oaks, A. (1994) Efficiency of nitrogen utilization in C$_3$ and C$_4$ cereals. *Plant Physiology* 106, 407–414.

Oaks, A., He, X. and Zoumadakis, M. (1990) Nitrogen use efficiency in C$_3$ and C$_4$ cereals. In: Sinha, S.K. (ed.) *Proceedings International Congress of Plant Physiology, Delhi, India*, Vol. 2, pp. 1038–1045.

Ortiz-Monasterio, J.I., Sayre, K.D., Rajaram, S. and McMahom, M. (1997) Genetic progress in wheat yield and nitrogen use efficiency under four nitrogen rates. *Crop Science* 37, 898–904.

Oscarson, P., Lundborg, T., Larsson, M. and Larsson, C.M. (1995) Genotypic differences in nitrate uptake and nitrate utilization for spring wheat grown hydroponically. *Crop Science* 35, 1056–1062.

Paccaud, F.X., Fossati, A. and Cao, H.S. (1985) Breeding for yield and quality in winter wheat: consequences for nitrogen uptake and partitioning efficiency. *Zeitschrift fuer Pflanzenzuechtung* 94, 89–100.

Pace, G.M., Mackown, C.T. and Volk, R.J. (1982) Maximizing nitrate reduction during Kjeldahl digestion of plant tissue extracts and stem exudates: application to ^{15}N studies. *Plant Physiology* 69, 32–36.

Pan, W.L., Kamprath, E.J., Moll, R.H. and Jackson, W.A. (1984) Proliferation in corn: its effects on nitrate and ammonium uptake and utilization. *Soil Science Society of America Journal* 48, 1101–1106.

Peng, S., Cassman, K.G. and Kropff, M.J. (1995) Relationship between leaf photosynthesis and nitrogen content of field grown rice in the tropics. *Crop Science* 35, 1627–1630.

Perez, P., Martinez-Carrasco, R. and Sanchez de La Puente, L. (1983) Uptake and distribution of nitrogen in wheat plants supplied with different amounts of nitrogen after stem elongation. *Annals of Applied Biology* 102, 399–406.

Peuke, A.D., Jeschke, W.D., Dietz, K.J., Schreiber, L. and Hartung, W. (1998) Foliar application of nitrate or ammonium as sole nitrogen supply in *Ricinus communis*: I. Carbon and nitrogen uptake and inflows. *New Phytologist* 138, 675–687.

Plassard, C., Meslem, M., Souche, G. and Jaillard, B. (1999) Localization and quantification of net fluxes of H^+ along maize roots by combined use of pH-indicator dye video-densitometry and H^+-selective microelectrodes. *Plant and Soil* 211, 29–39.

Plenet, D. and Lemaire, G. (1999) Relationships between dynamics of nitrogen uptake and dry matter accumulation in maize crops: determination of critical N concentration. *Plant and Soil* 216, 65–82.

Pollmer, W.G., Eberhard, D., Klein, D. and Dhillon, B.S. (1979) Genetic control of nitrogen uptake and translocation in maize. *Crop Science* 19, 82–96.

Poorter, H., Remkes, C. and Lambers, H. (1990) Carbon and nitrogen economy of 24 wild species differing in relative growth rate. *Plant Physiology* 94, 621–627.

Pouteau, S., Cherel, I., Vaucheret, H. and Caboche, M. (1989) Nitrate reductase mRNA regulation in *Nicotiana plumbaginifolia* nitrate reductase-deficient mutants. *Plant Cell* 1, 1111–1120.

Provan, F. and Lillo, C. (1999) Photosynthetic post-translational activation of nitrate reductase. *Journal of Plant Physiology* 154, 605–609.

Purcino, A.A.C., Arellano, C., Athwal, G.S. and Huber, S.C. (1998) Nitrate effect on carbon and nitrogen assimilating enzymes of maize hybrids representing seven eras of breeding. *Maydica* 43, 83–94.

Rajcan, I. and Tollenaar, M. (1999) Source–sink ratio and leaf senescence in maize. II. Metabolism of nitrogen and soluble carbohydrates during the grain filling period. *Field Crops Research* 60, 255–265.

Raun, W.R. and Johnson, G.V. (1999) Improving nitrogen use efficiency for cereal production. *Crop Science* 91, 357–363.

Reed, A.J., Below, F.E. and Hageman, R.H. (1980) Grain protein accumulation and the relationship between leaf nitrate reductase and protease activities during grain development in maize: I. Variation between genotypes. *Plant Physiology* 66, 164–170.

Richards, R.A., Lopez Castaneda, C., Gomez-Macpherson, H. and Condon, A.G. (1993) Improving the efficiency of water use by plant breeding and molecular biology. *Irrigation Science* 14, 93–104.

Salsac, L., Chaillou, S., Morot-Gaudry, J.F., Lesaint, C. and Jolivoe, E. (1987) Nitrate and ammonium nutrition in plants. *Plant Physiology and Biochemistry* 25, 805–812.

Scheible, W.R., Gonzalez-Fontes, A., Morcuende, R., Lauerer, M., Geiger, M., Glaab, J., Gojon, A., Schulze, E.D. and Stitt, M. (1997) Tobacco mutants with a decreased number of functional nia genes compensate by modifying the diurnal regulation of transcription, post-translational modification and turnover of nitrate reductase. *Planta* 203, 304–319.

Sehtiya, H.L. and Goyal, S.S. (2000) Comparative uptake of nitrate by intact seedlings of C_3 (barley) and C_4 (corn) plants: effect of light and exogenously applied sucrose. *Plant and Soil* 227, 185–190.

Sherchand, K. and Paulsen, G.M. (1985) Genotypic variation in partitioning of phosphorus in relation to nitrogen and dry matter during wheat grain development. *Journal of Plant Nutrition* 8, 1161–1170.

Siddiqi, M.Y., Glass, A.D.M., Ruth, T.J. and Rufty, T.W. (1990) Studies of the uptake of nitrate in barley. I. Kinetics of NO_3^- influx. *Plant Physiology* 93, 1426–1432.

Simmons, S.R. (1987) Growth, development, and physiology. In: Heyne, E.G. (ed.) *Wheat and Wheat Imrovement*, 2nd edn. Agronomy Monograph Series No. 13, American Society of Agronomy, Madison, Wisconsin, pp. 77–113.

Sinclair, T.R. and Horie, T. (1989) Leaf nitrogen, photosynthesis and radiation use efficiency – a review. *Crop Science* 29, 90–98.

Sinclair, T.R. and Muchow, R.C. (1995) Effect of nitrogen supply on maize yield. I. Modeling physiological responses. *Agronomy Journal* 87, 632–641.

Sinclair, T.R. and Muchow, R.C. (1999) Radiation use efficiency. *Advances in Agronomy* 65, 215–265.

Singh, Y. and Singh, B. (2001) Efficient management of primary nutrients in the rice–wheat system. *Journal of Crop Production* 4, 23–85.

Sivapalan, S., O'Brien, L., Ortiz-Ferrara, G., Hollamby, G.J., Barclay, I. and Martin, P.J. (2000) An adaptation analysis of Australian and CIMMYT/ICARDA wheat germplasm in Australian production environments. *Australian Journal of Agricultural Research* 51, 903–915.

Slafer, G.A., Andrade, F.H. and Feingold, F.E. (1990) Genetic improvement of bread wheat (*Triticum aestivum* L.) in Argentina: relationship between nitrogen and dry matter. *Euphytica* 50, 63–71.

Sofield, I., Wardlaw, I.F., Evans, L.T. and Zee, S.Y. (1977) Nitrogen, phosphorus and water contents during grain development and maturation in wheat. *Australian Journal of Plant Physiology* 4, 799–810.

Spiertz, J.H.J. and Ellen, J. (1978) Effects of nitrogen on crop development and grain growth of winter wheat in relation to assimilation and utilization of assimilates and nutrients. *Netherlands Journal of Agricultural Science* 26, 210–231.

Stitt, M. (1999) Nitrate regulation of metabolism and growth. *Current Opinion in Plant Biology* 2, 178–186.

Tabachow, R.M., Jeffrey Pierce, J. and Richter, D.D. (2001) Biogeochemical models relating soil nitrogen losses to plant-available N. *Environmental Engineering Science* 18, 81–89.

Taub, D.R. and Lerdau, M.T. (2000) Relationship between leaf nitrogen and photosynthetic rate for three NAD-ME and three NADP-ME C4 grasses. *American Journal of Botany* 87, 412–417.

Taylor, A.R. and Bloom, A.J. (1998) Ammonium, nitrate, and proton fluxes along the maize root. *Plant, Cell and Environment* 21, 1255–1263.

Teyker, R.H., Moll, N.A. and Jackson, N.A. (1989) Divergent selection among maize seedlings for nitrate uptake. *Crop Science* 29, 879–884.

Tindall, T.A., Stark, J.C. and Brooks, R.H. (1995) Irrigated spring wheat response to topdress nitrogen as predicted by flag leaf nitrogen concentration. *Journal of Production Agriculture* 8, 46–52.

Tischner, R. (2000) Nitrate uptake and reduction in higher and lower plants. *Plant, Cell and Environment* 23, 1005–1024.

Tollenaar, M. (1991) The physiological basis of the genetic improvement of maize hybrids in Ontario from 1959 to 1988. *Crop Science* 31, 119–124.

Tsai, C.Y., Dweikat, I., Huber, D.M. and Warren, H.L. (1992) Interrelationship of nitrogen nutrition with maize (*Zea mays*) grain yield, nitrogen use efficiency and grain yield. *Journal of Science of Food and Agriculture* 58, 1–8.

Uhart, S.A. and Andrade, F.H. (1995) Nitrogen deficiency in maize: I. Effects on crop growth, development, dry matter partitioning, and kernel set. *Crop Science* 35, 1376–1383.

Van der Werf, A., Van Neuenen, M., Visser, A.J. and Lambers, H. (1993) Contribution of physiological and morphological plant traits to a species' competitive ability at high or low nitrogen supply: a hypothesis for inherently fast- and slow-growing species. *Oecologia* 94, 434–440.

Van Sanford, D.A. and MacKown, C.T. (1987) Cultivar differences in nitrogen remobilization during grain fill in soft red winter wheat. *Crop Science* 27, 295–300.

Von Wiren, N., Gazzarrini, S., Gojon, A. and Frommer, W.F. (2000) The molecular physiology of ammonium uptake and retrieval. *Current Opinion in Plant Biology* 3, 254–261.

Vose, P.B. (1990) Screening techniques for plant nutrient efficiency: philosophy and methods. In: El Bassam, N., Dambroth, M. and Loughman, B.C. (eds) *Genetic Aspects of Plant Mineral Nutrition*. Kluwer Academic Publishers, Dordrecht, pp. 283–289.

Walch-Liu, P., Neumann, G., Bangerth, F. and Engels, C. (2000) Rapid effects of nitrogen form on leaf morphogenesis in tobacco. *Journal of Experimental Botany* 51, 227–237.

Wang, M.Y., Siddiqi, M.Y., Ruth, T.J. and Glass, A.D.M. (1993) Ammonium uptake by rice roots. II. Kinetics of $^{13}NH_4^+$ influx across the plasmalemma. *Plant Physiology* 103, 1259–1267.

Wang, R., Guegler, K., LaBrie, S.T. and Crawford, N.M. (2000) Genomic analysis of a nutrient response in *Arabidopsis* reveals diverse expression patterns and novel metabolic and potential regulatory genes induced by nitrate. *Plant Cell* 12, 1491–1509.

Wang, X. and Below, F.E. (1992) Root growth, nitrogen uptake, and tillering of wheat induced by mixed-nitrogen source. *Crop Science* 32, 997–1002.

Westcott, B. (1986) Some methods of analyzing genotype–environment interactions. *Heredity* 56, 243–253.

Whitfield, D.M. and Smith, C.J. (1989) Effects of irrigation and nitrogen on growth, light interception and efficiency of light conversion in wheat. *Field Crops Research* 20, 279–295.

Yamasaki, H. (2000) Nitrite-dependent nitric oxide production pathway: implications for involvement of active nitrogen species in photoinhibition *in vivo*. *Philosophical Transactions of the Royal Society of London B Biological Sciences* 355, 1477–1488.

Yin, Z.H., Kaiser, W., Heber, U. and Raven, J.A. (1998) Effects of gaseous ammonia on intracellular pH values in leaves of C_3- and C_4-plants. *Atmospheric Environment* 32, 539–544.

Ying, J., Peng, S., Yang, G., Zhou, N., Visperas, R.M. and Cassman, K.G. (1998) Comparison of high-yield rice in tropical and subtropical environments. II. Nitrogen accumulation and utilization efficiency. *Field Crops Research* 57, 85–93.

19 Biplot Analysis of Multi-environment Trial Data

Weikai Yan and L.A. Hunt

*Department of Plant Agriculture, University of Guelph, Guelph,
Ontario N1G 2W1, Canada*

Introduction

Regional multi-environment trials (MET) are conducted every year for all major crops throughout the world, constituting a costly but essential step towards new crop genotype release and cultivar recommendation. MET are essential because the presence of genotype–environment interaction (GE), i.e. differential genotype responses in different environments, complicates cultivar evaluation. Some important concepts, such as ecological region, ecotype, mega-environment, specific adaptation, stability, etc., all originate from GE. Were there no GE, a single cultivar would prevail all over the world and a single trial would suffice for cultivar evaluation (Gauch and Zobel, 1996). GE constitutes a major challenge to cultivar improvement, and MET data analysis constitutes an important aspect of plant breeding. Because of this, improvement in the methods used for MET data analysis should be of interest to the plant-breeding community. This chapter deals with the biplot method, which has been receiving attention in recent years.

Utilities of multi-environment trial data analysis

The primary objective of MET is, of course, to identify superior cultivars. The most common practice used to achieve this end is to compare the mean yield of genotypes across test environments (usually year–location combinations) represented in the MET. The validity of this practice is, however, based on the usually unstated assumption that the environments in the MET belong to a single mega-environment, defined as a group of locations in which the same set of cultivars perform best across a number of years. Although usually unstated, cultivar evaluation is always specific to single mega-environments. If the test environments are sufficiently heterogeneous, the cultivars that are selected based on mean yield may not be the best in some of the test environments; in extreme cases, they may even not be the best in any of the environments. Thus, a second utility of MET data analysis, prior to cultivar evaluation, should be to investigate the relationships among the test environments and the possibility of mega-environment differentiation within the target environment. Identification of mega-environments would allow exploitation of the GE that is repeatable across years.

For a given mega-environment, genotypes should be evaluated for mean yield (or, in more general terms, mean performance) and stability across test environments. The ideal cultivar should be one that is both high-yielding and stable. Mean performance is simply the mean across all environments, whereas stability is a measure of variability across environments. Most research has focused on quantification of stability, and numerous stability measures have been proposed (Lin *et al.*, 1986; Lin and Binns, 1994; Kang, 1998). For a given mega-environment and parallel to cultivar evaluation, individual test environments should be evaluated for their ability to provide data that allow for discrimination among genotypes and, at the same time, for the extent to which they represent the target mega-environment.

The ultimate reason for differential stability among genotypes and for differential results from various test environments is non-repeatable GE. Since this type of GE cannot be effectively exploited, it must be avoided. A fourth utility of MET data analysis should be the development of a better understanding of the causes of GE. Such an understanding may help to avoid confounding plant responses to specific and rare conditions with overall cultivar evaluation.

To summarize, MET data analysis should, and potentially can, fulfil four functions: (i) investigation of possible mega-environment differentiation in the target environment; (ii) selection of superior cultivars for individual mega-environments; (iii) selection of better test environments; and (iv) development of a better understanding of the causes of GE. An ideal MET data-analysis system should accomplish all four tasks so that the information contained in the MET is maximally exploited and utilized.

Visualization of multi-environment trial data

With the belief that 'a picture is worth a thousand words', many attempts have been made to graphically present MET data. The general pattern of such a graphical display of MET data is to plot the mean yield of each genotype against a measure of stability, which can be any parameter that is listed in Lin *et al.* (1986), among others.

Another popular presentation of MET data is based on the Finlay and Wilkinson

(1963) model, in which the yield of each genotype is plotted against the mean yield of each environment and in which each genotype is represented by a fitted straight line. Philosophically, this type of graphical display of MET data is very attractive, since it clearly indicates differential genotype responses to test environments. The problem with this method is that the environmental means are not always a good, and are frequently a poor, measure of environments, such that the fitted lines in most cases only account for a small fraction of the total GE (Zobel *et al.*, 1988).

A visualization method that is similar to that of Finlay and Wilkinson (1963) but which explains more GE was developed by Gauch and Zobel (1997). In this method, the nominal yields of genotypes are plotted against the first interaction principal component (IPC1) scores of environments, so that each genotype is represented by a line with the mean yield as the intercept and the genotype IPC1 score as the slope. Such a plot indicates the 'which-won-where' patterns of the data, provided that the IPC1 explains most of the GE.

The recently developed GGE-biplot method (Yan *et al.*, 2000, 2001) provides a more elegant and useful display of MET data. It effectively addresses both the issue of mega-environment differentiation and the issue of genotype selection for a given mega-environment based on mean yield and stability. It also allows environments to be evaluated just as well as genotypes. In addition, it facilitates interpretation of GE as genotypic factor by environmental factor interactions (Yan and Hunt, 2001). In the rest of the chapter, we shall describe the rationale and applications of the GGE-biplot methodology in MET data analysis.

The GGE-biplot Methodology

The GGE-biplot methodology consists of two concepts: biplot and 'GGE'. Both components are discussed below.

The concept of biplot

The concept of biplot was first proposed by Gabriel (1971). The main ideas follow. Any two-way table or matrix X that contains n rows and m columns can be regarded as the product of two matrices A, with n rows and r columns, and B, with r rows and m columns. Therefore, matrix X can always be decomposed into its two component matrices A and B. If r happens to be 2, matrix X is referred to as a rank-two matrix. Each row in matrix A has two values that can be displayed as a point in a two-dimensional plot. Similarly, each column in matrix B has two values and can also be displayed as a point in a two-dimensional plot. When both the n rows of A and the m columns of B are displayed in a single plot, the plot is called a 'biplot'. Therefore, the biplot of a rank-two matrix contains $n + m$ points, as compared with $n \times m$ values in the matrix *per se*, and yet contains all the information of the matrix.

One interesting property of a biplot is that each of the $n \times m$ values can be precisely recovered by viewing the $n + m$ points on the biplot. Assume that we have three-genotype × three-environment data on yield and that it is a rank-two matrix. After decomposition of the data into its two component matrices, the three genotypes and three environments can be presented in a biplot, as shown in Fig. 19.1. The yield of genotype i in environment j, Y_{ij}, can be recovered by the following formula:

$$Y_{ij} = \overline{OE}_j \cos \alpha_{ij} \overline{OG}_i = \overline{OE}_j \overline{OP}_{ij}$$

where $\overline{OG}_i$ (or OG_i) is the absolute distance from the biplot origin O to the marker of the genotype i, $\overline{OE}_j$ (or OE_j) is the absolute distance from the biplot origin O to the marker of environment j, α_{ij} is the angle between the vectors $\overline{OG}_i$ and $\overline{OE}_j$ and $\overline{OP}_{ij}$ (or OP_{ij}) = cos $\alpha_{ij}\overline{OG}_i$ is the projection of the marker of genotype i to the vector of environment j. To compare yields of the three genotypes in environment E_1, we have

$$Y_{11} = (OE_1)(\cos\alpha_{11})(OG_1) = (OE_1)(OP_{11})$$
$$Y_{21} = (OE_1)(\cos\alpha_{21})(OG_2) = (OE_1)(OP_{21})$$
$$Y_{31} = (OE_1)(\cos\alpha_{31})(OG_3) = (OE_1)(OP_{31})$$

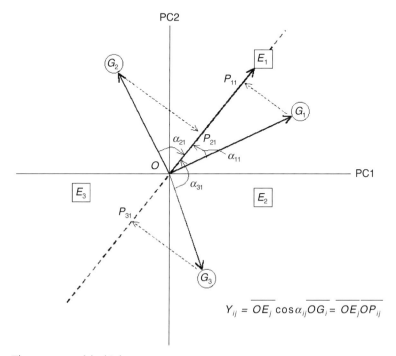

$$Y_{ij} = \overline{OE}_j \cos\alpha_{ij}\overline{OG}_i = \overline{OE}_j\overline{OP}_{ij}$$

Fig. 19.1. The geometry of the biplot.

where OP_{11}, OP_{21} and OP_{31} are the projections of the markers of the genotypes on to the vector or its extension of environment E_1. Since OE_1 is non-negative and common to all genotypes, comparisons among Y_{11}, Y_{21} and Y_{31} can be performed by simply visualizing OP_{11}, OP_{21} and OP_{31}. In our example (Fig. 19.1), it is obvious that $OP_{11} > OP_{21} > OP_{31}$, and therefore, $Y_{11} > Y_{21} > Y_{31}$. Note that OP_{11} and OP_{21} are above average, whereas OP_{31} is below average, since $\cos \alpha_{11}$ and $\cos \alpha_{21}$ are positive whereas $\cos \alpha_{31}$ is negative.

Approximation of any two-way table using a rank-two matrix

A biplot is obviously an elegant display of a rank-two matrix. In reality, however, it is rare that a two-way data set is exactly a rank-two matrix. Nevertheless, if a two-way data set, e.g. the yield data of a number of genotypes tested in a number of environments, can be approximated by a rank-two matrix, the latter can then be displayed in a biplot (Gabriel, 1971). The process of decomposing matrix X into its component matrices A and B is called 'singular value decomposition' (SVD), the result of which is r principal components (r equals the smaller of n and m). If the first two principal components (PC1 and PC2) explain a large proportion of the total variation of X, X is said to be sufficiently approximated by a rank-two matrix and can be approximately displayed in a biplot.

The concept of GGE

The concept of GGE originates from analysis of MET of crop cultivars. The yield of a genotype (or any other measure of genotype performance) in an environment is a mixed effect of genotype main effect (G), environment main effect (E) and GE. In normal MET, E accounts for 80% and G and GE each account for about 10% of the total variation. For the purpose of cultivar evaluation, however, only G and GE are relevant (Gauch and Zobel, 1996). Furthermore, both G and GE must be considered in cultivar evaluation: hence the term 'GGE' (Yan et al., 2000). Simultaneous examination of G and GE is, thus, an important principle in cultivar evaluation.

Models for constructing a GGE biplot

The GGE biplot displays the GGE part of a MET data set. Compared with other types of biplots, a GGE biplot has the advantage in that it: (i) displays most information that is relevant to cultivar evaluation; and (ii) displays only the information that is relevant to cultivar evaluation. A GGE biplot can be generated based on SVD of: (i) environment-centred data; (ii) environment-centred and within-environment standard deviation-scaled data; and (iii) environment-centred and within-environment standard error-scaled data.

Singular value decomposition of environment-centred data

The model for a GGE biplot based on SVD of environment-centred data is:

$$Y_{ij} - \overline{Y}_j = \lambda_1 \xi_{i1} \eta_{j1} + \lambda_2 \xi_{i2} \eta_{j2} + \varepsilon_{ij} \quad (19.1)$$

where:

Y_{ij} is the mean yield of genotype i in environment j

$\overline{Y}_j$ is the mean yield across all genotypes in environment j

λ_1 and λ_2 are the singular values for the first and second principal components, PC1 and PC2, respectively

ξ_{i1} and ξ_{i2} are the PC1 and PC2 scores, respectively, for genotype i

η_{j1} and η_{j2} are the PC1 and PC2 scores, respectively, for environment j

ε_{ij} is the residual of the model associated with genotype i in environment j

To display the PC1 and PC2 in a biplot, the equation is rewritten as:

$$Y_{ij} - \overline{Y}_j = \xi_{i1}^* \eta_{j1}^* + \xi_{i2}^* \eta_{j2}^* + \varepsilon_{ij}$$

where $\xi_{in}^{*} = \lambda_{n}^{1/2}\xi_{in}$ and $\eta_{jn}^{*} = \lambda_{n}^{1/2}\eta_{jn}$, with $n = 1$, 2. Although all scaling methods are equally valid, this method has the advantage that PC1 and PC2 have the same unit, which is the square root of the original unit, such as t ha^{-1} for yield. This property is important when genotypes are visually evaluated by both mean yield and stability (discussed later).

A GGE biplot is generated by plotting ξ_{j1}^{*} and η_{j1}^{*} against ξ_{i2}^{*} and η_{j2}^{*}, respectively. Although this type of biplot has been used previously in MET data analysis (e.g. Cooper *et al.*, 1997), methods for the utilization of the information contained in a biplot to its fullest extent became available only recently (Yan, 1999; Yan *et al.*, 2000).

Singular value decomposition of within-environment standard deviation-scaled data

The second model that can be used to generate a GGE biplot is:

$$\left(Y_{ij} - \overline{Y}_{j}\right) / s_{j} = \lambda_{1}\xi_{i1}\eta_{j1} + \lambda_{1}\xi_{i1}\eta_{j2} + \varepsilon_{ij} \quad (19.2)$$

where s_{j} is the standard deviation for genotype means for environment j, and all other parameters are the same as in Equation 19.1. This model removes the units of the data and assumes an equal ability of all environments to discriminate among genotypes, which may be an undesired property for genotype–environment data analysis. It is useful for analysing genotype–trait data, however, in which different traits use different units.

Singular value decomposition of within-environment standard error-scaled data

The third model is based on:

$$\left(Y_{ij} - \overline{Y}_{j}\right) / z_{j} = \lambda_{1}\xi_{i1}\eta_{j1} + \lambda_{1}\xi_{i1}\eta_{j2} + \varepsilon_{ij} \quad (19.3)$$

where z_{j} is the standard error for environment j. Since z_{j} can be estimated only with replicated data, this model can only be used when replicated data are available. It is preferred for all types of two-way data when replicated observations are available, since it adjusts any heterogeneity among testers, which can be environments, traits, etc.

Alternative models for generating a GGE biplot

To make sure that the abscissa of the GGE biplot represents the mean yield of the genotypes, Yan *et al.* (2001) proposed the following model:

$$Y_{ij} - \overline{Y}_{j} = b_{j}\alpha_{i} + \lambda_{1}\xi_{i1}\eta_{j1} + \varepsilon_{ij} \quad (19.4)$$

where α_{i} is the main effect of genotype i, b_{j} is the regression coefficient of the environment-centred yield of genotypes in environment j against the genotype main effects, λ_{1} is the singular value for the PC1 from subjecting the residue of the regressions to SVD, ξ_{i1} and η_{j1} are the scores for genotype i and environment j on PC1, respectively, and ε_{ij} is the residual associated with genotype i in environment j. The regressions $b_{j}\alpha_{i}$ in Equation 19.4 correspond to PC1 in Equation 19.1, and the PC1 in Equation 19.4 corresponds to PC2 in Equation 19.1. Equations 19.2 and 19.3 can also have their counterparts of Equation 19.4.

Biplots based on Equation 19.4 are more interpretative than Equation 19.1, since its abscissa of the biplot represents exactly the genotype main effects and, therefore, its ordinate is a measure of stability (or variability, or GE). This advantage, however, is offset by the explanation of a smaller percentage of the total GGE variation (Yan *et al.*, 2001). A recent finding is that biplots based on Equation 19.1 can also be used to approximately indicate the main effects and stability of the genotypes through axis rotation (discussed later). Therefore, the alternative models will not be further discussed in this chapter.

Biplot Analysis of Multi-environment Trial Data: an Example

This section exemplifies biplot analysis of MET data using the 1993 Ontario winter-wheat performance trial data. Efforts will be made to demonstrate how a GGE biplot can be used to address the four major utilities of MET data analysis.

The steps in biplot analysis

The sample data are presented in Table 19.1, which contains the mean yield of 18 winter-wheat genotypes tested in nine Ontario locations in 1993. The trials were replicated four to six times at each location, but we present only the mean data for the purpose of illustration. Generating a GGE biplot based on Equation 19.1 from Table 19.2 data involves the following steps:

1. Centring the data, i.e. subtracting the respective environmental means from each of the cells.

Table 19.1. Yield data (t ha^{-1}) of 18 genotypes in nine environments.

Genotypes	Environments									Mean
	BH93	EA93	HW93	IN93	KE93	NN93	OA93	RN93	WP93	
ANN	4.5	4.2	2.9	3.1	5.9	4.5	4.4	4.0	2.7	4.0
ARI	4.4	4.8	2.9	3.5	5.7	5.2	5.0	4.4	2.9	4.3
AUG	4.7	4.6	3.1	3.5	6.1	5.0	4.7	3.9	2.6	4.2
CAS	4.7	4.7	3.4	3.9	6.2	5.3	4.2	4.9	3.5	4.5
DEL	4.4	4.6	3.5	3.9	5.8	5.4	5.2	4.1	2.8	4.4
DIA	5.2	4.5	3.0	3.8	6.6	5.0	4.0	4.3	2.8	4.4
ENA	3.4	4.2	2.7	3.2	5.3	4.3	4.2	4.1	2.0	3.7
FUN	4.9	4.7	4.4	4.0	5.5	5.8	4.2	5.1	3.6	4.7
HAM	5.0	4.7	3.5	3.4	6.0	4.9	5.0	4.5	2.9	4.4
HAR	5.2	4.7	3.6	3.8	5.9	5.3	3.9	4.5	3.3	4.5
KAR	4.3	4.5	2.8	3.4	6.1	5.3	4.9	4.1	3.2	4.3
KAT	3.2	3.0	2.4	2.4	4.2	4.3	3.4	4.1	2.1	3.2
LUC	4.1	3.9	2.3	3.7	4.6	5.2	2.6	5.0	2.9	3.8
M12	3.3	3.9	2.4	2.8	4.6	5.1	3.3	3.9	2.6	3.5
REB	4.4	4.7	3.7	3.6	6.2	5.1	3.9	4.2	2.9	4.3
RON	4.9	4.7	3.0	3.9	6.1	5.3	4.3	4.3	3.0	4.4
RUB	3.8	5.0	3.4	3.4	4.8	5.3	4.3	4.9	3.4	4.3
ZAV	4.2	4.7	3.6	3.9	6.6	4.8	5.0	4.4	3.1	4.5

Table 19.2. PC1 and PC2 scores for each genotype and each environment used in constructing the GGE biplot.

Genotypes	PC1	PC2	Environments	PC1	PC2
ANN	−0.14	−0.44	RN93	0.19	0.72
ARI	0.17	−0.22	NN93	0.44	0.63
AUG	0.19	−0.41	WP93	0.54	0.59
CAS	0.43	0.31	IN93	0.66	0.36
DEL	0.32	−0.23	HW93	0.77	0.32
DIA	0.31	−0.07	BH93	0.97	0.23
ENA	−0.60	−0.51	EA93	0.76	0.08
FUN	0.51	0.79	KE93	1.11	−0.62
HAM	0.40	−0.22	OA93	0.85	−0.96
HAR	0.37	0.39			
KAR	0.17	−0.32			
KAT	−1.35	−0.18			
LUC	−0.73	0.86			
M12	−0.94	0.10			
REB	0.20	0.06			
RON	0.31	0.05			
RUB	−0.11	0.42			
ZAV	0.48	−0.37			

2. Subjecting the environment-centred data to SVD, which results in singular values – genotype and environment scores for each of the *n* principal components, *n* being the number of environments. SVD is a complex mathematical operation that decomposes a matrix into two component matrices using the least-squares method. Fortunately, it becomes a routine function in all major statistical analysis systems. The SAS package (SAS Institute, 1996) has an SVD function in the IML or MATRIX procedure, so that performing the SVD of a matrix takes no more than a single statement. The PRINCOMP procedure of SAS, which performs principal-component analysis, gives outputs in which the singular values are tied with the genotype (row) eigenvectors.

3. Partitioning the singular value into genotype and environment scores for each of the principal components. Theoretically, the singular value can be partitioned in any proportion, but symmetrical partitioning is preferred because it results in the same units for both the genotype scores and the environment scores and for all principal components.

4. Plotting the PC1 scores against the PC2 scores to generate a biplot. Biplots using other principal components are also possible. The plotting can be done using a spreadsheet, but the abscissa and ordinate must be drawn to scale.

5. Labelling the biplot with the genotype and environment names, which can be a very tedious job.

6. Adding supplementary lines to facilitate visualization and interpretation of the biplot.

As can be seen, although the biplot is an elegant tool for visualizing MET data, the process is tedious, if not difficult, even for well-trained biometricians. Fortunately, a Windows application, GGEbiplot, was recently created (Yan, 2001), which fully automated the biplot analysis process. All biplots presented below are the direct outputs of this software. In these biplots, the genotypes are labelled with lower-case letters and the environments with upper-case letters.

Visualizing the performance of different genotypes in a given environment

This is a direct application of the biplot theory described in Fig. 19.1 and associated descriptions. To visualize the performance of different genotypes in a given environment, say, BH93, draw a line that passes through the biplot origin and the marker of BH93; this may be called the BH93 axis. The genotypes will be ranked according to their projections on to the BH93 axis (Fig. 19.2). Thus, the order of yields of the genotypes in BH93 was: kat < m12 < ena < luc < ann < . . . < har ≈ cas < fun. The line passing through the biplot origin and perpendicular to the BH93 axis separates genotypes that yielded below the mean (kat, m12, ena, luc and ann) from genotypes that yielded above the mean (all other genotypes) in BH93.

Visualizing the relative adaptation of a given genotype in different environments

Analogous to the above, to visualize the relative performance of a given genotype, say, rub, in different environments, draw a line that passes through the biplot origin and the marker of rub, which may be called the rub axis. The environments would be ranked along the rub axis in the direction towards the marker of rub (Fig. 19.3). Thus, the relative performance of rub in different environments was: RN93 > NN93 > WP93 > IN93 > BH93 > EA93 > KE93 > OA93. The line passing through the biplot origin and perpendicular to the rub axis separates environments in which rub yielded below the mean (OA93, KE93 and EA93) from environments in which rub yielded above the mean (all other environments, except BH93). Environment BH93 was right on the perpendicular line, implying that rub yielded near the mean in BH93.

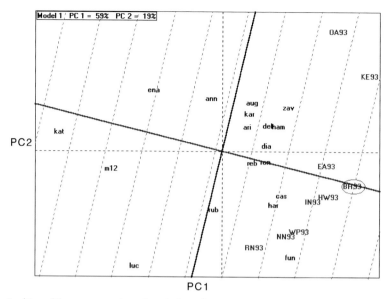

Fig. 19.2. Ranking of the genotypes based on their performance in environment BH93.

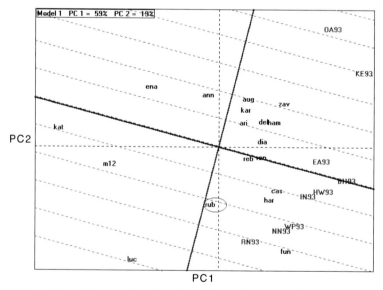

Fig. 19.3. Ranking of the environments based on the relative performance of genotype rub.

Visual comparison of two genotypes in different environments

Biplot comparison of two genotypes is an extension of the basic biplot principle. To compare two genotypes, connect the two genotypes to be compared, say, aug and rub, with a straight line (called a connector line) and draw a line that is perpendicular to the

connector line and passes through the biplot origin (Fig. 19.4). This perpendicular line separates environments where aug yielded better than rub from environments where rub yielded better than aug. Thus, Fig. 19.4 indicates that aug was better than rub in OA93, KE93, EA93 and BH93, and rub was better than aug in the other five environments. Based on the basic principle

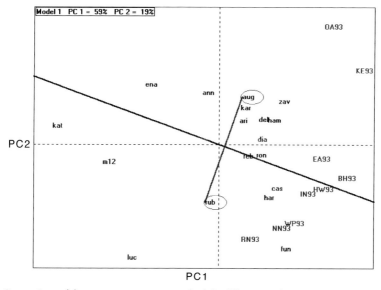

Fig. 19.4. Comparison of the two genotypes aug and rub in different environments.

of biplot geometry described earlier, the two genotypes would yield exactly the same in environments whose markers fall on the perpendicular line. If all environments fall on the same side of the perpendicular line, the genotype with the environments on its side would yield better than the other genotype in all environments. If the two genotypes are spatially close, they are likely to have yielded similarly in all or most of the environments.

Visual identification of the best genotype(s) for each environment

A further extended application of the biplot geometry is to visually identify the highest-yielding genotypes for each of the environments in a single step. For this purpose, the genotypes that are located far away from the biplot origin are connected with straight lines so that a polygon or vertex hull is formed with all other genotypes contained within the vertex hull (Fig. 19.5). The vertex genotypes in our example are fun, zav, ena, kat and luc. These genotypes are the most responsive genotypes; they are either the best or the poorest genotypes in some or all of the environments. Perpendicular lines to

the sides of the vertex hull are drawn, starting from the biplot origin, to divide the biplot into five sectors or quadrants, each having a vertex genotype. The beauty of Fig. 19.5 is this: the vertex genotype for each quadrant is the one that gave the highest yield for the environments that fall within that quadrant. Thus, genotype fun gave the highest yield in environments RN93, NN93, WP93, IN93, HW93, BH93 and EA93 and genotype zav gave the highest yield in environments OA93 and KE93. The other vertex genotypes, i.e. ena, kat and luc, did not give the highest yield in any of the environments. Actually, they were the poorest genotypes in some or all of the environments.

Now we explain why the above statements are valid. According to the section 'Visual comparison of two genotypes in different environments', the line perpendicular to the polygon side that connects genotypes luc and fun facilitates the comparison between luc and fun; fun yielded higher than luc in all environments because all environments are on the side of fun. Likewise, the line perpendicular to the polygon side that connects genotypes zav and fun facilitates the comparison between zav and fun; fun yielded higher than zav in seven

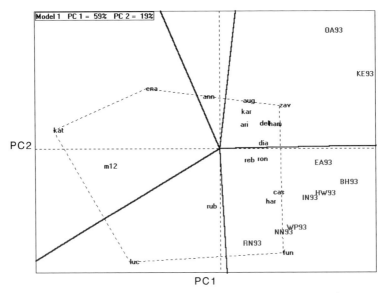

Fig. 19.5. The polygon view of the GGE biplot indicating the best genotype(s) in each environment and groups of environments.

environments that fall into the fun sector because they are on the side of fun. Within the fun sector, fun has the longest vector (distance from biplot origin to the marker of a genotype); it therefore gave higher yields than other genotypes in these seven environments, for reasons discussed in the section 'Visualizing the performance of different genotypes in a given environment'. Collectively, fun gave the highest yield in environments that fell in its sector. Using the same reasoning, zav was the best genotype in environments KE93 and OA93.

Visualizing groups of environments

Another utility of Fig. 19.5 is that the environments are grouped based on the best genotypes and we have two groups of environments: KE93 and OA93 as one group, with zav being the highest-yielding genotype, and the other seven environments as another group, with fun being the highest-yielding genotype.

The environment groups suggest different mega-environments. In our example, KE93 and OA93 represent eastern Ontario and the other environments represent western and southern Ontario. The hypothesis that eastern Ontario is a different mega-environment from the rest of Ontario for winter-wheat production was tested and confirmed using 1989–2000 Ontario winter-wheat performance trial data (Yan, 1999). Assuming two mega-environments, the variance component for genotype–mega-environment interaction explained 80% of the total GE (Yan, 1999).

Visualizing the mean performance and stability of genotypes

Once mega-environments are defined, cultivar selection should be specific to individual mega-environments. For a given mega-environment, genotypes are evaluated based on mean performance (such as mean yield) and stability across environments. Assuming that the nine environments in our example belong to a single mega-environment, a 'mean' environment can be defined in the biplot, using the mean-environment PC1 and PC2 scores of all environments. The mean yield of the genotypes can then be approximated by nominal yields of the genotypes in that mean environment.

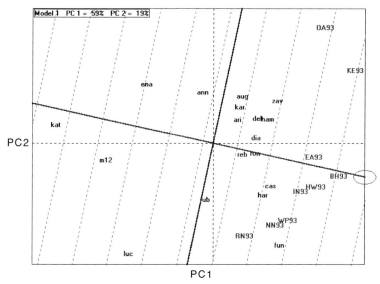

Fig. 19.6. The mean-environment coordination showing the mean yield and stability of each of the genotypes.

In Fig. 19.6, a line is drawn that passes through the biplot origin and the mean environment, which is marked by an oval at its positive end. This line will be called the mean-environment axis. Another line is drawn that passes through the biplot origin and is perpendicular to the mean-environment axis. These two lines constitute 'the mean-environment coordination'.

The projections of the genotypes to the mean-environment axis approximate the mean yield of the genotypes. Thus, the mean yield of the genotypes is in the following order: fun > cas ≈ har > . . . > rub > ann > luc > ena > m12 > kat. This order is highly consistent with the actual mean yield of the genotypes (Table 19.1). The parallel lines in Fig. 19.6 facilitate ranking the genotypes based on their predicated mean yield. Since the biplot contains both G and GE and since the two axes of the mean-environment coordination are orthogonal, if projections of the genotypes to the mean-environment axis approximate the mean yield of the genotypes, projections of the genotypes on to the perpendicular axis must approximate the GE associated with the genotypes. The longer the projection of a genotype, regardless of direction, the greater the GE associated with the genotype, which is a measure

of variability or instability of the genotype across environments. Thus, the performance of genotypes luc and fun is highly variable (less stable), whereas genotypes ron and reb are highly stable.

It should be pointed out that stability *per se* is not necessarily a positive factor. High stability is desirable only when associated with a high mean yield. A genotype with high stability is highly undesirable if it is associated with a low mean yield; it is simply a genotype that is consistently poor. It is even less desirable than genotypes with low stability but high mean yield.

An ideal genotype is one that has both high mean yield and high stability. The centre of the concentric circles in Fig. 19.7a represents the position of an 'ideal' genotype, which is defined by a projection on to the mean-environment axis that equals the longest vector of the genotypes that had above-average mean yield and by a zero projection on to the perpendicular line (zero variability across environments). A genotype is more desirable if it is closer to the 'ideal' genotype. Thus, genotypes cas and har are more desirable than genotype fun, even though the latter had the highest mean yield. The low-yielding genotypes kat, m12, luc, ena and ann, are, of course, undesirable because they are far away from the 'ideal' genotype.

Visualizing the discriminating ability and representativeness of environments

Although MET are conducted primarily for genotype evaluation, they can also be used in evaluating environments. An ideal environment should be highly differentiating of the genotypes and at the same time representative of the target environment. Assuming that the test environments used in the MET are representative samples of the target environment, the ideal environment should be located on the mean-environment axis. The centre of the concentric circles represents the ideal environment, which has the longest vector of the test environments that had positive projections onto the mean environment axis (Fig. 19.7b). An environment is more desirable if it is closer to the 'ideal' environment. Therefore, BH93, EA93, HW93 and IN93 were relatively desirable test environments, whereas OA93 and RN93 were relatively undesirable test environments.

Discussion and Conclusions

Strength of the GGE-biplot approach

The GGE-biplot approach graphically displays genotype main effect and genotype–environment interaction of a MET, which are the only two parts of yield variation that are relevant to genotype evaluation and mega-environment identification. Assuming that the GGE of a MET is sufficiently approximated by the first two principal components, all individual genotype–environment relationships in the MET should be displayed by the GGE biplot. Such a biplot graphically addresses three of the four utilities of MET data analysis listed in the introduction of this chapter, namely: (i) investigating possible mega-environment differentiation in the target environment; (ii) selecting superior genotypes for individual

mega-environments; and (iii) selecting better test environments. In addition, the GGE biplot also facilitates pairwise genotype comparisons. The GGE biplot does not directly address the fourth utility of the MET data analysis, i.e. understanding the causes of GE. To fulfil this task, information other than yield *per se* is necessary. Once such information is available, the genotype and environment scores can be related to genotypic and environmental factors, so that the observed genotype–environment interactions can be explained in terms of interactions between genotypic factors and environmental factors (Yan and Hunt, 2001). Therefore, the GGE biplot is an ideal approach for MET data analysis.

Constraints of the GGE-biplot approach

All methods have their limitations. The limitations of the GGE biplot lie in four aspects. First, it requires balanced data; secondly, it may explain only a small portion of the total GGE; thirdly, it lacks a measure of uncertainty; and, fourthly, although elegant, GGE biplot analysis is tedious to perform using conventional tools. Now that the GGEbiplot software is available, the fourth constraint is no longer an issue. Once the data are prepared, all functions are just a 'mouse-click' away. All the figures presented in this chapter, along with many other options, are the direct outputs of this software.

Although quite common, unbalanced MET data are really not a problem of the GGE-biplot approach; they are a problem of experimental design and execution, which create problems for all kinds of analyses. The GGEbiplot software offers two options on this problem. It allows generation of a balanced subset, which can be used in GGE-biplot analysis; alternatively, missing cells are automatically replaced by mean yields of the respective environments. In either case, unbalancedness means that part of the information contained in the data cannot be utilized effectively. Therefore, experimental design and execution should

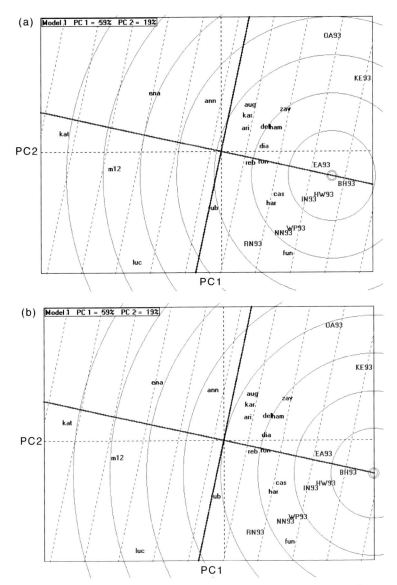

Fig. 19.7. Comprehensive evaluation of genotypes and environments. (a) Comparison of genotypes with the 'ideal' genotype for both mean yield and stability. (b) Comparison of environments with the 'ideal' environment based on both discriminating ability and representativeness of the target environment.

be improved to prevent missing cells as soon as possible.

The GGE biplot may explain only a small proportion of the GGE when the genotype main effect is considerably smaller than the GE and when the GE pattern is complex. In such cases, the GGE biplot consisting of PC1 and PC2 may not be sufficient to explain the GGE, even though the most important pattern of the MET is already displayed. To remedy this problem, the GGEbiplot software offers options for viewing biplots of PC3 vs. PC4, PC5 vs. PC6, etc.

Unlike conventional approaches, which allow calculation of probability for a particular hypothesis, the GGE-biplot approach does not have a measure of uncertainty. Therefore, the GGE biplot is better used as a hypothesis-generator rather than as a decision-maker (Yan et al., 2001), and hypotheses based on biplots should be tested using conventional statistical methods. For example, biplots based on individual years of Ontario winter-wheat performance trials suggested that eastern Ontario sites and other sites of Ontario belong to different mega-environments, and this hypothesis was tested and confirmed by variance component analysis (Yan, 1999). Sometimes, the biplot distance of two genotypes, relative to the biplot size, may be sufficiently informative about the significance of the difference between two genotypes or two environments.

Other applications of the GGE-biplot approach

The GGE-biplot methodology was developed for MET data analysis. It is a generic method, however. It has been successfully used in analysing genotype–trait data (Yan and Rajcan, 2002), diallel-cross data (Yan and Hunt, 2002), host genotype–pathogen race data (W. Yan, unpublished), etc. The GGE-biplot methodology and the GGEbiplot software described in this chapter should thus be useful for the graphical presentation of all types of two-way data that conform to an entry–tester data structure. A demo version of the GGEbiplot software can be downloaded at www.ggebiplot.com.

Acknowledgements

Dr Rich Zobel and Dr Hugh Gauch are acknowledged for their stimulating suggestions and critiques during the development of the GGE-biplot methodology. Dr Paul Cornelius and Dr Jose Crossa are acknowledged for their valuable encouragement and editing of our first paper on the GGE-biplot methodology.

References

Cooper, M., Stucker, R.E., DeLacy, I.H. and Harch, B.D. (1997) Wheat breeding nurseries, target environments, and indirect selection for grain yield. Crop Science 37, 1168–1176.

Finlay, K.W. and Wilkinson, G.N. (1963) The analysis of adaptation in a plant breeding program. Australian Journal of Agricultural Research 14, 742–754.

Gabriel, K.R. (1971) The biplot graphic display of matrices with application to principal component analysis. Biometrika 58, 453–467.

Gauch, H.G. and Zobel, R.W. (1996) AMMI analysis of yield trials. In: Kang, M.S. and Gauch, H.G. (eds) Genotype-by-Environment Interaction. CRC Press, Boca Raton, Florida, pp. 1–40.

Gauch, H.G. and Zobel, R.W. (1997) Identifying mega-environments and targeting genotypes. Crop Science 37, 311–326.

Kang, M.S. (1998) Using genotype by environment interaction for crop cultivar development. Advances in Agronomy 62, 199–252.

Lin, C.S. and Binns, M.R. (1994) Concepts and methods for analysis regional trial data for cultivar and location selection. Plant Breeding Reviews 11, 271–297.

Lin, C.S., Binns, M.R. and Lefkovitch, L.P. (1986) Stability analysis: where do we stand? Crop Science 26, 894–900.

SAS Institute Inc. (1996) Version 6, SAS/STAT User's Guide. SAS Institute, Cary, North Carolina.

Yan, W. (1999) Methodology of cultivar evaluation based on yield trial data – with special reference to winter wheat in Ontario. PhD dissertation, University of Guelph, Guelph, Ontario, Canada.

Yan, W. (2001) GGEbiplot – a Windows application for graphical analysis of multi-environment trial data and other types of two-way data. Agronomy Journal 93(5), 1111.

Yan, W. and Hunt, L.A. (2001) Genetic and environmental causes of genotype by environment interaction for winter wheat yield in Ontario. Crop Science 41(1), 19–25.

Yan, W. and Hunt, L.A. (2002) Biplot analysis of diallel data. Crop Science 41(1), 21–30.

Yan, W. and Rajcan, I. (2002) Biplot evaluation of test sites and trait relations of soybean in Ontario. Crop Science 41(1), 11–20.

Yan, W., Hunt, L.A., Sheng, Q. and Szlavnics, Z. (2000) Cultivar evaluation and mega-environment investigation based on the GGE biplot. *Crop Science* 40(3), 597–605.

Yan, W., Cornelius, P.L., Crossa, J. and Hunt, L.A. (2001) Two types of GGE biplots for analyzing multi-environment trial data. *Crop Science* 41(3), 656–663.

Zobel, R.W., Wright, M.J. and Gauch, H.G. (1988) Statistical analysis of a yield trial. *Agronomy Journal* 80, 388–393.

20 Linear–Bilinear Models for the Analysis of Genotype–Environment Interaction

J. Crossa[1] and P.L. Cornelius[2]

[1]Biometrics and Statistics Unit, International Maize and Wheat Improvement Center (CIMMYT), Lisboa 27, Apdo. Postal 6-641, 06600 Mexico DF, Mexico; [2]Departments of Agronomy and Statistics, University of Kentucky, Lexington, KY 40546-0091, USA

Introduction

The presence of genotype–environment interaction (GEI) in a multi-environment trial (MET) is expressed either as inconsistent responses of some genotypes with respect to others due to the alteration of the ordering of the genotypes from one environment to another (GEI with rank change or crossover interaction (COI)) or as changes in the absolute differences between genotypes without rank change (GEI without rank change or non-crossover interaction (non-COI)).

Early approaches to the analysis of GEI included the conventional, two-way fixed-effects (FE2W) model, with sum to zero constraints, and the simple linear regression of genotype yields on the environment means. The FE2W model expresses the empirical mean, $\bar{y}_{ij}$, of the ith genotype ($i = 1, 2, \ldots, g$) in the jth environment ($j = 1, 2, \ldots, e$) with n replications in each of the $g \times e$ cells as:

$$\bar{y}_{ij} = \mu + \tau_i + \delta_j + (\tau\delta)_{ij} + \bar{\varepsilon}_{ij} \qquad (20.1)$$

where μ is the grand mean across all genotypes and environments and is estimated by $\bar{y}_{..}$, τ_i is the additive effect of the ith genotype defined as a deviation of the genotype mean from the overall mean and is estimated by $\bar{y}_{i.} - \bar{y}_{..}$, which satisfies constraint $\sum_i \hat{\tau}_i = 0$. Similarly, δ_j is the additive effect of the jth environment estimated by $\bar{y}_{.j} - \bar{y}_{..}$, which satisfies constraint $\sum_i \hat{\delta}_j = 0$. Also, $(\tau\delta)_{ij}$ is the non-additivity, i.e. GEI, of the ith genotype and the jth environment estimated as the residual $\bar{y}_{ij} - \bar{y}_{i.} - \bar{y}_{.j} + \bar{y}_{..}$ after fitting the main effects. The $(\tau\delta)_{ij}$ satisfies constraints $\sum_i \sum_j (\tau\delta)_{ij} = \sum_i (\tau\delta)_{ij} = \sum_j (\tau\delta)_{ij} = 0$.

The $\bar{\varepsilon}_{ij}$ term is the mean of the errors contributing to measurements on the ith genotype in the jth environment. The $\bar{\varepsilon}_{ij}$ are assumed NID(0, σ^2/n) where σ^2 is the pooled within-environment error variance, assumed to be homoscedastic. For the complete random effects two-way (RE2W) model, τ_i, δ_j and $(\tau\delta)_{ij}$ are assumed to be normally and independently distributed, with variances σ_τ^2, σ_δ^2 and $\sigma_{\tau\delta}^2$, respectively. From the GEI perspective, the FE2W and RE2W models are unparsimonious, the analysis is uninformative and the $(g - 1) \times (e - 1)$ independent parameters of the GEI are difficult to interpret.

Fisher and MacKenzie (1923), who analysed data from an experiment evaluating 12 potato cultivars under each of six soil-fertilization treatments, were the first authors to propose breaking down the response into a series of multiplicative terms fitted by least squares; however, this work was apparently forgotten for many years. Yates and Cochran (1938) proposed breaking down the GEI into one multiplicative term and a deviation therefrom, examining whether the GEI is a linear function of the additive environmental component, that is, $(\tau\delta)_{ij} = \xi_i\delta_j + d_{ij}$, where $1 + \xi_i$ is the linear regression coefficient of yields of the ith genotype on the environmental mean and d_{ij} is a deviation. This regression approach expresses the GEI simply as heterogeneity of slopes and was later used by Finlay and Wilkinson (1963) and slightly modified by Eberhart and Russell (1966).

Tukey (1949) proposed a test for the GEI in which the $(\tau\delta)_{ij}$ term is a constant multiplied by the product of the main effects of genotypes and environments, $(\tau\delta)_{ij} = \lambda\tau_i\delta_j$. Mandel (1961) generalized Tukey's (1949) model by letting the GEI term be the product $(\tau\delta)_{ij} = \lambda\alpha_i\delta_j$ for regression of genotype simple effects on environment main effects or $(\tau\delta)_{ij} = \lambda\tau_i\gamma_j$ for regression of environment simple effects on genotype main effects. Either of these consists of a 'bundle of regression lines', which may be tested for concurrence (i.e. in the first case, whether the α_i are proportional to the τ_i or, in the second case, whether the γ_j are proportional to the δ_j) or non-concurrence. Proportionality of the α_i to the τ_i is a special case of regression of genotypes on the environmental mean, in which the regression lines all intersect at one point.

The Mandel (1961) tests for concurrent and non-concurrent regression lines partition the GEI into one degree of freedom (d.f.) for the concurrence of genotype (or environment) regressions on environment (or genotype) main effects (this is the same as Tukey's (1949) one d.f. for non-additivity), $g - 2$ d.f. for the non-concurrence of genotype regressions, $e - 2$ for the non-concurrence of site regressions and $(g - 2)$ $(e - 2)$ d.f. for the remainder of the GEI. Cornelius et al. (1996) suggested Mandel's (1961) analysis as a diagnostic for choice of multiplicative model form.

Freeman (1973) cited Williams (1952) as the first researcher to link the FE2W model with principal-component analysis (PCA)

and showed that the GEI term can be represented by the sum of eigenvalues of a matrix. Gollob (1968) and Mandel (1969, 1971) introduced the linear–bilinear model (LBM),

$$\bar{y}_{ij} = \mu + \tau_i + \delta_j + \sum_{k=1}^{t} \lambda_k\alpha_{ik}\gamma_{jk} + \bar{\varepsilon}_{ij} \quad (20.2)$$

Where λ_k is a scale parameter or singular value for the kth bilinear (multiplicative) component. The λ_k are ordered, i.e. $\lambda_1 \geq \lambda_2 \geq \ldots \geq \lambda_t$. Further, the α_{ik} and γ_{jk} are elements of the left and right singular vectors, respectively, contributing to the kth bilinear (multiplicative) term. The α_{ik} represents genotypic sensitivities to a hypothetical environmental factor the level of which, in the jth environment, is represented by the element γ_{jk} in the right singular vector for the kth component. Elements of the singular vectors for genotypes and environments (α_{ik} and γ_{jk}) are subject to normalization constraints, $\Sigma_i\alpha_{ik}^2 = \Sigma_j\gamma_{jk}^2 = 1$, and to orthogonality constraints, $\Sigma_i\alpha_{ik}\alpha_{ik'} = \Sigma_j\gamma_{jk}\gamma_{jk'} = 0$ for $k \neq k'$. When Equation 20.2 is saturated, the number of bilinear terms is $t = \min(g - 1, e - 1)$ and, for any smaller value of t, the model is said to be 'truncated'. The word 'truncated' here is used not in the sense of having a truncated distribution, but, rather, in the sense of truncating the string of bilinear terms at something less than the number of terms that will saturate the model.

Mandel (1971) computed the number of degrees of freedom associated with the sum of squares (SS) due to each of the first three bilinear terms in Equation 20.2 by a Monte Carlo study, and Johnson and Graybill (1972) found that Mandel's (1971) results were close to the exact values. Gabriel (1978) showed that a least-squares (LS) solution for model parameters in Equation 20.2 can be obtained by taking the estimates of the bilinear terms as the t largest components of the singular value decomposition (SVD) of the matrix $\mathbf{Z} = [z_{ij}] = [\bar{y}_{ij} - \bar{y}_{i.} - \bar{y}_{.j} + \bar{y}_{..}]$, with the additive (linear) effects μ, τ_i and δ_j estimated as we have previously given for their estimates in the FE2W model (Equation 20.1).

If the components of the SVD of $\mathbf{Z}$ are arranged in decreasing order with respect to the singular values, the first component gives a rank-one matrix that, in an LS sense, approximates matrix $\mathbf{Z}$; the first two components of the SVD give a rank-two matrix that approximates $\mathbf{Z}$, etc.

Zobel *et al.* (1988) and Gauch (1988) named Equation 20.2 the 'additive main effects and multiplicative interaction' (AMMI) model. They further introduced a data-splitting and cross-validation procedure for determining the number of multiplicative components to retain in a truncated AMMI model.

The AMMI model and four other LBMs and their LS estimates were described and unified in one general methodology by Cornelius *et al.* (1996). These authors described various statistical tests for the significance of the bilinear terms and mentioned the possibility of using shrinkage estimates of these models for improving the prediction of the $g \times e$ cells. The four models that can be derived from the AMMI model are:

genotypes regression model (GREG)

$$\bar{y}_{ij} = \mu_i + \sum\nolimits_{k=1}^{t} \lambda_k \alpha_{ik} \gamma_{jk} + \bar{\varepsilon}_{ij}$$

sites (i.e. environment) regression model (SREG)

$$\bar{y}_{ij} = \mu_j + \sum\nolimits_{k=1}^{t} \lambda_k \alpha_{ik} \gamma_{jk} + \bar{\varepsilon}_{ij}$$

completely multiplicative model (COMM)

$$\bar{y}_{ij} = \sum\nolimits_{k=1}^{t} \lambda_k \alpha_{ik} \gamma_{jk} + \bar{\varepsilon}_{ij}$$

shifted multiplicative model (SHMM)

$$\bar{y}_{ij} = \beta + \sum\nolimits_{k=1}^{t} \lambda_k \alpha_{ik} \gamma_{jk} + \bar{\varepsilon}_{ij}$$

The LS estimates of the additive effects of these models and the elements of the residual matrix $\mathbf{Z}$ are:

AMMI $\hat{\mu} = \bar{y}_{..},\ \hat{\tau}_i = \bar{y}_{i.} - \bar{y}_{..},\ \hat{\delta}_j = \bar{y}_{.j} - \bar{y}_{..},\ z_{ij} = \bar{y}_{ij} - \bar{y}_{i.} - \bar{y}_{.j} + \bar{y}_{..}$

SREG $\hat{\mu}_j = \bar{y}_{.j},\ z_{ij} = \bar{y}_{ij} - \bar{y}_{.j}$

GREG $\hat{\mu}_i = \bar{y}_{i.},\ z_{ij} = \bar{y}_{ij} - \bar{y}_{i.}$

COMM $z_{ij} = \bar{y}_{ij}$

SHMM $z_{ij} = \bar{y}_{ij} - \hat{\beta},\ \hat{\beta} = \bar{y}_{..} - \sum\nolimits_{k=1}^{t} \lambda_k \bar{\alpha}_k \bar{\gamma}_k$

Seyedsadr and Cornelius (1992) developed the SHMM model, which is a reparameterization of the Tukey (1949)

model for testing non-additivity. The singular vectors for genotypes and environments for the ordered components are called 'primary effects' $(\alpha_{i1}, \gamma_{j1})$, 'secondary effects' $(\alpha_{i2}, \gamma_{j2})$, and so on. The LS solution for $\hat{\beta}$ requires an iterative algorithm, because the solutions for the bilinear terms are the t largest components of the SVD of matrix $\mathbf{Z} = [z_{ij}]$, where, in this case, $z_{ij} = \bar{y}_{ij} - \hat{\beta}$, but $\hat{\beta} = \bar{y}_{..} - \sum\nolimits_{k=1}^{t} \lambda_k \bar{\alpha}_k \bar{\gamma}_k$, where $\bar{\alpha}_k = g^{-1} \Sigma_i \hat{\alpha}_{ik}$, and $\bar{\gamma}_k = e^{-1} \Sigma_j \hat{\gamma}_{jk}$. Thus, $\mathbf{Z}$ depends on $\hat{\beta}$, but $\hat{\beta}$ depends on the SVD of $\mathbf{Z}$. Consequently, the LS solution does not exist in closed form. Moreover, the value of $\hat{\beta}$ changes if the number of bilinear components, t, is changed.

Apparently, the SHMM model was the first LBM that, along with other statistical tools, was used for identifying subsets of genotypes or environments in which genotypic rank changes are negligible (Cornelius *et al.*, 1992, 1993b; Crossa and Cornelius, 1993; Crossa *et al.*, 1993, 1995). Later, the SREG model was suggested as a better model to use for identifying such subsets of environments (Crossa and Cornelius, 1997) (but not for identifying such subsets of genotypes). The SREG model is very appealing for breeders and agronomists because its multiplicative terms contain the main effects of genotypes plus the GEI, making it possible to assess both the general and the specific adaptation of genotypes. Crossa and Cornelius (1997) have used the SREG model for clustering sites without genotypic rank change under heterogeneity of within-site error variances. Yan *et al.* (2000) used the biplot of the first two bilinear components obtained from the SREG model to graphically identify specific 'winner' genotypes in certain subsets of environments.

The GREG model with $t = 1$ is a reparameterization and with $t > 1$ a generalization of the linear regression model of Yates and Cochran (1938), Finlay and Wilkinson (1963) and Eberhart and Russell (1966), except that we replace their estimator of δ_j with an LS solution and further impose orthonormality constraints, as in the AMMI model. Typically, the LS estimators of the γ_{j1} are very highly correlated with the

Finlay–Wilkinson/Eberhart–Russell estimator $\bar{y}_{\cdot j} - \bar{y}_{\cdot \cdot}$ of δ_j.

Cornelius and Seyedsadr (1997) defined the general linear–bilinear model (GLBM) as:

$$\bar{y}_{ij} = \sum_{k=1}^{m} \beta_k x_{kij} + \sum_{k=1}^{t} \lambda_k \alpha_{ik} \gamma_{jk} + \bar{\varepsilon}_{ij}$$

where the x_{kij} are known constants and the β_k parameters (regression coefficients) for the linear terms and the λ_k, α_{ik} and γ_{jk} in the bilinear terms are parameters to be estimated (α_{ik} and γ_{jk} subject to the previously defined orthonormality constraints).

In matrix notation, the GLBM can be expressed as:

$$\mathbf{Y} = \sum_{k=1}^{m} \beta_k \mathbf{X}_k + \mathbf{A} \Lambda \mathbf{G}' + \mathbf{E}$$

where $\mathbf{Y} = [\bar{y}_{ij}]$, $\mathbf{X}_k = [x_{kij}]$, $\mathbf{E} = [\bar{\varepsilon}_{ij}]$, $\Lambda = \mathrm{diag}(\lambda_k, k = 1, 2, \ldots, t)$, $\lambda_1 \geq \lambda_2 \geq \ldots \geq \lambda_t$, $\mathbf{A} = (\alpha_1, \ldots, \alpha_t)$, $\mathbf{G} = (\gamma_1, \ldots, \gamma_t)$ and $\mathbf{A}'\mathbf{A} = \mathbf{G}'\mathbf{G} = \mathbf{I}_t$. Define $\mathbf{Z} = \mathbf{Y} - \sum_{k=1}^{m} \hat{\beta}_{k(t)} \mathbf{X}_k$, where $\hat{\beta}_{k(t)}$ is the LS estimate of β_k when the fitted model contains t bilinear terms ($t \leq \mathrm{rank}(\mathbf{Z})$). Then the first t components of the SVD of $\mathbf{Z}$ provide the LS estimates of parameters in the bilinear terms. An LS solution for the linear effects (the $\hat{\beta}_k$) is given by any solution to the equation:

$$\mathbf{C}\hat{\beta} = \mathbf{T}$$

where the elements of $\hat{\beta}$ are the $\hat{\beta}_k$, the hkth element of $\mathbf{C}$ is $C_{hk} = \mathrm{tr}(\mathbf{X}_h' \mathbf{X}_k) = \sum_i \sum_j x_{hij} x_{kij}$

and the kth element of $\mathbf{T}$ is $T_k = \mathrm{tr}$ $\left[\mathbf{X}_k' (\mathbf{Y} - \hat{\mathbf{A}} \hat{\Lambda} \hat{\mathbf{G}}') \right] = \sum_i \sum_j x_{kij} \left(\bar{y}_{ij} - \sum_{u=1}^{t} \hat{\lambda}_u \hat{\alpha}_{ui} \hat{\gamma}_{uj} \right)$.

Here $\mathrm{tr}(\cdot)$ denotes the trace of the (square) matrix given as the argument.

An LBM is said to be 'balanced' (a BLBM) if $\mathbf{Z} = \mathbf{PYQ}$, where $\mathbf{P}$ and $\mathbf{Q}$ are projection matrices free of the bilinear effects. Under this condition T_k reduces to $\sum_i \sum_j x_{kij} \bar{y}_{ij}$

and, thus, in a BLBM, an LS solution for $\hat{\beta}$ ignoring the bilinear effects (i.e. for $t = 0$) is also a solution for $\hat{\beta}$, given any value for $t \leq \mathrm{rank}(\mathbf{Z})$. Provided there are no missing cells, AMMI, GREG, SREG and COMM are BLBMs (COMM actually being without any linear terms at all), but SHMM is not.

In this chapter, we review the use of the SHMM and SREG models for finding clusters of environments with negligible genotypic COI and examine some of the unconstrained and constrained non-COI solutions for finding the 'distance' between pairs of environments. We also summarize results of 'shrinkage' estimators of LBMs developed as analogues of best linear unbiased predictors (BLUPs) and justified by a Bayesian argument, and further show empirical evidence that shrinkage estimators are usually better predictors than the best truncated LBM and sometimes better than BLUPs of a RE2W model with interaction.

SHMM for Assessing COI

Since the early 1990s, theoretical and practical studies have shown the utility of the SHMM model for identifying subsets of environments and genotypes without genotypic rank change (Cornelius et al., 1992; Crossa and Cornelius, 1993; Crossa et al., 1993, 1995, 1996; Abdalla et al., 1997; Trethowan et al., 2001). Cornelius et al. (1992), observing results obtained with SHMM$_1$ (SHMM with one multiplicative term), defined sufficient conditions for the absence of significant genotype COI in a set of environments and/or genotypes.

1. SHMM with $t = 1$ (SHMM$_1$) must be an adequate model for fitting the data. This implies that the multiplicative components, beyond the first, are not significantly different from zero.

2. The primary effects of environments, γ_{j1}, are all of like sign.

The SHMM model satisfying the above condition 2 has the following two proportionality properties:

1. Differences between genotypes in any single environment are proportional to genotype differences in any other environment.

2. Differences between environments with respect to the performance of any single genotype are proportional to environmental differences with respect to performance of

any other genotype (but, for environment differences, proportionality constants can be negative).

The second proportionality restriction is irrelevant for the case of genotypic non-COI and is relaxed in the SREG$_1$ model (Crossa and Cornelius, 1997).

When SHMM$_1$-predicted values, $\hat{y}_{ij} = \hat{\beta} + \hat{\lambda}_1 \hat{\alpha}_{i1} \hat{\gamma}_{j1}$, are plotted against the primary effects of environments, $\hat{\gamma}_{j1}$, the graph consists of a set of regression lines, one for each genotype, all of which concur (i.e. intersect) at the point $(0, \hat{\beta})$. For a non-COI SHMM$_1$, the $\hat{\gamma}_{j1}$ are all of like sign (or zero) and thus the point of intersection is a point either at the boundary (if one $\hat{\gamma}_{j1} = 0$) or outside (left or right of) the region containing the plotted points.

If the $\hat{\gamma}_{j1}$ have different signs (some positive, some negative), then the point of intersection is within the region containing the plotted points and a complete reversal of rank order of genotypes is displayed on the right, as compared with the left, of the point of intersection. If the intersection point is far outside the region containing the plotted points, then the genotype regression lines appear very nearly parallel, implying that the data are essentially additive (provided that the SHMM$_1$ adequately fits the data).

SHMM clustering of environments with non-COI

Typically, when SHMM is fitted to the entire set of data from an MET, in addition to primary effects, one must include secondary and perhaps even higher-order effects if an adequate fit is to be achieved. The clustering strategy is to divide the environments into subsets such that significant variation captured as secondary, tertiary, etc., effects, when SHMM is fitted to the entire data set, can be expressed as primary effects in separate analyses of data from the subsets. In doing this clustering, the measure of 'distance' between two environments is taken as the residual mean square (RMS) after fitting SHMM$_1$ (RMS(SHMM$_1$))

to the data from the two environments subject to a non-COI constraint, namely, that both $\hat{\gamma}_{j1}$ must be either non-positive or non-negative. RMS(SHMM$_1$) is obtained as [RSS(SHMM$_1$)]/f, where RSS stands for residual sum of squares and f is the d.f., namely, $f = g - 2 + v$, where v is the number of additional constraints imposed to achieve a non-COI solution.

It is a property of SHMM that, if $e < g$, RSS(SHMM$_{e-1}$) = RSS(SREG$_{e-1}$) in unconstrained LS solutions (Seyedsadr and Cornelius, 1992). Thus, for a subset of $e = 2$ environments, this property provides a closed-form solution for the distance, provided that the two $\hat{\gamma}_{j1}$ values are of like sign. Otherwise a constrained solution must be computed. The constraint, if needed, is imposed by putting $\hat{\gamma}_{j1} = 0$ for one of the two environments and $\hat{\gamma}_{j1} = \pm 1$ for the other environment. Two different methods for doing this have been devised, namely, a constrained LS method and a constrained SVD method. In the constrained LS method, we put $\hat{\gamma}_{j1} = 0$ for the environment that has the smaller value of $\sum_i (\bar{y}_{ij} - \bar{y}_{.j})^2$ (moreover, this value becomes the distance value) and put $\hat{\gamma}_{j1} = \pm 1$ for the other environment. Other properties of the solution are $\hat{\beta} = \bar{y}_{.j}$ and thus $\hat{y}_{ij} = \hat{\beta} = \bar{y}_{.j}$ for the environment with $\hat{\gamma}_{j1} = 0$ and the quantities $\hat{\lambda}_1 \hat{\alpha}_{i1}$ are chosen such that $\hat{y}_{ij} = \hat{\beta} + \hat{\lambda}_1 \hat{\alpha}_{i1} \hat{\gamma}_{j1} = \bar{y}_{ij}$ for j, now representing the environment for which $\hat{\gamma}_{j1} = \pm 1$.

The constrained SVD solution chooses $\hat{\beta}$ such that the first right singular vector $(\hat{\gamma}_1)$ of $\mathbf{Z} = [z_{ij}] = [\bar{y}_{ij} - \hat{\beta}]$ is either $(\pm 1, 0)'$ or $(0, \pm 1)'$. A sufficient condition for this is that the two columns of $\mathbf{Z}$ must be orthogonal to one another. There are two solutions for $\hat{\beta}$ that will satisfy this. For either solution, the right singular vectors are $(\pm 1, 0)'$ and $(0, \pm 1)'$, either in that order or in the reverse order. The final choice among the four combinations of one of the two possible $\hat{\beta}$ values and one of the two possible environments for which to put $\hat{\gamma}_{j1} = 0$ will be the combination for which the quantity $\sum_i (\bar{y}_{ij} - \hat{\beta})^2$ is smallest. For further details, we refer the reader to Crossa *et al.* (1996). The residual d.f. for either the constrained LS or con-strained SVD solution fitted to data from two environments are $g - 1$. We doubt that the choice of method for computing constrained solutions will ever be a critically important issue in clustering environments or geno-types into non-COI groups.

After the distances for all possible pairs of environments have been computed, a dendrogram is constructed using the com-plete linkage (furthest neighbour) clustering method. The final step is to analyse the subsets of data for each of the clusters suggested by branches of the dendrogram for adequacy of fit of SHMM$_1$ (constrained if necessary to obtain a non-COI solution).

A constrained SVD non-COI SHMM$_1$ solution for a subset containing more than two environments can usually be computed by iteratively alternating between computa-

tion of $\hat{\beta} = \dfrac{\sum_i \hat{\alpha}_{i1} \bar{y}_{ih}}{\sum_i \hat{\alpha}_{i1}}$ and computation of the

$\hat{\alpha}_{i1}$ as elements of the first left singular vector of $\mathbf{Z} = [z_{ij}] = [\bar{y}_{ij} - \hat{\beta}]$, where the subscript h denotes the environment to have its primary effect $(\hat{\gamma}_{h1})$ put equal to zero and the $\bar{y}_{ij}$ used in computing $\mathbf{Z}$ are only those for the particular subset being analysed. If what at first seems to be the most reasonable choice for environment h fails to give a non-COI solution (which occurs if the non-zero $\hat{\gamma}_{j1}$ in the constrained solution are not all of like sign), another choice for environment h may be tried. We doubt if it is possible to find a solution for $\hat{\beta}$ that will simultaneously constrain primary effects of more than one environment. If such a solution appears to be necessary, computation of a constrained LS solution may be mandatory. A Newton–Rapshon algorithm for computing a constrained LS solution for any number of environments to have primary effects put equal to zero can be found in Crossa et al. (1996). Despite the more complicated algorithm and under-lying mathematics, we prefer constrained LS solutions to constrained SVD solutions.

For a set (or subset) containing g geno-types and e environments, the d.f. of RSS (SHMM$_1$) is $ge - g - e$ in an unconstrained

solution, $ge - g - e + 1 = (g - 1)(e - 1)$ in a constrained SVD solution and $ge - g - e + v$ in a constrained LS solution, where v is the number of environments with their $\hat{\gamma}_{j1} = 0$ in the constrained LS solution.

The SREG Model and its Relationship with the COI

It has been shown that SREG with one multiplicative term (SREG$_1$) is a viable alternative to SHMM$_1$ as a model for identi-fying groups of environments without geno-type COI, because SREG$_1$, like SHMM$_1$, also displays proportionality of genotype differ-ences in different environments, but, unlike SHMM$_1$, SREG$_1$ does not impose proportion-ality of environmental differences with resp-ect to performance of genotypes (Crossa and Cornelius, 1997). Proportionality of envi-ronmental differences with respect to differ-ent genotypes is not relevant to the issue of genotype COI and SREG's relaxation of these constraints may allow larger non-COI clusters to be obtained. Furthermore, SREG can be quite satisfactorily used to deal with heterogeneity of within-environment error variances by the simple device of rescaling

the $\bar{y}_{ij}$ by dividing by $\sqrt{\dfrac{s_j^2}{n}}$, where s_j^2 is the

error mean square within the jth environ-ment. (So also can SHMM, but the result has the unpalatable property that SHMM fitted to the scaled data is no longer a SHMM when back-transformed to the original scale (Crossa and Cornelius, 1997).)

In the SREG$_1$ model, it is the deviations of genotype yields $(\bar{y}_{ij})$ from environment means $\bar{y}_{.j}$ that are modelled by the bilinear term. The fitted bilinear effects, $\hat{\lambda}_1 \hat{\alpha}_{i1} \hat{\gamma}_{j1}$, can be plotted as a set of regression lines, one for each genotype, with the $\hat{\gamma}_{j1}$ as regressor variable and with zero intercepts. Because these regression lines all intersect at the zero point on the $\hat{\gamma}_{j1}$ scale, the graph, like the graph of SHMM$_1$, does not display genotype COI within the region of the plotted points $(\hat{\lambda}_1 \hat{\alpha}_{i1} \hat{\gamma}_{j1}, \hat{\gamma}_{j1})$ if, and only if, the $\hat{\gamma}_{j1}$ are either all non-negative or all non-positive. Addi-tion of the environment mean to the plotted

ordinates $(\hat{\lambda}_1\hat{\alpha}_{i1}\hat{\gamma}_{j1})$ gives the SREG$_1$ predic-
ted response $\hat{y}_{ij} = \bar{y}_{.j} + \hat{\lambda}_1\hat{\alpha}_{i1}\hat{\gamma}_{j1}$. If the $\hat{y}_{j1}$ are
plotted against the $\hat{\gamma}_{j1}$, the plotted points
for any given genotype no longer fall on a
straight line, but, if plotted points for adjacent values of $\hat{\gamma}_{j1}$ are connected with line
segments, the figure displays an overlaid set
of broken-line graphs, one for each genotype, which, although they are not straight
lines, nevertheless display no genotype COI
within the region of plotted points.

For SREG$_1$ clustering of environments,
the measure of distance between two environments is RMS(SREG$_1$). For a subset of
data with only two environments, RSS
(SREG$_1$) = RSS(SHMM$_1$) for both unconstrained and constrained LS non-COI
solutions. Consequently, provided that constrained LS solutions are used for non-COI
solutions (when needed), dendrograms for
SREG and SHMM clustering are identical.
Thus, subsets that the dendrogram suggests
as groups to be evaluated for acceptability
of fit of the one-term model are the same
subsets whether SREG$_1$ or SHMM$_1$ is used,
but the less parsimonious SREG$_1$ may sometimes give an acceptable fit to a subset to
which SHMM$_1$ does not acceptably fit.

Unlike SHMM$_1$, the constrained LS
non-COI SREG$_1$ solution for a subset containing more than two environments exists
in closed form (Crossa and Cornelius, 1997).
For a subset consisting of sub-subset S_1 containing e_1 environments that are to have their
$\hat{\gamma}_{j1} = 0$ and sub-subset S_2 containing e_2 environments that are to have their $\hat{\gamma}_{j1}$ unconstrained, the constrained LS SREG$_1$ solution
is to put $\hat{\mu}_j = \bar{y}_{.j}$ for all of the environments in
the subset, but obtain $\hat{\lambda}_1$, $\hat{\alpha}_{i1}$ and the non-
zero $\hat{\gamma}_{j1}$ as the first component of the SVD of
$\mathbf{Z}_2 = [z_{2ij}] = [\bar{y}_{ij} - \bar{y}_{.j}]$, where the $\bar{y}_{ij}$ and $\bar{y}_{.j}$
used in computing $\mathbf{Z}_2$ are only those from
environments in sub-subset S_2.

We have recently developed a constrained SVD solution for SREG$_1$ by putting
$\hat{\mu}_j = \bar{y}_{.j} + \hat{\beta}$, where $\hat{\beta}$ is a constant chosen to
force a particular $\hat{\gamma}_{j1} = 0$ (P.L. Cornelius and
J. Crossa, unpublished result). The useful-
ness of such constrained SVD SREG$_1$ solutions for the SREG clustering problem has
not been evaluated.

For a set (or subset) containing g genotypes and e environments, the d.f. of RSS
(SREG$_1$) is $ge - g - 2e + 2 = (g - 2)(e - 1)$ in
an unconstrained solution, $(g - 2)(e - 1) + 1$
in the above described constrained SVD
SREG$_1$ solution and $(g - 2)(e - 1) + v$ in a
constrained LS solution, where v is the
number of environments with their $\hat{\gamma}_{j1} = 0$ in
the constrained LS solution. Note that $v = e_1$,
where e_1 is as previously defined.

SHMM Clustering of Genotypes with Non-COI

Cornelius *et al.* (1993b) used SHMM clustering to group 41 winter-wheat (*Triticum
aestivum* L.) genotypes into non-COI
clusters. The MET included seven environments. Thirty-five of the genotypes were
grouped into nine clusters, leaving six genotypes unclustered because the procedure
did not enter them into any cluster to which
a non-COI SHMM$_1$ would give a satisfactory
fit. Constrained non-COI solutions, when
needed, were computed using constrained
SVD solutions. Other examples of SHMM
clustering of genotypes have been reported
by Crossa *et al.* (1996) and Abdalla *et al.*
(1997).

Crossa *et al.* (1996) compared constrained LS and constrained SVD solutions
when constrained solutions were needed
and found, for the particular example, that
choice of method for computing constrained
solutions made no difference with respect to
the dendrogram or final acceptable clusters
obtained. To our knowledge, this is the only
published example in which consequences
of the choice of method for computing
constrained non-COI solutions has been
studied.

SHMM clustering of genotypes is essentially by the same strategy as for clustering
environments. The distance between two
genotypes is defined as RMS(SHMM$_1$), using
a constrained solution, if necessary, when
SHMM$_1$ is fitted to the subset of data deriving from only those two genotypes evaluated
in the entire set of environments. Note that it
is still the $\hat{\gamma}_{j1}$ (and not the $\hat{\alpha}_{i1}$) that must be

either all non-positive or all non-negative to have a non-COI solution.

The unconstrained SHMM₁ solution for RSS(SHMM₁) can be obtained in closed form using the result RSS(SHMM₁) = RSS(GREG₁) if the set of data being analysed contains only two genotypes. Because when clustering genotypes the number of environments will always exceed two, constrained non-COI solutions will not exist in closed form. They can be computed as previously described for subsets containing more than two environments in the context of SHMM clustering of environments. The residual d.f. is $e - 2$ for an unconstrained solution, $e - 1$ for a constrained SVD solution and $e - 2 + v$ for a constrained LS solution, where v is the number of environments with their $\hat{\gamma}_{j1} = 0$ in the constrained LS solution. Further details can be found in Crossa et al. (1996).

Use of SREG as a model for identifying groups of genotypes without significant genotype COI is not practical because, for a pair of genotypes, the matrix $\mathbf{Z} = [z_{ij}] = [\bar{y}_{ij} - \bar{y}_{.j}]$ is of rank one and an unconstrained SREG₁ will fit the values exactly, resulting in the distance RSS(SREG₁) = 0. Thus, it is only for pairs of genotypes for which a constrained non-COI solution is needed that a non-zero distance will be obtained. Consequently, SREG₁ clustering of genotypes will, typically, not provide a unique starting-point for the clustering. When clustering genotypes, the d.f. of RSS (SHMM₁) in unconstrained and constrained solutions for a subset containing more than two genotypes will be by the same formulae as previously given for subsets containing more than two environments in the context of clustering environments.

Tests for Lack of Fit of SHMM₁ and SREG₁

Inadequacy of SHMM₁ or SREG₁ (either of these constrained, if necessary) for modelling data from a subset of environments and/or genotypes may be tested statistically using the F_R and/or F_{GH} (F_{GH1} or F_{GH2}) tests.

The F_R test (Cornelius et al., 1992, 1996) of RMS(SHMM₁) or RMS(SREG₁) is from the fit of SHMM₁ or SREG₁ (constrained if necessary) to the subset against the MET's pooled error mean square. The d.f. of RMS(SHMM₁) or RMS(SREG₁) are as given earlier in this chapter.

The F_{GH} tests are sequential tests of the bilinear components. Significance of one or more components beyond the primary effects implies inadequacy of inclusion of only one bilinear term. Letting SS_k denote the sequential sum of squares (on an observation basis) due to the kth bilinear term, the F_{GH2} test is constructed as:

$$F_{GH2} = \frac{SS_k}{u_{1k}s^2}$$

where s^2 is the pooled error mean square and $u_{1k} = E(SS_k/\sigma^2 \mid \lambda_k = 0,\ \lambda_{k-1}$ is large). The denominator d.f. are the pooled error d.f., and the numerator d.f. are approximated as $2u_{1k}^2 / u_{2ik}^2$, where $u_{2k}^2 = V(SS_k/\sigma^2 \mid \lambda_k = 0,\ \lambda_{k-1}$ is large).

A function that will closely approximate u_{1k} and u_{2k} for use in F_{GH} tests of bilinear components in SREG is given by Liu and Cornelius (2001). For doing so, always put u_{1k} and u_{2k} equal to their approximating functions for $E(\hat{\theta}_1)$ and $SD(\hat{\theta}_1)$, respectively, with their r and c defined as $r = \max(g - 1,\ e) - k + 1$ and $c = \max(g - 1,\ e) - k + 1$. The approximating functions are valid for $r \le 199$ and $c \le 149$. For AMMI, SREG, GREG and COMM, the approximating functions given by Liu and Cornelius (2001) supersede the functions previously given by Cornelius et al. (1996).

For tests of bilinear components in SHMM, use the SHMM approximating functions for u_{1k} and u_{2k} given in the appendix of Cornelius et al. (1996) if $\max(g, e) < 100$ and $\min(g, e) < 20$. For cases that violate either of these bounds, use the formulas for $E(\hat{\theta}_1)$ and $SD(\hat{\theta}_1)$ from Liu and Cornelius (2001), with r and c defined as $r = \max(g, e) - k + 1$ and $c = \max(g, e) - k + 1$. Use of functions given by Liu and Cornelius (2001) for SHMM analysis tests the SHMM sequential bilinear terms as if they were bilinear terms in

COMM. This should give sufficiently accurate results for SHMM if r or c is large.

The F_{GH1} derives from a method of moments approximation of the distribution of the quantity $1 + [SS_k/(\text{pooled error SS})]$ as the reciprocal of a beta random variable. The P value can be computed directly from the approximate beta distribution, but, because plant breeders will generally find the value of an F statistic more interpretable than the value of a beta statistic, we routinely transform the beta statistic to an F statistic (with d.f. equal to twice the values of the beta distribution). For details, see Cornelius *et al.* (1992, 1996) and Cornelius (1993). If the pooled error d.f. are large, as they generally are in a MET, P values for F_{GH1} and F_{GH2} are typically almost identical and thus there is ordinarily no need to compute both.

Example of the SHMM and SREG Clustering of Environments with Non-COI in Maize MET

The data come from an international maize (*Zea mays* L.) MET with nine genotypes ($g = 9$) evaluated in a randomized complete block design with four replicates in each of 20 environments ($e = 20$). The SHMM and SREG analyses showed that the first three components were statistically different from zero ($P < 0.05$) by the F_{GH1} test (Table 20.1, results for all environments). Since the second and third components were statistically significant, SHMM$_1$ will not adequately fit the data from the entire set of environments. Moreover, even the fitted SHMM$_1$ (unconstrained) has its point of concurrence within the region containing the plotted points and thus the fitted SHMM$_1$ itself displays genotype COI. This is observed in Fig. 20.1, where three environments have $\hat{\gamma}_{j1} < 0$ and all others have $\hat{\gamma}_{j1} > 0$. Genotype 8 performed worst in the environment with the largest primary effect, but was one of the best two genotypes in the environment with the smallest (most negative) primary effect.

Figure 20.2 depicts the dendrogram of the 20 environments when RMS(SHMM$_1$) was used as distance measurement and clustering was by the complete linkage (furthest neighbour) method. The dichotomous

Table 20.1. Probability values (P) for the F_R and F_{GH1} tests for the secondary and tertiary effects of the SHMM and SREG models for subsets of environments suggested by the dichotomous splitting of the dendrogram of Fig. 20.2.

Environments	Model form	F_R		F_{GH1}	
		Secondary effect	Tertiary effect	Secondary effect	Tertiary effect
All	SHMM	0.0000	0.0244	0.0000	0.0037
	SREG	0.0016	0.0685	0.0042	0.0322
{1, 3, 8, 10}	SHMM	0.0001	0.4757	0.0000	0.2532
	SREG	0.0552	0.4383	0.0481	0.2155
{1, 3, 10}	SHMM	0.6952	0.9984	0.4550	0.9984
	SREG	0.6595	0.9984	0.1245	0.4059
{2, 4, 5, 6, 7, 8, 9,	SHMM	0.0000	0.0418	0.0000	0.0076
11, 12, 13, 14, 15,	SREG	0.0027	0.1042	0.0040	0.0287
16, 17, 18, 19, 20}					
{2, 6, 9, 12, 13, 18}	SHMM	0.1106	0.3641	0.2229	0.4686
	SREG	0.2134	0.7129	0.1455	0.6899
{4, 5, 7, 11, 14, 15,	SHMM	0.0254	0.7313	0.0004	0.4335
16, 17, 19, 20}	SREG	0.2481	0.7067	0.1982	0.4839
{4, 5, 11, 14, 15, 16}	SHMM	0.3524	0.8925	0.1615	0.7020
	SREG	0.8113	0.9186	0.8642	0.8521
{7, 17, 19, 20}	SHMM	0.5342	0.9748	0.2717	0.9850
	SREG	0.4529	0.9581	0.2172	0.9717

F_R test of RMS(SHMM$_1$) or RMS(SREG$_1$) against the pooled error mean square. F_{GH1} test is a sequential test of the bilinear components.

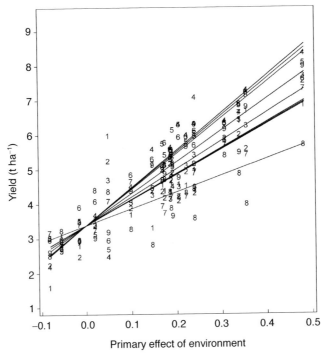

Fig. 20.1. SHMM₁ model fitted by least squares (unconstrained) to nine genotypes and 20 environments ($\hat{\beta}=3.4$). The scatter points are empirical cell means plotted using the digits identifying the genotype as plotting symbols and the regression lines, one for each genotype, plot the SHMM₁ predicted cell means. The regression lines from top to bottom in the region to the right of the point of concurrence $(0, \hat{\beta})$ are genotypes 4, 5, 6, 3, 2, 9, 7, 1 and 8. This rank order for SHMM-predicted yields is completely reversed to the left of the point of concurrence.

splitting suggested by Crossa *et al.* (1993) for finding subsets of environments with genotypic non-COI consisted of fitting SHMM₁ to subsets defined by the branches of the dendrogram, starting at the first split of the entire set of data into the two subsets of environments, {1, 3, 8, 10} and {2, 4, 5, 6, 7, 9, 11, 12, 13, 14, 15, 16, 17, 18, 19, 20}. For the first of these subsets, the lack of fit of SREG₁ is marginally significant ($P = 0.0552$) when tested by the F_R (Cornelius *et al.*, 1996) test, and the SREG secondary effect is significant at $P < 0.05$ when tested by the F_{GH1} test (Table 20.1). For the latter of these two subsets, there is highly significant lack of fit of both SHMM₁ and SREG₁, detected by both F_R and F_{GH1} tests (Table 20.1). Continuing with the dichotomous splitting, the adequacy of SHMM₁ for fitting subsets {1, 3, 10}, {2, 6, 9, 12, 13, 18} and {4, 5, 7, 11, 14, 15, 16, 17, 19, 20} is tested. According to the F_R and the

F_{GH1} tests, SHMM₁ is adequate for fitting the first two subsets, but not for the last one. However, SREG₁ did adequately fit all three subsets. Thus, we have here an example of a subset that is acceptably modelled by a non-COI SREG₁, but not by a non-COI SHMM₁.

In Fig. 20.3, the consistent response of the nine genotypes across the ten environments of subset {4, 5, 7, 11, 14, 15, 16, 17, 19, 20} is clearly depicted through the overlaid broken line SREG₁ graphs that do not cross over. The residuals for the two models, both with unconstrained solutions, were RMS (SHMM₁) = RSS(SHMM₁)/($ge - g - e$) = 14,006,110/71 = 197,269 and RMS(SREG₁) = RSS(SREG₁)/($g - 2)(e - 1$) = 9,795,021/63 = 155,476. Further splitting of {4, 5, 7, 11, 14, 15, 16, 17, 19, 20} gives subsets {4, 5, 11, 14, 15, 16} and {7, 17, 19, 20}, both of which can be adequately modelled by either

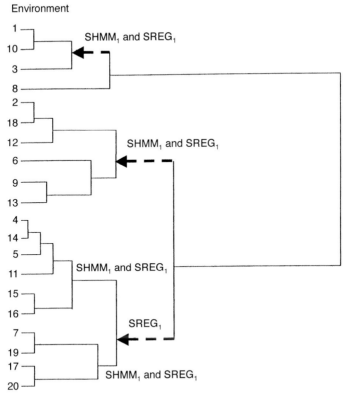

Fig. 20.2. Dendrogram of SHMM and SREG clustering of 20 environments using RMS(SHMM$_1$) = RMS(SREG$_1$) as distance measurement. Arrows denote the subsets of environments where SREG$_1$ gave an adequate fit. The subset indicated by the lower arrow on the figure had to be subdivided once more in order for SHMM$_1$ to give an adequate fit.

non-COI SHMM$_1$ or non-COI SREG$_1$ (Table 20.1).

The ten highest values of RMS(SHMM$_1$) (the distance) for pairs of environments with unconstrained and constrained LS and SVD non-COI solutions are shown in Table 20.2. As expected, based on the results obtained from the SHMM$_1$ and SREG$_1$ clustering (Fig. 20.2), environment 8 is the most frequently occurring environment in pairs of environments with large distance values.

To illustrate constrained and unconstrained solutions for a pair of environments needing a constrained solution for their distance, SHMM$_1$ fitted without constraint to the data from environment 8 ($\bar{y}_{.8} = 4027$, $\hat{\gamma}_{8,1} = -0.7285$) and environment 11 ($\bar{y}_{.11} = 5307$, $\hat{\gamma}_{11,1} = 0.6849$) has its point of concurrence within the region containing the plotted data points (Fig. 20.4). In the constrained

LS solution (Fig. 20.5), the environment that had a positive primary effect ($\hat{\gamma}_{11,1} = 0.6849$) in the unconstrained solution has its primary effect put equal to zero ($\hat{\gamma}_{11,1} = 0$), thus moving the point of concurrence to the right boundary of the region containing the plotted points. In this constrained LS solution, $\hat{\gamma}_{8,1} = -1$.

Both the SHMM and SREG clustering of environments have been extensively used for finding associations between international testing environments used by the International Maize and Wheat Improvement Center (CIMMYT) for maize, bread (*T. aestivum* L.) and durum (*Triticum turgidum* var. *durum*) wheat and triticale (*Triticosecale* Wittm.) METs. Abdalla *et al.* (1997) used SHMM clustering of environments and genotypes and found that durum-wheat genotypes with similar genetic backgrounds formed non-COI clusters, but COI more frequently occurred with genotypes derived from different genetic backgrounds, especially those with different levels of resistance to specific diseases. Consequently, genotypes from diverse genetic

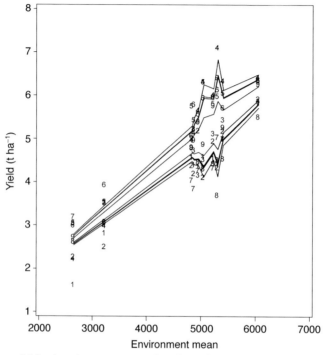

Fig. 20.3. SREG$_1$ model fitted to nine genotypes and a subset of 10 environments. The rank order of genotypes with respect to the overlaid broken-line graphs is 6, 4, 5, 7, 9, 3, 1, 2, 8.

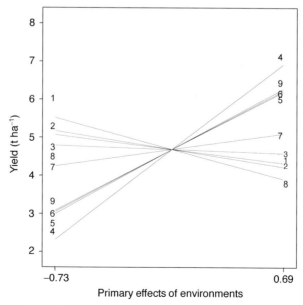

Fig. 20.4. Unconstrained SHMM$_1$ solution for nine genotypes in two environments (8 and 11).

backgrounds tended to cluster into different non-COI groups. Recently, Trethowan *et al.* (2001) used the SHMM and SREG clustering of environments to study long-term associations between international sites for a drought-tolerant bread-wheat MET. Results of this study demonstrated the usefulness of this approach for identifying key testing environments around the world.

Shrinkage Estimates of Linear–Bilinear Models Analogous to BLUPs

Simulation studies of Cornelius (1993) and Cornelius *et al.* (1996) for the AMMI model showed that the inter-

Table 20.2. The ten largest distances (RMS (SHHM$_1$)) for unconstrained and constrained SVD and LS solutions. The degrees of freedom for the unconstrained solutions are the number of genotypes minus two, whereas for the constrained solution they are the number of genotypes minus one.

Pair of environments		Degrees of freedom	Residual mean square
SVD non-COI SHMM$_1$ solution			
5	8	8	534,929.83
3	4	8	538,867.47
8	17	8	546,730.47
8	12	8	621,562.28
3	11	8	630,314.30
4	8	8	636,836.50
8	15	8	1,118,858.50
8	13	8	1,197,429.00
8	18	8	1,278,760.50
8	11	8	1,697,992.90
LS non-COI SHMM$_1$ solution			
3	17	7	445,676.58
15	18	7	473,768.72
4	8	8	474,787.63
3	8	7	486,411.11
3	15	7	491,284.81
8	12	8	600,516.22
8	15	8	809,543.62
8	18	8	861,644.18
8	13	8	1,137,120.80
8	11	8	1,327,651.00

SVD, singular value decomposition constrained solution; LS, least-squares constrained solution.

action mean squared errors (IMSE),

$$\sum_i \sum_j \left(\sum_{k=1}^{P} \hat{\lambda}_k \hat{\alpha}_{ik} \hat{\gamma}_{jk} - \sum_{k=1}^{P} \lambda_k \alpha_{ik} \gamma_{jk} \right)^2,$$

where $p = \text{rank}(\mathbf{Z})$, can be reduced if the LS estimate of λ_k, i.e. $\hat{\lambda}_k$, is replaced by a shrinkage estimate of the form $S_k \hat{\lambda}_k$. The authors found that the IMSE of the shrinkage estimates were smaller than the IMSE of the best AMMI truncated model. These shrinkage estimators are of the form *signal* divided by (*signal* + *noise*) and are similar to the functions of variance components that occur in computation of empirical BLUPs of cell means in a RE2W model and can be justified by a Bayesian argument.

The conventional RE2W model with interactions for the mean of the *i*th genotype in the *j*th environment is given in Equation 20.1 with the τ_i, δ_j and $(\tau\delta)_{ij}$ considered random. Under normality, independence and a balanced data set, it is easy to compute empirical BLUPs of the realized performance levels, $\mu_{ij} = \mu + \tau_i + \delta_j + (\tau\delta)_{ij}$, of the genotypes in the environments where they were tested (i.e. empirical BLUPs of cell means). The BLUP estimates of the main effects of τ_i, δ_j and $(\tau\delta)_{ij}$ for a balanced data set are:

$$\text{BLUP}(\tau_i) = \frac{n e \sigma_\tau^2}{\sigma^2 + n\sigma_{\tau\delta}^2 + n e \sigma_\tau^2}(\bar{y}_{i..} - \bar{y}_{...})$$

$$\text{BLUP}(\delta_j) = \frac{n g \sigma_\delta^2}{\sigma^2 + n\sigma_{\tau\delta}^2 + n g \sigma_\delta^2}(\bar{y}_{.j.} - \bar{y}_{...})$$

$$\text{BLUP}\left[(\tau\delta)_{ij}\right] = n\sigma_{\tau\delta}^2 \left[\frac{(\bar{y}_{i..} - \bar{y}_{...})}{\sigma^2 + n\sigma_{\tau\delta}^2 + n e \sigma_\tau^2} \right.$$
$$\left. + \frac{(\bar{y}_{.j.} - \bar{y}_{...})}{\sigma^2 + n\sigma_{\tau\delta}^2 + n g \sigma_\delta^2} + \frac{(\bar{y}_{ij.} - \bar{y}_{i..} - \bar{y}_{.j.} - \bar{y}_{...})}{\sigma^2 + n\sigma_{\tau\delta}^2} \right]$$

Not only do the ordinary LS estimates of the main effects of genotypes $(\bar{y}_{i..} - \bar{y}_{...})$ and environments $(\bar{y}_{.j.} - \bar{y}_{...})$ contribute to the BLUP of the *i*th genotypic main effect and of the *j*th environmental main effect, respectively, but each also contributes to the BLUP of the GEI. Then, for $\hat{\mu} = \bar{y}_{...}$, the BLUP of the cell mean is given by:

$$\hat{\mu} + \text{BLUP}(\tau_i) + \text{BLUP}(\delta_j) + \text{BLUP}\left[(\tau\delta)_{ij}\right] =$$

$$\hat{\mu} + \frac{n\sigma_{\tau\delta}^2 + n e \sigma_\tau^2}{\sigma^2 + n\sigma_{\tau\delta}^2 + n e \sigma_\tau^2}(\bar{y}_{i..} - \bar{y}_{...}) +$$

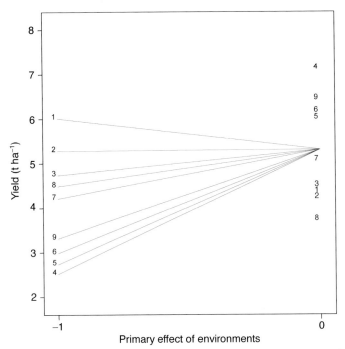

Fig. 20.5. Constrained LS SHMM$_1$ solution for nine genotypes in two environments (8 and 11).

$$\frac{n\sigma_{\tau\delta}^2 + ng\sigma_{\delta}^2}{\sigma^2 + n\sigma_{\tau\delta}^2 + ng\sigma_{\delta}^2}\left(\bar{y}_{.j.} - \bar{y}_{...}\right) + \frac{n\sigma_{\tau\delta}^2}{\sigma^2 + n\sigma_{\tau\delta}^2}$$

$$\left(\bar{y}_{ij.} - \bar{y}_{i..} - \bar{y}_{.j.} + \bar{y}_{...}\right) = \bar{y}_{...} + S_{\tau}\left(\bar{y}_{i..} - \bar{y}_{...}\right) +$$

$$S_{\delta}\left(\bar{y}_{.j.} - \bar{y}_{...}\right) + S_{\tau\delta}\left(\bar{y}_{ij.} - \bar{y}_{i..} - \bar{y}_{.j.} - \bar{y}_{...}\right)$$

In computing empirical BLUPs of cell means, the functions of the estimated variance components that multiply the LS estimates of the main effects of genotypes, environments and the GEI, namely $(\bar{y}_{i..} - \bar{y}_{...})$, $(\bar{y}_{.j.} - \bar{y}_{...})$ and $(\bar{y}_{ij.} - \bar{y}_{i..} - \bar{y}_{.j.} - \bar{y}_{...})$, respectively, are 'shrinkage' factors that can be estimated as:

$$S_{\tau} = \frac{n\hat{\sigma}_{\tau\delta}^2 + ne\hat{\sigma}_{\tau}^2}{\hat{\sigma}^2 + n\hat{\sigma}_{\tau\delta}^2 + ne\hat{\sigma}_{\tau}^2} = 1 - \frac{1}{F_{\tau}} =$$

$$1 - \frac{1}{\dfrac{MS(\text{Genotype})}{MS(\text{Error})}} = 1 - \frac{MS(\text{Error})}{MS(\text{Genotype})},$$

$$S_{\delta} = \frac{n\hat{\sigma}_{\tau\delta}^2 + ng\hat{\sigma}_{\delta}^2}{\hat{\sigma}^2 + n\hat{\sigma}_{\tau\delta}^2 + ng\hat{\sigma}_{\delta}^2} = 1 - \frac{1}{F_{\delta}} =$$

$$1 - \frac{1}{\dfrac{MS(\text{Environment})}{MS(\text{Error})}} =$$

$$1 - \frac{MS(\text{Error})}{MS(\text{Environment})},$$

and

$$S_{\tau\delta} = \frac{n\hat{\sigma}_{\tau\delta}^2}{\hat{\sigma}^2 + n\hat{\sigma}_{\tau\delta}^2} = 1 - \frac{1}{F_{\delta}} =$$

$$1 - \frac{1}{\dfrac{MS(\text{Genotype} \times \text{Environment})}{MS(\text{Error})}} =$$

$$1 - \frac{MS(\text{Error})}{MS(\text{Genotype} \times \text{Environment})}$$

where MS = mean square.

For a saturated AMMI model, the estimates of the interaction parameters are such that $\sum_{k=1}^{t}\lambda_k\alpha_{ik}\gamma_{jk} = \bar{y}_{ij.} - \bar{y}_{i..} - \bar{y}_{.j.} + \bar{y}_{...}$, i.e. the BLUP of a cell mean is a shrinkage estimate of AMMI. It is reasonable to suppose, however, that an optimum strategy for obtaining shrinkage estimates of AMMI, COMM, SREG, GREG and SHMM should include different shrinkage factors for the different bilinear components.

Cornelius and Crossa (1995, 1999) and Cornelius *et al.* (1993a, 1996) used a procedure of shrinking each bilinear component by an estimate of the *signal* to (*signal + noise*) ratio of its eigenvalue, i.e. the *k*th bilinear component $\hat{\lambda}_k \hat{\alpha}_{ik} \hat{\gamma}_{jk}$ is shrunk as $S_k \hat{\lambda}_k \hat{\alpha}_{ik} \hat{\gamma}_{jk}$. Then the sum of shrunken bilinear terms is $\sum_k S_k \hat{\lambda}_k \hat{\alpha}_{ik} \hat{\gamma}_{jk}$, where $\hat{\lambda}_k, \hat{\alpha}_{ik}$ and $\hat{\gamma}_{jk}$ are the corresponding LS estimates. The shrinkage factor S_k is computed as:

$$S_k = \left(\hat{\lambda}_k^2 - \ddot{u}_k s^2 / n \right) / \hat{\lambda}_k^2 = 1 - \frac{1}{\ddot{F}_k}$$

where s^2 is the error mean square, $\ddot{u}_k$ is a value equal to, or that estimates, the expected error variance absorption by the *k*th bilinear term, i.e. $nE\left[\left(\hat{\lambda}_k^2 - \lambda_k^2 \right) / \sigma^2 \right]$, and $\ddot{F}_k = n \hat{\lambda}_k^2 / \ddot{u}_k s^2$, provided that $\ddot{F}_k > 1$ and therefore $S_k > 0$; otherwise, put $S_k = 0$. The $\ddot{F}_k$ is a statistic similar in structure to the F_{GH2} statistic that Cornelius *et al.* (1996) used to test the null hypothesis $H_0 : \lambda_k = 0$, but F_{GH2} replaces $\ddot{u}_k$ with u_k, the latter defined as the conditional expectation $nE\left[\left(\hat{\lambda}_k^2 / \sigma^2 \right) | \lambda_k^2 = 0 \right]$, i.e. $\ddot{u}_k$ is an estimate of $nE\left[\left(\hat{\lambda}_k^2 | \lambda_1, \lambda_2, \ldots, \lambda_k, \ldots \right) \right]$. The value of $\ddot{u}_k$ in S_k is appropriate for the alternative hypothesis $H_1 : \lambda_k > 0$, but not for the null hypothesis. The above-described shrinkage estimators are appropriate for any of the balanced LBMs – AMMI, SREG, GREG and COMM. See Cornelius and Crossa (1999) for a mathematical justification for the above-described shrinkage estimators and for details of a scheme for computing shrinkage estimates of SHMM.

In practice, we obtain initial values of $\ddot{u}_k$ as the number of independent parameters in the *k*th bilinear component (i.e. number of parameters minus number of constraints). These initial $\ddot{u}_k$ are used to obtain an initial set of shrinkage estimates of the λ_k. Then, these initial shrinkage estimates are used as supposed 'true' values of the λ_k in a simulation (parametric bootstrap) scheme to obtain more accurate values of the $\ddot{u}_k$, which, in turn, are used to obtain a new set of shrinkage estimates of the λ_k. The scheme can be iterated as often as desired. The scheme has been observed to move the shrinkage estimates into a rather stable neighbourhood in about five iterations. For our scheme for computing shrinkage estimates of SHMM, see Cornelius and Crossa (1999).

Prediction accuracy of the shrinkage estimates of linear–bilinear models

It has been suggested that shrinkage estimates of LBMs will provide better estimates of the realized values of the cell means (μ_{ij}) than will the empirical cell means or LS solutions of parsimonious models with the number of multiplicative terms chosen by cross-validation or any test of statistical significance (Cornelius *et al.*, 1993a, 1996; Cornelius and Crossa, 1995). Cornelius and Crossa (1999) analysed five MET data sets using random data splitting and cross-validation. They evaluated the predictive accuracy of the shrinkage estimates of the LBMs and compared them with: (i) the best LS fitted truncated LBM; (ii) the empirical BLUPs of the cell means based on the RE2W model; and (iii) the empirical cell means. The root mean square predictive difference (RMSPD) was computed for judging the best predictive model, i.e. the one having the lowest RMSPD. The authors used data adjusted by replicate differences within environments to reduce the noise in the modelling and validation data that otherwise occurs as a consequence of ignoring block differences when randomly splitting the data.

Results showed that the worst predictors of the genotypic performance were the empirical cell means (Table 20.3). In all five experiments, shrinkage estimates of LBMs were better predictors than the best truncated model fitted by LS and, except in one experiment, also better than the BLUPs of the cell means. These results suggest that shrinkage estimates of LBMs eliminate the need for testing hypotheses and cross-validation to select an optimum number of bilinear terms. Results also indicate that predictive accuracy differs little among the five model forms if shrinkage estimators are used.

Experimental data set 3 of Cornelius and Crossa (1999) is the maize MET data, with

nine genotypes evaluated in 20 environments, used earlier in this chapter to illustrate SHMM and SREG clustering of environments. For this MET, clearly

Table 20.3. RMSPD (kg ha⁻¹) values for the best truncated least-squares fitted model and shrinkage estimates for linear–bilinear model forms AMMI, GREG, SREG, COMM and SHMM, for the BLUPs of cell means and for empirical cell means, all obtained by cross-validation in five MET data sets (Cornelius and Crossa, 1999).

Model form*	Truncated	Shrinkage	BLUP	Empirical cell mean
Experiment 1				
$AMMI_4$	637.45	627.65	–	–
$COMM_5$	636.40	626.92	–	–
$SREG_5$	641.12	628.04	–	–
$GREG_4$	638.90	628.17	–	–
$SHMM_5$	636.71	628.61	–	–
	–	–	637.40	671.10
Experiment 2				
$AMMI_8$	1322.60	1273.67	–	–
$COMM_8$	1316.89	1272.27	–	–
$SREG_8$	1318.83	1275.44	–	–
$GREG_8$	1316.28	1272.21	–	–
$SHMM_8$	1315.54	1275.35	–	–
	–	–	1284.04	1331.26
Experiment 3				
$AMMI_1$	816.41	800.36	–	–
$COMM_2$	813.39	799.42	–	–
$SREG_1$	816.30	800.48	–	–
$GREG_2$	808.03	798.24	–	–
$SHMM_2$	810.16	799.40	–	–
	–	–	817.64	849.45
Experiment 4				
$AMMI_{10}$	832.57	796.69	–	–
$COMM_{11}$	822.29	795.66	–	–
$SREG_6$	823.79	798.33	–	–
$GREG_{11}$	822.53	797.38	–	–
$SHMM_{11}$	823.61	799.55	–	–
	–	–	798.86	831.10
Experiment 5				
$AMMI_0$	677.49	668.50	–	–
$COMM_1$	675.79	668.14	–	–
$SREG_1$	684.56	671.60	–	–
$GREG_1$	675.08	668.09	–	–
$SHMM_1$	676.03	671.48	–	–
	–	–	663.95	715.84

*The subscripts on the model forms indicate the number of bilinear terms retained in the best truncated model. This subscript is not related to the shrinkage estimates.
RMSPD, root mean square predicted difference; AMMI, additive main effects and multiplicative interaction model; GREG, genotypes regression model; SREG, sites regression model; COMM, completely multiplicative model; SHMM, shifted multiplicative model; BLUP, best linear unbiased predictor; MET, multi-environment trials.

shrinkage estimates are better predictors than empirical cell means and BLUPs of cell means (Table 20.3). Crossa *et al.* (2002) computed SREG$_1$ clustering of the 20 environments in this MET, with the empirical cell means replaced with SREG shrinkage estimates as input to the procedure. Two of the final groups of environments obtained, namely, {2, 4, 5, 6, 9, 11, 12, 13, 14, 16, 18} and {7, 15, 17, 19, 20}, were different from those in Fig. 20.2. Only the environmental group {1, 3, 10} was the same. Whereas two of these groups did not agree with the clustering computed from empirical cell means, they did agree with groups of environments delineated by sectors of an SREG$_2$ biplot computed from deviations of empirical cell means from site means.

This is a very intriguing empirical result, suggesting that the methodology deserves study in more examples. Because of the separation of pattern from random error that appears to be achieved by the shrinkage estimators, we believe SREG clustering using shrinkage estimates of cell means as input has considerable promise as a routine procedure for the study of interaction patterns in an MET.

Software

To compute SHMM and SREG clustering of environments or SHMM clustering of genotypes, we first use an SAS® program (SAS/IML) to obtain a dendrogram. Then, adequacy of fit of SHMM$_1$ or SREG$_1$ (constrained, if necessary) is evaluated using the Fortran program EIGAOV. Enquiries concerning the availability of the software may be sent to P.L. Cornelius (corneliu@ms.uky.edu).

References

Abdalla, O.S., Crossa, J. and Cornelius, P.L. (1997) Results and biological interpretation of shifted multiplicative model clustering of durum wheat cultivars and test sites. *Crop Science* 37, 88–97.

Cornelius, P.L. (1993) Statistical tests and retention of terms in the additive main effects and multiplicative interaction model for cultivar trials. *Crop Science* 33, 1186–1193.

Cornelius, P.L. and Crossa, J. (1995) *Shrinkage Estimators of Multiplicative Models for Cultivar Trials.* Technical Report No. 352, Department of Statistics, University of Kentucky, Lexington, Kentucky, USA.

Cornelius, P.L. and Crossa, J. (1999) Prediction assessment of shrinkage estimators of multiplicative models for multi-environment cultivar trials. *Crop Science* 39, 998–1009.

Cornelius, P.L. and Seyedsadr, M. (1997) Estimation of general linear–bilinear models for two-way tables. *Journal of Statistical Computation and Simulation* 58, 287–322.

Cornelius, P.L., Seyedsadr, M. and Crossa, J. (1992) Using the shifted multiplicative model to search for 'separability' in crop cultivar trials. *Theoretical and Applied Genetics* 84, 161–172.

Cornelius, P.L., Crossa, J. and Seyedsadr, M. (1993a) Tests and estimators of multiplicative models for variety trials. In: *Proceedings of the 5th Annual Kansas State University Conference on Applied Statistics in Agriculture, Manhattan, Kansas*, pp. 156–169.

Cornelius, P.L., Van Sanford, D.A. and Seyedsadr, M. (1993b) Clustering cultivars into groups without rank-change interactions. *Crop Science* 33, 1193–1200.

Cornelius, P.L., Crossa, J. and Seyedsadr, M. (1996) Statistical tests and estimators of multiplicative models for cultivar trials. In: Kang, M.S. and Gauch, H.G., Jr (eds) *Genotype-by-Environment Interaction.* CRC Press, Boca Raton, Florida, pp. 199–234.

Crossa, J. and Cornelius, P.L. (1993) Recent developments in multiplicative models for cultivar trials. In: Buxton, D.R., Shibles, R., Forsberg, R.A., Blad, B.L., Asay, K.H., Paulsen, G.M. and Wilson, R.F. (eds) *International Crop Science I.* Crop Science Society of America, Madison, Wisconsin, pp. 571–577.

Crossa, J. and Cornelius, P.L. (1997) Sites regression and shifted multiplicative model clustering of cultivar trial sites under heterogeneity of error variance. *Crop Science* 37, 405–415.

Crossa, J., Cornelius, P.L., Seyedsadr, M. and Byrne, P. (1993) A shifted multiplicative model cluster analysis for grouping environments without genotypic rank-change. *Theoretical and Applied Genetics* 85, 577–586.

Crossa, J., Cornelius, P.L., Sayre, K. and Ortiz-Monasterio, I.J. (1995) A shifted multiplicative model fusion method for grouping environments without cultivar rank change. *Crop Science* 35, 54–62.

Crossa, J., Cornelius, P.L. and Seyedsadr, M. (1996) Using the shifted multiplicative model cluster methods for crossover genotype-by-environment interaction. In: Kang, M.S. and Gauch, H.G., Jr (eds) *Genotype-by-Environment Interaction.* CRC Press, Boca Raton, Florida, pp. 175–198.

Crossa, J., Cornelius, P.L. and Yan, W. (2002) Biplots of linear–bilinear models for studying crossover genotype × environment interaction. *Crop Science* (in press).

Eberhart, S.A. and Russell, W.A. (1966) Stability parameters for comparing varieties. *Crop Science* 6, 36–40.

Finlay, K.W. and Wilkinson, G.N. (1963) The analysis of adaptation in a plant breeding programme. *Australian Journal of Agriculture Research* 14, 742–754.

Fisher, R.A. and MacKenzie, W.A. (1923) Studies in variation II. The manurial response in different potato varieties. *Journal of Agricultural Science* 13, 311–320.

Freeman, G.H. (1973) Statistical methods for the analysis of genotype–environment interactions. *Heredity* 31(3), 339–354.

Gabriel, K.R. (1978) Least squares approximation of matrices by additive and multiplicative models. *Journal of the Royal Statistical Society, Series B* 40, 186–196.

Gauch, H.G., Jr (1988) Model selection and validation for yield trials with interaction. *Biometrics* 44, 705–715.

Gollob, H.F. (1968) A statistical model which combines features of factor analytic and analysis of variance. *Psychometrika* 33, 73–115.

Johnson, D.E. and Graybill, F.A. (1972) An analysis of a two-way model with interaction and no replication. *Journal of the American Statistical Association* 67, 862–868.

Liu, G. and Cornelius, P.L. (2001) Simulations and decimal approximations for the means and standard deviations of the characteristic roots of a Wishart matrix. *Communications in Statistics B. Simulation and Computation* (in press).

Mandel, J. (1961) Non-additivity in two-way analysis of variance. *Journal of the American Statistical Association* 56, 878–888.

Mandel, J. (1969) The partitioning of interaction in analysis of variance. *Journal of Research of the National Bureau of Standards, Series B* 73, 309–328.

Mandel, J. (1971) A new analysis of variance model for non-additive data. *Technometrics* 13, 1–18.

Seyedsadr, M. and Cornelius, P.L. (1992) Shifted multiplicative models for non-additive two-way tables. *Communications in Statistics B. Simulation and Computation* 21, 807–822.

Trethowan, R.M., Crossa, J., vanGinkel, M. and Rajaram, S. (2001) Relationships among bread wheat international yield testing locations in dry areas. *Crop Science* 41, 1461–1469.

Tukey, J.W. (1949) One degree of freedom for non-additivity. *Biometrics* 5, 232–242.

Williams, E.J. (1952) The interpretation of interactions in factorial experiments. *Biometrika* 39, 65–81.

Yan, W., Hunt, L.A., Sheng, Q. and Szlavnics, Z. (2000) Cultivar evaluation and mega-environment investigation based on the GGE biplot. *Crop Science* 40, 597–605.

Yates, F. and Cochran, W.G. (1938) The analysis of groups of experiments. *Journal of Agricultural Science* 28, 556–580.

Zobel, R.W., Wright, M.J. and Gauch, H.G., Jr (1988) Statistical analysis of a yield trial. *Agronomy Journal* 80, 388–393.

21 Exploring Variety–Environment Data Using Random Effects AMMI Models with Adjustments for Spatial Field Trend: Part 1: Theory

Alison Smith,[1] Brian Cullis[1] and Robin Thompson[2]
[1]Wagga Wagga Agricultural Institute, Private Mail Bag, Wagga Wagga, NSW 2650, Australia; [2]IACR-Rothamsted, Harpenden, Hertfordshire AL5 2JQ, UK

Introduction

The recommendation of new plant varieties for commercial use requires reliable and accurate predictions of the average yield of each variety across a range of target environments and knowledge of important interactions with the environment. This information is obtained from a series of plant variety trials, also known as multi-environment trials (MET). Each year, many millions of dollars are spent worldwide on the acquisition of such data. To maximize the cost-efficiency of this work, it is crucial that the data be scrutinized using an appropriate, informative statistical analysis. Two key aspects of the analysis are the assumptions associated with the variety effects and interactions and those associated with the field-plot errors from individual trials. For the latter, many approaches assume simple, often inappropriate, within-trial structures, such as randomized complete block (RCB), and a common error variance for all trials. Data from field trials often exhibit spatial variation, so called because it is a function of the location of the plots in the field. Gilmour et al. (1997) present a method of analysis in which spatial variation is modelled, resulting in estimates of treatment effects that have greater accuracy and precision than more traditional methods, such as RCB and IB (see, for example, Gleeson and Cullis, 1987). Cullis et al. (1998) adopted this approach in their mixed-model analysis of MET data. They allowed for a separate spatial covariance structure and error variance for each trial. This was a major step forward, resulting in more accurate estimates of overall variety means (see, for example, Smith et al., 2001).

Smith et al. (2001) extended the Cullis et al. (1998) approach with the use of multiplicative models for the variety–environment (V × E) effects. The model allows a separate genetic variance for each trial and provides a parsimonious and interpretable model for the genetic covariances between pairs of trials. The genetic model can be regarded as a random effects analogue of the additive main effects and multiplicative interaction (AMMI) model (Gauch, 1992). The Smith et al. (2001) approach, therefore, combines the interpretative strengths of AMMI with the advantages afforded by the mixed-model framework. The approach encompasses most current mixed-model approaches to the analysis of MET data (including Patterson et al., 1977; Piepho, 1997; Cullis et al., 1998).

This chapter is the first in a two-part series that explores the spatial multiplicative mixed model of Smith *et al.* (2001) in full detail. In the second section, we present the theory associated with the spatial analysis of a single field trial. In the third section, we describe the spatial analysis for MET data, incorporating multiplicative models for V × E effects. Topics covered include tests of goodness of fit and model interpretation. The methods are illustrated in Chapter 22 (Part 2: Applications).

Spatial Analysis of a Field Experiment

Gilmour *et al.* (1997) partition spatial variation into two types of smooth spatial trend (local and global) and extraneous variation. Local trend reflects, for example, small-scale soil depth and fertility fluctuations. Global trend reflects non-stationary trend across the field. Extraneous variation is often linked to the management of the trial. An example is the effect of harvesting in a serpentine manner up and down the rows in the field, with plots harvested in the 'up' direction being consistently lower/higher-yielding than plots harvested in the 'down' direction. Global trend and extraneous variation are accommodated in the model by including appropriate terms, such as design factors and polynomial functions of the spatial coordinates of the field plots. Local stationary trend is accommodated using a covariance structure. The decomposition of error variation provides a more plausible approach than the original spatial methodology of Gleeson and Cullis (1987) and Cullis and Gleeson (1991), in which error variation as a whole was modelled using a covariance structure.

It is assumed that an individual experiment consists of n plots that are laid out in the field as a rectangular array of r rows by c columns ($n = rc$). The data $y^{(n \times 1)}$ are ordered correspondingly (as rows within columns). The model for y is given by

$$y = X\tau + Zu + e \tag{21.1}$$

where $\tau^{(t \times 1)}$ and $u^{(b \times 1)}$ are the vectors of fixed and random effects, respectively. $X^{(n \times t)}$ and $Z^{(n \times b)}$ are associated design matrices, the former assumed to be of full column rank. The vector of residuals is given by $e^{(n \times 1)}$. It is assumed that the joint distribution of (u, e) is Gaussian, with zero mean and variance matrix:

$$\begin{bmatrix} G(\gamma) & 0 \\ 0 & R(\phi) \end{bmatrix}$$

where $G^{(b \times b)}$ and $R^{(n \times n)}$ are symmetric positive definite matrices that are functions of the vectors of variance parameters γ and ϕ, respectively. The distribution of the data is thus Gaussian, with mean $X\tau$ and variance matrix $H = ZGZ' + R$, where Z' denotes the transpose of Z.

The vector of errors e is assumed to follow a (second-order stationary) spatial process with $\text{Var}[e] = R = \sigma^2\Sigma(\alpha)$ where Σ is the spatial correlation matrix that is a function of parameters α and has associated variance σ^2. Possible forms for Σ are described in the section on covariance models for local spatial trend.

The model in Equation 21.1 is a mixed model, so the estimation strategy outlined in the section on Estimation can be used. The variance parameters to be estimated are $\kappa = (\gamma, \phi)$, where $\phi = (\sigma^2, \alpha)$.

Traditional methods of analysis, such as RCB and IB, are special cases of Equation 21.1. For an RCB analysis, u contains replicate effects and $\Sigma = I_n$, where I_n is the $n \times n$ identity matrix. The parameter σ^2 is simply the trial error variance, the residual maximum likelihood (REML) estimate of which is identical to the error mean square from an ordinary analysis of variance (ANOVA). For an IB analysis with recovery of interblock information, u contains effects for replicates and blocks within replicates and $\Sigma = I_n$. The parameter σ^2 is the within-block error variance. Treatment effects will be included in either τ or u, depending on the aims of the experiment.

The key to the Gilmour *et al.* (1997) approach to spatial analysis is the identification of an appropriate variance structure for the plot errors. There is no longer a dichotomy between spatial analysis and traditional methods, such as RCB and IB. The latter provide legitimate error variance models that would be adopted in the spatial approach if found to be consistent with the data. This is rarely the case, however.

Covariance models for local spatial trend

Local trend reflects the fact that, in the absence of design effects, data from plots that are close together are more similar than those that are further apart. Thus, the elements of e are correlated, the correlations being a function of the spatial distance between plots. Let $\Sigma = \{\rho_{ij}\}$, where $\rho_{ij} = \text{Cor}[e_i, e_j]$ is the spatial correlation between plots i and j. Since field experiments are arranged as rectangular arrays, a two-dimensional coordinate system is required to define the location of each plot. Let $s_i = (s_{ir}, s_{ic})$ denote the spatial location of ith plot in the field, where s_{ir} and s_{ic} are the row and column coordinates, respectively. The spatial correlation between $e_i = e_i(s_i)$ and $e_j = e_j(s_j)$ can then be written as:

$$\rho_{ij} = V(s_i, s_j; \alpha)$$

where the correlation function V depends on a vector of unknown parameters α. Since the process for e is second-order stationary, the correlation between two plots depends only on the distance between them. Thus:

$$V(s_i, s_j; \alpha) = V(l_{ij}; \alpha)$$

where $l_{ij} = (l_{ijr}, l_{ijc}) = s_i - s_j$. It is further assumed that the two-dimensional process is separable, so that the correlation function is given by the product of the correlation function for each dimension. The separability assumption is computationally convenient and appears to be reasonable for the two-dimensional spatial trend process associated with field trials (see, for example, Martin, 1990; Cullis and Gleeson, 1991). Thus:

$$V(l_{ij}; \alpha) = V_r(l_{ijr}; \alpha_r)V_c(l_{ijc}; \alpha_c)$$

where V_r and V_c are the correlation functions for rows and columns, respectively. Correspondingly, the variance matrix for e can be written as:

$$\text{Var}[e] = \sigma^2 \Sigma(\alpha) = \sigma^2 \Sigma_c(\alpha_c) \otimes \Sigma_r(\alpha_r)$$

where Σ_r and Σ_c are the $r \times r$ and $c \times c$ correlation matrices for rows and columns, respectively.

Many forms for V are possible. Zimmerman and Harville (1991) give examples used in geostatistical applications, including the exponential model, which, for a single dimension, is given by $V(l_{ij}; \alpha) = \exp(-\alpha |l_{ij}|^p)$ for some integer p. Extending to a separable two-dimensional process gives:

$$V(l_{ij}; \alpha) = \exp(-\alpha_r |l_{ijr}|^p - \alpha_c |l_{ijc}|^p) \quad (21.2)$$

The model with $p = 1$ is particularly important for field experiments.

In field experiments, plots are often of equal size and are laid out in a contiguous array, so that the distance between plots can be measured simply in terms of row and column numbers. Let l_{ijr}^* be the difference in row numbers between plots i and j, so that l_{ijr}^* has possible values $0, 1, \ldots (r - 1)$. Define l_{ijc}^* similarly. If d_r and d_c are the actual distances (in metres, say) between the centroids of plots in row and column directions, respectively, then $l_{ijr} = d_r l_{ijr}^*$, so that $\exp(-\alpha_r |l_{ijr}|) = \rho_r^{|l_{ijr}^*|}$, where $\rho_r = \exp(-\alpha_r d_r)$. The function in Equation 21.2 with $p = 1$ is then given by:

$$V(l_{ij}; \alpha) = \rho_r^{|l_{ijr}^*|} \rho_c^{|l_{ijc}^*|} \quad (21.3)$$

where ρ_r and ρ_c are, by definition, positive. If this restriction is lifted, Equation 21.3 is the correlation function for a separable autoregressive process of order 1 (hereafter denoted AR1 × AR1). Cullis and Gleeson (1991) proposed this as a plausible correlation structure for spatial trend. The parameters $\alpha = (\rho_r, \rho_c)$ are known as the autoregressive coefficients. Many other forms for Σ are possible (see, for example, Gleeson and Cullis, 1987). Experience has shown, however, that the AR1 × AR1 model (or a variant with an identity matrix for one of the dimensions) usually provides an adequate variance structure for local spatial trend.

The variogram

A tool that is widely used in repeated measures and geostatistical analyses to visualize temporal or spatial dependence is the variogram. For a general two-dimensional spatially correlated process, $\mathbf{E}(\cdot)$, the value of the variogram for two locations, $\mathbf{s}_i$ and $\mathbf{s}_j$, is defined as:

$$\omega(\mathbf{s}_i, \mathbf{s}_j) = \frac{1}{2} E\left[\left\{E(\mathbf{s}_i) - E(\mathbf{s}_j)\right\}^2\right]$$

If $\mathbf{E}(\cdot)$ has zero mean, this can be interpreted as half the variance of the difference between the two locations.

For the second-order stationary spatial trend process $\mathbf{e}$ in Equation 21.1, the theoretical variogram is given by:

$$\omega(\mathbf{s}_i, \mathbf{s}_j) = \omega(\mathbf{l}_{ij}) = \sigma^2\{1 - V(\mathbf{l}_{ij}; \alpha)\}$$

where V is the correlation function defined in the section on Covariance models for spatial trend.

The variogram for an independent process is therefore constant, irrespective of the distance $\mathbf{l}_{ij}$ between plots. As previously noted, the AR1 × AR1 process of Equation

21.3 is particularly important for field experiments. The associated theoretical variogram is given by:

$$\omega(\mathbf{l}_{ij}) = \sigma^2\left\{1 - \rho_r^{|l^*_{ijr}|}\rho_c^{|l^*_{ijc}|}\right\} \tag{21.4}$$

This increases monotonically in both the row and column directions as the separation between plots increases (and thus correlation between plots decreases). It reaches a plateau that is given by the variance σ^2. The greater the autoregressive correlation coefficients, the slower the rise to the plateau (see Fig. 21.1, for example).

The sample variogram

Given a set of data $\mathbf{y}$ that follows the model in Equation 21.1, the sample variogram is defined by:

$$v_{ij} = \frac{1}{2}\left\{e_i(\mathbf{s}_i) - e_j(\mathbf{s}_j)\right\}^2$$

Where $\mathbf{e} = \{e_i(\mathbf{s}_i)\} = \mathbf{y} - \mathbf{X}\tau - \mathbf{Zu}$. Clearly, this is unbiased for $w(\mathbf{s}_i, \mathbf{s}_j)$.

In practice, the vector of residuals is unknown and so is replaced by the

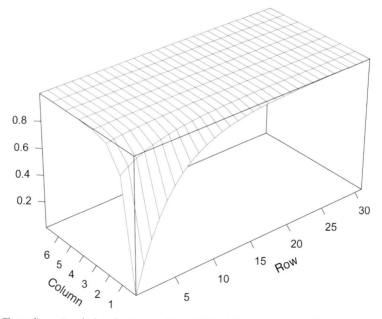

Fig. 21.1. Three-dimensional plot of variogram for an AR1 × AR1 process with $\sigma^2 = 1$, $\rho_r = 0.8$ and $\rho_c = 0.2$. The x and y ordinates are displacements in the row and column directions, respectively, measured as differences in row/column numbers.

estimate $\tilde{\boldsymbol{e}} = \{\tilde{e}_i(\boldsymbol{s}_i)\} = \boldsymbol{y} - \boldsymbol{X}\hat{\boldsymbol{\tau}} - \boldsymbol{Z}\tilde{\boldsymbol{u}}$. An estimate of the sample variogram is then obtained as:

$$\tilde{v}_{ij} = \frac{1}{2}\left\{\tilde{e}_i(\boldsymbol{s}_i) - \tilde{e}_j(\boldsymbol{s}_j)\right\}^2$$

With the assumption of equal plot sizes and a contiguous layout, there will be many values of $\tilde{v}_{ij}$ with the same absolute displacement $\boldsymbol{l}_{ij}^*$. The mean, denoted $\bar{v}_{ij}$, is calculated and the sample variogram is defined by the triple $(l_{ijr}^*, l_{ijc}^*, \bar{v}_{ij})$. This can be viewed graphically as a three-dimensional plot. Since at large displacements $\bar{v}_{ij}$ is based on only a few pairs of points, the graph should be truncated to exclude these ordinates.

As discussed in Gilmour *et al.* (1997), the use of estimated residuals introduces bias into the sample variogram. We propose the use of the sample variogram purely as an informal diagnostic tool so that this bias can be ignored. It has no adverse effect on the visual interpretation of the sample variogram.

Error model identification

The determination of an appropriate variance structure for the separable spatial trend process and the detection of extraneous effects is made possible through the use of graphical diagnostics. Two key graphs are of estimated residuals against row (column) number (hereafter referred to generically as 'residual plots') and the three-dimensional graph of the sample variogram. Future work is focused on the construction of score tests to determine the adequacy of a particular spatial model. This will remove possible ambiguities in the interpretation of the visual diagnostics.

Global trend

Non-stationary global trend in the row direction, say, will be displayed in the residual plots as a smooth trend (linear or non-linear) over row number for each column. A sample variogram that fails to reach a plateau in the row/column direction is evidence of global trend. Historically, non-stationarity of this type was corrected by differencing the data (see, for example, Gleeson and Cullis, 1987), but this complicates the analysis. Gilmour *et al.* (1997) recommend an alternative approach that involves the fitting of polynomial functions or cubic smoothing splines (Verbyla *et al.*, 1999) to the row and/or column coordinates of the plots. Thus, the non-stationary global trend is explicitly modelled.

Extraneous variation

Management of field trials involves procedures that are aligned with the rows and columns. Examples are the sowing and harvesting of plots. Certain procedures may result in row and column effects (systematic and/or random) in the data. Gilmour *et al.* (1997) label this extraneous variation to distinguish it from smooth trend.

Extraneous variation may be evident from an examination of the residual plots. It may be more clearly seen, however, using the sample variogram. For example, a variogram with a sawtooth appearance indicates the presence of cyclic row/column effects. Since this is a systematic effect, it can be accommodated in the model by fitting a fixed factor with number of levels corresponding to the length of the cycle.

There may also be non-systematic variation associated with rows and columns. This is accommodated by fitting random row/column effects in the model. The variogram can be used to diagnose the existence of random row and column effects. If there are random row effects, the variogram ordinates will be lower at zero row displacement compared with other row displacements, and similarly for random column effects.

Outliers

As with any data analysis, it is vital that erroneous data points be excluded. In a spatial analysis, potential outliers will be revealed on the residual plots. Work is in progress to construct formal tests of significance for outliers in a spatial analysis (Gogel, 1997). Consultation with the researcher will determine whether an outlier is erroneous and should therefore be omitted.

The Spatial Mixed Model for MET Data

Consistent with the notation in the section on Spatial Analysis of a Field Experiment, it is assumed that the jth trial (synonymous with environment), $j = 1 \ldots p$, consists of N_j plots laid out in a rectangular array with r_j rows and c_j columns ($N_j = r_j \times c_j$). The data vector $y_j^{(N_j \times 1)}$ is ordered correspondingly as rows within columns. The model for the combined vector of data across trials $y^{(n \times 1)} = \{y_j\}$, $n = \sum_{j=1}^{p} N_j$, is given by:

$$y = X\tau + Zu + e$$

where $\tau^{(t \times 1)}$ and $u^{(b \times 1)}$ are vectors of fixed and random effects, respectively. $X^{(n \times t)}$ and $Z^{(n \times b)}$ are the associated design matrices, the former assumed to be of full column rank. The vector of residuals is given by e. It is assumed that the joint distribution of $(u', e')'$ is Gaussian, with zero mean and variance matrix:

$$\begin{bmatrix} G(\gamma) & 0 \\ 0 & R(\phi) \end{bmatrix}$$

where γ and ϕ are vectors of variance parameters. The distribution of the data y is thus Gaussian, with mean $X\tau$ and variance matrix $H = ZGZ' + R$.

The errors e consist of subvectors $\{e_j\}$, where $e_j^{(N_j \times 1)}$ is the vector of plot error effects for the jth trial, which is a (second-order stationary) spatially dependent process. The errors from different trials are assumed to be independent. The error variance matrix for trial j is given by $R_j = \sigma_j^2 \Sigma_j(\alpha_j)$ where Σ_j is a spatial correlation matrix that is a function of parameters α_j and has associated variance σ_j^2. Models for Σ_j are described in the section on Covariance models for local spatial trend.

The random effects u consist of subvectors $\{u_i\}$, where $u_i^{(b_i \times 1)}$ is the vector of effects for the ith random term, $i = 1 \ldots q$. The matrix Z is partitioned conformably as $[Z_1 \ldots Z_q]$. The subvectors in u are assumed to be mutually independent. The variance

matrix G_i for the ith random term has many possible forms, including the standard variance component structure, namely $G_i = \sigma_i^2 I_{b_i}$. In the most general case, G_i could be completely unstructured, comprising $b_i(b_i + 1)/2$ parameters.

For the variety effects, we adopt the view that the yields from different environments constitute different traits for each variety (Falconer, 1952). It is therefore natural to consider genetic variances for each environment (reflecting the magnitude of variation between varieties in individual environments) and genetic correlations between pairs of environments (reflecting the agreement in variety rankings). This framework is synonymous with a statistical model in which the variety effects in each environment are regarded as random. Let u_g be the $mp \times 1$ vector of (genetic) effects for m varieties in each of p environments (ordered as varieties within environments). This represents a two-dimensional (variety by environment) array of effects, namely $U_g^{(m \times p)}$, where $u_g = \text{vec}[U_g]$. It is assumed that the associated variance structure is separable with:

$$\text{Var}[u_g] = G_e \otimes I_m \tag{21.5}$$

where $G_e = \{\sigma_{g_{jj'}}\}$ is the $p \times p$ symmetric genetic variance matrix. The diagonal elements are the genetic variances for individual environments and the off-diagonal elements are the genetic covariances between pairs of environments.

Finally, the spatial mixed model for MET data can be written as:

$$\begin{aligned} y &= X\tau + Zu + e \\ &= X\tau + Z_0 u_0 + Z_g u_g + e \end{aligned} \tag{21.6}$$

where the fixed effects τ include environment main effects and trial specific effects for extraneous field variation (as described in the section on Extraneous variation), u_g are the variety effects in each environment with associated design matrix $Z_g^{(n \times mp)}$ and variance matrix as in Equation 21.5 and u_0 comprise any additional random effects (including trial specific effects for extraneous variation). The latter have design matrix Z_0 and variance matrix G_0. If there are trials in which only a subset of the m

varieties are grown, then $\boldsymbol{Z}_g$ will contain zero columns. Thus, unbalanced data are easily accommodated. Note that, in Equation 21.6, the variety effects in each environment are specified as a single random term, that is, a single subvector of $\boldsymbol{u}$. The partitioning into variety main effects and V × E interactions is considered as a submodel.

Factor-analytic models for variety effects

There are many possible forms for the genetic variance matrix. The standard mixed model for MET data (see Patterson et al., 1977) involves the fitting of variety main effects $\boldsymbol{u}_v^{(m \times 1)}$ (assumed here to be random) and (random) V × E interactions. Thus, the variety effects in different environments are partitioned as:

$$\boldsymbol{u}_g = (1_p \otimes \boldsymbol{I}_m)\boldsymbol{u}_v + \boldsymbol{u}_{ve} \quad (21.7)$$

The main effects and interactions are assumed to be sets of independent effects with zero means and constant variances, given by σ_v^2 and σ_{ve}^2, respectively. The genetic variance matrix is then given by $\boldsymbol{G}_e = \sigma_v^2 \boldsymbol{J}_p + \sigma_{ve}^2 \boldsymbol{I}_p$, where $\boldsymbol{J}_p$ is a $p \times p$ matrix of ones, so that genetic variances, covariances and correlations are given by:

$$\sigma_{g_{jj}} = \sigma_v^2 + \sigma_{ve}^2, \ \forall j$$

$$\sigma_{g_{jj'}} = \sigma_v^2, \ \forall j \neq j'$$

$$\rho_{g_{jj'}} = \sigma_v^2 / (\sigma_v^2 + \sigma_{ve}^2), \ \forall j \neq j'$$

This so-called compound symmetry structure or repeatability model, therefore, implies that all environments have the same genetic variance and all pairs have the same genetic covariance (thence all pairs have the same genetic correlation). This rarely provides an adequate fit to the data. At the other extreme, we have a completely general or unstructured form for $\boldsymbol{G}_e$ that contains $p(p + 1)/2$ parameters. Estimation of such a structure may be inefficient or unstable for even moderately large values of p, so a more parsimonious representation is desirable. This can be achieved using multiplicative models for the variety effects in

each environment. Here we consider the multiplicative model associated with the multivariate technique of factor analysis (see, for example, Mardia et al., 1988).

Factor analysis is used to model the covariance structure among a set of p observed variates $x_1 \ldots x_p$. The aim is to account for the covariances of the x variates in terms of a much smaller number of hypothetical factors. As Lawley and Maxwell (1971) state:

in correlation terms . . . the first question that arises is whether any correlation exists . . . If there is correlation, the next question is whether there is a random variate f_1 such that all partial correlation coefficients between the x variates after eliminating the effect of f_1 are zero. If not, then two random variates f_1 and f_2 are postulated and the partial correlation coefficients after eliminating f_1 and f_2 are examined. The process continues until all partial correlations between the x variates are zero.

In the context of MET data, the factor-analysis approach can be used to provide a class of structures for the genetic variance matrix, $\boldsymbol{G}_e$. The model is postulated in terms of the (unobserved) variety effects in different environments:

$$u_{g_{lj}} = \sum_{r=1}^{k} \lambda_{jr} f_{lr} + \delta_{lj} \quad (21.8)$$

where $u_{g_{lj}}$ is the effect for variety l in environment j, f_{lr} is the score for variety l in factor r and λ_{jr} is the associated loading for environment j. The error, δ_{lj}, represents lack of fit of the model. This model is represented in vector notation as:

$$\boldsymbol{u}_g = (\lambda_1 \otimes \boldsymbol{I}_m)\boldsymbol{f}_1 + \ldots + (\lambda_k \otimes \boldsymbol{I}_m)\boldsymbol{f}_k + \boldsymbol{\delta}$$

$$= (\Lambda \otimes \boldsymbol{I}_m)\boldsymbol{f} + \boldsymbol{\delta} \quad (21.9)$$

where $\lambda_r^{(p \times 1)} = \{\lambda_{jr}\}$, $\boldsymbol{f}_r^{(m \times 1)} = \{f_{lr}\}$, $\boldsymbol{\delta}^{(mp \times 1)} = \{\delta_{lj}\}$, $\Lambda^{(p \times k)} = [\lambda_1 \ldots \lambda_k]$ and $\boldsymbol{f}^{(mk \times 1)} = (\boldsymbol{f}_1', \boldsymbol{f}_2' \ldots \boldsymbol{f}_k')'$. The joint distribution of $\boldsymbol{f}$ and $\boldsymbol{\delta}$ is given by:

$$\begin{pmatrix} \boldsymbol{f} \\ \boldsymbol{\delta} \end{pmatrix} \sim N \left[\begin{pmatrix} \boldsymbol{0} \\ \boldsymbol{0} \end{pmatrix}, \begin{pmatrix} \boldsymbol{I}_k \otimes \boldsymbol{I}_m & \boldsymbol{0} \\ \boldsymbol{0} & \Psi \otimes \boldsymbol{I}_m \end{pmatrix} \right]$$

where $\Psi = \text{diag} (\psi_1 \ldots \psi_p)$. The variance matrix for the variety effects in each environment is then given by:

$$\text{Var}\big[\boldsymbol{u}_g\big] = (\Lambda \otimes \boldsymbol{I}_m)\,\text{Var}\big[\boldsymbol{f}\big](\Lambda' \otimes \boldsymbol{I}_m) + \text{Var}[\boldsymbol{\delta}]$$
$$= (\Lambda\Lambda' + \Psi) \otimes \boldsymbol{I}_m \qquad (21.10)$$

Thus, the model for the variety effects in Equation 21.9 leads to a model for $\boldsymbol{G}_e$ in which:

$$\sigma_{g_{jj}} = \sum_{r=1}^{k} \lambda_{jr}^2 + \psi_j$$

$$\sigma_{g_{jj'}} = \sum_{r=1}^{k} \lambda_{jr}\lambda_{j'r}$$

$$e_{g_{jj'}} = \sum_{r=1}^{k} \lambda_{jr}\lambda_{j'r} \Big/ \sqrt{\Big(\sum_{r=1}^{k}\lambda_{jr}^2 + \psi_j\Big)\Big(\sum_{r=1}^{k}\lambda_{j'r}^2 + \psi_{j'}\Big)}$$

Note that Equation 21.9 has the form of a (random) regression on k environmental covariates, $\lambda_1 \dots \lambda_k$, in which all regressions pass through the origin. It may be more appropriate to allow a separate (non-zero) intercept for each variety. This is equivalent to the model with variety main effects, $\boldsymbol{u}_v$, and a k factor-analytic model for the V × E interactions. Using Equation 21.7 we then have:

$$\boldsymbol{u}_g = (1_p \otimes \boldsymbol{I}_m)\boldsymbol{u}_v + (\Lambda \otimes \boldsymbol{I}_m)\boldsymbol{f} + \boldsymbol{\delta} \quad (21.11)$$

where $\boldsymbol{u}_v$ has mean zero and variance $\sigma_v^2 \boldsymbol{I}_m$, say. The model can be written as:

$$\boldsymbol{u}_g = (\sigma_v 1_p \otimes \boldsymbol{I}_m)\boldsymbol{f}_0 + (\Lambda \otimes \boldsymbol{I}_m)\boldsymbol{f} + \boldsymbol{\delta}$$
$$= (\Lambda_v \otimes \boldsymbol{I}_m)\boldsymbol{f}_v + \boldsymbol{\delta} \qquad (21.12)$$

where $\Lambda_v^{p \times (k+1)} = [\sigma_v 1_p\ \Lambda]$, $\boldsymbol{f}_0 = \boldsymbol{u}_v/\sigma_v$ and $\boldsymbol{f}_v' = (\boldsymbol{f}_0', \boldsymbol{f}')$. Thus, the model with variety main effects and a k factor-analytic model for the V × E interactions is a special case of a $(k+1)$ factor-analytic model for the variety effects in each environment, in which the first set of loadings are constrained to be equal.

The feature that distinguishes Equations 21.9 and 21.11 from standard random regression problems is that both the covariates Λ and the regression coefficients $\boldsymbol{f}$ are unknown and, as such, must be estimated from the data. The model is, therefore, a multiplicative model of environment and variety coefficients (known as loadings and scores, respectively). Herein lies the analogy with approaches such as AMMI (Gauch, 1992). A key difference is that the multiplicative model in Equation 21.11 accommodates random effects, whereas AMMI is a fixed-effects model. (See the section on Connection with AMMI models for a detailed comparison.) Gogel et al. (1995) and Piepho (1997) proposed multiplicative mixed models for MET data, but assumed random-environment rather than random-variety coefficients. This leads to a factor-analytic model for the right-hand side of the separable structure in Equation 21.5 rather than the left-hand side. It is therefore assumed that the variance structure is associated with varieties rather than environments. In this chapter, the converse is used, as it is consistent with the logic of environments as traits. Our experience has shown a greater need to allow for heterogeneity of genetic variance and covariance between environments compared with varieties. This can, of course, be examined formally for each data set or chosen to suit the application.

Estimation

Full details regarding estimation of the mixed model in Equation 21.6 are given in Smith et al. (2001). Only a brief outline is presented here.

The estimation procedure leads to best linear unbiased estimates (BLUEs) of the fixed effects:

$$\hat{\boldsymbol{\tau}} = \big(\boldsymbol{X}'\boldsymbol{H}^{-1}\boldsymbol{X}\big)^{-1}\boldsymbol{X}'\boldsymbol{H}^{-1}\boldsymbol{y}$$

and best linear unbiased predictors (BLUPs) of the random effects:

$$\tilde{\boldsymbol{u}}_0 = \boldsymbol{G}_0\boldsymbol{Z}_0'\boldsymbol{P}\boldsymbol{y}$$
$$\tilde{\boldsymbol{u}}_g = (\boldsymbol{G}_e \otimes \boldsymbol{I}_m)\,\boldsymbol{Z}_g'\boldsymbol{P}\boldsymbol{y} \qquad (21.13)$$

where:

$$\boldsymbol{H} = \text{Var}[\boldsymbol{y}] = \boldsymbol{Z}_0\boldsymbol{G}_0\boldsymbol{Z}_0' + \boldsymbol{Z}_g\,(\boldsymbol{G}_e \otimes \boldsymbol{I}_m)\,\boldsymbol{Z}_g' + \boldsymbol{R}$$
$$\boldsymbol{P} = \boldsymbol{H}^{-1} - \boldsymbol{H}^{-1}\boldsymbol{X}\big(\boldsymbol{X}'\boldsymbol{H}^{-1}\boldsymbol{X}\big)^{-1}\boldsymbol{X}'\boldsymbol{H}^{-1}$$

For the factor-analytic model, BLUPs of the variety scores $\boldsymbol{f}$ and residuals $\boldsymbol{\delta}$ can be obtained in terms of $\tilde{\boldsymbol{u}}_g$ as:

$$\tilde{\boldsymbol{f}} = \text{Var}[\boldsymbol{f}]\big[\boldsymbol{Z}_g(\Lambda \otimes \boldsymbol{I}_m)\big]'\boldsymbol{P}\boldsymbol{y}$$
$$= \big[\Lambda'(\Lambda\Lambda' + \Psi)^{-1} \otimes \boldsymbol{I}_m\big]\tilde{\boldsymbol{u}}_g$$

and

$$\tilde{\boldsymbol{\delta}} = \big[\Psi(\Lambda\Lambda' + \Psi)^{-1} \otimes \boldsymbol{I}_m\big]\tilde{\boldsymbol{u}}_g$$

To calculate the estimates of fixed and random effects, we require estimates of the parameters in G_o, G_e and $\mathbf{R}$. In terms of the factor-analytic model, the variance parameters associated with G_e are Λ and Ψ. We estimate variance parameters using the method of REML (Patterson and Thompson, 1971). We use a scoring algorithm known as the average-information algorithm (Gilmour *et al.*, 1995), which is a modified Fisher scoring algorithm, in which the expected information matrix is replaced by an approximate average of the observed and expected information matrices.

The approach to estimation of the factor-analytic model in Smith *et al.* (2001) is computationally intensive. An alternative algorithm that uses sparse matrix methods is given in Thompson *et al.* (2001). This has been shown to reduce computing time substantially. It also accommodates cases where some (or all) specific variances need to be constrained equal to zero, thereby leading to a variance structure that has less than full rank. Current research focuses on alternatives to the average-information algorithm, in particular the expectation-maximization (Dempster *et al.*, 1977) and parameter-expanded expectation-maximization (EM) algorithms (Liu *et al.*, 1998).

Constraints on loadings

When $k > 1$, the number of free variance parameters in a k factor-analytic structure is not $pk + p$, as suggested by Equation 21.10, but only $pk + p - k(k-1)/2$. This arises from the fact that the distribution of $u_r = (\Lambda \otimes I_m) f$ is singular. It is necessary to impose $k(k-1)/2$ independent constraints on the elements of Λ to ensure uniqueness. The non-uniqueness of the loadings has implications both for the estimation and for the interpretation of the factor-analytic model.

For estimation purposes, Mardia *et al.* (1988) choose to constrain $\Lambda'\Psi^{-1}\Lambda$ to be diagonal. Jennrich and Schluchter (1986) suggest an alternative approach, stipulating that '$k(k-1)/2$ factor loadings in Λ . . . should

be . . . set equal to zero to fix the rotation'. A key point that is implicit in this statement but may easily be overlooked is that it is both the number and the position of the zeros that are important. One pattern that ensures uniqueness (and therefore fixes the rotation) is when all $k(k-1)/2$ elements in the upper triangle of Λ are zero, that is, $\lambda_{jr} = 0$ for $j < r = 2 \ldots k$. (The proof of uniqueness follows from theorem 14.5.5 in Harville (1997).) In terms of an algorithm for REML estimation of a k factor-analytic variance structure, this approach is more convenient than that of Mardia *et al.* (1988), since it constitutes a set of linear constraints. The solution of the score equations subject to these constraints can, therefore, be achieved using the approach described in Smith (1999).

Goodness of fit

The aim of the factor-analytic model for $V \times E$ effects is to account for the genetic covariances among p environments in terms of a much smaller number k of (unknown) factors $f_1 \ldots f_k$. Since it is fitted within a mixed-model framework, the adequacy of the factor-analytic model can be formally tested. The model with k factors, denoted FA(k), is nested within the model with $k + 1$ factors. An intermediate model is that with variety main effects and k factors for the $V \times E$ interactions. This will be denoted $V + FA(k)$. Residual maximum likelihood ratio tests (REMLRT) can be used to compare these models.

The standard test of goodness of fit of a factor-analytic model involves the comparison with an unstructured form for the variance matrix (see, for example, Mardia *et al.*, 1988). This can be carried out when the number of environments is small.

Connection with AMMI models

The AMMI model has become a popular method for analysing MET data. To use AMMI, the data must comprise a complete two-way factorial structure (variety × environment) and may or may not be replicated. The model is a fixed-effects model with (additive) main effects for varieties and environments and multiplicative terms for the interaction. The latter are obtained using a singular value decomposition (SVD) of the V × E interactions. Let U_{ve} denote the $m \times p$ matrix of V × E interactions. In AMMI, U_{ve} is decomposed as $U_{ve} = ALB^{*\prime}$, where A and B^* are $m \times t$ and $p \times t$ matrices, such that $A'A = I_t = B^{*\prime}B^*$, $L = \mathrm{diag}(l_1 \ldots l_t)$ and t is the rank of U_{ve}. Defining $B = B^*L$, the decomposition can be written as:

$$U_{ve} = AB'$$
$$= \sum_{r=1}^{t} a_r b_r' \qquad (21.14)$$

The columns of A $(a_r^{(m \times 1)})$ are called the variety scores and the columns of B $(b_r^{(p \times 1)})$ are the environment loadings.

As in factor analysis, the aim in the AMMI approach is to account for structure in the genetic effects using the minimum number, k, of multiplicative terms. Isolation of the first k terms in Equation 21.14 gives:

$$U_{ve} = \sum_{r=1}^{k} a_r b_r' + \sum_{r=k+1}^{t} a_r b_r'$$
$$= A_1 B_1' + A_2 B_2'$$

where A_1 and B_1 are $m \times k$ and $p \times k$ matrices, respectively. Thus, in the AMMI model, the V × E interactions are modelled as:

$$u_{ve} = (B_1 \otimes I_m)a + e_g \qquad (21.15)$$

where $a^{(mk \times 1)} = \mathrm{vec}[A_1] = (a_1' \ \ldots \ a_k')'$ and $e_g^{(mp \times 1)}$ are the residual V × E interactions that remain if not all t components of the SVD are used. The latter are assumed to be independent with constant variance.

There is a clear connection between Equation 21.15 and the k factor-analytic model for the V × E interactions, namely:

$$u_{ve} = (\Lambda \otimes I_m)f + \delta$$

There is a correspondence between the environment loadings for the two models $(B_1$ and $\Lambda)$ and the variety scores $(a$ and $f)$. Thus, the k factor-analytic model of

Equation 21.11 is a random-effects analogue of the AMMI model.

Model interpretation

Prediction of overall variety means

It is often of interest to obtain an overall mean (across environments) for each variety. One possibility is to obtain the prediction at the mean values of the loadings. By definition of the loadings, these are predictions of variety means for an environment that is 'average', in the sense of having average genetic covariance with all other environments. For a simple model in which there are no effects (fixed or random) for extraneous variation and in which τ is the vector of environment means, the prediction for variety l is given by:

$$\bar{\hat{\tau}} + \sum_{r=1}^{k} \bar{\hat{\lambda}}_{.r} \tilde{f}_{lr} \qquad (21.16)$$

where $\bar{\hat{\tau}}$ is the mean across environments of the estimated environment means and $\bar{\hat{\lambda}}_{.r}$ is the mean across environments of the estimated loadings for the rth factor.

An alternative is to form the two-way table of predicted variety means for each environment and then simply average across environments. For variety l, this is given by:

$$\frac{1}{p}\sum_{j=1}^{p}\left(\bar{\hat{\tau}}_j + \tilde{u}_{glj}\right) = \bar{\hat{\tau}} + \sum_{r=1}^{k} \bar{\hat{\lambda}}_{.r}\tilde{f}_{lr} + \bar{\tilde{\delta}}_{l.} \quad (21.17)$$

where $\bar{\tilde{\delta}}_{l.}$ is the mean across environments of the BLUPs of the residual effects for variety l.

Note that the difference between the two types of means is the inclusion of unexplained V × E effects (lack of fit from the factor-analytic model) in the second approach.

The calculation of overall variety means is the same, irrespective of the inclusion of variety main effects in the model. If they are included then $\hat{\lambda}_1$ in Equations 21.16 and 21.17 corresponds to the loadings for the main effects and so, from Equation 21.12, is given by $\hat{\sigma}_v$. The issue of interpretation of the variety main effects in Equation 21.12 is important. These are not main effects in the usual sense, namely, a measure of overall variety performance, but are merely intercepts in the regression. They therefore reflect variety performance in an environment that has zero values for the loadings. As in ordinary regression, a centring of the covariates (loadings) would result in the intercepts reflecting predictions at the average values of the covariates. In our application, this would provide variety main effects that are identical to the means in Equation 21.16.

Overall measures of performance must be used with caution. If the genetic correlations between environments are small, that is, if there are large changes in variety rankings from one environment to the next, then it is unwise to use an overall measure of performance to make broad selection decisions (see, for example, Cooper and DeLacy, 1994).

Interpretation of environment loadings

The non-uniqueness of Λ when $k > 1$ introduces ambiguity into the interpretation of both the environment loadings and the variety scores. The constrained form of Λ described in the section on Constraints on loadings is purely for computational ease and has no biological basis. It is therefore necessary to rotate the solution to obtain a meaningful interpretation of the loadings and scores. Lawley and Maxwell (1971) describe a number of rotations designed to aid with interpretation in the analysis of social-science data. We choose a rotation analogous to AMMI, namely, a principal-component representation of the loadings. If no variety main effects are included in the model, this means that the first factor accounts for the maximum amount of genetic covariance between environments, the second factor accounts for the next largest amount and is orthogonal to the first, and so on. If variety main effects are included, the rotation is applied to the loadings associated with the V × E interactions. The principal-component rotation then means that the first factor accounts for the maximum amount of V × E interaction in the data, and so on.

Consider the SVD of Λ, namely:

$$\Lambda = ALB'$$

where L is a diagonal matrix with elements given by the square roots of the eigenvalues of $\Lambda\Lambda'$ (arranged in decreasing order) and A and B are orthogonal matrices whose columns are the eigenvectors of $\Lambda\Lambda'$ and $\Lambda'\Lambda$, respectively. The required rotation is then $\Lambda^* = \Lambda B$.

Note that predictions of overall performance are unaffected by the rotation, provided that both the loadings and the scores have been rotated. Thus, if the rotated loadings Λ^* are used in Equation 21.16 or 21.17, the scores must correspond to this solution, namely:

$$\tilde{f}^* = \left[\Lambda^{*\prime}\left(\Lambda^*\Lambda^{*\prime} + \Psi\right)^{-1} \otimes I_m\right]\tilde{u}_g$$

Graphical representation

Graphs of loadings (columns of Λ^*) from factor-analytic models without variety main effects are particularly useful for clustering environments in terms of genetic correlations. Environment loadings from one multiplicative term are plotted against another and are displayed as vectors (lines from the origin). Being two-dimensional, the plot is most useful when only two or three multiplicative terms are needed in the model. If we consider the plot for the first two multiplicative terms (factors) from a k factor-analytic model without variety main effects, then:

1. The squared length of the vector for an environment is the genetic variance explained by the two factors.
2. The cosine of the angle between the vectors for two environments is the genetic correlation arising from the two factors.

Note that if the $k = 2$ factor model provides a reasonable fit (so that specific variances are small), then 1 approximates the genetic variance for the environment and 2 the genetic correlation between the pair of environments.

The multiplicative part of the k factor-analytic model (either with or without variety main effects) can also be displayed using biplots (see Gabriel, 1971). The 'bi' in the name refers to the fact that both environment loadings and variety scores are displayed on the same graph. This facilitates an examination of relationships among varieties, among environments and between varieties and environments. Detailed discussions of biplots and their interpretation can be found in Kempton (1984) and Meulman (1998). The data sets we routinely analyse often have so many environments and/or varieties that a joint display is uninformative. We therefore often choose to graph loadings and scores separately.

Discussion

The spatial multiplicative mixed-model approach of Smith et al. (2001) for the analysis of $V \times E$ data facilitates the modelling of important sources of variation associated with both variety and error effects. We have shown that the multiplicative mixed model they use to explain $V \times E$ variation is a random-effects analogue of the AMMI model (Gauch, 1992). Since a mixed model is used, the advantages over the AMMI model are numerous. They include:

- Within-trial spatial variation and between-trial error variance heterogeneity can be accommodated.
- Unbalanced data are easily handled.
- Variety effects and $V \times E$ interactions can be regarded as random (leading to better predictions).
- Goodness of fit of the model (that is, number of multiplicative terms needed) can be formally tested.

In the second chapter in this series (Chapter 22: Part 2: Applications), the importance of some of these points will be demonstrated.

Acknowledgements

We thank Arthur Gilmour for many helpful discussions. We gratefully acknowledge the financial support of the Grains Research and Development Corporation of Australia. The BBSRC Institute for Arable Crops receives grant-aided support from the Biotechnology and Biological Sciences Research Council.

References

Cooper, M. and DeLacy, I.H. (1994) Relationships among analytical methods used to study genotypic variation and genotype-by-environment interaction in plant breeding multi-environment experiments. *Theoretical and Applied Genetics* 88, 561–572.

Cullis, B.R. and Gleeson, A.C. (1991) Spatial analysis of field experiments – an extension to two dimensions. *Biometrics* 47, 1449–1460.

Cullis, B.R., Gogel, B.J., Verbyla, A.P. and Thompson, R. (1998) Spatial analysis of multi-environment early generation trials. *Biometrics* 54, 1–18.

Dempster, A.P., Laird, N.M. and Rubin, D.B. (1977) Maximum likelihood from incomplete data via the EM algorithm. *Journal of the Royal Statistical Society Series B* 39, 1–38.

Falconer, D.S. (1952) The problem of environment and selection. *The American Naturalist* 86, 293–298.

Gabriel, K.R. (1971) The biplot graphic display of matrices with application to principal components analysis. *Biometrika* 58, 453–467.

Gauch, H.G., Jr (1992) *Statistical Analysis of Regional Yield Trials: AMMI Analysis of Factorial Designs*. Elsevier, Amsterdam.

Gilmour, A.R., Thompson, R. and Cullis, B.R. (1995) AI, an efficient algorithm for REML estimation in linear mixed models. *Biometrics* 51, 1440–1450.

Gilmour, A.R., Cullis, B.R. and Verbyla, A.P. (1997) Accounting for natural and extraneous variation in the analysis of field experiments. *Journal of Agricultural, Biological, and Environmental Statistics* 2, 269–273.

Gleeson, A.C. and Cullis, B.R. (1987) Residual maximum likelihood (REML) estimation of a neighbour model for field experiments. *Biometrics* 43, 277–288.

Gogel, B.J. (1997) Spatial analysis of multi-environment variety trials. PhD thesis, Department of Statistics, University of Adelaide, South Australia.

Gogel, B.J., Cullis, B.R. and Verbyla, A.P. (1995) REML estimation of multiplicative effects in multi-environment variety trials. *Biometrics* 51, 744–749.

Harville, D.A. (1997) *Matrix Algebra from a Statistician's Perspective.* Springer-Verlag, New York.

Jennrich, R.L. and Schluchter, M.D. (1986) Unbalanced repeated measures models with structured covariance matrices. *Biometrics* 42, 805–820.

Kempton, R.A. (1984) The use of biplots in interpreting variety by environment interactions. *Journal of Agricultural Science, Cambridge* 103, 123–135.

Lawley, D.N. and Maxwell, A.E. (1971) *Factor Analysis as a Statistical Method*, 2nd edn. Butterworths, London.

Liu, C., Rubin, D.B. and Wu, Y.N. (1998) Parameter expansion to accelerate EM: the PX-EM algorithm. *Biometrika* 85, 755–770.

Mardia, K.V., Kent, J.T. and Bibby, J.M. (1988) *Multivariate Analysis.* Academic Press, London.

Martin, R.J. (1990) The use of time-series models and methods in the analysis of agricultural field trials. *Communications in Statistics* 19, 55–81.

Meulman, J.J. (1998) Optimal scaling methods for graphical display of multivariate data. In: *COMPSTAT98 Proceedings in Computational Statistics.* Physica-Verlag, Heidelberg, pp. 65–76.

Patterson, H.D. and Thompson, R. (1971) Recovery of interblock information when block sizes are unequal. *Biometrika* 31, 100–109.

Patterson, H.D., Silvey, V., Talbot, M. and Weatherup, S.T.C. (1977) Variability of yields of cereal varieties in UK trials. *Journal of Agricultural Science, Cambridge* 89, 238–245.

Piepho, H.-P. (1997) Analyzing genotype–environment data by mixed models with multiplicative terms. *Biometrics* 53, 761–767.

Smith, A.B. (1999) Multiplicative mixed models for the analysis of multi-environment trial data. PhD thesis, Department of Statistics, University of Adelaide, South Australia.

Smith, A.B., Cullis, B.R. and Thompson, R. (2001) Analyzing variety by environment data using multiplicative mixed models and adjustments for spatial field trend. *Biometrics* 57(4), 1138–1147.

Thompson, R., Cullis, B.R., Smith, A.B. and Gilmour, A.R. (2001) An efficient algorithm for fitting factor analytic and reduced rank models in complex experiments. *Australian and New Zealand Journal of Statistics* (submitted).

Verbyla, A.P., Cullis, B.R., Kenward, M.G. and Welham, S.J. (1999) The analysis of designed experiments and longitudinal data using smoothing splines (with discussion). *Journal of the Royal Statistical Society, Series C* 48, 269–312.

Zimmerman, D.L. and Harville, D.A. (1991) A random field approach to the analysis of field plot experiments. *Biometrics* 47, 223–239.

22 Exploring Variety–Environment Data Using Random Effects AMMI Models with Adjustments for Spatial Field Trend: Part 2: Applications

Alison Smith,[1] Brian Cullis,[1] David Luckett,[1] Gil Hollamby[2] and Robin Thompson[3]

[1]Wagga Wagga Agricultural Institute, Private Mail Bag, Wagga Wagga, NSW 2650, Australia; [2]Roseworthy Agricultural College, Roseworthy, South Australia, Australia; [3]IACR-Rothamsted, Harpenden, Hertfordshire AL5 2JQ, UK

Introduction

The first chapter in this series (Chapter 21: Part 1: Theory) contained the methodology for a spatial multiplicative mixed-model analysis for variety–environment (V × E) data. In the analysis, random V × E effects are modelled using the multiplicative model associated with factor analysis (Mardia et al., 1988). This is done simultaneously with the modelling of spatial field trend for individual trials (Gilmour et al., 1997). The current chapter illustrates the methodology using two examples. It is assumed that this chapter will be read in conjunction with Chapter 21.

The first example relates to a series of 16 plant-breeding trials from a single season in which the aim is to make varietal selections, either for retention for further yield testing or for commercial release. In this example, we illustrate in detail the model-fitting process. Predictions of overall variety means (across trials) are calculated for the purpose of selection. The impact of accommodating

spatial variation and error variance heterogeneity is demonstrated.

The second example relates to a series of 61 trials spanning several years and a range of geographical sites in which the varieties have been chosen as 'probes' for the sites. The aim is to identify sites that may be suitable for use as key testing sites in the early stages of the breeding programme. This example illustrates how the methodology can be applied to very large, complex data sets. The data set comprises 11,597 yield records and the total number of variance parameters estimated in the highest-order model is 485. An approximate solution to the problem of site selection is presented.

We present the results of a small simulation study to show the impact on prediction of ignoring error variance heterogeneity between trials and spatial variation within trials. We also briefly discuss the accommodation of spatial field trend in relation to the detection of quantitative trait loci (QTL) and marker-assisted selection (MAS).

Software

All analyses in this chapter were conducted using the software program ASREML (Gilmour et al., 1999), which is a FORTRAN program for mixed-model estimation. A wide range of models can be fitted. ASREML uses sparse matrix methods and the average information algorithm (Gilmour et al., 1995) for residual maximum likelihood (REML) (Patterson and Thompson, 1971) estimation of variance parameters. As a result, large and complex data sets can be efficiently analysed. Details on the availability of the ASREML program can be found on the web site ftp://ftp.res.bbsrc.ac.uk/pub/aar/

Example 1: NSW Lupins

Here we consider a series of trials from the New South Wales (NSW) Department of Agriculture lupin-breeding programme. In this programme, there are a number of stages of variety testing for yield, commencing with stage 1 (S1), in which a large number (between 100 and 150) of lines (cross-breds and standard commercial varieties) are grown in an unreplicated (grid) trial at a single location. If there is sufficient seed, two replicates will be sown. The most promising lines are selected to be grown in stage 2 trials in the following year. The process continues to the final stage (S4) of testing, in which a small number of élite lines (approximately 30) are grown in a diverse range of environments. Decisions are then made regarding the commercial release of new lines. Selection decisions are based on a number of traits, including disease resistance, quality parameters and yield. In this chapter, we focus on the analysis of yield data.

Description of data

The data set under study comprises all S2, S3 and S4 trials grown in the 2000 season for the narrow-leaf lupin species *Lupinus angustifolius*. A total of 12 trial locations

were used, with the locations for the S2 trials also being sown with an S3 trial. (See Table 22.1, in which the first two characters of the trial acronym specify the stage and the last four characters specify the location.) All locations are in NSW, apart from KATA, which is in Victoria. All trials were designed as randomized complete block (RCB) designs with neighbour balance (Coombes, 1999). Each had three replicates and was laid out as a contiguous rectangular array of plots, with numbers of rows and columns as given in Table 22.1. Replicates generally occupied one or several columns of the trial. A composite data set is used in preference to a separate analysis for each series, since there is commonality between series in terms of the varieties grown. Each S3 trial contains a complete set of S4 lines and all trials contain the set of nine standard commercial varieties. The combined analysis therefore maximizes the information used for selection decisions. The full data set comprised 2022 records on 108 varieties tested in 16 trials, with 674 variety–trial combinations observed out of the possible 1728 (39%).

Table 22.1. Summary statistics for lupin trials.

Trial	Rows	Columns	Varieties	Mean yield (t ha^{-1})	Missing values
S2ARDL	30	6	60	0.98	1
S2BURU	30	6	60	3.04	0
S2COWR	30	6	60	3.40	0
S2WARI	30	6	60	3.27	0
S3ARDL	22	6	44	0.99	0
S3BURU	22	6	44	3.11	0
S3COWR	22	6	44	3.69	2
S3HARD	44	3	44	2.80	5
S3WARI	22	6	44	3.29	0
S4CORO	31	3	31	1.28	0
S4GANM	31	3	31	2.45	0
S4KATA	28	3	28	3.00	0
S4KILR	31	3	31	3.30	0
S4SUNT	31	3	31	2.38	0
S4URAN	31	3	31	2.20	0
S4THUD	31	3	31	3.49	0

Modelling spatial variation

The first step in the analysis is to determine appropriate spatial models for each trial. For this purpose, the variety effects in different trials are regarded as independent. In the context of the model in Chapter 21, Equation 21.6, this is achieved using a genetic variance matrix of the form $G_e = \text{diag}(\sigma_{g_j})$, $j = 1 \ldots 16$, which is analogous to conducting 16 separate analyses. The use of a diagonal model for G_e in the first instance, rather than the more complex factor-analytic model, greatly reduces the computing load. The spatial models chosen can be verified when more realistic and complex forms for G_e are used.

The spatial models for each trial are determined using the approach in Chapter 21, the section on Spatial Analysis of a Field Experiment. The first choice is a separable autoregressive process of order 1 (AR1 × AR1) for local trend for each trial. In terms of the model in Chapter 21, Equation 21.6, there are 48 variance parameters in R, 26 fixed effects in X (to account for missing values, an overall mean, trial effects and measured covariates for two trials) and the term u_0 is omitted. After fitting this model, diagnostics – in particular, the sample variogram – provided evidence of global trend and extraneous variation (both systematic and random) associated with rows and

columns in some of the trials. For example, the variogram for the trial S2COWR shows evidence of non-stationarity in the column direction (Fig. 22.1a), with the variogram failing to reach a plateau. This was accommodated in the model by adding a fixed effect for the linear regression on column number for this trial. There is also evidence of random row effects, with lower variogram ordinates at zero row displacement compared with other row displacements. This was accommodated by adding random effects for the row factor for this trial.

After the examination of residual plots and sample variograms for all trials, the final model included six variance parameters in G_0 and an extra nine fixed effects in X to account for global and extraneous variation. Local trend was modelled using the AR1 × AR1 process for 12 of the trials, whereas the remaining four trials required a one-dimensional model for rows only (see Table 22.2). The sample variograms from these models showed no further evidence of global or extraneous variation and were much closer to the theoretical autoregressive forms (see Fig. 22.1b, for example).

For comparative purposes, an RCB analysis was conducted for each trial. The gain of spatial over RCB analysis is likely to be greatest for large trials with many varieties and few replicates. The trials here are relatively small. Despite this, the

(a)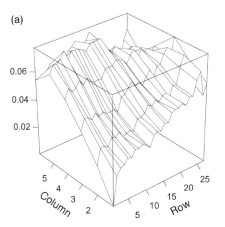

(b)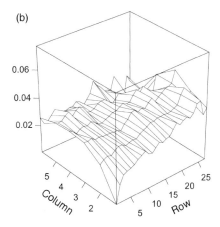

Fig. 22.1. Sample variograms for (a) initial model: AR1 × AR1; and (b) final model: AR1 × AR1 + lin(col) + ran(row) for lupin trial S2COWR. The x and y ordinates are displacements in the row and column directions, respectively, measured as differences in row/column numbers.

accommodation of spatial variation has resulted in reductions in effective error variance for all trials (Table 22.2).

Modelling variety effects

Factor-analytic forms for G_e were considered, maintaining the spatial models determined using the diagonal form for G_e (the spatial parameters being re-estimated, however). Table 22.3 contains the REML log-likelihoods and likelihood ratio tests (REMLRT) for successive models. The standard variety (V) main effects and V × E interaction model was fitted for comparative purposes. The FA(1) model provides a substantially better fit ($P < 0.001$). The V + FA(1) model was superior ($P < 0.001$) to the FA(1) but inferior to the FA(2) model ($P < 0.001$). In the V + FA(1) model, the estimate of the specific variance for trial 1 was on the boundary, that is, it was estimated as zero. In the FA(2) model, the estimates of the specific variances for four trials were on the boundary. A zero specific variance for a trial means that the variety effects for this trial are completely determined by the

Table 22.2. Spatial models for lupin trials.

Trial	Local trend model (row × col)	Extraneous variation*		Effective error mean square (ems)	RCB/spatial ems (%)
		Fixed	Random		
S2ARDL	AR1 × ID	lin(col)	ran(col)	0.0079	175
S2BURU	AR1 × ID	lin(col)	ran(row)	0.0241	127
S2COWR	AR1 × AR1	lin(col)	ran(row)	0.0309	162
S2WARI	AR1 × AR1	col2		0.0426	134
S3ARDL	AR1 × AR1	lin(col)		0.0084	174
S3BURU	AR1 × ID	lin(col)		0.0271	107
S3COWR	AR1 × AR1			0.0441	119
S3HARD	AR1 × AR1	lin(col), lin(row)		0.1155	290
S3WARI	AR1 × AR1			0.0483	102
S4CORO	AR1 × AR1		ran(col)	0.0189	159
S4GANM	AR1 × AR1	lin(col)	ran(col)	0.0553	114
S4KATA	AR1 × ID			0.2146	110
S4KILR	AR1 × AR1		ran(col)	0.1133	114
S4SUNT	AR1 × AR1			0.0434	106
S4URAN	AR1 × AR1			0.0421	124
S4THUD	AR1 × AR1			0.0344	265

ID, identity matrix.
*lin(col) represents fixed linear regression on column number; col2 represents 2 level fixed factor for cyclic column effects; ran(col) and ran(row) represent random column and row factors.

Table 22.3. REML log-likelihoods, ℓ_R, and REMLRT for the models fitted to the lupin data (spatial error structures fitted unless otherwise indicated).

Model for G_e	Variance parameters		ℓ_R	REMLRT (P value)	% Variance accounted
	G_e	Total			
Uniform (V + V × E)	2	52	1702.39		
FA(1)	32	82	1801.13	197.48 ($P < 0.001$)	48
V + FA(1)	33	83	1824.11	45.96 ($P < 0.001$)	54
FA(2)	47	97	1844.49	40.76 ($P < 0.001$)	77
FA(2), RCB*	47	49	1452.92		

*RCB error structure with common block and error variance for all trials; includes fixed trial effects, as for spatial analyses.

multiplicative part of the model. If more than one trial has a zero specific variance, this means that the genetic variance structure G_e has less than full rank. In our experience, this is a common occurrence. To estimate such a model, a special algorithm is required (see Thompson *et al.*, 2001). This has been implemented in ASREML (Gilmour *et al.*, 1999).

The final column in Table 22.3 gives the percentage of genetic variance accounted for by the multiplicative part of the model. This is calculated as the trace of the matrix $\Lambda\Lambda'$ divided by the trace of $(\Lambda\Lambda' + \Psi)$. The FA(2) model accounts for 77% of the genetic variance, which is sufficient for the purposes of the analysis, namely, variety selection. For this reason, higher-order models, such as V + FA(2) and FA(3), were not fitted.

The loadings (after the rotation described in Chapter 21, the section on the Spatial Mixed Model for MET Data) are displayed in Fig. 22.2. Since the FA(2) model provided a reasonable fit for the data, the cosine of the angle between vectors approximates the genetic correlation between the

pair of environments. Note that, for the location where both an S2 and an S3 trial were sown, the two trials are highly correlated, the exception being ARDL (also see Table 22.4). The trial in Victoria (S4KATA) is at one end of the spectrum in terms of genetic correlations.

Elements of the estimated genetic variance matrix, G_e, can be obtained from the estimated loadings and specific variances using the formulae following Equation 21.10 in Chapter 21. The portion of the matrix relating to the S2 and S3 lupin trials is shown in Table 22.4. The correlations in this matrix are well displayed in Fig. 22.2, e.g. the strong correlations between trials sown at the same location (correlations underlined in Table 22.4).

The final model in Table 22.3 was fitted with the error structure commonly used in the analysis of V × E data, namely, an RCB structure with common block and error variance for all trials. Such a structure is implicit in the additive main effects and multiplicative interaction (AMMI) (Gauch, 1992) approach, for example. To enable a

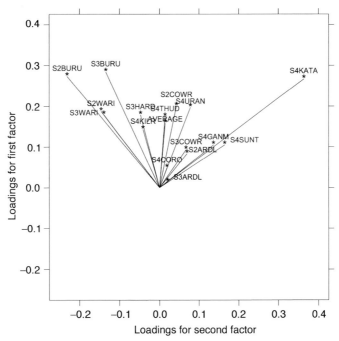

Fig. 22.2. Plot of environment loadings from FA(2) model for lupin data. Cosine of angle between vectors approximates genetic correlation. Centroid of loadings (the 'average' environment) is also marked.

direct comparison of residual log-likelihood with the spatial approach, the fixed effects included in the spatial analysis to account for global and extraneous variation were also fitted in the RCB model. The final two models in Table 22.3 are not nested, so a REMLRT cannot be used for comparison. Instead, we use the Akaike information criterion (AIC) (Akaike, 1974). The smaller the AIC value, the better the fit of the model. The model with spatial errors is far superior with an AIC value of −3494.98 compared with −2807.84 for the RCB model. Figure 22.3 presents a plot of the overall variety means from the two analyses. These were calculated as predictions at the mean values of the loadings and so reflect performance in an 'average' environment in terms of genetic

Table 22.4. Estimated genetic variance matrix for S2 and S3 lupin trials. Upper triangle contains covariances × 100; diagonals are variances × 100 (also in bold type); lower triangle contains correlations.

	S2 trials				S3 trials				
Trial	ARDL	BURU	COWR	WARI	ARDL	BURU	COWR	HARD	WARI
S2ARDL	**1.181**	0.672	1.973	0.541	0.264	1.432	1.246	1.161	0.522
S2BURU	0.172	**13.003**	4.481	8.507	−0.108	10.892	0.985	5.988	8.122
S2COWR	0.760	0.520	**5.705**	3.117	0.372	5.112	2.157	3.377	2.984
S2WARI	0.160	0.755	0.418	**9.757**	−0.049	7.295	0.753	4.042	5.395
S3ARDL	0.234	−0.029	0.150	−0.015	**1.076**	0.111	0.273	0.149	−0.046
S3BURU	0.408	0.936	0.663	0.723	0.033	**10.424**	1.735	5.722	6.970
S3COWR	0.998	0.238	0.785	0.210	0.229	0.468	**1.322**	1.344	0.725
S3HARD	0.425	0.661	0.563	0.515	0.057	0.705	0.465	**6.316**	3.863
S3WARI	0.167	0.785	0.435	0.602	−0.015	0.752	0.220	0.535	**8.241**

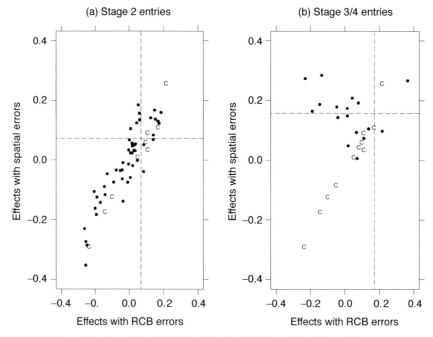

Fig. 22.3. Plot of overall variety effects for lupin trials from FA(2) model with spatial error structures against overall effects from model with RCB, common error structure. Vertical/horizontal lines mark cut-off points for the top 15 varieties in S2 and the top ten varieties in S3/S4. Commercial standard varieties are labelled 'C'.

covariances with other environments (see Chapter 21, the section on The Spatial Mixed Model for MET Data). This environment is indicated in Fig. 22.2. While there is reasonable agreement in overall performance for the varieties in the S3/S4 trials, there is substantial disagreement for the varieties in the S2 trials. Selection pressure for the S2 entries is lower than for S3/S4. If we consider identification of the top 15 lines (25%) in the S2 trials, there would be ten lines common to the two analyses (top right-hand quadrant in Fig. 22.3a). The agreement rate is thus only 67%. If we consider the identification of the top ten lines (18%) in the S3/S4 trials, the agreement rate is much higher, namely, 90%. A contributing factor to the levels of agreement in overall variety performance between the two analyses is the fact that most S3/S4 entries were in 12 trials, whereas S2 entries were in four trials only. With a smaller number of trials, the effect of spatial and error variance heterogeneity on the overall predictions is greater.

Example 2: SA Wheat

Here we consider a series of indicator trials conducted by the University of Adelaide (Roseworthy campus) wheat-breeding programme. The data set comprises 61 trials spanning the years 1994–1998. The trials were situated at 20 locations in the wheat-growing areas of South Australia (SA) and south-eastern Western Australia (WA). A total of 72 varieties were grown.

The purpose of the indicator trials was to obtain information that could be used to select trial locations for early-stage variety testing in the Roseworthy programme. It was important, therefore, to determine the extent of V × E interaction in target environments and to characterize and investigate relationships among trial locations. The 20 locations were thus chosen to be representative of the environments targeted by the Roseworthy wheat-breeding programme. The varieties were chosen to allow expression of V × E and provide a 'bioassay' of the influence of environmental conditions on variety

performance, that is, they were chosen on the basis of their known reactions to certain environmental factors encountered in the target environments.

Description of data

All trials were designed as RCB, with between three and six replicates, although there was sometimes unequal replication of varieties. Each trial was laid out as a contiguous rectangular array of plots with between ten and 25 rows and six and 15 columns. Blocks generally occupied one or several columns of the trial. The number of varieties in each trial varied from 39 to 50. Generally, the same varieties were used for all trials in the same year. A total of 30 varieties occurred in all trials, with 53 varieties occurring in more than ten trials.

A total of 20 locations (see Table 22.5) were used during the 5 years. Only six locations were used every year, whereas five locations were used in 1 year only. In 1996, there were three trials located at Roseworthy. One trial was sown with the normal seeder and at the normal sowing date. The other two varied either the seeder or the sowing date, so additional locations (named Wintersteiger and Roseworthy – Late, respectively) were defined. The number of trials per year varied from eight (1994) to 17 (1998). The trial mean yields varied from 0.48 to 4.92 t ha^{-1} (Table 22.5). Yields were lowest in 1994, whereas 1996 and 1997 were quite high-yielding years, particularly for trials sown in WA. The final data set comprised 11,592 records on 72 varieties tested in 61 trials, with 2781 variety–trial combinations observed out of the possible 4392 (63%).

Modelling spatial variation

The initial model fitted was as for the NSW lupins example, namely, with a diagonal form for G_e and an AR1 × AR1 process for local trend for each of the 61 trials. There were 183 variance parameters in R and 98 fixed effects in X (to account for missing values, an overall mean and trial effects). The use of diagnostics resulted in the addition of terms to the model to accommodate non-stationarity and extraneous variation. The final model included 64 variance parameters in G_0 and an extra 83 fixed effects in X. Local trend was modelled using the AR1 × AR1 process for 59 of the trials, whereas the remaining two trials required a one-dimensional model for rows only.

Table 22.5.　Summary of SA wheat trial mean yeilds (t ha⁻¹) for each location and year.

Location	1994	1995	1996	1997	1998	Overall*
Buckleboo (BUCB)					0.76	0.73
Buckley (BUCK)					3.17	3.14
Coomalbidgup, WA (COOM)			2.99	2.68	3.10	2.72
Coonalpyn (COON)		2.22		1.20	1.11	1.35
Fisher (FISH)	1.71	1.69	1.59	2.43	2.80	2.05
Hunt (HUNT)	1.79					2.52
Kapunda (KAPU)	1.51	3.73	2.84	2.60	3.28	2.79
Lake King, WA (LAKE)				3.01	2.33	2.49
Loxton (LOXT)		1.43	1.55	2.02	2.33	1.65
Minnipa (MINN)	0.51	0.54	0.88	0.76	1.07	0.75
Nelshaby (NELS)		2.28	1.83		1.39	1.70
Palmer (PALM)	1.43	1.46	1.62		1.19	1.51
Roseworthy (ROSE)	1.38	3.20	4.92	3.42	2.77	3.14
Roseworthy – Late (ROSL)			2.40			2.16
Stow (STOW)	1.96	2.43	2.70	2.69	3.07	2.57
Tuckey (TUCK)	0.54	1.21	1.81	2.82	0.48	1.37
Wilgoyne, WA (WILG)				1.74	2.16	1.78
Wintersteiger (WINT)			4.46			4.22
Wittenoom Hills, WA (WITT)			3.02	3.40	2.62	2.81
Yeelanna (YEEL)				3.22	1.64	2.25
Overall*	1.45	2.32	2.43	2.52	2.21	2.18

*Overall mean obtained from linear model with trial mean yields as data and year and location as explanatory factors.

Table 22.6.　Spatial models for wheat trials in 1996. Local trend modelled using AR1 × AR1 for all trials.

Trial	Number of			Extraneous variation*		Effective error mean square	RCB/spatial ems (%)
	Rows	Cols	Reps	Fixed	Random		
COOM	14	12	4	lin(row)	ran(col)	0.0585	360
FISH	14	15	5	lin(row)	ran(col)	0.1050	170
KAPU	21	12	6		ran(col)	0.0498	241
LOXT	22	12	6		ran(row)	0.0365	181
MINN	14	15	5		ran(row), ran(col)	0.0088	140
NELS	21	12	6	lin(row)	ran(col)	0.0209	140
PALM	21	12	6	lin(row)	ran(col)	0.0301	216
ROSL	21	12	6	lin(col)		0.0683	154
ROSE	21	12	6		ran(row), ran(col)	0.1282	151
STOW	21	12	6	lin(row)	ran(col)	0.0915	138
TUCK	14	15	5	lin(row)	ran(col)	0.0234	176
WINT	21	12	6	lin(row), lin(col)		0.2231	211
WITT	14	12	4	lin(row)	ran(col)	0.0549	173

*lin(col) and lin(row) represent fixed linear regressions on column/row number; ran(col) and ran(row) represent random column and row factors.

A summary of the spatial models fitted to the 13 trials conducted in 1996 is contained in Table 22.6. The local trend model for all these trials was AR1 × AR1. Individual trials were also analysed as RCB. The reductions in effective error mean square for the spatial analyses compared to the RCB analyses are substantial in most cases.

Modelling variety effects

Factor-analytic models for $\mathbf{G}_e$ were then considered. Since the main aim of this analysis is to investigate the genetic correlation structure, the sequence of models fitted (FA(1), FA(2) and FA(3)) did not include variety main-effect models, since these are intermediate. The best fitting of the models was the FA(3) (Table 22.7), which explained 71% of the genetic variance.

In an attempt to answer questions about site selection, we may examine plots of the (rotated) environment loadings. The aim is to choose a site suitable for selecting varieties with broad adaptation. In the absence of information about the frequency of target environments (see also the section Discussion below), we may consider sites that are consistently near average; that is, sites with loadings near to the mean values of the loadings for individual years. Figure 22.4 shows the loadings for the sites in 1996, plotted as the first against the second loading and the first against the third. The position of the environment with average loadings is also marked. Clearly, the trial at Fisher (FISH) was closest to average in this year. As a summary of the full set of plots of loadings, the Euclidean distance between individual sites and the average site for each year was calculated. The resultant ranking of sites in

Table 22.7. REML log-likelihoods, ℓ_R, and REMLRT for the models fitted to the wheat data.

Model for $\mathbf{G}_e$	Variance parameter		ℓ_R	REMLRT (*P* value)	% Variance accounted
	$\mathbf{G}_e$	Total			
FA(1)	122	367	10630.4		55
FA(2)	182	427	10858.4	456.0 (*P* < 0.001)	65
FA(3)	241	486	10961.1	205.4 (*P* < 0.001)	71

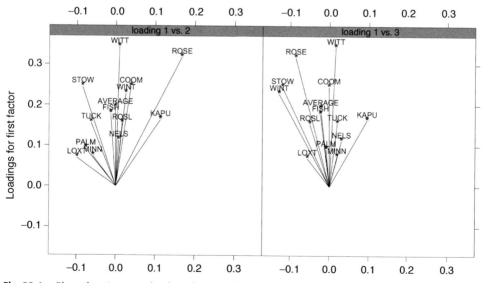

Fig. 22.4. Plots of environment loadings for 1996 from FA(3) model for wheat data. Centroid of loadings (the 'AVERAGE' environment) is also marked.

terms of proximity to the average is given in Table 22.8. This could be used as a guide for choosing a site.

We may also use the graphical display of loadings to help identify sites for selecting varieties with specific adaptation. In this case, we would investigate sites that deviate from the average. We must be certain, however, that the $V \times E$ interaction thus exhibited is repeatable (see, for example, Cooper et al., 1996).

Simulation Study

A simulation study was conducted to investigate the impact on variety predictions from a factor-analytic model of ignoring spatial variation within trials and error variance heterogeneity between trials. The simulation was based on the example data set used by Smith et al. (2001a) comprising 172 barley varieties tested in seven trials. Each trial was designed as an RCB and laid out as 43 rows by 12 columns with each block of four columns constituting a complete replicate. The simulations were

based on a model with (fixed) environment and (random) variety main effects and an FA(1) structure for the (random) $V \times E$ interactions. When fitted to the data in conjunction with a separate spatial covariance structure for the errors for each trial, namely, $AR1 \times AR1$, the estimated variety main-effect variance component was $\sigma_v^2 = 0.00292$. The estimated variance parameters associated with the factor-analytic structure, that is, loadings and specific variances, were as in Table 22.9. The estimated spatial parameters (row and column autocorrelations and spatial variance) were as in Table 22.10.

The genetic parameters in Table 22.9 and spatial parameters in Table 22.10 were used to generate 100 data sets, each comprising three replicates of 172 varieties in each of seven trials. The spatial configuration of rows, columns and replicates was as for the real data.

Three models were fitted to the data sets. The genetic model was the same in all cases. At the error level, the models were:

- Model 1: separate $AR1 \times AR1$ structure for each trial (implicitly includes a separate error variance for each trial).

Table 22.8. Site rankings for wheat data in terms of proximity to average site in each year (a rank of 1 is closest to average).

Site	1994	1995	1996	1997	1998
BUCB					13
BUCK					16
COOM			3	9	17
COON		1		5	2
FISH	3	4	1	8	12
HUNT	6				
KAPU	4	10	12	1	8
LAKE				4	14
LOXT		8	10	6	3
MINN	5	7	8	11	6
NELS		2	5		4
PALM	7	6	7		11
ROSE	1	5	13	7	10
ROSL			2		
STOW	2	9	9	3	7
TUCK	8	3	4	12	15
WILG				10	1
WINT			6		
WITT			11	13	9
YEEL				2	5

Table 22.9. Estimated factor-analytic parameters for $V \times E$ interactions from analysis of barley data.

Trial	λ_i	ψ_i
1	−0.062	0.00221
2	0.012	0.00244
3	−0.132	0.00141
4	−0.005	0.00279
5	−0.068	0.00267
6	−0.131	0.00380
7	−0.074	0.00477

Table 22.10. Estimated spatial parameters from analysis of barley data.

Trial	ρ_{col}	ρ_{row}	σ^2
1	0.072	0.549	0.0092
2	0.704	0.598	0.0134
3	0.096	0.604	0.0248
4	0.172	0.686	0.0200
5	0.183	0.477	0.0252
6	0.082	0.490	0.0279
7	0.063	0.653	0.0166

- Model 2: fixed block effects for each trial, common error variance for all trials.
- Model 3: fixed block effects for each trial, separate error variance for each trial.

The average per cent bias in estimates of the loadings and specific variances over the 100 simulations are shown in Figs 22.5 and 22.6. The average per cent bias for both the loadings and specific variances are similar (and small) for the models in which different error variances are allowed. The bias, when an RCB model with common error is assumed, is very high for most of the specific variances.

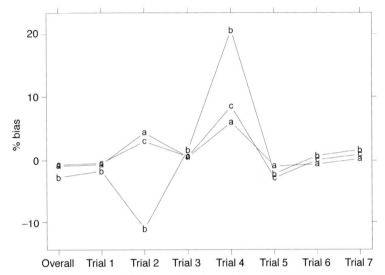

Fig. 22.5. Per cent bias in estimates of loadings for the V + FA(1) model with three error models: (a) spatial, (b) RCB same error and (c) RCB different error. The overall loading is the square root of the variety main effect variance.

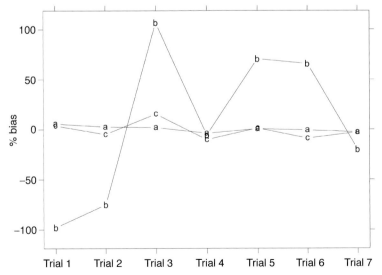

Fig. 22.6. Per cent bias in estimates of specific variances for V + FA(1) model with three error models: (a) spatial, (b) RCB same error and (c) RCB different error.

The models were compared in terms of the mean squared error of prediction (MSEP) for the overall variety effects and the variety effects for each trial. The overall effects were calculated as predictions at the mean values of the loadings (see Chapter 21, the section on The Spatial Mixed Model for MET Data). For ease of interpretation, the MSEP for variety effects for each trial have been expressed relative to the (true) genetic variance for the trial. For trial i the latter is obtained as:

$$\sigma_{gi}^2 = \sigma_v^2 + \lambda_i^2 + \psi_i$$

where the values for the parameters are taken from Table 22.9. The MSEP for overall variety effects have been expressed relative to the overall (true) genetic variance, which is obtained as:

$$\sigma_g^2 = \sigma_v^2 + \bar{\lambda}^2$$

where $\bar{\lambda}$ is the mean of the λ_i values.

The averages of these (relative) MSEP values over 100 simulations are shown in Fig. 22.7. Figure 22.8 shows the average correlation between true and predicted effects. The MSEP and correlations are consistently best for the spatial model. With respect to individual trials, the superiority of this model over the RCB models is greatest for the trials with the strongest spatial trend (trials 2

and 4). The spatial model is also superior in terms of the overall variety predictions.

We also generated 100 RCB data sets with fixed block effects and common error variance for all trials. For these data, there was no loss associated with fitting the spatial, heterogeneous error model (A.B. Smith, B.R. Cullis and R. Thompson, unpublished results). The per cent bias in estimated genetic variance parameters and the MSEP of prediction were very similar for the spatial, heterogeneous error model and the RCB, common error model.

Adjusting for Spatial Field Trend in QTL Detection and MAS

Another application in which the accommodation of spatial field trend is important is the detection of quantitative trait loci (QTL) and marker-assisted selection (MAS). Moreau et al. (1999) comment that statistical design and analysis issues are often neglected in QTL detection experiments. They conducted analyses for MAS for grain yield in maize and showed that when 'spatial field heterogeneity is considered through appropriate statistical models the accuracy of genetic value predictions is improved and the same genetic gain

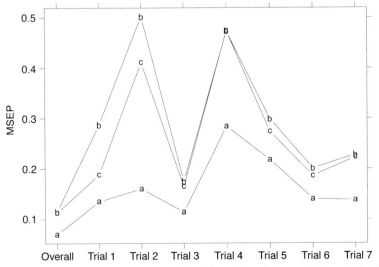

Fig. 22.7. (Relative) MSEP for variety effects (overall and for each trial) for V + FA(1) model with three error models: (a) spatial, (b) RCB same error and (c) RCB different error.

can be achieved with a reduced number of trials'. Standard QTL software for interval mapping does not allow for such models. Moreau *et al.* (1999) suggest that a two-stage approach could be used in which spatially adjusted means are obtained from an analysis of the raw plot yields and are then used as data for the standard mapping software. They comment that an integrated, one-step approach would be preferable. Such an approach is presented in Eckermann *et al.* (2001), who use a mixed model for QTL detection based on pairwise regressions on marker covariates, this being done simultaneously with the modelling of field trend using the spatial approach of Gilmour *et al.* (1997). Smith *et al.* (2001b) extend this approach for quality trait data that are characterized by two potential sources of error variation, namely, the field experiment and the measurement process in the laboratory. They conduct QTL detection for milling yield in wheat in several doubled haploid populations. In all cases, the accommodation of spatial field trend and/or variation in the laboratory had a large impact on the detection of QTL compared with the standard approach, in which raw doubled haploid means were used as data in a standard mapping programme. In the latter,

the significance of QTL effects was often overestimated. More importantly, an anomalous (apparently highly significant) QTL was detected in the standard approach. The spatial approach revealed that the effect was due purely to extraneous variation in the laboratory process. This highlighted the importance of not only adjusting for field and laboratory trend in the analysis but also accommodating these trends at the design stage, allocating varieties to field plots and 'positions' in the laboratory process in some optimal way.

Discussion

The analyses of the examples in this chapter clearly demonstrate the strengths of the spatial multiplicative mixed model of Smith *et al.* (2001a) for the analysis of V × E data. The models provided a good fit to the data, in terms of both V × E effects and error variation. Important information about overall variety performance and the nature of V × E interaction was obtained.

The aim of the analysis of the lupin data was to provide information for the purposes of identifying the highest-yielding lines for promotion to the next stage of

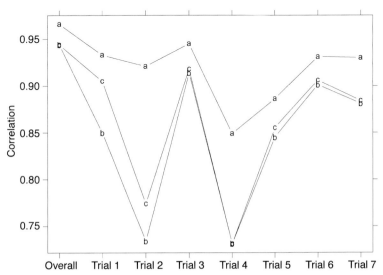

Fig. 22.8. Average correlation between true and predicted variety effects (overall and for each trial) for V + FA(1) model with three error models: (a) spatial, (b) RCB same error and (c) RCB different error.

testing. Factor-analytic models with spatial adjustments were used to obtain superior estimates of the variety performance in individual trials and important information about the genetic correlation structure for environments. To make selections based on yield, we require an index that combines variety means across trials in some meaningful way. We may choose to weight trials based on the importance of the target population of environments they represent (see, for example, Cooper and Byth, 1996). Devising an appropriate weighting scheme is a difficult issue. Long-term genetic correlations between test and target environments would be useful and could be obtained through the use of factor-analytic models on appropriate data sets. In the absence of such a scheme for the lupin data, we calculated predictions of variety performance at the mean values of the loadings, that is, performance in an 'average' environment in terms of genetic covariances with other environments. Overall variety performance calculated in this way is the current yield-selection criterion used in most NSW Department of Agriculture breeding programmes. Modifications may be made if the estimated genetic correlation matrix indicates unusual trials.

The aim of the analysis of the wheat data was to determine relationships among geographical locations (across years) to identify locations suitable for early-stage yield testing. In the earliest stage, varietal selection is for broad adaptation, so the location must consistently reflect average performance across the range of target environments. An approximate solution to the location selection problem was presented and was based on the loadings from a factor-analytic model on the two-way table of $V \times E$ effects. For a more exact solution, we must take account of the fact that environments are a factorial combination of locations (L) and years (Y). Thus, we would like to partition $V \times E$ interaction as $V \times L$, $V \times Y$ and $V \times L \times Y$. Technically, this raises no difficulties, but conceptually, there are issues regarding the type and compatability of structures fitted to the individual sources of $V \times E$.

In both examples, the gains in accuracy and precision of spatial analysis compared with RCB analysis of individual trials were clearly shown. We used the spatial approach of Gilmour *et al.* (1997), but other spatial approaches or analyses based on incomplete block designs will be superior to the RCB analysis and can be incorporated in the mixed model for $V \times E$ data. In terms of the combined analysis, the lupin example showed the impact on varietal selection of using spatial models and allowing a separate error variance for each trial compared with RCB models with common error variance. Despite the low levels of spatial trend in the lupin data, selection decisions for the S2 entries were very different under the two analyses, whereas for the S3/S4 entries they were almost identical. The key factor here was that the S2 entries were grown in fewer trials, so the effect of spatial and error variance heterogeneity on the overall predictions was greater. This has important consequences for the analysis of early-generation breeding data, which typically involve a large number of varieties but a small number of trials (possibly only two or three).

The simulation study provided additional evidence about the impact of ignoring error variance heterogeneity and spatial variation on variety predictions from a factor-analytic model. The spatial, heterogeneous error model had a lower mean squared error of prediction and higher correlation between true and predicted effects than the RCB models (common or heterogeneous error), both for variety effects for individual trials and variety effects overall. The impact of ignoring spatial and error variance heterogeneity is clearly related to the degree of heterogeneity in the data set under study. The level of error variance heterogeneity and strength of spatial correlation in the simulated data is low by Australian standards. The impact was non-ignorable, however. Additionally, the realized line mean heritability in the simulated data was high and it is expected that, with lower heritabilities, the choice of error model will have a larger influence on the predictions.

Acknowledgements

We thank Arthur Gilmour for implementation of the methodology into ASREML. We gratefully acknowledge the financial support of the Grains Research and Development Corporation of Australia. The BBSRC Institute for Arable Crops receives grant-aided support from the Biotechnology and Biological Sciences Research Council.

References

Akaike, H. (1974) A new look at statistical model identification. *IEEE Transactions on Automatic Control* 19, 716–722.

Coombes, N.E. (1999) *SpaDes, a Spatial Design Generator*. Biometric Bulletin, NSW Agriculture, Wagga Wagga, New South Wales.

Cooper, M. and Byth, D.E. (1996) Understanding plant adaptation to achieve systematic applied crop improvement – a fundamental challenge. In: Cooper, M. and Hammer, G.L. (eds) *Plant Adaptation and Crop Improvement*. CAB International, Wallingford, UK, pp. 5–24.

Cooper, M., Brennan, P.S. and Sheppard, J.A. (1996) A strategy for yield improvement of wheat which accommodates large genotype by environment interactions. In: Cooper, M. and Hammer, G.L. (eds) *Plant Adaptation and Crop Improvement*. CAB International, Wallingford, UK, pp. 487–512.

Eckermann, P.J., Verbyla, A.P., Cullis, B.R. and Thompson, R. (2001) The analysis of quantitative traits in wheat mapping populations. *Australian Journal of Agricultural Research* 52, 1195–1206.

Gauch, H.G., Jr (1992) *Statistical Analysis of Regional Yield Trials: AMMI Analysis of Factorial Designs*. Elsevier, Amsterdam.

Gilmour, A.R., Thompson, R. and Cullis, B.R. (1995) AI, an efficient algorithm for REML estimation in linear mixed models. *Biometrics* 51, 1440–1450.

Gilmour, A.R., Cullis, B.R. and Verbyla, A.P. (1997) Accounting for natural and extraneous variation in the analysis of field experiments. *Journal of Agricultural, Biological, and Environmental Statistics* 2, 269–273.

Gilmour, A.R., Cullis, B.R., Welham, S.J. and Thompson, R. (1999) *ASREML Reference Manual*. Biometric Bulletin No. 3, NSW Agriculture, Orange, New South Wales.

Mardia, K.V., Kent, J.T. and Bibby, J.M. (1988) *Multivariate Analysis*. Academic Press, London.

Moreau, L., Monod, H., Charcosset, A. and Gallais, A. (1999) Marker-assisted selection with spatial analysis of unreplicated field trials. *Theoretical and Applied Genetics* 98, 234–242.

Patterson, H.D. and Thompson, R. (1971) Recovery of interblock information when block sizes are unequal. *Biometrika* 31, 100–109.

Smith, A.B., Cullis, B.R. and Thompson, R. (2001a) Analyzing variety by environment data using multiplicative mixed models and adjustments for spatial field trend. *Biometrics* 57, 1138–1147.

Smith, A.B., Cullis, B.R., Appels, R., Campbell, A., Cornish, G., Martin, D. and Allen, H. (2001b) The statistical analysis of quality traits in plant improvement programs with application to the mapping of milling yield in wheat. *Australian Journal of Agricultural Research* 52, 1207–1219.

Thompson, R., Cullis, B.R., Smith, A.B. and Gilmour, A.R. (2002) An efficient algorithm for fitting factor analytic and reduced rank models in complex experiments. *Australian and New Zealand Journal of Statistics* (in press).

23 Applications of Mixed Models in Plant Breeding

Mónica Balzarini

Faculdad Cs. Agropecuarias, Estadistica y Biometria, Universidad Nacional de Córdoba, Av. Valparaiso s/n, 5000 Córdoba, Argentina

Introduction

Statistical modelling is based on the specification of expected values and variance–covariance structures of observed data. The traditional fixed linear model, coupled with ordinary least-squares estimation procedures, is too restrictive to perform satisfactory data analyses for the typical data structure of most breeding programmes because of the independence assumption. Error structure in 'real-world' experiments is often more complex than that used in standard linear models for conventional data analysis (Stroup, 1989).

In contrast, the general linear mixed model can easily accommodate covariances among observations. The mixed model handles correlated data by incorporating random effects and estimating their associated variance components to model variability over and above the residual error (Wolfinger and Tobias, 1998). Because of the estimation procedures usually involved, mixed-model approaches can circumvent the problems associated with unbalanced and incomplete data.

Mixed-model analysis applies particularly to research involving factors with a few levels that usually can be controlled by the researcher (fixed) as well as factors with levels that are beyond the researcher's control (random). These random factors vary from experiment to experiment and may be interpreted in the context of a symmetric probability function. Most breeding trials have some mixed-model aspect. At crossing, genetic effects may be reasonably assumed as normal random variables (Henderson, 1990). During the early stages of a selection programme, the nature of genotypic effects may still be regarded as random. At later selection stages, genotypes might be assumed to be fixed, but environmental and/or genotype–environment interaction (GEI) effects may be considered as random variables, since they represent a larger target population (Stroup, 2000).

Regular mixed-model applications in plant breeding have focused on variance component estimation and identification of appropriate error terms to test fixed-effect hypotheses; rarely have they been used for the most general purpose of modelling the underlying covariance structure and predicting random effects. This chapter discusses applications of mixed-model theory to predict cross performance and to analyse multi-environmental trials (METs).

Background: a Mixed-model Framework

The general form of a linear mixed model is:

$$Y = X\beta + Zu + e$$

where $\mathbf{Y}$ is the response vector (data), $\mathbf{X}$ and $\mathbf{Z}$ are known design matrices, $\boldsymbol{\beta}$ is a vector of fixed parameters, and $\mathbf{u}$ (random effects) and $\mathbf{e}$ (error terms) are unobservable random vectors. The $E(\mathbf{u})$ and $E(\mathbf{e})$ are usually assumed to be zero. Assumptions regarding the structure of $\mathbf{G}$ – the variance–covariance matrix of the random effects in $\mathbf{u}$ – and $\mathbf{R}$ – the variance–covariance matrix of the random error terms in $\mathbf{e}$ – will define a particular mixed model. Different models for the variance–covariance of the data, $\mathbf{V} = \mathbf{ZGZ'} + \mathbf{R}$, are obtained by specifying the structure of $\mathbf{Z}$, $\mathbf{G}$ and $\mathbf{R}$.

The simplest form for $\mathbf{G}$ and $\mathbf{R}$ is one that arises from the independence and constant variances of the random effects and the error terms. Independence in the random effects does not imply that the observations are independent. On the contrary, one sets up a common correlation among all observations having the same level of $\mathbf{u}$. Laird and Ware (1982) consider the unstructured model for a covariance matrix, i.e. the more general case where all elements of the matrix are allowed to be different. Intermediate structures for $\mathbf{G}$ and $\mathbf{R}$ are more efficient in plant breeding. They allow for modelling correlations with a smaller number of co-variance parameters than the unstructured one. In general, genetic correlations may be introduced into the model trough $\mathbf{G}$ and experimental correlations among observations may be modelled by the off-diagonal elements of $\mathbf{R}$. When data are indexed in space (georeferenced data), the covariances in $\mathbf{R}$ may reflect correlations due to spatial arrangement of the experimental units.

Mixed-model solutions can be written as:

$$\hat{\boldsymbol{\beta}} = \left(\mathbf{X'V^{-1}X}\right)^{-1}\mathbf{X'V^{-1}y}$$

$$\hat{\mathbf{u}} = \mathbf{GZ'V^{-1}}\left(\mathbf{y} - \mathbf{X}\hat{\boldsymbol{\beta}}\right)$$

If $\mathbf{G}$, $\mathbf{R}$ and $\mathbf{Z}$, and hence $\mathbf{V}$ are known, the generalized least-squares solution for $\boldsymbol{\beta}$ is the best linear unbiased estimator (BLUE) and the solution for the prediction of random effects is the best linear unbiased predictor (BLUP). The BLUP, as a technique for predicting random effects (Harville, 1990; Robinson, 1991), should be understood as a subject-specific mixed-model prediction. It represents the conditional expectation of the random effects, given the observed data, and is also a Bayesian estimator under normal priors. The BLUP of a linear combination of fixed and random effects is the linear combination of the BLUE of fixed effects and the BLUP of random effects.

Theoretically, BLUPs have the smallest mean squared error of prediction among all linear unbiased predictors, provided the assumed model holds and the parameters of the model are known (Searle *et al.*, 1992). In practice, however, $\mathbf{V}$ is usually unknown. Therefore, estimation of covariance parameters usually comes prior to estimation of $\boldsymbol{\beta}$ and $\mathbf{u}$. Assuming normality, the restricted maximum likelihood method (REML) (Patterson and Thompson, 1971) or related versions are usually preferred for estimating the variance components in a mixed model. To indicate that $\mathbf{G}$ and $\mathbf{R}$ have been estimated prior to getting the BLUPs, the term 'empirical BLUPs' (EBLUPs) is frequently used instead of BLUPs to refer to $\hat{\mathbf{u}}$.

Consider a simple model with one random effects vector, $\mathbf{u}$, representing genotypic effects and the response vector, $\mathbf{Y}$, containing phenotypic data for $j = 1, \ldots, g$ genotypes. The prediction equation for the mean performance of genotype j, $\mu_j = \mu + u_j$, under a mixed model is $\hat{\mu}_j = \hat{\mu} + w\left(y_j - \mu\right)$, where μ is the population mean and w is a weighting or shrinkage factor.

If $\mathbf{G}$ and $\mathbf{R}$ have the simplest structures ($\mathbf{G} = \sigma_u^2\mathbf{I}$, $\mathbf{R} = \sigma_e^2\mathbf{I}$), the elements of $\mathbf{GZ'V^{-1}}$, which define the weights w, are functions of $\sigma_u^2 / \left(\sigma_u^2 + \sigma_e^2\right)$. Thus, the weights may represent the heritability of the trait. Thus, a BLUP is a centred, fixed-effects estimate shrunken towards μ, with more shrinkage taking place for smaller value ratios of the estimated variance components in w. Alternative models for $\mathbf{G}$ and $\mathbf{R}$ lead to different BLUPs. Therefore, the term BLUP is quite general and a precise identification of the underlying model is needed to avoid confusion.

To do model selection, the log-likelihood ratio test criterion can be used with nested mixed models. The procedure demands the evaluation of the restricted log-likelihood

(LL$_R$) for the reduced model (model with smaller number of parameters) and for the full model (model with higher number of parameters). The test criterion for the likelihood ratio test is:

$$L = -2\{LL_R \text{ (reduced model)} - LL_R \text{ (full model)}\}$$

For the null hypothesis that the reduced model is not different from the full model, under normality, the likelihood ratio statistic is distributed as χ^2, with degrees of freedom equal to the difference in the number of parameters of both models. If the fixed part of the two mixed models under comparison is the same, the test compares the covariance structures of the two models. Information criteria, such as the Akaike's information criterion (AIC) (Wolfinger, 1993), are used to compare any set of mixed models; thus they are suitable for non-nested models.

Mixed Models for Cross Prediction

The ultimate goal of plant breeding is to generate productive cultivars improved for one or more traits. The breeding process begins with the selection of parents that possess the desired attributes. The choice of parents and hybrid combinations affect the quality of the progeny. Parents are typically derived from advanced stages of selection or they are recognized commercial lines or cultivars. Selecting superior genotypes in the early generations might be highly ineffective for several crops (Gopal *et al.*, 1992). Several researchers have demonstrated the gain in efficiency of selection by using cross prediction trials or progeny tests for family selection (Simmonds, 1996). Progeny tests are commonly employed at the beginning of each breeding cycle in clonally propagated species (DeSousa-Vieira and Milligan, 1999). In hybrid crops, the advanced material for further testing or commercialization depends on which inbreds to cross and the development of new inbreds depends on F$_2$ × tester crosses. However, only a few of all potential single crosses are actually made and evaluated in progeny tests, since the amount of resources needed for testing all possible crosses is large.

Regularly, the performance of a cross combination is predicted by calculating mid-parent values (MPVs) of the raw or scaled parental mean. However, it is important to note that parental generations are rarely discrete. They commonly originate from different selection and crossing series (for example, in sugarcane (*Saccharum* spp.)). Because the genotypes that represent potential parents often derive from different stages of selection, the amount and precision of data may vary dramatically. Improved estimates of parental means for a trait are often obtained with some form of additive linear model. Such models adjust observed values for non-genetic effects (Panter and Allen, 1995). A classic method of obtaining parental genetic effects is by combining data across progeny tests and considering all effects in the model as fixed (White *et al.*, 1986). Unfortunately, cross appraisal databases are typically incomplete and unbalanced, which creates theoretical concerns about the fixed linear model underlying the MPV prediction (Henderson, 1973). Mixed models provide alternative analytical approaches that may overcome limitations of the fixed analytical approach for cross prediction (Henderson, 1974).

BLUP of genetic merit has been used for selecting tested material in plant breeding (Bridges, 1989; Chang and Milligan, 1992). Mixed-model-based prediction, i.e. BLUP, has also been proposed for predicting the performance of untested crosses in maize (*Zea mays* L.) (Bernardo, 1995, 1996), soybean (*Glycine max* L. Merr.) (Panter and Allen, 1995) and sugarcane (Balzarini, 2000). In maize, validation results have indicated that BLUP is useful, not only for a routine selection of single crosses but also for choosing F$_2$ populations for inbred development (Bernardo, 1999).

The mixed-model prediction of untested crosses relies on performance data of tested crosses and the genetic relationships among tested and untested crosses. BLUP uses data about related individuals to yield information for a target individual prediction (an untested cross). If data are not available for the target cross but are available for two crosses genetically related to the target cross, the predicted performance of the target cross is obtained by a linear combination of tested cross data weighted by the degree of relatedness of these two crosses and the target, as well as by the heritability of the trait.

To quantify the degree of relatedness among individuals, coefficients of coancestry might be calculated from pedigree data (Falconer, 1989). The infinitesimal model commonly used in traditional breeding does not use the knowledge about location and number of genes that determine a trait. Molecular genetics provides new strategies for modelling genetic covariances. It is possible to analyse progeny test data knowing the parent genotypes and their genetic relationship from molecular-marker information. Bernardo (1994) predicted maize single-cross performance using restriction fragment length polymorphism (RFLP) marker data to assess genetic relationships among parental inbreds. At present, it is feasible to study the differences among individuals on the basis of knowledge about their genotype gained from molecular data (Bernardo, 1997). By incorporating a new design matrix related to gene random effects, the mixed-model approach allows deeper understanding of how important traits are inherited. The covariance between two of the random effects can be expressed as a function of their conditional covariance for the given marker genes, i.e. a variance component.

To predict hybrid performance in maize, Bernardo (1996) used a BLUP procedure that requires: (i) available data on all single crosses (heterotic group 1 × heterotic group 2 of parental inbreds) that have been evaluated in yield trials – say, n_t crosses; and (ii) coancestry coefficients among parental inbreds of each group. The linear mixed model used to represent a single-cross performance contains yield-trial fixed effects in β and three vectors of random genetic effects. The random vector $\mathbf{u}_1$ involves general combining-ability effects of parental inbreds in group 1, $\mathbf{u}_2$ involves general combining-ability effects of parental inbreds in group

2 and $\mathbf{u}_3$ is a vector of specific combining-ability effects.

After the component variances (genetic and residual variances) related to each type of random effect have been estimated from the tested cross data, the performance of n_u untested or new crosses is predicted as $\mathbf{Y}_u = \mathbf{CV}^{-1}\mathbf{Y}_t$, where $\mathbf{Y}_u$ is an n_u-dimensional vector containing the predicted performance of untested crosses (BLUPs); $\mathbf{C}$ is an $n_u \times n_t$ matrix of genetic covariances between the untested single crosses and the tested single crosses (obtained from coancestry coefficients or molecular-marker data), $\mathbf{V}$ is the $n_t \times n_t$ phenotypic variance–covariance matrix among the tested crosses and $\mathbf{Y}_u$ is an n_t-dimensional vector containing the mean performance of tested crosses corrected for fixed trial effects, i.e.:

$$\mathbf{Y}_t = (\mathbf{Z}'\mathbf{Z})^{-1}\mathbf{Z}'(\mathbf{Y} - \mathbf{X}\hat{\beta})$$

It is important to note that the cross predictor is not only based on the genetic relationship among single crosses but also involves $\mathbf{V}$, which depends on the underlying mixed model for the tested cross responses. Therefore, various BLUPs can be obtained by modelling $\mathbf{V}$.

Models involving female and male random parental effects and genetic covariances among parents are not a unique strategy for cross predictions. Chang (1996) suggested that using genetic covariances among the parents of sugarcane crosses to modify predictions would not be fruitful. He speculated that parents, having been obtained through a highly selective process, might vitiate the value of such covariances, because the genetic covariances estimated from pedigree analysis assume randomly selected parents. The offspring used in a progeny test are not selected and hence using the genetic covariances among crosses may enhance the predictive value of BLUPs.

Alternative models to predict single-cross performance were simultaneously assessed in sugarcane (Balzarini, 2000). Among others, the traditional MPV (fixed-model prediction) was compared against a predictor derived from a one-way classification model involving a random cross effect to model the genetic portion of the response

(C-BLUP). The additive genetic relationship among crosses was used to set up covariances among tested and untested cross effects. By tracking the pedigree of each cross and assuming parents are not related, covariances among crosses within and between trials are easy to obtain based on the cross parentage. I used coancestry coefficients to establish these genetic relationships. After variance components were obtained from such a mixed model for the tested crosses, BLUP of the untested cross effects was obtained using the expression, $\mathbf{Y}_u = \mathbf{CV}^{-1}\mathbf{Y}_t$. To obtain MPV, female and male parent means, after adjusting for fixed trial effects, were calculated for each clone that was used as a parent in crosses evaluated in cross appraisal trials. For each potential new cross, corresponding female and male parent means were averaged.

Using sugarcane data from the Louisiana Sugarcane Variety Development Program (LSVDP), these predictors, among other versions of BLUPs, were compared by cross-validation procedures. Information from 719 parental combinations obtained after 5 years of cross appraisal trials for five traits (stalk diameter, stalk height, stalk weight, stalk number and cane yield) was used. Models with BLUP-based predictors showed smaller mean-square prediction error and higher fidelity of top cross identification compared with MPV for all traits evaluated. The C-BLUP was better than the MPV, but not consistently the best one among the BLUPs assessed. A simpler BLUP-based predictor calculated as a function of the female and male BLUP (MP-BLUP) of a two-way mixed model was better. Possible factors for the behaviour of BLUPs based on cross effects with regard to MP-BLUP might be related to an insufficient number of related crosses per cross, equal weighting of female and male parent-related crosses (the control of experimental errors related to identification of female and male parents is different when crossing sugarcane) and dominance variance. Those specific aspects of the sugarcane data, however, do not exclude a BLUP based on cross effects and genetic relationship as a valid alternative to the BLUP based on parental effects.

All BLUP-based predictors were better than MPV.

The rank correlations between predicted and observed performances of untested crosses ranged from non-significant values for stalk number to a mean correlation of 0.52 ($P = 0.001$) for stalk diameter. The maximum expected correlation between the predicted and the observed value is not unity but depends on the heritability of the trait (Bernardo, 1992). This is because, in the cross-validation procedure, we are correlating predicted genotype with phenotypic values. Heritability for cane yield is not higher than 0.30; thus the correlation between genotype and phenotype, $(0.30)^{1/2} = 0.55$, is the upper bound for an observed correlation. Bernardo (1999) reported that correlations between the predicted and the observed performance of untested crosses, obtained by cross-validation across 16 heterotic patterns, ranged from 0.46 to 0.77 for maize yield and were 75–85% of the maximum expected value. Bernardo (1992) clarified that a correlation between predicted and true genetic value of 0.60 would allow a breeder to select the top 20 out of 100 single crosses, while ensuring at least an 80% chance of retaining the best hybrid in the selected group. Therefore, BLUP-based predictors have been shown to be effective for predicting the performance of untested crosses in more than one crop.

Mixed Models for Multi-environment Trials

Most important traits of commercial crops are controlled by polygenes with various kinds of genetic effects that are affected by the environment. Variety trials commonly involve several environments. Replicated yield trials involving several environments or METs are used in late stages of breeding programmes to select genotypes based on yield and other economically important traits. Broad (across environments) inference, narrow (environment-specific) inference and GEI are the focus in MET (Kang and Gauch, 1996; Littell *et al.*, 1996; Stroup, 2000).

The traditional analytical approach for broad inferences is based on genotype means that are subjected to multiple pairwise comparisons, and it does not take into account GEI. Narrow inference relies on comparisons of genotypic means in specific environments. Unfortunately, this procedure does not use all the available information. It is only possible to infer performance in a specific environment for genotypes that have been tested in that environment.

The stability approach addressing GEI (Crossa, 1990; Lin and Binns, 1994) quantifies a genotype's contribution to the overall GEI from a fixed-effects model. The additive main effects and multiplicative interaction (AMMI) models (Gauch, 1988), coupled with biplots for visual representation (Gabriel, 1971), have been broadly used to study GEI. They work under a fixed-model framework and use least-squares estimation procedures to estimate the interaction parameters. The estimation procedure has limited its application to cases where the original two-way table of genotype–environment (GE) data is complete. The information related to variety trials is often incomplete, however, because not all genotypes may be evaluated in all environments. The assumption of homogeneity of variances for the error terms is important to obtain clear GEI patterns when least-squares estimation is involved.

The mixed-model theory provides a more relaxed approach to dealing with MET data. Closely related mixed models related with different models for **G** and **R** can be used for genotype mean separations and GEI studies. Under a unified mixed-model approach, stability parameters are integrated into broad and narrow inferences about genotype performance. Environment-specific inferences are obtained from the conditional mean of the jth genotype in the ith environment. The likelihood-based techniques involved in mixed-model estimation provide a more flexible analytical approach for the analysis of METs because balanced data are not required.

The typical model for a random variable, Y_{ijk}, occurring in row i, column j and replication k within of a two-way table

(with rows representing environments and columns representing genotypes) is:

$$Y_{ijk} = \mu + E_i + R_{k(i)} + G_j + GE_{ji} + \varepsilon_{ijk}$$

where Y_{ijk} is the kth observation for the jth genotype in the ith environment, μ, E_i, G_j, $R_{k(i)}$, and GE_{ji} denote the overall mean, the environmental effect ($i = 1, \ldots, s$), the genotype effect ($j = 1, \ldots, g$), the replication-within-environment effect ($k = 1, \ldots, r$), and the GEI effect, respectively; and ε_{ijk} is the error term associated with Y_{ijk}.

Only a few highly selected genotypes are involved at late stages of a breeding cycle. The genotype effects are seldom treated as random effects, whereas environments and/or GEI may be regarded as random variables. A random approach for environment and GEI effects allows the modelling of correlation and heterogeneous variances (Cullis et al., 1996; Magari and Kang, 1997; Piepho, 1997, 1999). Thus, the GE terms are regarded as normal random effects with zero means but with a variance–covariance matrix not necessarily implying independence and homogeneity of variances. Modelling the variances and covariances of the random GE terms should permit a comparison of the mean performance in a more realistic manner and produce stability parameters and GE analysis as a by-product of the mixed-model approach.

For simplicity, I shall assume independence of error terms and also independence of environment and replications within environment effects. This means that the covariance between the random terms of every pair of different environments is zero or, in other words, the environments provide independent information. Although environment effects are uncorrelated, the response means within a given environment are not. Therefore, for observations in the same environment, Cov $(Y_{ij}, Y_{ij'}) = \text{Var}(E_i) + \text{Cov}$ $(GE_{ij}, GE_{ij'})$ for $j \neq j'$, whereas for those in different environments, $\text{Cov}(Y_{i'j}, Y_{ij'}) = 0$ for $i \neq i'$.

Different models can be hypothesized for the variance–covariance structure of the random effects (**G**) to model yield variability among environments. The regular mixed analysis of variance (ANOVA) assumed that all the GE terms have the same variance and are independent. Magari and Kang (1997) used a mixed model involving a diagonal **G** matrix with heterogeneous variances (by genotype) for the GE random terms. Thus, the model assumes that all GE terms involving a particular genotype have the same GE variance, and there will be as many different GE variance components as the number of genotypes. The REML variance components, assignable to each genotype, estimate the same parameters as Shukla's stability variance (Shukla, 1972).

By further modelling the variance–covariance structure of environment and interaction random effects, several well-known stability measures can be expressed as parameters of closely related mixed models (Piepho, 1999). The common regression approach for studying genotype sensitivities to environmental changes with multiplicative models for the GE terms (Finlay and Wilkinson, 1963; Eberhart and Russell, 1966; Gauch, 1988; Zobel *et al.*, 1988) can be handled by integrating a factor-analytic variance–covariance structure (Jenrich and Schluchter, 1986) into a mixed model for the observed yield (Piepho, 1998). A multiplicative interaction model can be as follows:

$$GE_{ij} = \sum_{m=1}^{M} \lambda_{mj} x_{mi} + d_{ij}$$

where $\sum_{m=1}^{M} \lambda_{mj} x_{mi}$ is the sum of multiplicative terms used to explain interaction signals and d_{ij} is the residual interaction term. The values λ_{mj} and x_{mi} are the mth scores ($m = 1, \ldots, M$; $M \le \min(g-1, s-1)$) for the jth genotype and the ith environment of a regression model, respectively.

Each multiplicative term represents a linear regression model of the residuals from the main effect model for the jth genotype on a latent unobservable variable related to the ith environment. A sum of multiplicative terms is used to model the GE variability pattern in more than one dimension. The subscript m indexes the axis of variability on which the fixed genotype and random environment scores are obtained. Thus, for each axis of variation, the genotypic score,

λ_j, can be interpreted as the response of the jth genotype to changes in some latent environmental variable with value x_i in the ith environment. The model for the GE terms resembles the non-additive part of the traditional AMMI models, but in the fixed AMMI models environment scores are fixed. A model analogous to the Eberhart and Russell (1966) regression model can be obtained in the mixed-model framework by excluding the environmental main effect from the model and using one multiplicative term for the GE random effect:

$$GE_{ij} = \lambda_j x_i + d_{ij}$$

where λ_j is the sensitivity of the jth genotype to a non-observed environmental variable x_i and d_{ij} is the unexplained part of the GEI. The deviations, d_{ij}, are allowed to have a separate variance for each genotype, $\sigma^2_{d(j)}$. The absolute value of λ_j indicates genotype sensitivity to environmental changes expressed in x_i.

Multiplicative interaction models provide a useful tool for analysing GEI in plant breeding. Typically, they have been used in a fixed-effects model framework in the analysis of complete GE data sets. In the traditional fixed-model approach, the multiplicative term(s) is(are) part of the expected value of the response, whereas, under the mixed model, it belongs to the model covariance structure.

By assuming a factor-analytic structure for the **G** matrix of the GE terms in a given environment, the genotype scores on each multiplicative term, λ_{mj}, for $j = 1, \ldots, g$, are estimated as covariance parameters. The homoscedastic factor-analytic model for the variance–covariance matrix of the GE terms in environment i, assuming unitary variance for x_{mi}, is:

$$\sum_{(GE/i)} = \Lambda\Lambda' + \sigma^2_d \mathbf{I}_g$$

where Λ is a $g \times M$ matrix of constants, λ_{mj} ($m = 1, \ldots M$); each column of Λ contains the genotype scores for one of the M multiplicative terms. By analogy with factor analysis, the genotypic scores are also called 'factor loadings'. If only one environmental latent variable is considered, the model will contain only one multiplicative term, and Λ

will be a $g \times 1$ column vector carrying the factor loading of each genotype in the only multiplicative term involved; the structure of $\Sigma_{(GE/i)}$ under a model with one multiplicative term is called 'factor-analytic 1' or FA(1). If more than one, but not all, multiplicative terms are involved, Λ is a matrix of dimension $g \times m$, with m equal to the number of multiplicative terms, and the structure of $\Sigma_{(GE/i)}$ is called FA(m). By running several mixed AMMI models with different numbers of multiplicative components and keeping, without changes, the fixed portion of the model, it is possible to obtain sequential likelihood ratio tests to determine the number of terms that should be retained to explain the GE pattern in the matrix of residuals from the additive model.

The factors, x_{mi}, representing the environmental scores, are non-observable random variables. However, they can be predicted by BLUP. The BLUP of the environmental score vector, $\mathbf{x}_i$, contains the scores for each of the multiplicative terms of a given environment. The environmental scores of the ith environment are a $M \times 1$ vector containing M environmental scores for the ith environment, which can be expressed as:

$$\mathrm{EBLUP}(\mathbf{x}_i) = \hat{\Lambda}'\hat{\Sigma}_{\iota}^{-1}(\mathbf{Y}_i - \mathbf{Xb}_i)$$

where the hats over the matrix of factor loadings and the covariance matrix of $\mathbf{Y}_i$ indicate estimates of the covariance parameters, and $\mathbf{Xb}_i$ represents the generalized least-squares estimates of the fixed parameters in environment i.

Balzarini (2000) compared the biplots obtained by plotting on the mth axis λ_{mj} as genotype scores and the mth element of the scaled EBLUP($\mathbf{x}_i$) as the score for the ith environment against the traditional biplot obtained from a fixed model. Biplots under both approaches were obtained for several complete data sets of variety trials. The different procedures to obtain the biplots, under both approaches, showed the same interaction pattern. An important advantage of the mixed-models framework is that biplots can still be obtained with incomplete data.

Plant-cane data of LSVDP outfield trials from 1996 to 1998 (Quebedeaux *et al.*, 1996,

1997; Guillot *et al.*, 1998) were used to compare prediction accuracy of several mixed models against a fixed-model approach. Regular outfield tests involve ten to 12 genotypes per trial. Each year, trials are conducted at several (seven to ten) commercial farms distributed throughout the 158,000 ha crop region. Each trial is laid out in a randomized complete-block design with three replications. A prediction accuracy measure (mean square prediction error) of cane yield (Mg ha^{-1}) was obtained by a 'leave-one-block-out' cross-validation procedure. The fixed-model approach consistently produced larger prediction errors in cane-yield narrow inferences than any of the mixed models discussed here. Stroup (2000) showed that a type I error-rate inflation should be expected when testing genotype performance in the presence of even a small magnitude of GE variation under a fixed-model approach (with or without GE terms).

Summary and Conclusions

Identifying the best cross combinations among those that cannot be tested in progeny tests should improve the likelihood of producing élite progeny without increasing field costs. Large crossing databases are common in all breeding programmes, so data on tested crosses are already available. By choosing a particular mixed model to estimate variance components and to link the tested and untested cross information, BLUPs can be obtained from any software that provides for mixed-model analyses.

If it is possible to regard environments and GE terms of an MET as random variables, breeders can visualize genotype mean performance and GE measures as parameters of a general mixed model. By using the mixed-model approach properly, genotype broad performance can be assessed through mean separations, taking into account genotype stability variances or sensitivities to environmental changes, narrow inferences can be predicted by BLUP of genotype performance in specific environments that include GE effects, even though not all genotypes are evaluated in all environments, biplots to visualize interaction patterns can be obtained from incomplete data and, by running several mixed models with different covariance structure, it is possible to obtain sequential likelihood ratio tests or criteria such as the AIC to do model selection; thus the selection of stability and GE measures will be more objective.

Issues concerning computation in mixed models are no longer a true concern; many statistical systems allow the fitting of complex mixed models. Software is available from the author as a set of statistical analysis system (SAS) macros that use Proc Mixed, which can fit any class of model described in this chapter.

The literature on predicting genotype effects at different plant-breeding stages contains a bewildering variety of apparently unrelated statistical procedures. A mixed-model-based approach allows conceptualization of the complex correlation structures of plant-breeding data in terms of random effects and underlying variance components. It provides a unified strategy for analysing data that facilitate the handling of genetically and/or experimentally correlated information, heterogeneous variances and incomplete databases. Additionally, the unified mixed-model framework gives the prospect of incorporating useful breeding measures in inference and offers statistical tools for model selection. This chapter laid out strategies under a mixed linear model that can be used to analyse plant-breeding data. The singular characteristics of each stage in the breeding process demand different statistical modelling strategies. BLUP is discussed at two stages of a plant-breeding programme, crossing-progeny tests (to predict cross performance of untested crosses in the hybridization stage) and late-selection stages or multi-environment yield trials (to predict genotype performance and study GEI).

Acknowledgements

The author is grateful for the contributions of Drs Scott Milligan and Manjit Kang, distinguished professors and advisers of the research work involved in this manuscript. The author's participation was made possible through the Louisiana State University Chapter of Sigma Xi, the Scientific Research Society and the support provided by the University of Córdoba, Argentina. I sincerely compliment Dr Manjit Kang for his efficient work in organizing the Quantitative Genetics and Plant Breeding Symposium.

References

Balzarini, M. (2000) Biometrical models for predicting future performance in plant breeding. PhD dissertation, Louisiana State University, Baton Rouge, Louisiana.

Bernardo, R. (1992) Retention of genetically superior lines during early-generation test-crossing of maize. *Crop Science* 32, 933–937.

Bernardo, R. (1994) Prediction of maize single-cross performance using RFLPs and information from related hybrids. *Crop Science* 34, 25–30.

Bernardo, R. (1995) Genetic models for predicting maize single-cross performance in unbalanced yield trial data. *Crop Science* 35, 141–147.

Bernardo, R. (1996) Best linear unbiased prediction of maize single-cross performance. *Crop Science* 36, 872–876.

Bernardo, R. (1997) RFLP markers and predicted testcross performance of maize sister inbreds. *Theoretical and Applied Genetics* 95, 655–659.

Bernardo, R. (1999) Best linear unbiased predictor analysis. In: *The Genetics and Exploitation of Heterosis in Crops.* American Society of Agronomy, Crop Science Society of America, Soil Science Society of America, Madison, Wisconsin, pp. 269–276.

Bridges, W.C., Jr (1989) Analysis of a plant breeding experiment with heterogeneous variance using mixed model equations. In: *Applications of Mixed Models in Agriculture and Related Disciplines.* Southern Cooperative Series Bulletin No. 343, Louisiana Agricultural Experiment Station, Baton Rouge, Louisiana, pp. 145–154.

Chang, Y.S. (1996) Assessment of genetic merits for sugarcane parents. *Taiwan Sugar Research Institute* 153, 1–9.

Chang, Y.S. and Milligan, S.B. (1992) Estimating the potential of sugarcane families to produce elite progeny using univariate cross prediction methods. *Theoretical and Applied Genetics* 84, 662–671.

Crossa, J. (1990) Statistical analyses of multilocation trials. *Advances in Agronomy* 44, 55–85.

Cullis, B.R., Thompson, F.M., Fisher, J.A., Gilmour, A.R. and Thompson, R. (1996) The analysis of the NSW wheat variety database. II. Variance component estimation. *Theoretical and Applied Genetics* 92, 28–39.

DeSousa-Vieira, O. and Milligan, S.B. (1999) Intrarow plant spacing and family × environment interaction effects on sugarcane family evaluation. *Crop Science* 39, 358–364.

Eberhart, S.A. and Russell, W.A. (1966) Stability parameters for comparing varieties. *Crop Science* 6, 36–40.

Falconer, D.S. (1989) *Introduction to Quantitative Genetics*. Ronald Press, New York. Third edition, John Wiley & Sons, New York.

Finlay, K.W. and Wilkinson, G.N. (1963) The analysis of adaptation in a plant breeding programme. *Australian Journal of Agricultural Research* 14, 742–754.

Gabriel, K.R. (1971) Biplot display of multivariate matrices with application to principal components analysis. *Biometrika* 58, 453–467.

Gauch, H.G. (1988) Model selection and validation for yield trials with interaction. *Biometrics* 44, 705–715.

Gopal, J., Gaur, P.C. and Rana, M.S. (1992) Early generation selection for agronomic characters in a potato breeding program. *Theoretical and Applied Genetics* 84, 709–713.

Guillot, D.P., Milligan, S.B., Bischoff, K.P., Quebedeaux, K.L., Gravois, K.A., Garrison, D.D., Jackson W.R. and Waguespack, H.L. (1998) *1998 Outfield Variety Trials: Sugarcane Research Annual Progress Report.* Louisiana State University Agricultural Center, Louisiana Agricultural Experiment Station, Baton Rouge, Louisiana.

Harville, D.A. (1990) BLUP: best linear unbiased prediction and beyond. In: Gianola, D. and Hammond, K. (eds) *Advances in Statistical Methods for Genetic Improvement of Livestock*. Springer-Verlag, New York, pp. 239–276.

Henderson, C.R. (1973) Sire evaluation and genetic trends. In: Harvey, W.D. (ed.) *Proceedings of Animal Breeding and Genetics Symposium in Honor of J.L. Lush*. Virginia Polytechnic Institute and State University and American Society of Animal Science and Dairy Science Association, Champaign, Illinois, pp. 10–41.

Henderson, C.R. (1974) General flexibility of linear model techniques for sire evaluation. *Journal of Dairy Science* 57, 963.

Henderson, C.R. (1990) Statistical methods in animal improvement: historical overview. In: Gianola, D. and Hammond, K. (eds) *Advances in Statistical Methods for Genetic Improvement of Livestock*. Springer-Verlag, New York, pp. 1–14.

Jenrich, R.L. and Schluchter, M.D. (1986) Unbalanced repeated-measures models with structured covariance matrices. *Biometrics* 42, 805–820.

Kang, M.S. and Gauch, H.G. (eds) (1996) *Genotype-by-Environment Interaction*. CRC Press, Boca Raton, Florida.

Laird, N. and Ware, J.H. (1982) Random-effects models for longitudinal data. *Biometrics* 38, 963–974.

Lin, C.S. and Binns, M.R. (1994) Concepts and methods for analyzing regional trial data for cultivar and location selection. *Plant Breed Reviews* 12, 271–297.

Littell, R.C., Milliken, G.A., Stroup, W.W. and Wolfinger, R.D. (1996) *SAS® System for Mixed Models*. SAS Institute, Cary, North Carolina.

Magari, R. and Kang, M.S. (1997) SAS-STABLE, stability analysis of balanced and unbalanced data. *Agronomy Journal* 89, 929–932.

Panter, D.M. and Allen, F.L. (1995) Using best linear unbiased predictions to enhance breeding for yield in soybean, I. Choosing parents. *Crop Science* 35, 397–405.

Patterson, H.D. and Thompson, R. (1971) Recovery of interblock information when block sizes are unequal. *Biometrika* 58, 545–554.

Piepho, H.P. (1997) Analyzing genotype–environment data by mixed models with multiplicative effects. *Biometrics* 53, 761–766.

Piepho, H.P. (1998) Empirical best linear unbiased prediction in cultivar trials using factor-analytic variance–covariance structures. *Theoretical and Applied Genetics* 97, 195–201.

Piepho, H.P. (1999) Stability analysis using the SAS System. *Agronomy Journal* 91, 154–160.

Quebedeaux, K.L., Milligan, S.B., Martin, F.A., Garrison, D.D., Jackson, W.R. and Waguespack, H., Jr (1996) *1996 Outfield Variety Trials: Sugarcane Research Annual Progress Report*. Louisiana State University Agricultural Center, Louisiana Agricultural Experiment Station, Baton Rouge, Louisiana.

Quebedeaux, K.L., Milligan, S.B., Martin, F.A., Garrison, D.D., Jackson, W.R. and

Waguespack, H., Jr (1997) *1997 Outfield Variety Trials: Sugarcane Research Annual Progress Report.* Louisiana State University Agricultural Center, Louisiana Agricultural Experiment Station, Baton Rouge, Louisiana.

Robinson, G.K. (1991) The BLUP is a good thing, the estimation of random effects. *Statistical Science* 6, 15–51.

Searle, S.R., Casella, G. and McCulloch, C.H. (1992) *Variance Components.* John Wiley & Sons, New York.

Shukla, G.K. (1972) Some statistical aspects of partitioning genotype–environment components of variability. *Heredity* 29, 237–245.

Simmonds, N.W. (1996) Family selection in plant breeding. *Euphytica* 90, 201–208.

Stroup, W.W. (1989) Why mixed models? In: *Applications of Mixed Models in Agriculture and Related Disciplines.* Southern Cooperative Series Bulletin No. 343, Louisiana Agricultural Experiment Station, Baton Rouge, Louisiana, pp. 104–112.

Stroup, W.W. (2000) Mixed model issues in the analysis of multi-location trials. In: *The XXth International Biometrics Conference,* Vol. II. University of California, Berkeley, pp. 135–142.

White, T.L., Hodge, G.R. and De Lorenzo, M.A. (1986) Best linear prediction of breeding values in forest tree improvement. In: *Workshop of the Genetics and Breeding of Southern Forest Trees.* Southern Region Information Exchange Group 40, Gainesville, Florida, pp. 99–122.

Wolfinger, R.D. (1993) Covariance structure selection in general mixed linear models. *Communications in Statistics A, Theory and Methods* 22, 1079–1106.

Wolfinger, R.D. and Tobias, R. (1998) Joint estimation of location, dispersion, and random effects in robust design. *Technometrics* 40(1), 62–71.

Zobel, R.W., Wright, M. and Gauch, H. (1988) Statistical analysis of a yield trial. *Agronomy Journal* 80, 388–393.

24 Defining Adaptation Strategies and Yield-stability Targets in Breeding Programmes

Paolo Annicchiarico

Istituto Sperimentale per le Colture Foraggere, viale Piacenza 29, 26900 Lodi, Italy

Introduction

Genotype–environment (GE) interactions are generally considered to be a hindrance to crop improvement in a target region (Kang, 1998). Moreover, the GE interaction effects may be added to environmental effects in determining the temporal and spatial instability of crop yields. Temporal stability, in particular, negatively affects, farmers' income and, for staple crops in developing countries, contributes to food insecurity at the national and household level. GE interactions may also, however, offer opportunities, e.g. selecting and using genotypes that show positive interaction with the location and its prevailing environmental conditions (exploitation of specific adaptation) or genotypes characterized by low frequency of low yields or crop failure (exploitation of yield stability) (Simmonds, 1991; Ceccarelli, 1996; Annicchiarico, 2002).

The growing awareness of the importance of GE interactions has increased the role of multi-environment, regional testing for making cultivar recommendations or in the final stages of selection among élite breeding material. Analysing the GE interaction effects by proper techniques, rather than ignoring them, is useful for exploring the opportunities and/or limiting the disadvantages that these effects present. The

information provided by these trials may also help breeding programmes in: (i) gaining a better understanding of the type and size of GE interaction that can be expected in a given region and the reasons underlying its occurrence; and (ii) defining, if necessary, a breeding strategy to cope successfully with GE interactions. Decisions on adaptation and yield-stability targets, genetic resources, variety type, breeding plan and selection procedures (selection environments, indirect selection criteria, presence and extent of participatory plant breeding, etc.) may represent the components of this strategy (Ceccarelli, 1996; Annicchiarico, 2002). In particular, plant selection may be unique for the target region (wide-adaptation strategy) or specific for distinct areas of the region (specific-adaptation strategy), whereas greater yield stability may be considered or neglected as a target. There is a need for consistency between the components of the breeding target. For instance, the inconsistency of a wide-adaptation prospect with the use of genetic resources and selection procedures generating material specifically adapted to favourable environments has made partly unsuccessful a number of breeding programmes carried out in the Green-Revolution context (Simmonds, 1991; Ceccarelli, 1994).

The extensive application of genetic-engineering techniques would not eliminate the need for breeding programmes to cope with GE interactions, because hardly any cultivar could possess genes conferring superior performance in all environments within a relatively large region. This derives from genetically based trade-offs between yield potential and tolerance to major stresses, e.g. drought (Ludlow and Muchow, 1990), as well as from the need to choose between incompatible levels of a key adaptive trait, such as earliness of cycle (Wallace *et al.*, 1993). Also the possible use of marker-assisted selection for yield may require a preliminary definition of an adaptation strategy and yield-stability targets, because a substantial portion of useful markers is environment-specific (Paterson *et al.*, 1991; Hayes *et al.*, 1993).

Adaptation and Yield Stability

In an evolutionary biology context, adaptation is a process, adaptedness is the level of adaptation of the plant material to a given environment and adaptability is the ability to show good adaptedness in a wide range of environments. In plant breeding, the first two terms relate to a condition rather than a process, designating the ability of the material to be high-yielding in a given environment or in given conditions (to which it is adapted) (Cooper and Byth, 1996). Breeding for wide adaptation, i.e. adaptability, or for specific adaptation means having as a goal a variety that is well-performing in nearly all environments or in a specific subset of environments, respectively, within a target region.

Various authors (e.g. Barah *et al.*, 1981; Lin and Binns, 1988) have applied the adaptation concept only to consistency of genotype performance in space, using the yield-stability concept for consistency in time. In fact, only repeatable genotype–location (GL) interaction effects can be exploited by selecting and growing specifically adapted genotypes. More generally, adaptation as a workable concept for breeders may conveniently be restricted to the investigation of responses to locations, geographical areas, farming practices or other

aspects that can be controlled or predicted prior to sowing. The knowledge of positive GE interaction effects relative to genotype–year (GY) or genotype–location–year (GLY) interactions, as estimated by analysis of variance (ANOVA) models, cannot be exploited in future years, because the climatic conditions that generate year-to-year environmental variation are not known in advance. The GLY interaction, expressing non-repeatable GL interaction effects, represents the error term for testing the significance of GL interaction under the usual assumption of year as a random factor (independently of the definition of genotype and location factors as random or fixed) (Annicchiarico, 2002, ch. 4). These models can be represented as follows (omitting the possible block factor):

$$R_{ijkr} = m + G_i + L_j + Y_k + GL_{ij} + \\ GY_{ik} + LY_{jk} + GLY_{ijk} + e_{ijkr} \quad (24.1)$$

where R_{ijkr} is the yield of the genotype i at the location j, year k and plot r, m is the grand mean and e_{ijkr} is the random error. Likewise, GY interaction effects within locations in the following ANOVA models, holding the time factor random and nested into location (particularly useful when locations differ for test years):

$$R_{ijkr} = m + G_i + L_j + Y_k (L_j) + \\ GL_{ij} + GY_{ik} (L_j) + e_{ijkr} \quad (24.2)$$

cannot be exploited by specific adaptation and they act as the error term for the GL interaction (Annicchiarico, 2002, ch. 4) (these GY effects include the GY and GLY effects of the model in Equation 24.1). The same view can hold when adaptation focuses on genotype responses to environmental factors. Material can be selected with specific adaptation to environmental conditions prevailing in a given area, provided that these conditions are not highly variable between years. Focusing on GE rather than GL effects for analysis of adaptation may fail to provide the most useful information. For instance, genotype variation in response to rainfall amount may appear important when analysing GE interaction and negligible when analysing GL interaction when test locations have similar mean and large year-to-year variation for the environmental variable. The relatively frequent assumption that GL interaction and GY interaction within locations are affected by the same environmental factors within a given region is not supported by data sets analysed by Lin and Binns (1988) or Annicchiarico (2002, ch. 8), in which no clear relationship emerges between wide adaptation and yield stability of genotypes in time. Concentrating on GL interaction effects for analysis of adaptation offers the additional advantage of simplifying the analysis, because adaptation patterns that are remarkably complex when evaluated on a GE basis (requiring three or more dimensions for a convenient multivariate representation) become relatively simple (requiring only one or two dimensions) (Annicchiarico, 1997a).

Assessing the value of a specific-adaptation strategy is of obvious interest for the globally orientated breeding program-mes of large seed companies or international research centres, where the target region may encompass a number of countries and widely diversified environments. The portion of the target region that is the object of specific breeding, termed hereafter the subregion, may include several countries in this case. Specific adaptation, however, may also prove a valuable target for national breeding programmes, for which the yield gain derived from exploitation of GL inter-action effects within the country can help withstand the increasing competition exerted on local seed markets by international seed companies. For public institutions, the breeding of diversified, specifically adapted germ-plasm can be a major element of a research policy enforcing sustainable agri-culture by: (i) maximizing the potential of different areas by fitting cultivars to an environment, instead of altering the envi-ronment (by inputs, such as fertilizers, water, pesticides, etc.) to fit widely adapted cultivars; and (ii) safeguarding the bio-diversity of cultivated material (Bramel-Cox et al., 1991; Ceccarelli, 1996). This strategy, which can also contribute to food security for staple crops, may further enhance its impact by integrating participatory breeding schemes (McGuire et al., 1999). In any case,

specific adaptation may not imply a duplica-tion of breeding stations and the relative costs, because crossing and hybridization operations could be centralized at a single station that provides each subregion with material on which local selection is performed.

Subregions may be identified not only within large and/or transnational regions (e.g. DeLacy et al., 1994) but also within relatively small regions, as suggested by results from Syria (Ceccarelli, 1996), Italy (Annicchiarico and Perenzin, 1994) and northern Italy (Annicchiarico, 1992), New South Wales (Seif et al., 1979; Basford and Cooper, 1998), south-western Canada (Saindon and Schaalje, 1993) and Ontario (Yan et al., 2000). Some of these studies also suggest that different subregions may possess similar mean yield levels, although specific breeding has mostly been envisaged with regard to a high-yielding, favourable area, on the one hand, and a low-yielding, drought-prone or nutrient-deficient area, on the other hand (Bramel-Cox et al., 1991; Ceccarelli, 1994). Indeed, GL interaction may also occur between unfavourable or moderately favourable areas that differ for the prevailing abiotic or biotic stress(es) or for the pattern of one major stress (Annicchiarico, 1997a; Basford and Cooper, 1998).

Repeatable GL interaction effects can be either exploited by breeding specifically adapted germ-plasm or minimized by breeding widely adapted material. Also the remaining GE interactions can be either exploited by breeding material that tends to maintain the yield constant across environ-ments (i.e. responding relatively better in unfavourable years) or minimized by breed-ing genotypes with no marked deviation from their expected mean in each environ-ment (i.e. no GE interaction). These contrast-ing features relate to two concepts of yield stability, termed, respectively, static and dynamic by Becker and Léon (1988) and referred to as type 1 and type 2 by Lin et al. (1986). Lin and Binns (1988) have defined a type 4 concept of stability that is strictly related to the former concept. The difference is that type 4 relates to consistency of yield exclusively in time (across years or crop cycles) within locations, rather than across indefinite environments (belonging to the same or different sites). The former concept of stability (type 1 or type 4) is more useful than the latter in a food-security context, and the repeatability of its measures, on the whole, tends to be somewhat higher (although rarely high in absolute terms) (Brancourt-Hulmel *et al.*, 1997; Annicchiarico, 2002, ch. 7). Whatever the pursued concept and the related measure of yield stability, the direct selection for this trait requires many test environments (at least eight to ten) because of the high sampling error (Kang, 1998). The choice of parental germ-plasm with recognized yield stability and, if possible, that of a convenient variety type can be important elements of a breeding strategy in this context (Becker and Léon, 1988; Brancourt-Hulmel *et al.*, 1997; Kang, 1998).

The practical interest of facilitating the simultaneous selection for high mean yield and high yield stability has led to the development of the yield-reliability concept. A reliable genotype is characterized by a consistently high yield across environments and/or cropping seasons. Different measures of yield reliability are available that relate to either concept of stability (Barah *et al.*, 1981; Eskridge, 1990; Kang and Pham, 1991; Annicchiarico, 2002, ch. 7).

Analysis of Adaptation

Techniques for the analysis of adaptation have mainly been developed for three objectives: (i) defining an adaptation strategy for breeding programmes; (ii) targeting genotypes and/or making variety recommendations; and (iii) identifying optimal test or selection locations. Different objectives may coexist in the analysis of one data set, but they may require partly different analytical approaches. The first, in particular, focuses on the responses of a set of genotypes to generate predictions relative to future breeding material that may be produced from the genetic base, of which the tested genotypes are assumed to be a representative sample. Genotypes do not strictly need to be

randomly chosen (a group of carefully chosen élite varieties or breeding lines may well represent the genetic base of major interest), but they should adequately represent the relevant germ-plasm types and/or provenances for local breeding. The sample of test locations (probably never fewer than seven or eight) should be representative of the pool of sites and relative variation in environmental and agronomical factors within the target region (Cooper and Hammer, 1996). The adoption of a proportional allocation criterion would imply that areas of greater importance are more represented. For annual crops, the trials should be conducted for at least 2 years and, possibly, 3, in order to distinguish repeatable GL interaction effects from non-repeatable ones. Reports from various countries (e.g. Annicchiarico and Perenzin, 1994; Weber and Westermann, 1994; Sneller and Dombek, 1995) have shown that a reliable assessment of GL effects is hardly possible from data of only 1 year, because the estimation is inflated by non-repeatable effects. For perennials, the repetition in time may not be strictly needed. Results for lucerne in Italy suggest that the variation in environmental factors encountered by genotypes across a 3-year crop cycle is wide enough to act as a buffer against the occurrence of sizeable non-repeatable GL interaction effects (Annicchiarico, 1992).

Adaptation patterns relative to test locations are of limited interest *per se*, as the sample of sites is necessarily very small relative to the target region. Moreover, specific breeding can only be directed to areas, i.e. it cannot realistically be so fine-tuned as to exploit positive interaction effects of genotypes with individual locations. However, sites that are similar for the response of genotypes can be grouped by different methods, and each group may identify a cropping area for specific breeding. Such areas have been termed by different authors as subregions, subzones, subareas, macroenvironments, or mega-environments. Their definition is not just geographical; it may encompass farming practices as well (e.g. irrigated or rain-fed cropping). Additional information on the climatic, soil and biotic variables that are closely related to the occurrence of GL interaction can help locate geographical boundaries for subregions, besides contributing to our understanding of causal factors for the interaction (Bidinger *et al.*, 1996; Annicchiarico, 2002). There are many ordination and classification techniques that can jointly be useful for the zoning process (DeLacy *et al.*, 1996). The most popular and simplest to apply and/or least demanding of additional information are considered in detail below.

Modelling of GL_{ij} effects for yield can be attempted via the following techniques:

1. Joint regression model:

$$GL_{ij} = \beta_i L_j + d_{ij}$$

where β_i is the genotype regression coefficient according to Perkins and Jinks' model, equal to $(b_i - 1)$ in Finlay and Wilkinson's model; L_j is the location main effect; and d_{ij} is the deviation from the model, i.e. the residual GL interaction (Becker and Léon, 1988).

2. AMMI model:

$$GL_{ij} = \Sigma u_{in} v_{jn} l_n + d_{ij}$$

where u_{in} and v_{jn} are the eigenvectors (scaled as unit vectors) of genotypes and locations, respectively, and l_n is the square root of the eigenvalue, for $n = 1, 2, \ldots$ N axes of a double-centred principal-components (PC) analysis performed on the matrix of GL effects (Gauch, 1992, ch. 3). For trials repeated in time, testing of PC axes can be done with respect to the appropriate error term for the GL interaction in the ANOVA (Annicchiarico, 1997b).

3. Factorial regression model:

$$GL_{ij} = \Sigma \beta_{in} X_{jn} + d_{ij}$$

where β_{in} is the genotype regression coefficient for the environmental covariate n and X_{jn} is the effect of the covariate for the location. The equation may also be expressed as a function of the site value of the covariate (Piepho *et al.*, 1998). Again, testing of individual components of the GL interaction can be performed with respect to the appropriate error term for this interaction in the ANOVA.

For model comparison, both efficacy (high proportion of GL interaction sum of squares) and parsimony (low number of GL interaction degrees of freedom) are important (Gauch, 1992, ch. 4). These features can be combined into a unique evaluation criterion equal to the sum of the estimated variances of the significant components of the GL interaction (Annicchiarico, 2002, ch. 5). For instance, an additive main effects and multiplicative interaction (AMMI) model with two significant PC axes (AMMI-2) outperforms the joint regression model if the sum of the variances of the two PC axes is higher than the variance of the heterogeneity of genotype regression component. An alternative criterion for model assessment has been proposed by Brancourt-Hulmel *et al.* (1997).

Modelling of GL effects is just a step of the analysis, which, in the current context, should be followed by classification of locations based on their similarity for GL interaction effects. Various methods based on hierarchical cluster analysis have been proposed in this respect, among which Ward's incremental sum of squares and average linkage (with a squared Euclidean distance as the dissimilarity measure) are generally recommended (DeLacy *et al.*, 1996). When used in combination with AMMI analysis, the performance of cluster analysis on the significant GL interaction PC scores of locations (e.g. Annicchiarico, 1992) has the advantage of retaining only the 'pattern' portion of the GL interaction variation in the assessment of site similarity, since the 'noise' portion is excluded from the model (Gauch, 1992, ch. 4). Cluster analysis can be performed on mean yields of locations when combined with joint regression and on site mean values of the significant environmental covariates when combined with factorial regression. In the latter case, covariates may be assigned a weight in proportion to their importance as estimated from the partial regression sum of squares in the model (Annicchiarico, 2002, ch. 5). The lack of significant GL interaction within groups of locations, verified by a separate ANOVA at each clustering stage, may provide a simple truncation criterion for definition of groups

(Ghaderi et al., 1980). An alternative criterion for site classification has been proposed by Singh et al. (1999) in the framework of joint regression analysis. It contemplates the estimation of the value of site mean yield for which crossover interactions (i.e. changes in ranks) between genotypes reach the highest frequency. This main crossover point serves as a cut-off for defining two groups of locations. The same approach has been extended by Annicchiarico (2002, ch. 5) to the estimation of the main crossover point in AMMI-1 models, in which nominal yields of genotypes (i.e. yields from which the location main effect has been removed) can be expressed as a function of the site score on the first GL interaction PC axis (Gauch and Zobel, 1997), as well as in factorial regression models including only one covariate. Relative to cluster analysis, this approach can identify only two groups of locations, and it cannot be applied when adaptation patterns require a multidimensional description.

One further method for location classification is represented by pattern analysis (DeLacy et al., 1996), which, for trials repeated also in time, implies the application of a hierarchical cluster analysis on genotype yields averaged across time at each location and preliminarily standardized (subtracting location mean, and dividing by location standard deviation of genotype values). The use of Ward's method allows for the classification of locations into groups that reflect the opportunities to exploit indirect selection among locations (Cooper et al., 1996). The method is usually complemented by an ordination technique based on PC analysis or principal coordinates analysis (DeLacy et al., 1996) and may include the modelling of genotype standardized yields. Compared with other methods, pattern analysis is more suitable for application to largely unbalanced data sets (e.g. DeLacy et al., 1994), but there is no attempt to separate pattern from noise in GL effects prior to classification of locations. Other techniques, such as canonical variates analysis (Seif et al., 1979) and the shifted multiplicative model (Cornelius et al., 1992), can also be used for site classification.

Groups of locations represent provisional subregions in breeding for specific adaptation. Also the analysis of adaptation aimed at targeting/recommendation of genotypes may contemplate the definition of subregions (Gauch and Zobel, 1997; Annicchiarico, 2002, ch. 5). However, that use of the analysis differs from the current one in at least two major aspects. First, the relevant GL interaction effects in that context are those of the crossover type between top-ranking genotypes, each subregion grouping the sites with the same best-yielding genotype(s). Conversely, any GL effect relating to lack of genetic correlation between locations, i.e. not resulting from heterogeneity of genotypic variance between sites (Cooper et al., 1996), is relevant for the current assessment of site similarity. Secondly, large heterogeneity of genotypic variance between locations may represent a problem only in the current context. The heterogeneity may arise from the aforementioned main crossover point for genotype adaptive responses, as well as from the trend of yield differences between genotypes to increase in increasing the site mean yield (Yau, 1991). While the former is intrinsic to the adaptation patterns and can be accepted (also in view of its potential usefulness for definition of subregions), the latter should be corrected because: (i) it introduces unwanted GL interaction pattern as a result of a scale effect that implies no change of relative merit between genotypes; and (ii) it causes cluster analysis results also to be affected by the main effect of locations (Fox and Rosielle, 1982; Annicchiarico, 2002, ch. 5). A convenient correction may be suggested by the regression of the within-location phenotypic variance of genotype mean yields (s_p^2) on the mean yield of locations (m_{loc}), with both terms expressed on a logarithmic scale. A regression slope $b \approx 2$ (implying the relationship: $s_p \approx k \, m_{loc}$) supports a logarithmic transformation of plot yield data; $b \approx 1$ (implying the relationship: $s_p^2 \approx k \, m_{loc}$) supports a square-root transformation; and $b \approx 0$ (implying no relationship of s_p^2 with m_{loc}) discourages any transformation. Inspection of various data sets suggests that the transformation may be needed only in the presence of wide variation in location mean yield and, when needed, it tends to stabilize experiment error variances as well (Annicchiarico, 2002, ch. 5). In particular, the transformation is required for the data set from Algeria, which is considered hereafter as the first case-study ($b = 1.92$, $P \leq 0.01$), whereas it is unnecessary for several data sets from Italy. The data transformation can be of interest for joint regression, factorial regression and AMMI analysis techniques. For pattern analysis, the adopted data standardization within sites removes any heterogeneity of genotypic variance, including that originating from the main crossover point. This feature makes the modelling of genotype adaptation patterns less accurate by this technique (McLaren, 1996; Annicchiarico, 2002, ch. 5), which is a possible limitation when the analysis is aimed at targeting/recommending genotypes.

An Analytical Flow Chart

The main analytical steps involved in the definition of an adaptation strategy and yield-stability targets on the basis of yield trials repeated also in time are summarized in Fig. 24.1. There are six different, major conclusions that can be reached, implying a wide-adaptation or a specific-adaptation strategy and, in both cases, the inclusion or exclusion of increased yield stability as a target. Within the wide-adaptation strategy, indications may or may not urge the choice of selection locations that contrast for GL interaction effects.

At first, it is useful to estimate the variance components for genotype, GE interaction across environments (as location–year or location–crop-cycle combinations) and the two determinants of the GE interaction variance, i.e. the lack of genetic correlation among genotype values and the heterogeneity of genotypic variance, as described by Cooper *et al.* (1996). The latter determinant relates to both GL and other GE effects that originate from a scale effect of the environment and inflate the estimated genotypic variation for adaptation pattern and yield stability, respectively. The larger size of this variance component relative to the former

suggests its reduction through a suitable data transformation, as outlined, in particular, in respect of analysis of adaptation. In any case, the low size of the lack of genetic correlation component (say, less than 25–30%) relative to the genotypic variance reveals the limited extent of GE interaction effects relevant to breeding and supports, without further analyses, the selection for wide adaptation with no regard for yield stability (Fig. 24.1). Otherwise, different genotypic and genotype–environmental components of variance can be estimated through the models in Equations 24.1 or 24.2 of ANOVA. An analysis of adaptation is not justified if the GL interaction variance is low (e.g. less than 30% relative to the genotypic variance; Fig. 24.1).

Following an appropriate modelling of GL effects and/or the site classification based on similarity for these effects, subregions are provisionally identified that should lend themselves to a practical definition on a geographical basis and/or according to farming practices. Those that cannot be characterized as distinct from each other can be merged at this stage. Likewise, subregions that are too small to be of practical interest can be merged with larger ones. The availability of a geographical information system for major environmental variables can facilitate the characterization of subregions and the possible up-scaling of results both spatially (in relation to non-test sites) and temporally (in relation to long-term values of environmental data). Some procedures of possible interest in this context are described by Annicchiarico (2002, ch. 5).

Wide- vs. specific-adaptation strategies can be compared in terms of yield gains predicted on the basis of selection theory applied to the same yield data, as described in the following section. If there are several candidate subregions, specific adaptation may contemplate other possibilities besides targeting each subregion (e.g. merging of some subregions; neglecting some subregion of minor importance; etc.). The assessment is strictly valid only for materials, and under conditions, similar to those of the data set. A final decision on the adaptation strategy may be made at this stage or, especially in the absence of clear-cut indications, be

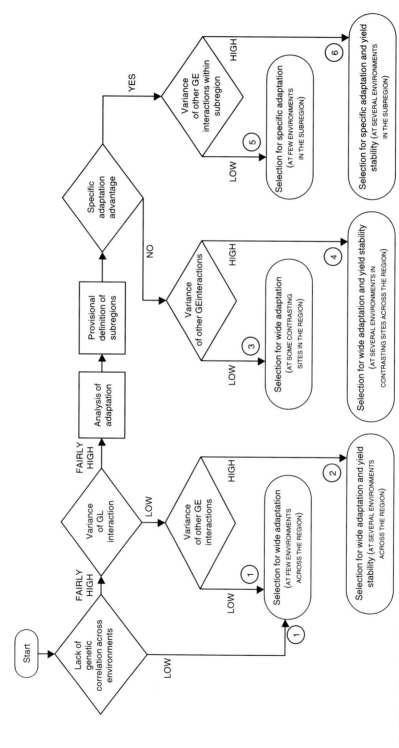

Fig. 24.1. Flow chart of steps for definition of an adaptation strategy and yield stability targets in the analysis of multilocational yield trials repeated in time (GL = genotype–location and GE = genotype–environment interactions; environment as location–year or location–crop cycle combination; see text for further details) (from Annicchiarico, 2002).

postponed to further research comparing the best options on the basis of actual yield gains (e.g. Ceccarelli *et al.*, 1998).

A wide-adaptation strategy may arise from two events, i.e. small GL interaction, or sizeable GL interaction but no clear advantage of specific breeding (Fig. 24.1). In the latter case, selection should be conducted across contrasting sites capable of reproducing the GL interaction effects and, thereby, the genotype mean responses across the target region. The provisional identification of subregions helps locate these locations, representative of different subregions and usable for parallel selection of material (Lin and Butler, 1988; Cooper *et al.*, 1996). Early stages of selection can ideally be devised at the site that reproduces with greatest accuracy the genotypic mean responses over the region (wide-adaptation prospect) or the subregion (specific prospect), maximizing the relative yield gain predicted on the basis of indirect selection theory (Cooper *et al.*, 1996). However, evaluation of selection locations based on just a few test years may be largely biased by unusually high or low within-site GY interaction and experimental error values.

The selection for wide or specific adaptation may also be attempted through artificial and/or managed selection environments capable of reproducing genotype responses across the target region, as shown by Cooper *et al.* (1995) and Annicchiarico and Mariani (1996) for wheat, and in the current, second, case-study for lucerne.

Decisions on the yield-stability issue depend essentially on the size of other GE interaction variance components (Fig. 24.1). Breeding for yield stability is probably justified only when the overall estimated variance of the relevant effects (for annual crops, either GY and GLY interactions, according to the ANOVA model in Equation 24.1, or average GY interaction within locations, according to the model in Equation 24.2) is large relative to the genotypic variance (say, 150–200%). In the specific-adaptation scenario, the indications may differ depending on the subregion, suggesting yield stability as a target in some cases only. Selection

for yield stability requires in all cases a higher number of selection environments (Fig. 24.1) and the definition of the concept of stability to breed for.

Comparison of Adaptation Strategies

Breeding for specific adaptation tends to imply greater genetic gains, as well as greater costs. The genetic gains derive from exploitation of GL interaction effects via useful adaptive traits (Bidinger *et al.*, 1996), as well as increased heritability of yield as a consequence of decreased GL interaction (Kang, 1998). The greater costs may be due to increased field testing rather than to duplication of breeding stations, since a single station could provide each subregion with novel germ-plasm. Under this scenario, a comparison of wide vs. specific adaptation that does not imply substantial differences in costs can be made on the basis of predicted yield gains obtained from the same number of selection environments (as in location–year combinations for annual crops: e.g. four sites by 2 years = eight environments, in a wide context, and two sites by 2 years = four environments for each of two subregions, in a specific context). The gain from wide adaptation is not necessarily lower in this case, because more environments allow for greater precision in the estimation of genotypic values (mainly in relation to random variation due to GE interactions). The use of selection theory for prediction of yield gains obtained from selection of genotypes is envisaged here for comparison of selection strategies for relative merit, rather than estimation of actual selection gains that can be expected. The latter use is inappropriate, unless the multi-environment testing relates to a random sample of élite breeding lines (as may be the case for self-pollinated species).

In general, the mean predicted yield gain across *E* environments can be estimated as (Falconer, 1989, ch. 11; Cooper *et al.*, 1996):

$$\Delta G = i\,h^2\,s_p \qquad (24.3)$$

where i = standardized selection differential, h^2 = estimated broad-sense heritability on a genotype mean basis, and s_p = square root of the estimated phenotypic variance across environments. In particular:

$$h^2 = s_g^2/[s_g^2 + (s_{ge}^2/E) + (s_e^2/E\,R)] \quad (24.4)$$

where s_g^2, s_{ge}^2 and s_e^2 are estimates of the components of variance for genotype, GE interaction and pooled error, respectively, and R = number of replicates. The s_p term is equal to the square root of the denominator in Equation 24.4. In the following formulae, h^2 values for prediction of yield gains are computed from components of variance estimated from ANOVAs including test environments of all locations or their subsets, whereas E and R values are set by the user as hypothesized for the selection work. E, in particular, is usually different from that of the data set. R and i are set to constant values in the assessment. Hereafter, E_A and E_B represent the number of selection environments for two subregions A and B, respectively, in a specific-adaptation scenario, whereas $E_{AB} = E_A + E_B$ represents the number of selection environments used, in a wide-adaptation scenario, for parallel selection across the subregions. P_A and P_B represent the proportion of the target region that is occupied by the subregions A and B, respectively, with $P_A + P_B = 1$. The proportion relates to the cropping area of each subregion (as indicated by the number of test locations assigned to subregions in the analysis of adaptation or, possibly, by the relative size of subregions following the spatial and temporal up-scaling of results). The relative number of selection environments for each subregion should be roughly proportional to the relative extent of the subregion (e.g. if $E_{AB} = 6$ and $P_A = 0.64$, then $E_A = 4$ = two sites by 2 years, and $E_B = 2$ = one site by 2 years). An exact match between proportion of selection environments and P value for each subregion in parallel selection for wide adaptation could be obtained by a weighted selection procedure that down-weights the contribution of environments that are over-represented, and vice versa (Podlich et al., 1999). The average predicted gain (per

unit area) provided by a wide-adaptation strategy is:

$$\Delta G_W = i\,h_{AB}^2\,s_{p(AB)} \quad (24.5)$$

where h_{AB}^2 and $s_{p(AB)}$ are obtained from Equation 24.4 after estimating the components of variance from the combined ANOVA including all test environments, and inserting E_{AB} and R values as appropriate in the formula. The average predicted yield gain over the target region provided by breeding for specific adaptation (ΔG_S) arises from a weighted mean of the gains ΔG_A and ΔG_B predicted for the subregions A and B, respectively:

$$\Delta G_A = i\,h_A^2\,s_{p(A)} \quad (24.6)$$

$$\Delta G_B = i\,h_B^2\,s_{p(B)} \quad (24.7)$$

$$\Delta G_S = [(\Delta G_A\,P_A) + (\Delta G_B\,P_B)]/(P_A + P_B) = (\Delta G_A\,P_A) + (\Delta G_B\,P_B) \quad (24.8)$$

where heritability and phenotypic variance values are obtained from Equation 24.4 after estimating the components of variance from the combined ANOVA, including only environments of the test locations grouped in subregion A (values h_A^2 and $s_{p(A)}$) or B (h_B^2 and $s_{p(B)}$), and inserting E_A or E_B and R values in the same equation as appropriate. The same procedure described for two subregions can easily be extended to the case of three or more subregions, comparing predicted yield gains over the region in a wide-adaptation scenario with those obtained from specific selection for each subregion.

In fact, a third scenario may also be envisaged for two subregions, which contemplates selection in only one subregion, so that yield gains in the other derive from correlated responses in an indirect selection context (Falconer, 1989, ch. 19; Cooper et al., 1996). The predicted yield gain in subregion B deriving from indirect selection in subregion A is:

$$\Delta G_{B/A} = i\,h_A\,h_B\,r_{g(AB)}\,s_{p(B)} \quad (24.9)$$

where h_A and h_B are square roots of broad-sense heritability values for each subregion previously estimated, together with $s_{p(B)}$, through Equation 24.4; and $r_{g(AB)}$ is the genetic correlation coefficient for genotype yields between the subregions. The predicted yield gain from direct selection in subregion A can be computed by Equation 24.6. However, the comparison with the other scenarios on the basis of the same number of selection environments suggests a modification of the current computation of h_A^2 and $s_{p(A)}$ values used for estimation of ΔG_A, because all selection environments would be attributed to subregion A for direct selection now. For instance, if $E_A = 4$ and $E_B = 2$ in previous formulae for direct selection in each subregion, $E_A = 4 + 2 = $ six environments assigned to subregion A and hence used for estimation of h_A^2 and its introduction in Equation 24.6 in the present context. The mean predicted gain over the region provided by breeding only for subregion A arises from a weighted mean of the gains ΔG_A and $\Delta G_{B/A}$ predicted for the subregions A and B, respectively:

$$[(\Delta G_A\, P_A) + (\Delta G_{B/A}\, P_B)]/(P_A + P_B) = (\Delta G_A\, P_A) + (\Delta G_{B/A}\, P_B) \quad (24.10)$$

Extending the case of two environments to two subregions (Burdon, 1977), an estimate of $r_{g(AB)}$ in Equation 24.9 can be provided by the following formula:

$$r_{g(AB)} = r_{p(AB)}/(h_A'\, h_B') \quad (24.11)$$

where $r_{p(AB)}$ is the phenotypic correlation coefficient between subregions for genotype yields (averaged across environments in each subregion), and h_A' and h_B' are square roots of the broad-sense heritability on a genotype mean basis estimated for subregions A and B, respectively. The difference of h_A' and h_B' from previous estimates (h_A and h_B) lies in the fact that E and R in Equation 24.4 are the actual numbers of test environments per subregion and experiment replicates, respectively, in the analysed data set.

Other, more complex scenarios may be assessed by a combination of the above procedures. For instance, direct selection could be devised specifically for two subregions, in either of which indirect selection is also performed for a third subregion.

The comparison between adaptation strategies based on predicted yield gains tends to underestimate the possible advantage of a specific prospect, when the material with markedly specific adaptation is underrepresented among the tested varieties or breeding lines for various reasons (e.g. previous selection for wide adaptation by local breeding; large representation in the sample of foreign, widely adapted material). Likewise, the potential gain of specific breeding for unfavourable areas may be underestimated when most tested genotypes have been selected in favourable environments. Especially in these cases, the results can be exploited for definition of the most promising specific-adaptation scenario to be compared with wide adaptation on the basis of actual yield gains.

The advantage of specific adaptation may be even greater if it also implies the utilization of a distinct genetic base for each subregion. For public breeding programmes, an additional advantage of specific breeding that is difficult to quantify is its contribution to the stability of production by increasing the number of varieties and, hence, the biodiversity of the material under cultivation (Ceccarelli, 1996).

Case-study 1: Durum Wheat in Algeria

In the framework of a bilateral cooperation project between the governments of Algeria and Italy, carried out by the Institut Technique des Grandes Cultures of Algeria and the Istituto Agronomico per l'Oltremare of Italy, 24 durum wheat cultivars were grown for 2 years at 17 Algerian locations for: (i) optimizing, for this fundamental staple crop, the variety recommendation across the country; and (ii) supporting the decisions on adaptation and yield-stability targets of the national breeding programme. Based on results published elsewhere (Annicchiarico *et al.*, 2002), the present study makes a comparison of various analytical methods used for definition of subregions and the relative yield gains predicted for each specific-adaptation scenario relative to wide adaptation. It is assumed that a valuable method can identify subregions that offer the best opportunity for a specific-adaptation strategy.

The estimated lack of genetic correlation component of variance proved about as large as the genotypic variance, justifying further analyses (Fig. 24.1). This component accounted for a lower portion (43%) of the GE interaction variance than the heterogeneity of genotypic variance component (57%), suggesting the need for transforming data. The aforementioned criterion, based on the relationship between variation in genotype mean values and mean yield of locations, supported the adoption of a logarithmic transformation, which, indeed, substantially reduced the proportion of GE interaction variance accounted for by the heterogeneity of genotypic variance component (34%). The compared methods differ in the type of analysed data (original or transformed), the adopted technique for modelling of GL effects and the criterion for grouping of sites (Table 24.1). The assessment was limited to the scenario with two subregions because this is the only one compatible with the

main crossover criterion for location classification. Joint regression, factorial regression and the AMMI-1 model proved adequate in all cases for original and transformed data, as indicated by the lack of significance of the residual GL interaction term (data not shown). The estimation of predicted gains relied upon the hypothesis of six selection environments (three sites by 2 years). Additional hypotheses are reported in Table 24.1.

Only two methods provided the same grouping of locations (those based on joint regression analysis of transformed data). The estimated gain from specific adaptation relative to wide adaptation ranged from −9.7% to +4.0%, depending on the method (Table 24.1). On the whole, the results suggest: (i) the usefulness of the logarithmic transformation of data and the criterion underlying its adoption; (ii) no consistent advantage of cluster analysis over the main crossover criterion; and (iii) the superiority of two methods, i.e. pattern analysis, and AMMI

Table 24.1. Comparison of analytical methods for definition of two subregions for durum-wheat breeding in Algeria. Predicted yield gain over the region from a specific-adaptation strategy (ΔG_S) relative to wide adaptation (ΔG_W) (analysis on data from Annicchiarico et al., 2002).

			Subregion A[b]			Subregion B[b]			
Data	Model	Grouping of sites	P_A	E_A	ΔG_A (t ha^{-1})	P_B	E_B	ΔG_B (t ha^{-1})	ΔG_S (t ha^{-1})	$\Delta G_S/\Delta G_W$ ratio (%)[c]
Original	JR	CA	0.71	4	0.178	0.29	2	0.498	0.271	91.1
Original	JR	CP	0.18	2	0.108	0.82	4	0.319	0.281	94.6
Original	AMMI	CA	0.82	4	0.179	0.18	2	0.843	0.298	100.4
Original	AMMI	CP	0.12	2	0.115	0.88	4	0.311	0.288	96.9
Original	FR	CA–nw	0.71	4	0.170	0.29	2	0.507	0.268	90.3
Original	FR	CA–w	0.71	4	0.191	0.29	2	0.448	0.266	89.6
Log$_{10}$-transformed	JR	CA	0.23	2	0.081	0.77	4	0.353	0.291	97.9
Log$_{10}$-transformed	JR	CP	0.23	2	0.081	0.77	4	0.353	0.291	97.9
Log$_{10}$-transformed	AMMI	CA	0.41	2	0.118	0.59	4	0.440	0.308	103.7
Log$_{10}$-transformed	AMMI	CP	0.23	2	0.117	0.77	4	0.352	0.298	100.2
Log$_{10}$-transformed	FR	CA	0.65	4	0.149	0.35	2	0.514	0.277	93.3
Log$_{10}$-transformed	FR	CP	0.18	2	0.155	0.82	4	0.328	0.297	100.0
Standardized	PA	CA	0.41	2	0.125	0.59	4	0.437	0.309	104.0

[a]JR = joint regression; AMMI = model with one GL interaction PC axis; FR = factorial regression holding as environmental variables winter mean temperature alone (transformed data) or with annual rainfall (original data). CA = cluster analysis (-nw = not weighted, and -w = weighted, environmental covariates in the FR-based approach); CP = main crossover point.
[b]P = proportion of the target region; E = number of selection environments; and ΔG = predicted yield gain per selection cycle (four experiment replicates; selection intensity = 10%, applied to 20 élite breeding lines).
[c]ΔG_W = 0.297 t ha^{-1}, from selection in $E = E_A + E_B$ = six environments.

analysis of transformed data complemented by cluster analysis. The subregion definition by these methods was very similar. With reference to Fig. 24.2, the indications differed only for locations 7 and 14 (attributed to subregions A and B, respectively, by the former method and to subregions B and A, respectively, by the latter). These sites may represent a transitional zone separating the high-elevation, cold- and drought-prone subregion A from the warmer, partly drought-prone subregion B. While site mean yield is mainly associated with rainfall amount, GL interaction and genotype adaptive responses relate mainly to winter cold (Annicchiarico *et al.*, 2002), just as for bread and durum wheat in Italy (Annicchiarico, 1997a), because of the importance of an appropriate matching of genotype phenology with the level and extent of cold stress in winter.

In this case, the relative advantage predicted for the specific-adaptation strategy was probably underestimated by the fact that most tested cultivars (15 out of 24) are modern varieties bred at international research centres or in south European countries and, as such, selected for wide adaptation and/or adaptation to areas quite different from Algerian ones, especially the stressful subregion A. The present definition of subregions could conveniently be used for extending the comparison of adaptation strategies to actual yield gains obtained from selection of breeding lines within and across the subregions.

Case-study 2: Lucerne in Northern Italy

In an earlier study (Annicchiarico, 1992), adaptation patterns of 11 lucerne (*Medicago sativa* L.) cultivars across 12 sites of northern Italy were modelled by AMMI analysis and joint regression. The former model was largely more adequate than the latter and, in combination with cluster analysis, allowed for the classification of lowland test sites into three subregions (Fig. 24.3). After eliminating two mountain locations that had clustered alone and were related to GL interaction variation on a second PC axis, adaptation patterns could conveniently be represented on a unidimensional basis by an AMMI-1 model. This is shown in Fig. 24.4a in respect of best-performing material, together with results of site classification producing the three subregions. Subregions A and C are sharply in contrast in relative responses of varieties, whereas subregion B is somewhat intermediate in adaptive responses (Fig. 24.4a) and geographically (Fig. 24.3). The site score on the first PC axis was correlated positively with soil clay content and negatively with summer water received by the crop (Annicchiarico, 1992). Subregion A is mostly characterized by sandy-loam soils, higher rainfall and/or irrigated cropping of lucerne; subregion C by clay soils, somewhat lower rainfall, and rainfed cropping; and subregion B by intermediate features.

A second study (Annicchiarico, 2000) was based on a novel set of trials, including partly different cultivars and test locations.

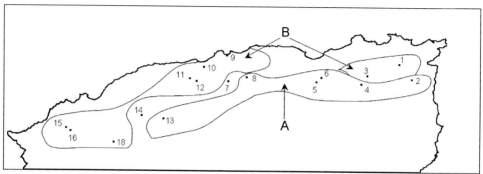

Fig. 24.2. Geographical position of test locations for durum wheat in Algeria, and definition of two subregions for a specific adaptation strategy based on pattern analysis and AMMI + cluster analysis results (analysis on data in Annicchiarico *et al.*, 2001).

Its results, partly summarized in Fig. 24.4b, showed that: (i) adaptation patterns across the region are largely repeatable; and (ii) the classification of locations is consistent with the previous definition of subregions (Fig. 24.3), which is therefore also useful in relation to other germ-plasm and locations in the region.

Based on data of the first study, wide- vs. specific-adaptation strategies were compared for predicted yield gains deriving from selection of populations (Table 24.2). Scenarios contemplated either one selection location per subregion or one for each of the contrasting subregions A and C. In the latter case, yield gains for subregion B in a specific-adaptation prospect resulted from selection in the subregion that could maximize the indirect selection gain, namely, subregion A. Selection for specific adaptation to three or two subregions was estimated to be at least six times more effective than selection for wide adaptation. The present comparison relative to selection of populations is, however, rather out of the context of selection schemes for open-pollinated populations, in which selection is basically performed on individual plants. Therefore, its conclusions should be taken with caution and validated for actual yield gains relative to phenotypic or genotypic selection.

The breeding station in Lodi, corresponding to the coded site 'k' in Fig. 24.3, is placed in subregion A. Therefore, selection at this site (implying irrigation) is expected to produce material with specific adaptation to this subregion (like, for instance, the variety 'Equipe' in Fig. 24.4b). As a public breeding programme for the region, widening the adaptation of our varieties and/or also producing varieties adapted to subregion C are highly desirable objectives, which are, however, hindered by the cost of additional selection sites. Therefore, an attempt has been made to reproduce at Lodi in artificial environments the adaptation patterns occurring across the lowland locations of the region. Based on previous information on relevant environment variables, the variation among test locations along the first PC axis of Fig. 24.4a,b was reproduced by four artificial environments created by the factorial combination of type of soil (sandy loam or clay) and level of drought stress in summer (limited; high). Local sandy-loam soil, or clay soil imported from subregion C, filled large (24.0 m × 1.6 m × 0.8 m deep), bottomless containers in concrete laid in a field. Irrigated or rainfed cropping during an ordinary summer season were simulated in each environment by irrigation under rain-shelter equipment. Results of the AMMI analysis of the GE

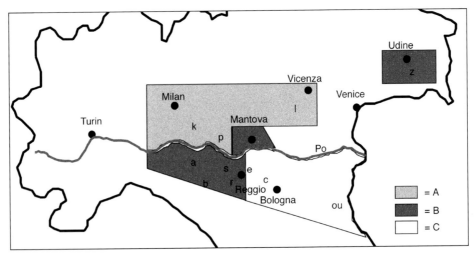

Fig. 24.3. Geographical position of test locations for lucerne in northern Italy, and definition of three subregions for a specific adaptation strategy based on AMMI + cluster analysis results (from Annicchiarico, 1992, 2000).

interaction are illustrated in Fig. 24.4c for a set of 'probe' varieties of known adaptation and a subset of farm landraces that were also included in the experiment. The responses of the varieties from the no-stress/sandy-loam soil environment (simulating sub-region A) through the stress/clay soil environment (simulating subregion C) agreed

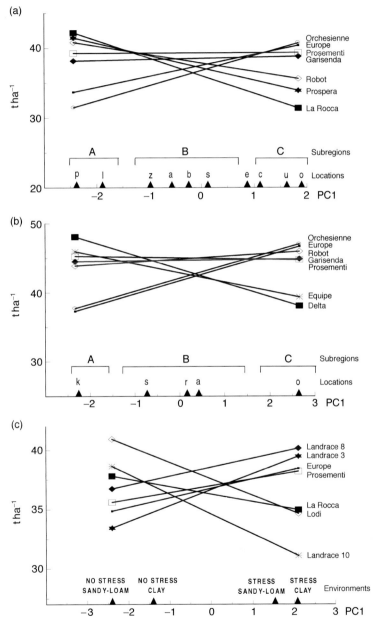

Fig. 24.4. Nominal yields of best-performing lucerne varieties as a function of the score on the first GL interaction principal-component (PC) axis of test locations (graphs a and b) or the first GE interaction PC axis of artificial environments (graph c), and location classification based on cluster analysis of site PC scores (see Fig. 24.3 for geographical position of locations).

well with the responses across the region reported in Fig. 24.4a,b. Landraces '10', '8' and '3', originating in subregions A, B and C, respectively, showed responses consistent with their respective subregion of origin, highlighting the role of evolutionary adaptation in this context. The ordination of environments along the PC axis suggested that the drought-stress level has a greater impact on the occurrence of GE interaction than soil type. These artificial environments may allow for the adoption of a wide- or a specific-adaptation strategy. Work is in progress for comparing these strategies in terms of actual yield gains obtained from phenotypic plant selection, as well as for elucidating the morphophysiological traits contributing to the adaptive responses.

Conclusions

Most breeding programmes have devoted a limited effort to investigation of GE interactions and their possible implications, despite the apparently considerable importance of the issue, the frequently large investment by public and private institutions in multi-environment testing and the ever-growing

number of statistical methods proposed for this target (Cooper and Byth, 1996). A substantial inversion of this trend can be expected to occur in so far as ordinary breeders are increasingly being sought as the main users of these methods (also through the availability of user-friendly and inexpensive software), the analyses are finalized to answer practical, crucial questions concerning adaptation strategies and yield-stability targets and, most of all, the documented application of results of these analyses to breeding programmes provides increasing evidence of their contribution to the attainment of higher and more stable crop yields.

Acknowledgements

I thank Dr K. Feliachi of the Institut Technique des Grandes Cultures of Algeria and Dr A. Perlini of the Istituto Agronomico per l'Oltremare of Italy for permission to use the durum-wheat data for the first case study. The work on lucerne in artificial environments reported in the second case study was carried out with the contribution of Dr E. Piano.

Table 24.2. Predicted yield gain over the lowland northern Italy region from selection of lucerne populations specifically adapted to three or two subregions relative to selection for wide adaptation ($\Delta G_S/\Delta G_W$ ratio) (analysis on data by Annicchiarico, 1992; see Fig. 24.3 for subregion definition).

Adaptation strategy	Selection within subregion	E	P	h^2	s_p	ΔG (t ha^{-1}) for subregions	ΔG (t ha^{-1}) over the region	$\Delta G_S/\Delta G_W$ ratio (%)
Specific	–	3	–	–	–	–	2.77	605
Subregion A	Direct	1	0.250	0.72	3.13	3.70	–	–
Subregion B	Direct	1	0.375	0.48	1.79	1.43	–	–
Subregion C	Direct	1	0.375	0.77	2.75	3.49	–	–
Wide	–	3	–	0.18	1.55	–	0.46	–
Specific	–	2	–	–	–	–	2.53	650
Subregion A	Direct	1	0.250	0.72	3.13	3.70	–	–
Subregion B	Indirect (from A)	0	0.375	0.48	1.79	0.80	–	–
Subregion C	Direct	1	0.375	0.77	2.75	3.49	–	–
Wide	–	2	–	0.13	1.84	–	0.39	–

E, number of selection environments; P, proportion of the target region; h^2, broad-sense heritability on a population mean basis; s_p, phenotypic standard deviation, estimated for E environments and four experiment replicates; ΔG, predicted yield gain (over a 3-year crop cycle) per selection cycle (selection intensity = 10%, applied to 20 élite populations). Estimated genetic correlations for subregion B are: $r_g = 0.46$ with subregion A, $r_g = -0.11$ with subregion C.

References

Annicchiarico, P. (1992) Cultivar adaptation and recommendation from alfalfa trials in northern Italy. *Journal of Genetics and Breeding* 46, 269–278.

Annicchiarico, P. (1997a) Joint regression vs. AMMI analysis of genotype–environment interactions for cereals in Italy. *Euphytica* 94, 53–62.

Annicchiarico, P. (1997b) Additive main effects and multiplicative interaction (AMMI) of genotype–location interaction in variety trials repeated over years. *Theoretical and Applied Genetics* 94, 1072–1077.

Annicchiarico, P. (2000) Variety × location interaction and its implications in breeding of lucerne: a case study. In: Veronesi, F. and Rosellini, D. (eds) *Lucerne and Medics for the XXI Century*. University of Perugia, Italy, pp. 35–43.

Annicchiarico, P. (2002) *Genotype × Environment Interactions: Challenges and Opportunities for Plant Breeding and Cultivar Recommendations.*. FAO, Rome (in press).

Annicchiarico, P. and Mariani, G. (1996) Prediction of adaptability and yield stability of durum wheat genotypes from yield response in normal and artificially drought-stressed conditions. *Field Crops Research* 46, 71–80.

Annicchiarico, P. and Perenzin, M. (1994) Adaptation patterns and definition of macro-environments for selection and recommendation of common-wheat genotypes in Italy. *Plant Breeding* 113, 197–205.

Annicchiarico, P., Chiari, T., Bellah, F., Doucene, S., Yallaoui-Yaïci, N., Bazzani, F., Abdellaoui, Z., Belloula, B., Bouazza, L., Bouremel, L., Hamou, M., Hazmoune, T., Kelkouli, M., Ould-Said, H. and Zerargui, H. (2002) Response of durum wheat cultivars to Algerian environments. I. Yield. *Journal of Agriculture and Environment for International Development* (in press).

Barah, B.C., Binswanger, H.P., Rana, B.S. and Rao, G.P. (1981) The use of risk aversion in plant breeding: concept and application. *Euphytica* 30, 451–458.

Basford, K.E. and Cooper, M. (1998) Genotype × environment interactions and some considerations of their implications for wheat breeding in Australia. *Australian Journal of Agricultural Research* 49, 153–174.

Becker, H.C. and Léon, J. (1988) Stability analysis in plant breeding. *Plant Breeding* 101, 1–23.

Bidinger, F.R., Hammer, G.L. and Muchow, R.C. (1996) The physiological basis of genotype by environment interaction in crop adaptation. In: Cooper, M. and Hammer, G.L. (eds) *Plant Adaptation and Crop Improvement*. CAB International, Wallingford, UK, pp. 329–347.

Bramel-Cox, P.J., Barker, T., Zavala-Garcia, F. and Eastin, J.D. (1991) Selection and testing environments for improved performance under reduced-input conditions. In: Sleper, D., Bramel-Cox, P.J. and Barker, T. (eds) *Plant Breeding and Sustainable Agriculture: Considerations for Objectives and Methods*. CSSA Special Publication 18, ASA, CSSA, SSSA, Madison, Wisconsin, pp. 29–56.

Brancourt-Hulmel, M., Biarnès-Dumoulin, V. and Denis, J.B. (1997) Points de repère dans l'analyse de la stabilité et de l'interaction génotype–milieu en amélioration des plantes. *Agronomie* 17, 219–246.

Burdon, R.D. (1977) Genetic correlation as a concept for studying genotype–environment interaction in forest tree breeding. *Silvae Genetica* 26, 168–175.

Ceccarelli, S. (1994) Specific adaptation and breeding for marginal conditions. *Euphytica* 77, 205–219.

Ceccarelli, S. (1996) Positive interpretation of genotype by environment interaction in relation to sustainability and biodiversity. In: Cooper, M. and Hammer, G.L. (eds) *Plant Adaptation and Crop Improvement*. CAB International, Wallingford, UK, pp. 467–486.

Ceccarelli, S., Grando, S. and Impiglia, A. (1998) Choice of selection strategy in breeding barley for stress environments. *Euphytica* 103, 307–318.

Cooper, M. and Byth, D.E. (1996) Understanding plant adaptation to achieve systematic applied crop improvement – a fundamental challenge. In: Cooper, M. and Hammer, G.L. (eds) *Plant Adaptation and Crop Improvement*. CAB International, Wallingford, UK, pp. 5–23.

Cooper, M. and Hammer, G.L. (1996) Synthesis of strategies for crop improvement. In: Cooper, M. and Hammer, G.L. (eds) *Plant Adaptation and Crop Improvement*. CAB International, Wallingford, UK, pp. 591–623.

Cooper, M., Woodruff, D.R., Eisemann, R.L., Brennan, P.S. and DeLacy, I.H. (1995) A selection strategy to accommodate genotype-by-environment interaction for grain yield of wheat: managed-environments for selection among genotypes. *Theoretical and Applied Genetics* 90, 492–502.

Cooper, M., DeLacy, I.H. and Basford, K.E. (1996) Relationships among analytical methods used to study genotypic adaptation in multi-environment trials. In: Cooper, M. and Hammer, G.L. (eds) *Plant Adaptation and Crop Improvement*. CAB International, Wallingford, UK, pp. 193–224.

Cornelius, P.L., Seyedsadr, M.S. and Crossa, J. (1992) Using the shifted multiplicative model to search for 'separability' in crop cultivar trials. *Theoretical and Applied Genetics* 84, 161–172.

DeLacy, I.H., Fox, P.N., Corbett, J.D., Crossa, J., Rajaram, S., Fischer, R.A. and Van Ginkel, M. (1994) Long-term association of locations for testing spring bread wheat. *Euphytica* 72, 95–106.

DeLacy, I.H., Basford, K.E., Cooper, M., Bull, J.K. and McLaren, C.G. (1996) Analysis of multi-environment data – an historical perspective. In: Cooper, M. and Hammer, G.L. (eds) *Plant Adaptation and Crop Improvement*. CAB International, Wallingford, UK, pp. 39–124.

Eskridge, K.M. (1990) Selection of stable cultivars using a safety-first rule. *Crop Science* 30, 369–374.

Falconer, D.S. (1989) *Introduction to Quantitative Genetics*, 3rd edn. Longman, New York, 438 pp.

Fox, P.N. and Rosielle, A.A. (1982) Reducing the influence of environmental main-effects on pattern analysis of plant breeding environments. *Euphytica* 31, 645–656.

Gauch, H.G. (1992) *Statistical Analysis of Regional Yield Trials: AMMI Analysis of Factorial Designs*. Elsevier, Amsterdam, 278 pp.

Gauch, H.G. and Zobel, R.W. (1997) Identifying mega-environments and targeting genotypes. *Crop Science* 37, 311–326.

Ghaderi, A., Everson, E.H. and Cress, C.E. (1980) Classification of environments and genotypes in wheat. *Crop Science* 20, 707–710.

Hayes, P.M., Liu, B.H., Knapp, S.J., Chen, F., Jones, B., Blake, T., Franckowiak, J., Rasmusson, D., Sorrells, M., Ulrich, S.E., Wesenberg, D. and Kleinhofs, A. (1993) Quantitative trait locus effects and environmental interaction in a sample of North American barley germplasm. *Theoretical and Applied Genetics* 87, 392–401.

Kang, M.S. (1998) Using genotype-by-environment interaction for crop cultivar development. *Advances in Agronomy* 62, 199–252.

Kang, M.S. and Pham, H.N. (1991) Simultaneous selection for high yielding and stable crop genotypes. *Agronomy Journal* 83, 161–165.

Lin, C.S. and Binns, M.R. (1988) A method for analysing cultivar × location × year experiments: a new stability parameter. *Theoretical and Applied Genetics* 76, 425–430.

Lin, C.S. and Butler, G. (1988) A data-based approach for selecting locations for regional trials. *Canadian Journal of Plant Science* 68, 651–659.

Lin, C.S., Binns, M.R. and Lefkovitch, L.P. (1986) Stability analysis: where do we stand? *Crop Science* 26, 894–900.

Ludlow, M.M. and Muchow, R.C. (1990) A critical evaluation of traits for improving crop yields in water-limited environments. *Advances in Agronomy* 43, 107–153.

McGuire, S., Manicad, G. and Sperling, L. (1999) *Technical and Institutional Issues in Participatory Plant Breeding – Done from a Perspective of Farmer Plant Breeding*. Working Document No. 2, CIAT, Cali, Colombia, 88 pp.

McLaren, C.G. (1996) Methods of data standardization used in pattern analysis and AMMI models for analysis of international multi-environment variety trials. In: Cooper, M. and Hammer, G.L. (eds) *Plant Adaptation and Crop Improvement*. CAB International, Wallingford, UK, pp. 225–242.

Paterson, A.H., Damon, S., Hewitt, J.D., Zamir, D., Rabinowitch, H.D., Lincoln, S.E., Lander, E.S. and Tanksley, S.D. (1991) Mendelian factors underlying quantitative traits in tomato: comparison across species, generations, and environments. *Genetics* 127, 181–197.

Piepho, H.P., Denis, J.B. and Van Eeuwijk, F.A. (1998) Predicting cultivar differences using covariates. *Journal of Agricultural, Biological, and Environmental Statistics* 3, 151–162.

Podlich, D.W., Cooper, M. and Basford, K.E. (1999) Computer simulation of a selection strategy to accommodate genotype–environment interactions in a wheat recurrent selection programme. *Plant Breeding* 118, 17–28.

Saindon, G. and Schaalje, G.B. (1993) Evaluation of locations for testing dry bean cultivars in western Canada using statistical procedures, biological interpretation and multiple traits. *Canadian Journal of Plant Science* 73, 985–994.

Seif, E., Evans, J.C. and Balaam, L.N. (1979) A multivariate procedure for classifying environments according to their interaction with genotypes. *Australian Journal of Agricultural Research* 30, 1021–1026.

Simmonds, N.W. (1991) Selection for local adaptation in a plant breeding programme. *Theoretical and Applied Genetics* 82, 363–367.

Singh, M., Ceccarelli, S. and Grando, S. (1999) Genotype × environment interaction of cross-over type: detecting its presence and estimating the crossover point. *Theoretical and Applied Genetics* 99, 988–995.

Sneller, C.H. and Dombek, D. (1995) Comparing soybean cultivar ranking and selection for yield with AMMI and full-data performance estimates. *Crop Science* 35, 1536–1541.

Wallace, D.H., Zobel, R.W. and Yourstone, K.S. (1993) A whole-system reconsideration of paradigms about photoperiod and temperature control of crop yield. *Theoretical and Applied Genetics* 86, 17–26.

Weber, W.E. and Westermann, T. (1994) Prediction of yield for specific locations in German winter-wheat trials. *Plant Breeding* 113, 99–105.

Yan, W., Hunt, L.A., Sheng, Q. and Szlavnics, Z. (2000) Cultivar evaluation and mega-environment investigation based on GGE biplot. *Crop Science* 40, 597–605.

Yau, S.K. (1991) Need of scale transformation in cluster analysis of genotypes based on multi-location yield data. *Journal of Genetics and Breeding* 45, 71–76.

Index